Student Study Guide for

BIOLOGY
CONCEPTS & CONNECTIONS

Richard M. Liebaert
Linn-Benton Community College
Albany, Oregon

The Benjamin/Cummings Publishing Company, Inc.
Redwood City, California • Menlo Park, California
Reading, Massachusetts • New York • Don Mills, Ontario
Wokingham, U.K. • Amsterdam • Bonn • Sydney • Singapore
Tokyo • Madrid • San Juan

Sponsoring Editor: Edith Beard Brady (1990–1993); Don O'Neal (1993–)
Assistant Editor: Korinna Sodic
Editorial Assistants: Chris Ruotolo, Tabinda Khan
Production Supervisor: Larry Olsen
Art Coordinator: Carol Ann Smallwood
Artists: Stacy J. Lund, Denise Schmidt
Copyeditor: Melissa Andrews
Proofreader: Adrienne Goss
Text Design and Composition: Detta Penna
Cover Designer: Yvo Riezebos

Cover Photo: © Art Wolfe.
Photo 8.12A courtesy of Carolina Biological Supply/Phototake/NYC.
Figure 16.2 copyright © 1986 *Discover Magazine,* Biruta Hansen.
Photo 31.3a copyright © Ed Reschke.

ISBN 0-8053-0921-7

5   6   7   8   9   10—CRK—98   97   96

The Benjamin/Cummings Publishing Company, Inc.
390 Bridge Parkway
Redwood City, California 94065

Contents

A Note to Students

A teacher I know once said that he likes to see a textbook that is tattered and torn because then he knows it is being used. I could think of no better end for this book than to have it tattered and frayed—blanks filled in, notes and sketches in the margins, phrases highlighted, and corners of pages turned down. It is a book that is meant to be used.

How To Use This Book

This study guide is intended to be interactive. I wrote it to prompt you to interact with the textbook, your lecture notes, your teacher, other students, and the biological environment. Each study guide chapter accompanies a chapter in *BIOLOGY: Concepts and Connections.* Here is an example of how you might use the text and study guide together: First, read the one-page story at the beginning of the text chapter, skim over the module headings, and read the summary. Next, read the modules, alternating between the textbook and the study guide exercises. Or, you may choose to read the entire text chapter first and then complete the study guide exercises.

Each chapter in the study guide has five parts:

1. *Introduction.* This short paragraph gets you started.

2. *Organizing Your Knowledge.* This section, which is the heart of the study guide, contains activities such as questions, diagrams, crossword puzzles, problems, and fill-in-the-blank stories that will help you master the concepts in the text. Some exercises in this section relate to single modules; others connect the information and concepts of several modules. (Module numbers are given at the beginning of each exercise.) If you get stuck, refer to the text and your lecture notes. Check your work by looking in the Answers section at the back of the book. Do not simply look up the answers—the value in doing the exercises is in actually thinking about them. If you answer questions incorrectly, you may want to review relevant modules.

3. *Testing Your Knowledge.* These multiple choice and essay questions require you to recall information from the chapter. When doing the multiple choice questions, look at all the possible answers and choose the best one. Only one correct answer is given. Some answers are tempting but wrong; others are only partially correct. The essay questions are worth your effort, as exams often include essay questions. Check your answers with those at the back of the study guide, but again, do not simply look up the answers.

4. *Applying Your Knowledge.* The multiple choice and essay questions in this section test the depth of your understanding by asking you to apply the material from the chapter to new situations, make comparisons, and put ideas together to solve biological problems.

5. *Using Your Knowledge.* This section suggests activities that you might pursue to apply what you have learned. These exercises are optional and open-ended, with a range of possible outcomes, so there are no right or wrong answers in the back of the book.

How To Study Biology

You will learn more in your biology course and enjoy it more if you have an effective study plan. Here are a few suggestions:

• *Use lecture notes to help you focus on important material.* Facts are different from concepts, and not all facts are equal. Facts are names, statistics, and simple relationships: The upper arm bone is called the humerus; a human cell contains 46 chromosomes. A concept is a generalization, a set of rules, an explanation: Living things naturally vary, and those variants best adapted to their environments produce the most offspring and become more numerous. Use your lecture notes as a guide to the facts and concepts your instructor wants you to focus on.

• *Understand how things work.* You cannot sufficiently understand a biological concept by simply recalling what you read in the textbook or heard in lecture. In fact, you do not need to be able to name and define every part of something like the heart, for example, to understand how it works. Focus on how it works first, then learn the terms you need to describe it. If you read a description of the heart and look at the illustrations, you can comprehend how it works and then learn the names for its parts. In fact, the names will come easily as you use them.

• *Write things in your own words.* The process of writing helps you organize your thoughts, and looking at what you have written helps you evaluate your understanding. Making outlines, concept maps, and diagrams is a good way to see how ideas connect. (Concept maps are discussed in Chapter 3 of the study guide.)

• *Study the illustrations.* Study the textbook illustrations carefully, and try to learn in visual images as well as words. Can you describe how a bicycle works? Chances are you can because you remember what a bicycle looks like, not because you have memorized a list of parts or a verbal description of how a bicycle works. This same approach works in biology. Remember that even tiny things like molecules and genes are real, just as real as a bicycle, and thinking in pictures is the way to remember how they work.

• *Relate biology to your own life.* Finding points of contact between biology and the rest of your life will help you understand many of the concepts you study in biology. Consider the function of your body (breathing, seeing, walking) and connections with other organisms and the nonliving environment (the food you eat, air you breathe, and microorganisms that make you sick). In addition, a multitude of important issues, including AIDS, global warming, nutrition, and population growth, have biological ramifications.

• *Participate in a study group.* Many students find it useful to study together in pairs or small groups. Talking about biology aloud gets another "learning channel" involved, and other members of the group can evaluate your understanding. A study group is a great way to hear different points of view and to find new ways to study and approach the subject.

•*Ask for help.* If you have questions or difficulties, make an appointment to talk with your professor during office hours. Teaching assistants and laboratory instructors are also valuable resources. Many colleges and universities have tutorial services, learning assistance or study skills centers, and computer labs with review programs that provide additional assistance.

I hope this study guide contributes to your understanding of life and to success in your biology course. If you have comments or suggestions, I would like to hear from you c/o The Benjamin/Cummings Publishing Company.

Acknowledgments. I would like to thank the staff at The Benjamin/Cummings Publishing Company—especially Edith Beard Brady, who convinced me that I could write this book, Lisa Donohoe, who got the project off the ground, and Korinna Sodic, who edited the study guide with considerable enthusiasm, skill, and patience. Larry Olsen skillfully guided the book through production. Many thanks to the three authors of *BIOLOGY: Concepts and Connections,* who had confidence in my ability to write the study guide and whose writing set a high standard. I would also like to thank my friend and mentor Bob Thornton, who inspired me and taught me how to teach. A number of reviewers made important suggestions; I hope the study guide meets their expectations. Thanks also to my students and colleagues at Linn-Benton Community College, from whom I learn new things every day. Steve Lebsack of LBCC read the entire manuscript and made many useful suggestions and corrections. Thanks to my friends who did not forget me while I spent many weekends and evenings at the computer. Finally, I would like to thank my wife, Ann Greig, who has been patient, supportive, and encouraging from the start.

Richard Liebaert

Reviewers

Jane Aloi, *Saddleback College*

M.C. Barnhart, *Southwest Missouri State University*

William E. Barstow, *University of Georgia, Athens*

Stephen Dina, *Saint Louis University*

Charles Duggins, Jr., *University of South Carolina*

Martin Hahn, *William Peterson College*

Lazlo Hanzley, *Northern Illinois University*

Jean Helgeson, *Collin County Community College*

George A. Hudock, *Indiana University*

Kris Hueftle, *Pensacola Junior College*

Ursula Jander, *Washburn University of Topeka*

Alan Jaworski, *University of Georgia, Athens*

Steve Lebsack, *Linn-Benton Community College*

Melanie Loo, *California State University, Sacramento*

Joseph Marshall, *West Virginia University*

Presley F. Martin, *Drexel University*

James E. Mickle, *North Carolina State University*

Henry L. Mulcahy, *Suffolk University*

Kathryn Stanley Podwall, *Nassau Community College*

Robert Salisbury, *State University of New York College at Oswego*

Gary Smith, *Tarrant County Junior College*

Mary Colavito-Shepansky, *Santa Monica College*

Gerald Summers, *University of Missouri, Columbia*

Edmund Wodehouse, *Skyline College*

Uko Zylstra, *Calvin College*

Introduction

We are living in a golden age of biology. More biologists (and people who have taken a biology course) are alive today than have lived in all of history. Biology has always been important to human life—from the earliest times we have had to understand our bodies, the plants and animals in our surroundings, and how to obtain food—but in recent years biological issues have taken on a new urgency. Issues such as global warming, overpopulation, AIDS, the challenges of genetic engineering, the green revolution, and depletion of global biodiversity demand that we apply more sophisticated knowledge to biological questions and issues. Just as "war is too important to be left to the generals," so also biology is too important to be left to the biologists. This chapter introduces the subject of biology and its importance in all our lives.

Organizing Your Knowledge

Exercise 1   (Introduction and Module 1.1)

Review the five kingdoms of life and their characteristics by matching each statement on the right with one or more of the kingdoms on the left.

A. Monera
B. Protista
C. Plantae
D. Fungi
E. Animalia

_____ 1. The most widespread of all organisms are in this kingdom
_____ 2. Three kingdoms of multicellular organisms
_____ 3. Protozoans and algae
_____ 4. More named species are in this kingdom than in any other
_____ 5. Multicellular decomposers
_____ 6. Mushrooms, molds, and yeasts
_____ 7. Multicellular organisms that move actively in search of food
_____ 8. These three kingdoms include photosynthetic organisms
_____ 9. Two kingdoms of single-celled organisms
_____ 10. Multicellular photosynthetic organisms
_____ 11. Have simple cells that lack nuclei

Exercise 2   (Modules 1.2 – 1.3)

Fill in the blanks in the following story to review the concepts of evolution.

While investigating the insect life of the rainforest canopy, a zoologist captured several specimens of a previously unknown species of butterfly. The butterfly was mostly black but had conspicuous red and yellow stripes on its wings. It rested on bare tree limbs in plain view; the zoologist was surprised she had not seen it before. The butterfly was very similar in structure to that of a much less conspicuous all-black species found in the same general area, so the zoologist figured that the two species were closely
[1]_____. (Insects are delicate and unlikely to be preserved, so biologists usually must rely on comparisons between living species, rather than on
[2]_____, to reconstruct evolutionary relationships.)

Biologists have long marveled at the diversity of insect life in the tropics. ³_____, the English biologist who wrote *On the Origin of Species*, was surprised by the large number of insect species he encountered in the rain forests of South America. In fact, biologists estimate that most species of living things are rainforest insects.

Like Darwin, the zoologist concluded that the black butterfly and the new species looked alike because they were both descended from a common ⁴_____. But why the difference in color pattern? She speculated that the red and yellow stripes were an ⁵_____, a beneficial feature that evolved by means of natural selection. But how could a bright color pattern be of any possible benefit? Wouldn't brightly colored butterflies be attacked by predators?

The next day the zoologist saw the red and yellow winged butterfly resting on a tree limb. A predatory bird landed nearby and peered at the butterfly. The butterfly responded by rapidly flapping its wings, displaying their striped pattern, and the bird flew off.

The zoologist now suspected that the bright wing pattern was an example of "warning coloration," often seen in harmful or bad-tasting animals—for example, the conspicuous yellow and black stripes of bees and wasps. How could such a color pattern have evolved in this species of butterfly? The zoologist speculated that at one time a ⁶_____ of black butterflies existed in this area, breeding among themselves but not with other members of their species. These butterflies exhibited ⁷_____ traits—slightly different wing shapes, sizes, behaviors, and so on. They also may have tasted different. Perhaps some were able to make a bad-tasting substance or store a bad-tasting substance obtained from food plants. Just as Darwin reasoned, the zoologist realized that ⁸_____ variation must be present in the population for natural selection to operate. In the population of butterflies, good-tasting butterflies were more likely to be eaten by ⁹_____ than bad-tasting ones before they had a chance to ¹⁰_____. The surviving bad-tasting butterflies were able to reproduce, and they passed their ability to make the bad-tasting chemical on to their ¹¹_____. Over time, the bad-tasting butterflies became more numerous. Among the bad-tasting butterflies, there may have been variation in wing coloration. Butterflies with bright colors on their wings were easier for predators to remember and avoid. They had more offspring than less-conspicuous individuals, and eventually bright-colored, bad-tasting butterflies became the norm in the population. In this situation, as in others explained by Charles Darwin, ¹²_____ occurs as heritable ¹³_____ are exposed to ¹⁴_____ factors that favor the ¹⁵_____ success of some individuals over others.

The zoologist concluded that, over a long period, the changes in palatability, wing pattern, and other characteristics must have combined, and a new ¹⁶_____ of butterfly came into existence. According to Darwin, the ¹⁷_____ of new species results from the accumulation of minute changes resulting from natural selection over ¹⁸_____. This short story is just one illustration of evolution, biology's central ¹⁹_____.

Exercise 3   (Modules 1.4 – 1.5)

Review the scientific method by reading the following story. Match each of the statements in italic type with some aspect of the scientific method. Write your answers in the parentheses. Choose from **question, variable, hypothesis, theory, repetition, experimental group, test, controlled experiment, observation, prediction, control group**.

The zoologist *noticed that predatory birds avoided the brightly colored butterflies* (1_____), and she wanted to find out why this occurred. *Did the red and yellow stripes on the butterflies' wings really deter predators? And did the butterflies really taste bad?* (2_____) She *guessed that the butterflies did taste bad and that their stripes served as a warning* (3_____), and she decided to *try out this guess* (4_____) in the laboratory.

She captured several insect-eating birds native to the area and put them in cages at a nearby tropical field station. Then she netted a number of the brightly striped butterflies and their black cousins. For her first experiment, she allowed birds to choose between a black butterfly and a striped one. They invariably chose the black butterflies and avoided the striped ones. This confirmed her field observations.

Next she used black paint to cover the bright stripes of a number of the striped butterflies. These *altered butterflies* (5_____) were offered to predators, along with some *"normal" striped butterflies as a comparison* (6_____). Actually, the normal striped butterflies were also painted, but with a clear paint, so that *only one factor—wing pattern—would differ between the two groups* (7_____). It is important to set up *an experiment like this, so that it is two parallel tests with only one factor differing between the two* (8_____). The researcher speculated that, *if the wing pattern really did warn predators, they would be fooled* (9_____) by the painted wings and try to eat the black-painted butterflies.

Just as the zoologist suspected, the birds chose the black-painted butterflies in every trial. This showed that the wing pattern did indeed function as a warning. Also, most of the birds quickly spit out the black-painted butterflies, and those who swallowed the butterflies became ill. The zoologist performed another experiment in which she painted the wings of the edible black-winged butterflies. The birds ate them with gusto, demonstrating that the paint itself was not distasteful and produced no ill effects.

The zoologist *performed the experiment several times* (10_____), verifying her technique and conclusions. Then she wrote a short paper for *The Journal of Tropical Entomology* describing this new example of warning coloration. The evolution of bad-tasting and brightly colored insects is another example that illustrates natural selection, Charles Darwin's *powerful idea that explains biological adaptations of many kinds* (11_____).

These modules discuss the hierarchy of structural levels into which life is organized. Each of these levels has unique emergent properties, which arise from the organization of its component parts. Review this structural hierarchy by completing the chart below.

Level	Description
Ecosystem	1.
2.	All the populations of organisms in a particular area
Population	3.
4.	An individual living thing
5.	Organs that work together to perform particular functions
Organ	6.
7.	A group of cells with a specific function
8.	A unit of living matter, separated from its environment by a membrane
9.	Atoms and molecules

Life is characterized by a particular set of emergent properties. Module 1.8 lists six characteristics that we associate with the state of being alive. Write a short paragraph that shows how you, a pet, or a house plant illustrates these six characteristics.

Exercise 6   (Modules 1.9 – 1.13)

These modules discuss the emergent properties seen at various levels in life's structural hierarchy—molecules, cells, organisms, and ecosystems. Review these properties and the hierarchy by completing the crossword puzzle.

Across
   3. The cellular level is the lowest level where we see ____.
   6. Each level in life's hierarchy has ____ properties.
   9. ____ structure and arrangement underlies all the properties of life.
  10. A central theme in biology is the connection between form and ____.
  11. The ____ is the highest level in the hierarchy of life.
  13. Most cells of multicellular organisms are highly ____.
  15. Differences in organisms reflect differences in their ____ nucleotide sequences.

Down
   1. DNA contains ____ information.
   2. Natural ____ shapes structures that allow an organism to perform certain functions.
   4. DNA information is in its sequence of ____.
   5. There is a complex web of ____ among organisms in an ecosystem.
   7. ____ and protists are single-celled organisms.
   8. Bacteria, fungi, and soil animals ____ the remains of dead organisms.
  12. ____ make most of the food in a forest ecosystem.
  14. The ____ is life's basic unit of structure and function.

Exercise 7 (Module 1.14)

On a separate sheet of paper, list six ways in which you think the science of biology and its technological applications may affect society in the next decade. Which of these are primarily scientific? Which are technological? Which do you think will have the most effect on you personally?

Testing Your Knowledge

Multiple Choice

1. Approximately how many species of living things have been identified and named?
   a. 250,000
   b. 750,000
   c. 1.5 million
   d. 5 million
   e. 30 million

2. Scientists think that most rainforest animals have not yet been identified and named. What is the primary reason described in the text for this lack of knowledge of the rain forest?
   a. Very few scientists are working in the rain forest.
   b. Most species live in the forest canopy, which is hard to study.
   c. Most species live in the rich tropical soil, which is of little interest to biologists.
   d. The heat and humidity of the tropics make it difficult to study the rain forest.
   e. The political situation in tropical countries makes it hard to work there.

3. Biologists group living things into ___ kingdoms.
   a. 2
   b. 3
   c. 4
   d. 5
   e. about 10

4. A bacterium and an amoeba are placed in different kingdoms because
   a. a bacterium is single-celled.
   b. an amoeba is photosynthetic.
   c. an amoeba can move.
   d. a bacterial cell is much simpler.
   e. an amoeba is single-celled.

5. Most of the organisms in Kingdom _____ are photosynthetic.
   a. Animalia
   b. Protista
   c. Plantae
   d. Fungi
   e. Monera

6. Which of the following provides clues about evolutionary relationships among organisms?
   a. DNA nucleotide sequences
   b. comparison of body structures
   c. fossils
   d. all of the above
   e. b and c only

7. An educated guess posed as a tentative explanation is called a
   a. theory.
   b. control.
   c. variable.
   d. prediction.
   e. hypothesis.

8. Which of the following is the simplest thing displaying the characteristics we associate with life?
   a. a population of deer
   b. a DNA molecule in a bone cell of a deer
   c. a deer
   d. the heart of a deer
   e. a deer skin cell

9. The ____ is the highest level in life's structural hierarchy.
   a. ecosystem
   b. cell
   c. organism
   d. population
   e. molecule

10. Which of the following properties of life is not seen at the organism level?
    a. ability to take in and use energy
    b. ability to evolve
    c. precise structural organization
    d. ability to reproduce
    e. capacity for growth and development

11. All of the properties of life arise from
    a. the organization of chemicals.
    b. the organism level in the structural hierarchy.

   c. relationships within the ecosystem.

   d. life's ability to reproduce.

   e. the properties of cells.

12. There are many interdependencies in an ecosystem. Bacteria and fungi play an important role in the ecosystem primarily because they
   a. cause diseases that keep populations in check.
   b. trap water, which is then used by other organisms.
   c. decompose the remains of dead organisms.
   d. are responsible for producing energy.
   e. do photosynthesis, which makes the food for other species.

Essay

1. Name the five kingdoms of life, identify the organisms in each, and briefly describe the criteria that separate each kingdom from the others.

2. Explain the difference between a hypothesis and a theory.

3. What kinds of questions can be answered and what kinds of problems can be solved by science? What kinds of questions are outside the realm of science?

4. Explain how the information in DNA relates to the emergent properties that characterize life.

Applying Your Knowledge

Multiple Choice

1. A crop scientist noted that, over a period of 10 years, a beetle species that feeds on rice gradually became resistant to insecticide. Which of the following best explains this in terms of natural selection?
   a. The insecticide mutated the beetles exposed to the biggest doses.
   b. Some beetles learned to tolerate the insecticide and passed this ability to their offspring.
   c. Beetles learned to avoid the spray and passed the knowledge to their offspring.
   d. The insecticide caused the beetles to reproduce more quickly than normal.
   e. Those beetles with natural resistance to the insecticide had the most offspring.

2. Researchers testing new drugs usually give the drug to one group of people and give placebos, "sugar pills," to another group. The group receiving the sugar pills
   a. constitutes the experimental group.
   b. is needed so that the test will be repeated enough times.
   c. is the control group.
   d. is a backup in case some of the people getting the drug drop out of the test.
   e. is the experimental variable.

3. ____ has emergent properties that result from the organization of its ____ .
   a. A population . . . ecosystems
   b. A tissue . . . organs
   c. A cell . . . tissues
   d. An organism . . . organ systems
   e. A molecule . . . cells

4. An ecologist studied the effect of nutrients and predators on the population growth of bacteria on the bottom of a pond. His study of bacteria could not involve which of the following levels in life's structural hierarchy?
   a. ecosystem
   b. organ
   c. organism
   d. population
   e. chemical

5. A rainforest primate called an aye-aye has a long middle finger that it uses to probe for insects in cracks and crevices in tree bark. This connection between structure and function developed gradually as a result of
   a. reproduction.
   b. population growth.
   c. natural selection.
   d. DNA replication.
   e. energy exchange.

6. Which of the following does *not* illustrate technology?
   a. DNA research is used to cure an inherited disease.
   b. Scientists develop a bacterium that destroys toxic wastes.
   c. A biologist identifies a new species of monkey.
   d. A chemical that slows cell division is used to treat cancer.
   e. Biologists breed a disease-resistant kind of corn.

Essay

1. Choose a familiar wild animal—a squirrel, a toad, or a duck, for example—and describe some of the web of relationships that connect it with other organisms in its ecosystem.

2. Beavers are descendants of land-dwelling rodents similar to rats. Explain how natural selection could have shaped the beaver's flat tail and webbed feet, which it uses for swimming.

3. Jason tried a new fertilizer called MegaGro on his garden. He said, "I used it on all my tomato plants this year, and they grew much better than they did last year! MegaGro is fantastic!" Was Jason's test of MegaGro scientifically valid? Why or why not?

4. Biologists think that oilbirds might be able to avoid obstacles in their caves and in the tangled growth of the rain forest much the way bats do—by making sounds and listening to the echoes. Describe a controlled experiment to test this hypothesis.

5. Explain how each of the following shows the connection between form and function: your hand, a leaf, a hawk's beak, a frog's hind legs.

6. Describe the hierarchy of structural levels of which your body is composed. Give a specific example of an emergent property at one level that is not seen in the parts that make it up.

7. A camera sends back pictures of purple gelatinous blobs in near-boiling water near volcanic vents on the ocean floor. What properties should scientists look for to determine whether the blobs represent life?

8. How is human DNA different from the DNA of a chimpanzee? The DNA of a goldfish?

Using Your Knowledge

1. Take a short walk outside, or look out a window, and describe how what you see relates to life's five kingdoms, life's hierarchy of structural organization, emergent properties, and the characteristics of life.

2. For one week, keep a list of biological issues in the news—the newspaper, magazines, television, radio, and conversations with friends and family. How many different issues did you encounter during the week? What were the major categories of biological issues in the news?

The Chemical Basis of Life

2

You may have read something like "The value of all the chemicals in the human body is about 97 cents." Putting a price tag on our chemicals misses the point. This book was written on a microcomputer. If you hit the computer with a sledgehammer, the chemicals in the pile of junk would probably be worth about 97 cents. But the value of the computer has less to do with the kinds of materials that go into it than how the materials are arranged. Like computers, living things are made from a few kinds of atoms obtained from water, soil, and air. These atoms, following a few basic laws, form the molecules and cells of a human being or a pine tree. What is special about life is not the chemicals themselves but the way they are organized to do what living things do—from the DNA in every cell to the valves that guide blood through the heart. It is easy to believe that a human being is made up of 97 cents' worth of chemicals; the truly amazing thing is that 97 cents' worth of chemicals can make a human being. This chapter is about the chemicals of life and the rules they follow in shaping a living thing.

Organizing Your Knowledge

Exercise 1 (Introduction)

Review contrasting approaches to the study of life by stating whether each of the following represents the reductionist approach (study of parts) or a holistic approach (study of the functioning whole).

_____ 1. A scientist studying global warming collects data about logging, agriculture, industry, transportation, and studies of natural ecosystems.

_____ 2. Microbiologists take an AIDS virus apart in hopes of finding out how it enters cells.

_____ 3. Researchers studying nerve repair remove cells from a spinal cord and watch them grow in a nutrient solution.

_____ 4. Having studied the behavior of individual birds, zoologists then try to predict how a flock will behave.

Exercise 2   (Module 2.1)

These diagrams illustrate the same biological levels of organization discussed in connection with the aspen tree in the introduction and in Module 2.1. In this case, the organism depicted is a zebra, but the concepts are the same. Name and number the levels, starting with the smallest. Use the names **tissue, molecule, cell, organism, system,** and **organ**.

1._____

2._____

3._____

4._____

5._____

6._____

Exercise 3   (Module 2.2)

Write the chemical symbol for each of the following elements, and state whether it is used by living things in large (L), moderate (M), or trace (T) amounts.

Symbol	Amount	Element	Symbol	Amount	Element
		1. Magnesium			7. Carbon
		2. Oxygen			8. Calcium
		3. Zinc			9. Phosphorus
		4. Hydrogen			10. Nitrogen
		5. Copper			11. Sodium
		6. Iodine			12. Iron

Exercise 4   (Module 2.4)

This module introduces atoms. It is most important to know what the subatomic particles are, where they are located in an atom, and that atoms of different elements differ because they contain different numbers of protons. Some atoms not covered in the module are compared below. Use the concepts in the module to fill in the blanks. Then sketch each atom, labeling and coloring **protons** blue, **neutrons** green, and **electrons** red. Also use + and – signs to remind yourself of the electrical charges carried by the particles.

Element	Symbol	Atomic Number	Mass Number	Number of Protons	Number of Neutrons	Number of Electrons
1. Carbon-12	C	6	12	6	6	6
2. Nitrogen-14	___	7	14	___	___	___
3. Chlorine-35	___	___	35	17	___	___
4. Oxygen-16	___	___	___	___	___	8
5. Oxygen-17	___	___	___	___	___	___

Exercise 5   (Modules 2.6 – 2.8)

The atoms of four elements important to life are diagrammed below. Pay particular attention to their electron shells. Remember that atoms with incomplete outer electron shells participate in chemical reactions that allow them to attain complete outer shells: 2 electrons for a hydrogen atom, 8 electrons for most other elements important to life.

NITROGEN (N)
Atomic number = 7

CHLORINE (Cl)
Atomic number = 17

HYDROGEN (H)
Atomic number = 1

CALCIUM (Ca)
Atomic number = 20

1. Given the information and diagrams above, show how electrons would be transferred between calcium and chlorine atoms to form calcium and chloride ions, which would then attract each other to form calcium chloride, $CaCl_2$. (Hint: An atom can gain or lose more than one electron.)

2. Using the information and diagrams above, show how nitrogen could form covalent bonds with several hydrogen atoms, forming a molecule of ammonia. What would be the molecular formula for ammonia?

Exercise 6   (Modules 2.9 – 2.14)

Review the properties of water by filling in the blanks in the following story.

When Amy came through the door, she found Liz poised over a glass of water, ready to drop a needle into the glass. Amy asked, "Liz, what are you trying to do?"

"We're studying the [1]_____ basis of life in my biology class," Liz replied. "I don't believe some of this stuff she's teaching us, so I need to do some experiments to figure it out."

Liz gently placed the needle on the water surface. "Watch this," she said. The needle rested in a dimple on the surface of the liquid.

"How did you do that?"

"I didn't. The water did. Water molecules have a tendency to stick together, which is called [2]_____. The water molecules are stuck together so tightly at the surface that they form a film that can support the weight of the needle. Bugs can walk on it. It's called [3]_____."

Amy was getting interested. She asked, "So how do the water molecules do it? What's so special about water?"

Liz explained, "A water molecule is H_2O, right? It is made of one [4]_____ atom and two [5]_____ atoms. The atoms stay together because they [6]_____ electrons. This holds them together. A shared pair of electrons forms a chemical bond called a [7]_____ bond between each hydrogen and the oxygen. Now if the electrons were shared evenly, the bond would be called a [8]_____ covalent bond. But they are not shared evenly. The oxygen tends to 'hog' the electrons away from the hydrogens. It has a greater attraction for electrons; it is more [9]_____ than hydrogen."

"So what does this have to do with floating needles?"

"Well, because the oxygen atom attracts the electrons more strongly, the shared electrons are closer to the oxygen than to the hydrogens, giving the oxygen a slight [10]_____ charge. Because the electrons are pulled away from the protons in the nuclei of the hydrogen atoms, the hydrogens are left with slight [11]_____ charges. So the bonding electrons are shared unevenly, producing a [12]_____ covalent bond between each hydrogen and the oxygen. In fact, the whole water molecule is polar, even though the molecule as a whole is electrically [13]_____."

Amy was getting impatient. "So what does that have to do with surface tension? And what's the biology connection?"

Liz went on, "Well, it is their polarity that causes water molecules to stick together. The [14]_____-charged oxygen of one water molecule is attracted to the [15]_____-charged hydrogens of other water molecules. These special bonds between water molecules are called [16]_____ bonds. These bonds form a network at the water's surface, creating surface tension strong enough to support the needle. Each water molecule is connected to [17]_____ others. Hydrogen bonds give water some other peculiar properties. For example, water is the only common substance on earth that naturally exists in all three states of matter—[18]_____,

[19]_____, and [20]_____. And lots of things will dissolve in water; it is a versatile [21]_____. Blood plasma, for example, is an [22]_____ solution containing many different [23]_____, or dissolved substances, such as salt and blood sugar."

Amy got up and opened the bathroom door, looked inside, and said, "It's steamy in there. Are you going to take a bath?"

Liz replied, "No, that's just another experiment. I'm trying to figure out the difference between heat and temperature."

"Are they different?"

"Yes. [24]_____ is the total amount of energy resulting from the movement of molecules in a body of matter, like a bathtub full of water. [25]_____ measures the intensity of movement. I compared the amount of heat in a cup of water at 98°C and a bathtub of water at 45°C. In the [26]_____, the intensity of movement of water molecules was greater, but the [27]_____ held more heat energy. I knew it did because the bathtub of water added more heat to the room as it cooled, warming up the room more than the cup of hot water did.

"Water has a great capacity to store heat, by the way. When water is heated, a lot of the energy goes into breaking the [28]_____ between water molecules before the molecules can move faster. For instance, if you had a kilogram of water and a kilogram of rock, the same amount of heat would raise the temperature of the water [29]_____ than the temperature of the rock. This means water can soak up a lot of heat, and its temperature will go up only a few degrees."

"And when water cools a few degrees, it [30]_____ a lot of heat."

"Correct. And since animals are mostly water, this helps us control our body temperature. It also stabilizes the temperatures of the ocean and coastal areas. In the summer, the ocean [31]_____ heat, and in the winter, it [32]_____ heat."

Amy's eyes narrowed. "So why do we sweat when we are hot? Wouldn't we want to hang onto all that good water?"

Liz was ready with an answer. "No, not necessarily. Because of their strong hydrogen bonds, it takes a lot of heat energy to get a water molecule moving [33]_____ enough to [34]_____ , to separate from its neighbors. This gives water an unusually high [35]_____, but it also makes [36]_____ cooling possible. The hottest—or fastest moving—water molecules evaporate first, taking a lot of heat energy with them and leaving the cooler—slower—molecules behind. So sweating cools you off on a hot day."

Amy looked at the clock and said, "It's 3:30. I told Sara I'd meet her at the ice rink at 3:30."

This inspired Liz anew. "Ice. Now, ice is very interesting. In ice the water molecules are locked into a crystal, linked by hydrogen bonds, but farther apart than they are in liquid water. This means that ice is [37]_____ dense than liquid water, so it [38]_____. This is important to life, because . . . "

But Amy was already out the door. Liz had a puzzled expression on her face as she got up and slid an ice tray out of the freezer.

Exercise 7   (Modules 2.15 – 2.16)

Practice using the pH scale by giving the approximate pH of each of the following. Some are listed in the modules; others you can estimate from the information given.

____ 1. Tomato juice

____ 2. Human blood

____ 3. Vinegar (moderately acidic)

____ 4. Pure water

____ 5. Cola (mildly acidic)

____ 6. Household ammonia

____ 7. Concentrated nitric acid (very acidic)

____ 8. Acid precipitation

____ 9. Drain cleaner (very basic)

____10. Antacid pills (mildly basic)

____11. Urine

____12. Gastric juice

Exercise 8   (Module 2.17)

This module introduces chemical reactions, chemical processes that change matter. A common chemical reaction in many cells is one that changes hydrogen peroxide (H_2O_2) into water and oxygen gas:

$$2H_2O_2 \rightarrow 2H_2O + O_2$$

Hydrogen peroxide is a harmful by-product of many reactions. Cells get rid of it by carrying out the above reaction, converting it to harmless water and oxygen. Note that the equation for a chemical reaction must be "balanced." Since atoms cannot be created or destroyed in a chemical reaction—only rearranged—the numbers of atoms on both sides must be equal. In this example, there are four hydrogen atoms in the two hydrogen peroxide molecules on the left. After the reaction occurs, the four hydrogen atoms reappear in the two water molecules on the right. Similarly, the four oxygen atoms reappear on the right. Your task in this exercise is simple: Label the reactant(s) and the product(s).

Exercise 9 (Summary)

Review basic chemical terminology by completing this crossword puzzle.

Across

2. ____ is the energy due to movement of molecules in a body of matter.

5. An ____ is a subatomic particle that circles an atom's nucleus.

7. The smallest particle of an element is called an ____.

9. An ____ is a charged atom or molecule.

12. Acid ____ is caused by pollutants that combine with water in the air.

16. ____ is anything that occupies space and has mass.

17. Two or more atoms held together by chemical bonds form a ____.

19. Neutrons and protons are found in an atom's ____.

22. The cohesion of water molecules is responsible for surface ____.

23. Variant forms of elements with different numbers of neutrons are called ____.

24. ____ is the tendency of water molecules to stick together.

Down

1. A ____ is a subatomic particle with no electrical charge.

3. ____ measures the intensity of heat.

4. Electrons are shared unequally in a ____ covalent bond.

5. There are 92 naturally occurring ____.

6. A ____ contains two or more elements in a fixed ratio.

8. Weak bonds between water molecules are called ____ bonds.

10. An ____ donates H^+ ions to solutions.

11. A ____ bond is formed when two atoms share electrons.

13. A ____ is a positively charged particle from the nucleus of an atom.

14. In a solution, the dissolving agent is called the ____.

15. A ____ is a liquid containing a homogeneous mixture of substances.

18. When two ions of opposite charges attract each other, an ____ bond forms.

20. The ____ is the substance dissolved in a solution.

21. A ____ accepts H^+ ions and removes them from solution.

Testing Your Knowledge

Multiple Choice

1. Which of the following is a trace element, required only in small amounts by most living things?
   a. oxygen
   b. iron
   c. nitrogen
   d. carbon
   e. hydrogen

2. An acid is a substance that
   a. dissolves in water.
   b. forms covalent bonds with other substances.
   c. donates hydrogen ions to solutions.
   d. is a versatile solvent.
   e. removes hydrogen ions from solutions.

3. How an atom behaves when it comes into contact with other atoms is determined by its
   a. nucleus.
   b. size.
   c. protons.
   d. neutrons.
   e. electrons.

4. Most of water's unique properties result from the fact that water molecules
   a. are very small.
   b. tend to repel each other.
   c. are extremely large.
   d. tend to stick together.
   e. are in constant motion.

5. Atoms of different phosphorus isotopes
   a. have different atomic numbers.
   b. have different numbers of neutrons.
   c. react differently with other atoms.
   d. have different numbers of electrons.
   e. have different numbers of protons.

6. An ion is formed when an atom
   a. forms a covalent bond with another atom.
   b. gains or loses an electron.
   c. becomes part of a molecule.
   d. gains or loses a proton.
   e. gains or loses a neutron.

7. The smallest particle of water is
   a. an atom.
   b. a crystal.
   c. an element.
   d. a compound.
   e. a molecule.

8. Why are biologists so interested in chemistry?
   a. Chemicals are the fundamental parts of all living things.
   b. Most chemicals are harmful to living things.
   c. They know little about life except the chemicals it is made from.
   d. If you understand the chemistry of life, you can make a lot of money.
   e. Everything about life can be known by understanding its chemistry.

9. Molecules are always moving. Some molecules move faster than others; _____ is a measure of their average velocity of movement.
   a. polarity
   b. heat
   c. temperature
   d. electronegativity
   e. density

10. Which of the following holds atoms together in a molecule?
    a. ionic bonds between atoms
    b. transfer of protons from one atom to another
    c. sharing of electrons between atoms
    d. loss of neutrons by atoms
    e. sharing of protons between atoms

11. Ice floats because
    a. it is colder than liquid water.
    b. its molecules are moving faster than in liquid water.
    c. it is more dense than liquid water.
    d. its hydrogen molecules bond to the water surface film.
    e. its water molecules are farther apart than in liquid water.

12. Adding acid tends to ____ of a solution.
    a. increase the hydrogen ion concentration and raise the pH
    b. increase the hydrogen ion concentration and lower the pH
    c. decrease the hydrogen ion concentration and raise the pH
    d. decrease the hydrogen ion concentration and lower the pH
    e. **c** or **d**, depending on the original acidity

Essay

1. List the four elements needed by living things in large amounts, two others needed in moderate amounts, and two elements needed in trace amounts.

2. Explain why the smallest particle of iron is an atom, but the smallest particle of water is a molecule.

3. Explain the following statement: The temperature of the water in a tea kettle is higher than the temperature of a swimming pool, but the swimming pool contains more heat.

4. How does acid precipitation form? How does it injure animals? Plants?

5. Explain why water molecules are polar, how this makes them tend to bond to each other, and how this causes water to have a large heat-storage capacity.

6. Explain how evaporation of water from your skin cools you on a hot day.

Applying Your Knowledge

Multiple Choice

1. Which of the following best states a reductionist point of view?
   a. You can understand something by taking it apart.
   b. Small things are more complex than large ones.
   c. A team can accomplish tasks individual members cannot.
   d. A system has functions more complex than its parts.
   e. If you look at something long enough, you will understand it.

2. An atom that normally has _____ in its outer shell would tend *not* to form chemical bonds with other atoms.
   a. 1 electron
   b. 3 electrons
   c. 4 electrons
   d. 6 electrons
   e. 8 electrons

3. Researchers studying the effects of toxic wastes knew that animals were poisoned by the heavy metal cadmium, but they wanted to know where cadmium accumulated in the body. They could find out by
   a. tracing the movement of cadmium isotopes in test animals.
   b. measuring the size of cadmium atoms.
   c. finding out whether cadmium atoms form ionic or covalent bonds.
   d. finding out whether cadmium is acidic in water.
   e. determining the number of bonds formed by cadmium atoms.

4. Changing the number of _____ would change it into an atom of a different element.
   a. bonds formed by an atom
   b. electrons circling the nucleus of an atom
   c. protons in an atom
   d. particles in the nucleus of an atom
   e. neutrons in an atom

5. A glass of grapefruit juice, at pH 3, contains _____ as much H^+ as a glass of tomato juice, at pH 4.
   a. one-tenth as much
   b. half as much
   c. twice as much
   d. three times as much
   e. ten times as much

6. Fluorine atoms tend to take electrons from any atoms that come near. As a result, fluorine atoms
   a. tend to become positively charged.
   b. are nonpolar.
   c. do not react readily with other atoms.
   d. tend to form ionic bonds.
   e. are not very electronegative.

7. Tim added 10 milliliters (mL) of hydrochloric acid and 10 mL of water (pH 7) to a beaker containing 100 mL of water. The pH of the resulting solution was 4. Next he is going to add 10 mL of hydrochloric acid and 10 mL of pH 7 buffer to a different beaker containing 100 mL of water. What do you think will happen?
   a. The resulting pH will be less than 4.
   b. The resulting pH will be between 4 and 7.
   c. The resulting pH will be 7.
   d. The resulting pH will be between 7 and 11.
   e. The resulting pH will be greater than 11.

8. A sodium atom has a mass number of 23. Its atomic number is 11. How many electrons does it have (if it is not an ion)?
   a. 11
   b. 12
   c. 22
   d. 23
   e. 34

9. Which of the following is the smallest in volume?
   a. nucleus of an oxygen atom
   b. water molecule
   c. proton
   d. ice crystal
   e. electron cloud of an oxygen atom

10. Potassium chloride consists of potassium ions (K^+) and chloride ions (Cl^-) in a crystal. If potassium chloride is placed in water, what do you think happens?
    a. The K^+ ions are attracted to the oxygen atoms of water molecules.
    b. It will not dissolve.
    c. The Cl^- ions are attracted to the oxygen atoms of water molecules.
    d. It acts as an acid.
    e. The K^+ ions are attracted to the hydrogen atoms of water molecules.

Essay

1. A sulfur atom has 6 electrons in its outer shell. How many covalent bonds is it likely to form with other atoms? Why? What do you think the formula for hydrogen sulfide would be?

2. Plants carry out chemical reactions that make sugars, which contain carbon, hydrogen, and oxygen, from carbon dioxide (CO_2) and water (H_2O). Researchers want to know whether the oxygen atoms in sugar come from the carbon dioxide or the water. How could they use radioactive tracers to find out? What would they look for in terms of results?

3. If you drop a hot 10-kg rock into 10 kg of cold water, the rock cools off and the water warms up, but the final temperature of both is much closer to the starting temperature of the water than that of the rock. Why?

4. A sugar molecule contains carbon, hydrogen, and oxygen atoms. The oxygen atoms tend to steal electrons from the other atoms in the molecule, giving the oxygen atoms negative charges and leaving other parts of the sugar molecule positively charged. The molecules in oil, however, consist only of hydrogen and carbon, which share electrons equally. Oil molecules do not have areas of positive and negative charge. Using this information and what you know about water, explain why sugar mixes with water but oil does not.

5. Why are the ratios of elements in molecular formulas always fixed—H_2O instead of H_5O, and CH_4 instead of CH_6? Explain this in terms of the characteristics of the atoms making up the molecules.

6. On a typical July day in Seattle, Washington, the high temperature is around 75°F, the low around 55°F. In January, the average high is around 45°F, the low 35°F. In Minneapolis, Minnesota (at roughly the same latitude as Seattle), July highs average around 83°F, lows around 60°F; January highs average around 20°F, lows close to 0°F. Explain the difference in terms of the concepts discussed in this chapter.

7. If you heat up a mixture of water and alcohol, the alcohol evaporates first, leaving most of the water behind. Why do you think this happens?

Using Your Knowledge

1. Do you take vitamin pills? The minerals in a vitamin and mineral supplement are most of the trace elements your body needs to function. Take a look at a vitamin bottle and list these trace elements. What other elements does your body need besides these? In smaller or larger amounts? Why is it not necessary to take carbon or sodium pills?

2. Most acid precipitation results from pollutants produced by electric power plants. How might you reduce acid precipitation by recycling items like aluminum cans? Even if you live in a part of the country where electricity is generated in other ways, such as hydropower, how could your recycling reduce acid precipitation somewhere else?

3. Have you ever gone scuba diving or snorkeling? When you release air under water, the bubbles that form are always spherical, never square or pyramidal. Have you ever wondered why? Can you explain it in terms of hydrogen bonding and surface tension?

The Molecules of Cells

3

Now that human beings have visited the moon, and now that space shuttles and space stations constantly orbit the Earth, some people hope that our goal in space for the next century will be to send people to Mars. One of the attractions of the red planet is the slim possibility that life exists there. The Viking missions of the 1970s tested Martian soil for the presence of life, and the results were at best inconclusive. Most scientists believed then, and still believe, that if we find life elsewhere—on Mars, or perhaps on Titan, a moon of Saturn, or maybe by listening for radio transmissions from other star systems—the life we find will be based on the chemistry of carbon. On Earth, only carbon seems able to form the variety and complexity of stable compounds that can perform the myriad of activities needed to produce life. This chapter is about carbon and carbon compounds in life.

Organizing Your Knowledge

Exercise 1 (Module 3.1)

The great variety of organic compounds results from the ability of carbon atoms to bond with four other atoms, forming branching chains of different lengths. Several hydrocarbon molecules, consisting only of carbon and hydrogen, are shown in Module 3.1. Practice seeing the versatility of carbon by sketching some hydrocarbon molecules of your own, as suggested below.

1. Sketch a hydrocarbon molecule that is a straight chain, containing five carbon atoms and twelve hydrogen atoms, molecular formula C_5H_{12}:

   Question: Why does each carbon bond to four other atoms?

2. Now sketch a shorter hydrocarbon chain, with only four carbon atoms:

   Question: What is the molecular formula of the above molecule?

3. Sketch another five-carbon hydrocarbon, but this time include one double bond:

   Question: What is the molecular formula of this molecule?

4. Sketch a five-carbon hydrocarbon molecule that is branched (and contains no double bonds):

   Question: What is the molecular formula of this molecule? What is the term for its relationship to molecule 1 (in this module)?

5. Sketch two five-carbon hydrocarbon molecules in the form of rings, one without double bonds and one with a single double bond.

   Question: How many hydrogen atoms are in each of these molecules?

Exercise 2   (Module 3.2)

Circle the functional groups that are discussed in this module in the molecules below. Label an example of each of the following: **hydroxyl group, carbonyl group, carboxyl group, amino group**. There are a total of ____ hydroxyl groups, ____ carbonyl groups, ____ carboxyl groups, and ____ amino groups. (The properties of the molecules are described at the right.)

 Formaldehyde is the starting point for making many chemicals.

 Formic acid is in ant venom.

 Lactic acid builds up as a waste product in exercising muscles and makes them feel tired.

Ethylene glycol is in automobile antifreeze.

 Acrolein is produced when meat is heated; it is the barbecue smell.

 Serine is part of many protein molecules.

Urea is a waste product in urine.

 Putrescene's name is descriptive; it is produced by rotting flesh.

Exercise 3   (Module 3.3)

There are four main classes of macromolecules. Most are polymers, assembled from smaller monomers in a process called dehydration synthesis. Hydrolysis breaks the polymers back down to monomers. State whether each of the following relates to dehydration synthesis or hydrolysis.

_____ 1. Connects monomers to form a polymer.
_____ 2. Produces water as a by-product.
_____ 3. Breaks up polymers, forming monomers.
_____ 4. Water is used to break bonds between monomers.
_____ 5. Joins amino acids to form a protein.
_____ 6. Glycerol and fatty acids combine this way to form a fat.
_____ 7. Occurs when polysaccharides are digested to form monosaccharides.
_____ 8. —H and —OH groups form water.
_____ 9. Nucleic acid breaks up to form nucleotides.
_____10. Water breaks up, forming —H and —OH groups on separate monomers.

Exercise 4   (Modules 3.3 – 3.7)

After reading these modules, review carbohydrates by filling in the blanks in the following story.

Carbohydrates are a class of molecules ranging from the simplest sugars, called
^{1}_____, to giant molecules called ^{2}_____, built of many sugars. Carbohydrates are the main fuel molecules for cellular work.

Plants make their own carbohydrates, but humans, like all animals, must obtain them from plants or other animals. Imagine eating a piece of whole-wheat bread spread with strawberry jam. It contains a mixture of carbohydrates, along with other macromolecules like ^{3}_____ and ^{4}_____. Much of the carbohydrate in the bread itself is in the form of a polysaccharide called ^{5}_____, which is simply a chain of ^{6}_____ monomers. The monomers were linked together in the wheat plant in a process called ^{7}_____ synthesis. As the glucose units joined, ^{8}_____ was produced as a by-product. When you swallow a bite of bread, digestive juices in the intestine separate the monomers in the opposite reaction, called ^{9}_____. In the intestine, this is actually a two-step process; secretions from the pancreas first break the starch down to maltose, a type of carbohydrate called a ^{10}_____, which consists of two glucose monomers. Secretions from the walls of the intestine complete the process, breaking each maltose molecule down to two individual glucose molecules.

There are other carbohydrates in the bread and jam. Whole wheat flour contains the tough coats of the wheat seeds. These contain a lot of ^{11}_____, the fibrous polysaccharide that makes up plant cell walls. Like starch, it is made of glucose monomers, but these monomers are ^{12}_____ in a different orientation. The human digestive tract is not capable of ^{13}_____ cellulose, so it passes through the digestive tract unchanged, in the form of ^{14}_____. Sucrose, a ^{15}_____ refined from sugar cane or sugar beets, is used to sweeten the strawberry jam. Each sucrose molecule is hydrolyzed in the small intestine to form one molecule of ^{16}_____ and one molecule of ^{17}_____. The jam naturally contains a small amount of fructose, a [18]_____that is produced by strawberries and is considerably sweeter than sucrose.

Now that all the carbohydrates have been hydrolyzed to small monosaccharides, they can be absorbed by the body. Glucose and fructose pass through the wall of the intestine and into the bloodstream, which carries them to the liver. Here the fructose is converted to glucose. This process is relatively easy because glucose and fructose are [19]_____, having the same molecular formula, [20]_____, but slightly different structures. Glucose circulates around the body as "blood sugar" and is taken up by the cells for fuel as needed. Extra glucose molecules are taken up by liver and muscle cells and linked together by [21]_____ synthesis to form a polysaccharide called [22]_____. This molecule is similar to [23]_____, except it is more branched. Later the glycogen can be hydrolyzed to release [24]_____ into the blood.

Exercise 5   (Modules 3.8 – 3.10)

Review the structures and functions of lipids by completing the following crossword puzzle.

Across

3. _____ means that hydrogen has been added to unsaturated fats.
4. _____ is a steroid common in cell membranes.
5. A _____ is similar to a fat; found in cell membranes.
8. A fat molecule is composed of _____ and 3 fatty acids.
9. Glycerol and 3 _____ acids make a triglyceride.
10. _____ is another name for "fat."
11. A _____ forms a waterproof coat that keeps a fruit or insect from drying out.
15. Olive and corn _____ are examples of unsaturated fats.
16. Fats with double bonds are said to be _____.
17. _____ is a lipid-containing deposit in a blood vessel.
18. _____ are grouped together because they do not dissolve in water.

Down

1. _____ is when lipid-containing deposits block blood vessels.
2. Female and male sex hormones are examples of _____.
6. Animal fats are said to be _____.
7. Lipids are water-avoiding, or _____ substances.
12. Unsaturated fats contain more _____ bonds than saturated fats.
13. A _____ is a large molecule whose main function is energy storage.
14. _____ steroids are dangerous variants of testosterone.

Exercise 6   (Module 3.11)

Everything a cell does involves proteins. Seven classes of proteins are discussed in Module 3.11. Match each of the classes with one of the descriptions below.

_____ 1. Hemoglobin carries oxygen in the blood.
_____ 2. A protein in muscle cells enables them to move.
_____ 3. Antibodies fight disease-causing bacteria.
_____ 4. Collagen gives bone strength and flexibility.
_____ 5. Insulin signals cells to take in and use sugar.
_____ 6. Proteins in seeds provide food for plant embryos.
_____ 7. A protein called sucrase promotes the chemical conversion of
                        sucrose into monosaccharides.

Exercise 7   (Modules 3.12 – 3.13)

Three amino acids not shown in the module are diagrammed below.

1. Draw a box around the unique R group of each, and label it **R group**.

2. Draw a red circle around the amino group of each, and label it **amino group**.

3. Draw a blue triangle around the acid group of each, and label it **acid group**.

Alanine   Threonine   Asparagine

4. In the space below, sketch the three amino acids to show how they would join to form a tripeptide. What is this chemical reaction called? How many molecules of water would be formed? Show where the water would come from.

Exercise 8   (Modules 3.14 – 3.18)

Identify each of the levels of protein structure in the diagrams. Then choose the descriptions from the list below that go with each of the levels.

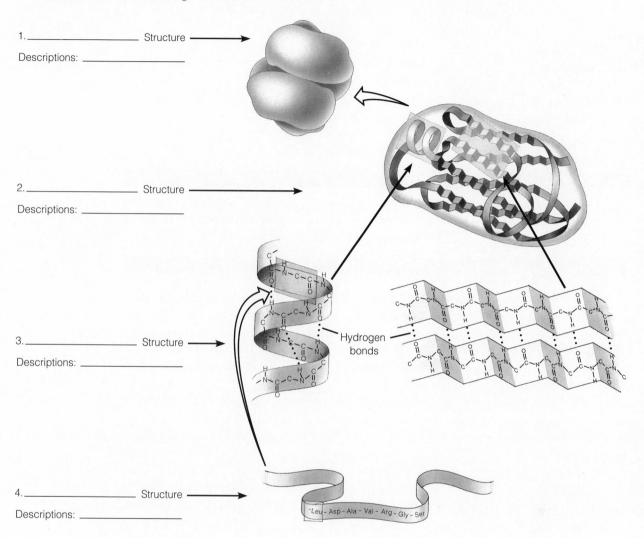

1._____ Structure ⟶

Descriptions: _____

2._____ Structure ⟶

Descriptions: _____

3._____ Structure ⟶

Descriptions: _____

Hydrogen bonds

4._____ Structure ⟶

Descriptions: _____

Leu – Asp – Ala – Val – Arg – Gly – Ser

Choose from these descriptions:

A. Overall three-dimensional shape
B. Amino acid sequence
C. Even a slight change in this can alter tertiary structure.
D. This level occurs in proteins with more than one polypeptide subunit.
E. Coiling and folding produced by hydrogen bonds between —NH and C=O groups
F. Not present in all proteins
G. Level of structure that is held together by peptide bonds
H. Alpha helix and pleated sheet
I. Stabilized mostly by hydrogen bonds and ionic bonds
J. "Globular" or "fibrous" might describe this level of structure.

Exercise 9   (Module 3.20)

Nucleic acids are the fourth group of macromolecules discussed in this chapter. Review their structures and functions by matching each of the phrases on the right with a word or phrase from the list on the left. Answers may be used more than once.

A. Phosphate group
B. Deoxyribose
C. A, T, C, G
D. DNA
E. Nucleotide
F. A, U, C, G
G. Double helix
H. Ribose
 I. Nitrogenous base
J. RNA

_____ 1. Sugar in RNA
_____ 2. Overall structure of DNA
_____ 3. Short for ribonucleic acid
_____ 4. Passed on from parent to offspring
_____ 5. Nitrogenous bases of RNA
_____ 6. Sugar in DNA
_____ 7. Nitrogenous bases of DNA
_____ 8. Short for deoxyribonucleic acid
_____ 9. DNA works through this intermediary.
_____ 10. Nucleotide is sugar, phosphate, and this.
_____ 11. Sugar of one nucleotide bonds to this of next nucleotide.
_____ 12. Monomer of nucleic acids

Exercise 10   (Summary)

You may find that making a concept map is a useful way to organize your knowledge. Such a map for the topic of carbohydrates is shown at the top of the next page. A concept map shows how key ideas are connected. Making a concept map can help you learn because it causes you to focus on main concepts and how they are related. It helps you to sort out what is important from unimportant details and helps you tie your knowledge together into a more meaningful and useful whole.

To make a concept map, you must first decide which ideas are most important. Place the biggest, or most inclusive, concept at the top of the page. Just a word or phrase is enough. Cluster subconcepts around it, and cluster sub-subconcepts around them. Draw lines between the concepts to show how they are connected, and describe these connections next to the lines. Again, use only a word or two.

If the topics or connections are not clear, perhaps they are unimportant, or perhaps you are not clear on how they connect, or perhaps they do not really connect. Remember, clarifying relationships is the purpose of making the map. Generally, maps that are more "branched" are more useful than ones with many long straight "chains" of boxes, but there is no one "correct" map for a particular topic.

Focus on the process of making the map, rather than on the map itself. More learning will take place while you are making the map than when you look at the finished product. You might want to "tune up" your maps by comparing them with maps made by other students. After reviewing the following concept maps, on separate paper, try making your own concept maps for proteins and nucleic acids. Keep them simple at first. Remember, "Practice makes perfect!" Also, keep the concept map idea in mind for upcoming chapters.

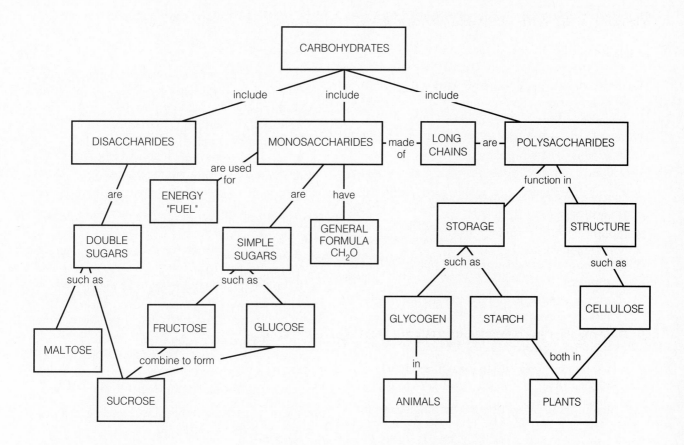

Practice working with a concept map by filling in the blanks on this map for lipids.

Testing Your Knowledge

Multiple Choice

1. Cellulose is a _____ made of many _____.
   a. polypeptide . . . monomers
   b. carbohydrate . . . fatty acids
   c. polymer . . . glucose molecules
   d. protein . . . amino acids
   e. lipid . . . triglycerides

2. In a hydrolysis reaction, _____, and in this process water is _____ .
   a. a polymer breaks up to form monomers . . . consumed
   b. a monomer breaks up to form polymers . . . produced
   c. monomers are assembled to produce a polymer . . . consumed
   d. monomers are assembled to produce a polymer . . . produced
   e. a polymer breaks up to form monomers . . . produced

3. The four main categories of macromolecules in a cell are
   a. proteins, DNA, RNA, and steroids.
   b. monosaccharides, lipids, polysaccharides, and proteins.
   c. proteins, nucleic acids, carbohydrates, and lipids.
   d. nucleic acids, carbohydrates, monosaccharides, and proteins.
   e. RNA, DNA, proteins, and carbohydrates.

4. The characteristic that all lipids have in common is
   a. they are all made of fatty acids and glycerol.
   b. they all contain nitrogen.
   c. none of them is very high in energy content.
   d. they are all acidic when mixed with water.
   e. none of them dissolves in water.

5. A flower's color is determined by the genetic instructions in its
   a. proteins.
   b. lipids.
   c. carbohydrates.
   d. nucleic acids.
   e. all of the above.

6. The most concentrated source of stored energy is a molecule of
   a. DNA.
   b. cellulose.
   c. fat.
   d. protein.
   e. glucose.

7. In some places a protein molecule may twist or fold back on itself. This is called _____ and the coils or folds are held in place by _____.
   a. tertiary structure . . . hydrogen bonds
   b. primary structure . . . covalent bonds
   c. secondary structure . . . peptide bonds
   d. tertiary structure . . . covalent bonds
   e. secondary structure . . . hydrogen bonds

8. A hydrophobic amino acid R group would be found where in a protein?
   a. forming a peptide bond with the next amino acid in the chain
   b. on the outside of the folded chain, in the water
   c. on the inside of the folded chain, away from water
   d. forming hydrogen bonds with other R groups
   e. only at one end of a protein chain

9. The overall three-dimensional folding of a protein molecule is called the
   a. double helix.
   b. primary structure.
   c. secondary structure.
   d. tertiary structure.
   e. quaternary structure.

10. How many different *kinds* of protein molecules are there in a typical cell?
    a. four
    b. twenty
    c. about a hundred
    d. thousands
    e. billions

11. Estrogen, cholesterol, and other steroids are examples of
    a. polysaccharides.
    b. lipids.
    c. polypeptides.
    d. triglycerides.
    e. nucleic acids.

12. The "building blocks" of nucleic acid molecules are called
    a. polysaccharides.
    b. amino acids.
    c. fatty acids.
    d. nucleotides.
    e. DNA and RNA.

Essay

1. Briefly describe the various functions of proteins in the cell and body.

2. Animal fats tend to be solid at room temperature, plant oils more liquid. Explain how a difference in the chemical structure of their molecules causes this physical difference.

3. What forces and bonds maintain the three-dimensional folded shape of a protein molecule? How does this relate to the sensitivity of proteins to changes in their environment?

4. Using circles to represent monosaccharides, show the difference between glucose, maltose, and starch. Maltose is an example of what kind of carbohydrate? Starch is an example of what kind of carbohydrate?

5. Sketch a protein molecule, using squares connected by lines to represent amino acids connected by peptide bonds. Does your protein display primary, secondary, and tertiary structure? Where?

Applying Your Knowledge

Multiple Choice

1. Citric acid makes lemons taste sour. Which of the following is a functional group that would cause a molecule like citric acid to be acidic?
   a. hydroxyl
   b. hydrocarbon
   c. amino
   d. carbonyl
   e. carboxyl

2. Which of the following do nucleic acids and proteins have in common?
   a. They are both made of amino acids.
   b. Their structures contain sugars.
   c. They are hydrophobic.
   d. They are large polymers.
   e. They each consist of four basic kinds of subunits.

3. A biochemist is analyzing a potato plant for the disaccharide sucrose. Where would he be most likely to find it?
   a. in cell membranes
   b. in grains in the cells of underground tubers (potatoes)
   c. in the nuclei of potato cells
   d. in the sap of the potato plant
   e. in the walls of the potato plant cells

4. Which of the following ranks the molecules in the correct order by size?
   a. water . . . sucrose . . . glucose . . . protein
   b. protein . . . water . . . glucose . . . sucrose
   c. water . . . protein . . . sucrose . . . glucose
   d. protein . . . sucrose . . . glucose . . . water
   e. glucose . . . water . . . sucrose . . . protein

5. How does glucose differ from sucrose, cellulose, and starch?
   a. It is a carbohydrate.
   b. It is larger.
   c. The others are polysaccharides.
   d. It is a monosaccharide.
   e. It contains carbon, hydrogen, and oxygen.

6. Steve noticed that his friend Jon had gained a little weight during the holidays. He commented, "Storing up some _____ for the winter, I see."
   a. polysaccharides
   b. triglycerides
   c. nucleotides
   d. polypeptides
   e. steroids

7. How does DNA differ from RNA?
   a. DNA is larger.
   b. One of their nitrogenous bases is different.
   c. They contain different sugars.
   d. DNA consists of two strands in a double helix.
   e. All of the above are differences.

8. Certain fatty acids are said to be essential because the body cannot make them itself; they must be obtained in the diet. If your diet were deficient in these essential fatty acids, you would not be able to make certain
   a. fats.
   b. glycerol molecules.
   c. monosaccharides.
   d. proteins.
   e. You would not be able to make any of the above.

9. Hydrolysis of a protein would produce
   a. amino acids.
   b. monosaccharides.
   c. polysaccharides.
   d. peptide bonds.
   e. nucleotides.

10. Glucose and hexanoic acid each contain six carbon atoms, but they have completely different properties. Glucose is necessary in food; hexanoic acid is poisonous. Their differences must be due to different
    a. monomers.
    b. macromolecules.
    c. hydrolysis.
    d. quaternary structures.
    e. functional groups.

11. Which of the following would probably *not* be affected when a protein is denatured?
    a. primary structure
    b. secondary structure
    c. hydrogen bonds
    d. tertiary structure
    e. All of the above must be affected for the protein to be denatured.

12. Palm oil and coconut oil are more like animal fats than other plant oils. Because they _____ than other plant oils, they can contribute to cardiovascular disease.
    a. contain fewer double bonds
    b. are less saturated
    c. contain more sodium
    d. are less soluble in water
    e. contain less hydrogen

Essay

1. Briefly explain why all starch molecules are pretty much the same, but there are millions of kinds of protein molecules.

2. Specific enzymes in your intestine enable you to break down starch and use the glucose molecules produced by this process. But you cannot break down cellulose. Explain why, in terms of both carbohydrate structure and protein shape.

3. Slight heating is often enough to render a protein nonfunctional. But a polysaccharide such as starch must literally be boiled in acid before it is significantly affected. Explain why.

4. Sketch the structural formulas of two hydrocarbon molecules that are isomers. Be sure the C and H atoms form the correct numbers of bonds. What are the molecular formulas of the molecules? What is identical about the molecules? How do the molecules differ?

5. Fred suffers from a disease that makes it difficult for his cells to produce glycogen. For him, three meals a day are not enough; he needs to snack constantly. Explain why.

6. A tripeptide, a molecule consisting of three amino acids, is a very small protein. Yet a huge variety of tripeptides is possible. Assume that the first, second, and third amino acids can be any of 20 choices. How many different tripeptides could there be? (Hint: Imagine you are stringing beads. If you have 20 colors to choose from, how many three-bead sequences are possible?) How could you calculate the number of possible polypeptides 100 amino acids long? (You probably won't want to actually work it out—it's a *very* large number.)

Using Your Knowledge

1. Most labels on packaged foods and household products list the chemicals they contain. You can use your knowledge of basic biological chemistry to figure out what the ingredients are. For example, vitamin pills contain folic acid. What makes it an acid? A shampoo contains hydroxypropyl methyl*cellulose* and hydrolyzed soy protein. What kinds of polymers are these? Pretzels are made with hydrogenated cottonseed oil. What does this mean? Poke around the kitchen cupboards and bathroom cabinet for other examples.

A Tour of the Cell

4

A sampling of the diversity of life on Earth includes a redwood tree, a jellyfish, a bacterium, a tiger, and a mushroom. At one time, living things seemed so varied that the only characteristic they were thought to have in common was a mysterious "vital force" that made them all alive. Then, with the invention of the microscope, biologists discovered cells. By the late nineteenth century, they realized that all living things are made of cells and that an organism is alive because its cells are alive. Even though the life of a redwood tree and the life of a jellyfish seem quite different, they look and work similarly on the cellular level. Now we have electron microscopes, and we can zoom in on the intricate structures within a single cell. We can take cells apart and analyze their chemistry, or probe them with radioactive isotopes, antibodies, lasers, or fluorescent dyes. This chapter describes what these techniques have revealed about the life of a cell.

Organizing Your Knowledge

Exercise 1   (Modules 4.1 – 4.2)

Use the information in the two modules and the chart in Module 4.2 to complete the following table comparing microscopes and the unaided human eye.

	Unaided Eye	*Light Microscope*	*Electron Microscope (SEM or TEM)*
Kind of radiation (beam) used	1.	2.	3.
Parts that focus beam	4.	5.	6.
Maximum magnification	7.	8.	9.
Smallest objects visible	10.	11.	12.
Size of smallest objects visible (resolving power)	13.	14.	15.
Limitations	16.	17.	18.

Exercise 2   (Module 4.3)

We need to use a microscope to see cells because cells are so small. Why can't a cell be as big as a house or at least as big as a baseball? Compare the two cells diagrammed below. For each cell, calculate the surface area, volume, and ratio of surface area to volume. Then answer the questions.

Cell 1

Surface area: $s = 6 \times (l \times l) =$
Volume: $v = l \times l \times l =$
Surface/volume: $s/v =$

Cell 2

Surface area: $s = 6 \times (l \times l) =$
Volume: $v = l \times l \times l =$
Surface/volume: $s/v =$

1. Which cell has the greatest surface area?

2. Which cell has the greatest volume?

3. Which cell has the greatest ratio of surface area to volume?

4. In which cell would the surface area of the membrane most efficiently service the cytoplasm?

Exercise 3   (Module 4.4)

Label the following on this diagram of a prokaryotic (bacterial) cell: **capsule, cell wall, plasma membrane, nucleoid region, cytoplasm, ribosome, bacterial flagella, pili**. Briefly state the function of each structure next to its label.

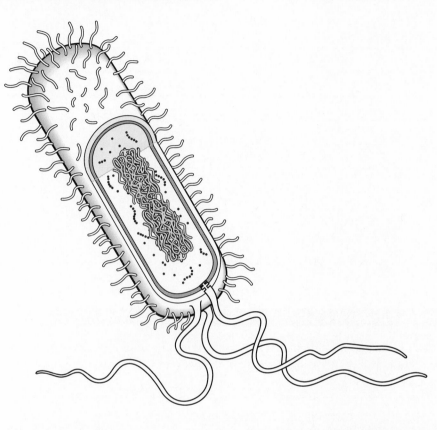

Exercise 4   (Modules 4.4 – 4.5)

Examine the diagrams in the text, and then compare the structures of bacterial, plant, and animal cells by checking off their characteristics below. You may want to revise or refer to this checklist as you complete the chapter.

Characteristic	Bacterial Cell	Plant Cell	Animal Cell
Prokaryotic			
Eukaryotic			
Relatively large size			
Relatively small size			
Extensive internal membranes			
Plasma membrane			
Cell wall			
Cytoplasm			
Ribosomes			

Characteristic	Bacterial Cell	Plant Cell	Animal Cell
Bacterial flagellum			
Nucleus			
Rough endoplasmic reticulum			
Smooth endoplasmic reticulum			
Golgi apparatus			
Lysosome			
Microbody			
Mitochondrion			
Chloroplast			
Central vacuole			
Cytoskeleton			
Flagellum			
Centriole			

Exercise 5   (Modules 4.6 – 4.14)

Review the nucleus and the various structures that make up the endomembrane system by matching each phrase on the right with a structure from the list on the left. Answers can be used more than once.

A. Nucleus
B. Transport vesicle
C. Central vacuole
D. Smooth ER
E. Lysosome
F. Golgi apparatus
G. Rough ER
H. Contractile vacuole
I. Ribosome

____ 1. Lipids manufactured here
____ 2. Small structure that makes protein
____ 3. Contains chromatin
____ 4. Sac of enzymes that digest things
____ 5. Carries secretions for export from cell
____ 6. Breaks down drugs and toxins in liver
____ 7. Makes cell membranes
____ 8. Cell control center
____ 9. Numerous ribosomes give it its name
____ 10. "Ships" products to plasma membrane, outside, or other organelles
____ 11. May store water, needed chemicals, wastes, pigments in plant cell
____ 12. Buds off from Golgi apparatus
____ 13. Defective in Pompe's disease and Tay-Sachs disease
____ 14. Proteins made here for secretion from cell
____ 15. Pumps out excess water from some cells
____ 16. Nonmembranous organelle
____ 17. Takes in transport vesicles from ER and modifies their contents
____ 18. Digests food, wastes, foreign substances
____ 19. Surrounded by double layer of membrane with pores
____ 20. How proteins, other substances get from ER to Golgi apparatus

Exercise 6   (Modules 4.7 – 4.14)

Sketch and label the endomembrane system on this diagram. Include **rough ER, smooth ER, ribosomes, Golgi apparatus, lysosome,** and **transport vesicles**. (1) Trace the path of a protein from its site of manufacture to the outside of the cell with a red arrow. (2) Trace the path of a protein incorporated into a lysosome in blue. (3) Trace the path of the protein incorporated into the plasma membrane in green. (4) Trace lipid secreted from the cell in yellow.

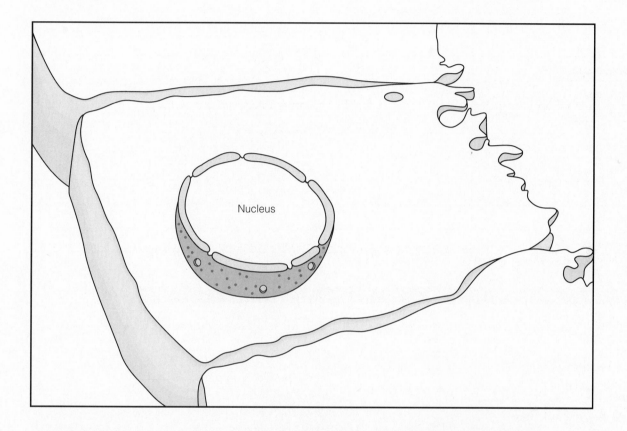

Nucleus

Exercise 7   (Modules 4.15 – 4.16)

Both mitochondria and chloroplasts are energy converters, but their functions are quite different. Compare them by filling in the chart below.

	Chloroplast	*Mitochondrion*
Found in the following organisms . . .		
Carries out process of . . .		
Converts energy of . . .		
Into chemical energy in . . .		

Exercise 8   (Modules 4.17 – 4.18)

Compare two main components of the cytoskeleton by indicating which of the following are characteristics of microfilaments and microtubules.

	Microfilaments	*Microtubules*
Hollow tubes		
Solid rods		
Made of tubulin		
Made of actin		
Help cell change shape		
Act in muscle cell contraction		
Move chromosomes		
Act as tracks for organelles		
Give cell rigidity, shape		
In cilia		
In flagella		
In centrioles		

Exercise 9   (Module 4.19)

Match each of the cell surface characteristics or structures on the left with a phrase on the right.

A. Tight junction
B. Plasmodesma
C. Anchoring junction
D. Cell wall
E. Communicating junction

_____ 1. Channel between animal cells
_____ 2. Made of cellulose
_____ 3. Link animal cells in leakproof sheet
_____ 4. Channel between plant cells
_____ 5. Connects animal cells, leaving space between them

Exercise 10 (Module 4.20 and Summary)

Label the organelles and other structures listed in Module 4.20 on these diagrams of animal and plant cells. (If you get stuck, refer to Module 4.5.) Try to group your labels according to the functional categories in Module 4.20 so that you can circle and label each category. Complete your diagrams by putting red boxes around the names of structures found only in animal cells but not plant cells. Put green boxes around the names of structures found only in plant cells but not animal cells.

Testing Your Knowledge

Multiple Choice

1. To enter or leave a cell, substances must pass through
   a. a microtubule.
   b. the Golgi apparatus.
   c. a ribosome.
   d. the nucleus.
   e. the plasma membrane.

2. Which of the following would *not* be considered part of a cell's cytoplasm?
   a. a ribosome
   b. the nucleus
   c. a mitochondrion
   d. a microtubule
   e. fluid between the organelles

3. Which of the following consist of prokaryotic cells?
   a. plants and animals
   b. bacteria
   c. plants, fungi, and bacteria
   d. animals
   e. plants and bacteria

4. Organelles involved in energy conversion are the
   a. rough ER and Golgi apparatus.
   b. nucleus and smooth ER.
   c. nucleus and chloroplast.
   d. lysosome and ribosome.
   e. mitochondrion and chloroplast.

5. The maximum size of a cell is limited by
   a. its need for enough surface area for exchange with its environment.
   b. the number of organelles that can be packed inside.
   c. the materials needed to build it.
   d. the amount of flexibility it needs to be able to move.
   e. the amount of food it needs to survive.

6. You would expect a cell with an extensive Golgi apparatus to
   a. make a lot of ATP.
   b. secrete a lot of material.
   c. move actively.
   d. perform photosynthesis.
   e. store large quantities of food.

7. Which of the following correctly matches an organelle with its function?
   a. mitochondrion . . . photosynthesis
   b. nucleus . . . cellular respiration
   c. ribosome . . . manufacture of lipids
   d. lysosome . . . movement
   e. central vacuole . . . storage

8. Cellular metabolism is
   a. a type of cell division.
   b. the process by which certain parts cause a cell to "self-destruct."
   c. the chemical activity of a cell.
   d. movement of a cell.
   e. control of the cell by the nucleus.

9. Which of the following breaks down toxins and drugs in liver cells?
   a. mitochondria
   b. smooth ER
   c. the Golgi apparatus
   d. contractile vacuoles
   e. rough ER

10. Of the following organelles, which group is involved in manufacturing substances needed by the cell?
    a. lysosome, vacuole, ribosome
    b. ribosome, rough ER, smooth ER
    c. vacuole, rough ER, smooth ER
    d. smooth ER, ribosome, vacuole
    e. rough ER, lysosome, vacuole

11. The internal skeleton of a cell is composed of
    a. microtubules, intermediate filaments, and microfilaments.
    b. cellulose and intermediate filaments.
    c. cellulose, microtubules, and centrioles.
    d. microfilaments.
    e. microfilaments and cellulose.

Essay

1. What are the advantages of an electron microscope over a light microscope? For what tasks would it be preferable to use a light microscope?

2. Briefly describe the major differences between prokaryotic and eukaryotic cells.

3. Name the structures present in plant cells but lacking in animal cells, and describe their functions.

4. Explain the advantages eukaryotic cells derive from being compartmentalized by many internal membranes.

5. Compare the functions of chloroplasts and mitochondria in a plant cell.

Applying Your Knowledge

Multiple Choice

1. A cell has mitochondria, ribosomes, smooth ER, and other parts. Based on this information, it could *not* be
   a. a cell from a pine tree.
   b. a grasshopper cell.
   c. a yeast (fungus) cell.
   d. a bacterium.
   e. Actually, it could be any of the above.

2. Dye injected into a cell might be able to enter an adjacent cell through a
   a. tight junction.
   b. microtubule.
   c. vacuole.
   d. plasmodesma.
   e. lysosome.

3. If a cell's chromatin were damaged, the cell would
   a. swell up and burst.
   b. run out of energy needed for its activities.
   c. go out of control.
   d. not be able to absorb light.
   e. divide immediately.

4. A researcher made an interesting observation about a protein made by the rough ER and eventually used to build a cell's plasma membrane. The protein in the membrane was actually slightly different from the protein made in the ER. The protein was probably changed in the
   a. Golgi apparatus.
   b. smooth ER.
   c. mitochondrion.
   d. nucleus.
   e. chloroplast.

5. If the nucleus is a cell's "control center," and chloroplasts its "solar collectors," which of the following might be called the cell's combination "food processor" and "garbage disposer"?
   a. lysosome
   b. Golgi apparatus
   c. flagellum
   d. ribosome
   e. nucleolus

6. When elongated, tube-shaped cells from the lining of the intestine are treated with a certain chemical, the cells sag and become round blobs. The internal structures disrupted by this chemical are probably
   a. cell junctions.
   b. microtubules.
   c. smooth and rough ER.
   d. mitochondria.
   e. microfilaments.

7. The electron microscope has been particularly useful in studying bacteria, because
   a. electrons can penetrate tough bacterial cell walls.
   b. bacteria are so small.
   c. bacteria move so quickly they are hard to photograph.
   d. they aren't really alive, so it doesn't hurt to "kill" them for viewing.
   e. their organelles are small and tightly packed together.

8. A cell possesses ribosomes, a plasma membrane, a cell wall, and other parts. It could not be
   a. a bacterium.
   b. a cell from a fungus.
   c. a cell from a mouse.
   d. an oak tree cell.
   e. It could not be **a** or **d**.

9. A mutant plant cell unable to manufacture cellulose would be unable to
   a. build a cell wall.
   b. divide.
   c. capture sunlight.
   d. move.
   e. store food.

10. A plant cell was grown in a test tube containing radioactive nucleotides, the parts from which DNA is built. Later examination of the cell showed the radioactivity to be concentrated in the
    a. rough ER.
    b. Golgi apparatus.
    c. smooth ER.
    d. central vacuole.
    e. nucleus.

Essay

1. Explain whether you would use a light microscope, a transmission electron microscope (TEM), or a scanning electron microscope (SEM) to perform each of the following tasks, and explain why: examining fine structural details within cell organelles, observing how a cell changes shape as it moves, studying tiny bumps on the cell surface, filming changes in the shape of the nucleus as a cell prepares to divide.

2. Imagine a cell shaped like a cube, 5 μm on each side. (Cells are not perfect cubes, but this assumption simplifies the question.) What is the surface area of the cell, in square μm ? What is its volume, in cubic μm ? What is the ratio of surface area to volume for this cell? (Sketches may help.) Now imagine a second cell, this one 10 μm on each side. What are its surface area, volume, and surface/volume ratio? Compare the surface/volume ratios of the two cells. How is this comparison significant to the functioning of cells?

3. An enzyme (a type of protein) called salivary amylase is manufactured in the cells of your salivary glands and secreted as part of saliva. Explain how these parts of the cell cooperate to produce and secrete salivary amylase: transport vesicles, rough ER, plasma membrane, nucleus, Golgi apparatus, ribosomes.

4. A chemical that acts specifically on mitochondria was found to interfere with the movement of cilia, slow down protein synthesis, reduce the frequency of cell division, and slow down manufacture of lipids. Explain how this chemical could affect so many different cell activities.

5. When you work harder, your muscle cells work harder and increase in size. How might various organelles in a muscle cell increase in size, number, or activity to respond to the challenge of an increased workload?

Using Your Knowledge

1. Analogies can be helpful in learning and understanding a new topic. For example, the cell can be thought of as a chemical factory, and the nucleus as its main office or control center. Can you compare the functions of other cell parts to the parts of a factory? What might be the factory doors? The power plant that provides energy for running the factory? The warehouse where raw materials or products are stored? Can you think of other comparisons? How does the cell differ from a factory?

2. Basic research is scientific investigation aimed at figuring out how something works, without thought for the immediate usefulness of the information obtained. The goal of applied research, on the other hand, is to put scientific knowledge to work. Much of cell biology is basic research—simply figuring out what the parts of a cell are and how they work. Nevertheless, the recent explosion of knowledge about cells has been immensely useful. For example, understanding lysosomes has helped us to understand disorders such as Tay-Sachs disease. Can you think of other examples where our understanding of cells has made our lives more healthy, comfortable, or productive?

The Working Cell

5

You can read these words because enzymes and membranes enable your cells to use energy. The light energy that bounces off the page enters your eyes and stimulates pigments held in special cell membranes. Enzymes make these pigments and convert them to a form that can absorb light. The eye cells can transmit signals through nerve cells to the brain because the membranes of these cells can selectively absorb and pump out charged particles. The energy for moving these particles comes from processes that make ATP. These processes take place through the action of enzymes on and between cell membranes. Every activity—not just reading, but walking, laughing, and thinking—depends on energy produced by processes that involve enzymes and membranes. Energy, enzymes, and membranes are the subjects discussed in this chapter.

Organizing Your Knowledge

Exercise 1 (Modules 5.1 – 5.5)

After reading Modules 5.1 – 5.5, review energy, chemical reactions, and the function of enzymes by filling in the blanks in the following story.

If you were to stop eating, you would probably starve to death in weeks or months. If you were unable to breathe, you would die in minutes. Animals need the energy that is released when food and oxygen combine. This energy is used not only to move the body but also to keep it from falling apart.

Energy is the ability to perform [1]_____. The sun is the source of the energy that sustains living things. Sunlight is pure [2]_____ energy, energy of movement that is actually doing work. In the process of photosynthesis, plants are able to use the energy of sunlight to produce food molecules. This process obeys the laws of [3]_____, the principles that govern energy transformations. Plants do not make the energy in food. According to the [4]_____ law of thermodynamics, energy can be [5]_____ or transferred, but it cannot be created or destroyed. In photosynthesis, no energy is created. Rather, the plant transforms the energy of sunlight into stored energy, called [6]_____ energy, stored in molecules of sugar.

No energy change is 100% efficient, and the changes that occur in photosynthesis are no exception to this rule. Some of the energy of sunlight is not stored in glucose, but rather is converted to [7]_____, which is random molecular motion. The [8]_____ law of thermodynamics says that energy changes are always accompanied by an increase in [9]_____, a measure of disorder. One of the reasons living things need a constant supply of energy is to counter this natural tendency toward disorder.

The products of photosynthesis contain [10]_____ potential energy than the reactants. This means that, overall, photosynthesis is an [11]_____ reaction. Such a reaction consumes energy, which in photosynthesis is supplied by the sun.

Photosynthesis produces food molecules, such as glucose, which store energy. An animal might obtain this food by eating a plant or an animal that has eaten a plant. The food molecules enter the animal's cells, where their potential energy is released in the process of cellular respiration. The products of this chemical reaction (actually a series of reactions) contain less potential energy than the reactants. Therefore, cellular respiration is an [12]_____ process; it [13]_____ energy. In fact, this is the same overall change that occurs when glucose in a piece of wood or paper burns in air. When paper burns, the energy escapes as the heat and light of the flames. In a cell, the reaction occurs in a more controlled way, and some of the energy is captured for use by the cell.

Energy released by the exergonic "burning" of glucose in cellular respiration is used to make a substance called [14]_____ . A molecule of [15]_____ and a [16]_____ group are joined to form each molecule of ATP. This is an endergonic reaction, because it takes energy to assemble ATP. The covalent bond connecting the phosphate group to the rest of the ATP molecule is unstable and easily broken. This arrangement of atoms stores [17]_____ energy. The [18]_____ of ATP is an exergonic reaction. When ATP undergoes hydrolysis, a [19]_____ is removed, ATP becomes [20]_____ , and energy is released. Thus, ATP is a kind of energy "currency" that can be used to perform cellular [21]_____ . Most cellular activities depend on ATP energizing other molecules by transferring its phosphate group to them—a process called [22]_____ . It should be noted that energy is not destroyed when ATP is used to do work. When an ATP molecule is hydrolyzed to make muscles move, some of its energy moves the body, and some ends up as random molecular motion, or [23]_____ .

A less obvious but important function of ATP is supplying the energy for fighting the natural tendency for a system to become disordered. A cell constantly needs to manufacture molecules to replace ones that are used up or damaged. Building a large molecule from smaller parts is an [24]_____ reaction. Energy released by the exergonic hydrolysis of ATP is used to drive essential endergonic reactions. The linking of exergonic and endergonic processes is called energy [25]_____ , and ATP is the critical connection between the processes that release energy and processes that consume it.

What prevents a molecule of ATP from breaking down until its energy is needed? Molecules can break down spontaneously; that is why ATP energy is needed to repair them. Fortunately for living things, it takes some additional energy, called energy of [26]_____, to get a chemical reaction started. This creates an energy [27]_____ that prevents molecules from breaking down spontaneously. Energy barriers exist for both exergonic and endergonic reactions. Most of the time, most molecules in a cell lack the extra energy needed to clear the barrier, so chemical reactions occur slowly, if at all.

So what enables the vital reactions of metabolism to occur when and where they are needed, at a rate sufficient to sustain life? This is where enzymes come in. An enzyme is a special [28]_____ molecule that acts as a biological [29]_____ . It [30]_____ the rate of a chemical reaction without being [31]_____ by it. An enzyme holds reactants in such a way as to [32]_____ the energy barrier that prevents them from reacting. Even though reactants would not normally possess the activation energy needed to start the reaction, the enzyme creates conditions that make the reaction possible. Enzymes enable the cell to carry out vital chemical changes when and where they are needed, enabling it to control the many chemical reactions that make up cellular [33]_____ .

Exercise 2   (Modules 5.1 – 5.5)

Briefly summarize the differences between the words or phrases in each of the following pairs.

1. Kinetic energy and potential energy

2. Exergonic reactions and endergonic reactions

3. Reactants and products

4. ATP and ADP

5. A reaction without an enzyme and a reaction with an enzyme

6. Photosynthesis and cellular respiration

7. First and second laws of thermodynamics

Review enzyme action by completing the activities below.

1. Complete the diagram below so that it shows the cycle of enzyme activity. Imagine that the reaction carried out by this enzyme is splitting a substrate molecule into two parts. Color the diagram as suggested and label the items in **boldface** type. Color the **enzyme** blue. Sketch the **substrate** as a dark green shape. Sketch the **products** and color them light green. Also label the **active site**.

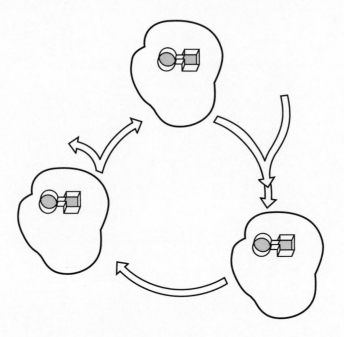

2. Make a sketch showing how heat or change in pH might change the above enzyme and alter its ability to catalyze its chemical reaction. Color and label the **enzyme**, its **active site**, and its **substrate**, as above.

3. On the left side of the space below, make a sketch showing how a competitive inhibitor might interfere with activity of the enzyme. Label the **competitive inhibitor** and color it orange. On the right side, make a sketch showing how a noncompetitive inhibitor might interfere with activity of the enzyme. Label the **noncompetitive inhibitor** and color it red.

Exercise 4 (Modules 5.10 – 5.13)

Review fluid mosaic membrane structure by coloring and labeling this diagram. It is a composite based on the figures in Modules 5.12 and 5.13. Label the items in **boldface** type: Start with the **cytoplasm** and **extracellular fluid**. In the membrane, color **phospholipids** yellow, **protein** molecules red, **carbohydrate I.D. tags** on **glycoprotein** and **glycolipid** molecules green, and **cholesterol** molecules purple. Also show the functions of certain proteins by labeling them **enzyme, receptor protein**, and **transport protein**.

Exercise 5 (Modules 5.14 – 5.20)

Review diffusion and the function of cell membranes by matching each of the phrases on the right with the appropriate mechanisms from the list on the left. Two questions require more than one answer.

A. Diffusion
B. Active transport
C. Osmosis
D. Phagocytosis
E. Passive transport
F. Facilitated diffusion
G. Pinocytosis
H. Receptor-mediated endocytosis
I. Exocytosis

_____ 1. Diffusion across a biological membrane
_____ 2. Moves solutes against concentration gradient
_____ 3. Any spread of molecules from area of higher concentration to area of lower concentration
_____ 4. Diffusion with help of transport protein
_____ 5. Three types of endocytosis
_____ 6. Engulfing of fluid in membrane vesicles
_____ 7. Movement of water through selectively permeable membrane, from hypotonic to hypertonic solution
_____ 8. Transport molecules need ATP to function
_____ 9. Enables cell to engulf bulk quantities of specific large molecules
_____10. How oxygen and carbon dioxide enter and leave cells
_____11. Two types of passive transport
_____12. Engulfing of particle in membrane vesicle
_____13. Fusion of membrane-bound vesicle with membrane, and dumping of contents outside cell
_____14. How a cell might capture a bacterium

Exercise 6   (Modules 5.15 – 5.16)

Osmosis is an important process that has many effects on living things. Test your understanding of osmosis by predicting in each of the following cases whether water will enter the cell (*In*) or leave the cell (*Out*), or whether there will be no net movement of water (*None*). Assume that the plasma membrane is permeable to water but not solutes.

_____ 1. Cell is exposed to hypertonic solution.

_____ 2. Cell is placed in salt solution whose concentration is greater than cell contents.

_____ 3. Due to disease, solute concentration of body fluid outside cell is less than solute concentration of cells.

_____ 4. Cell is in isotonic solution.

_____ 5. Single-celled organism is placed in drop of pure water for examination under microscope.

_____ 6. Cell is immersed in solution of sucrose and glucose whose individual concentrations are less than concentration of cytoplasm, but whose combined concentration is greater than concentration of cytoplasm.

_____ 7. Solute concentration of cell is greater than solute concentration of surrounding fluid.

_____ 8. Cell is exposed to hypotonic solution.

_____ 9. Concentration of solutes in cytoplasm is equal to solute concentration of extracellular fluid.

_____ 10. Cytoplasm more dilute than surrounding solution.

Exercise 7   (Modules 5.10 – 5.20)

Try to picture membranes and their functions close-up by completing the following story.

Your first mission as a Bionaut requires you to enter a blood vessel and observe the structure and functions of cell membranes. You step into the water-filled chamber of the microtron, which quickly shrinks you to a size much smaller than a red blood cell.

You tumble through the tunnel-like needle and into a blood vessel in the arm of a volunteer. Huge, rubbery red blood cells slowly glide past. Floating in the clear, yellowish blood plasma, you switch on your headlamp and examine the epithelial cells of the vessel wall. Their plasma membranes seem made of millions of small balloons. These are the polar "heads" of the [1]_____ molecules that make up most of the membrane surface. Through the transparent surface, you can see their flexible, [2]_____ tails projecting inward toward the interior of the membrane, and beyond them an inner layer of [3]_____ molecules with their tails pointing toward you. Here and there there are globular [4]_____molecules embedded in the membrane; some rest lightly on the surface, but most project all the way into the interior of the cell. The membrane is indeed a [5]_____ mosaic; the proteins are embedded like the pieces of a picture, but you can see that they are free to move around. You push on one of the proteins, and it bobs like an iceberg. Some of the phospholipids and proteins have chains of sugar molecules attached to them, forming [6]_____ and [7]_____. These are the molecules that act as cell [8]_____ tags. You notice that one of the proteins has a dimple in its surface. Just then a small, round molecule floating in the plasma nestles in the dimple. The molecule is a hormone, a chemical signal, and the dimpled protein is the [9]_____ that enables the cell to respond to it.

In your light beam, you can see the sparkle and shimmer of many molecules, large and small, in the blood and passing through the cell membrane. Oxygen is moving from the plasma, where it is more concentrated, to the cell interior, where it is less concentrated. This movement is [10]_____; when it occurs through a biological membrane, it is called [11]_____ transport. Similarly, carbon dioxide is flowing out of the cell, down its [12]_____ gradient, from the cell interior where it is [13]_____ concentrated to the blood where it is [14]_____ concentrated.

You note that water molecules are passing through the membrane equally in both directions. The total concentration of solutes in the cell and in the blood must be equal; the solutions must be [15]_____. You signal the control team to inject a small amount of concentrated salt solution into the blood, making the blood slightly [16]_____ relative to the cell contents. This causes water to flow [17]_____ the cell, until the two solutions are again in equilibrium. This diffusion of water through a [18]_____ permeable membrane is called [19]_____.

Some sugar molecules floating in the blood are simply too large to pass easily through gaps in the membrane like the much smaller water molecules can. The sugar molecules simply bounce off, unless they happen to pass through pores in special [20]_____ proteins. This is a type of passive transport, because the molecules move down a concentration gradient without the expenditure of [21]_____. Because transport proteins help out, it is called [22]_____ diffusion.

Your chemscanner detects a high concentration of potassium ions inside the cell. Transport proteins here and there in the membrane are able to move potassium into the cell against the concentration gradient. This must be [23]_____ transport; the cell expends [24]_____ to provide energy to "pump" the potassium into the cell.

Suddenly there is a tug at your foot. You look down to see your flipper engulfed by a rippling membrane. A white blood cell the size of a building quickly pins you against the vessel wall. The phospholipids of its membrane are pressed against your face mask. The cell is engulfing you, protecting the body from a foreign invader! Taking in a substance in this way is called [25]_____, more specifically [26]_____, if the substance is a solid particle. Suddenly the pressure diminishes, and you are inside the white blood cell, floating free in a membrane-enclosed bag, or [27]_____. Another sac is approaching; it is a [28]_____, full of digestive enzymes. You manage to get your legs outside of the vacuole and move it back toward the inner surface of the cell membrane. As the vacuole fuses with the membrane, you tear your feet free and swim away from the voracious cell, realizing that [29]_____ expelled you almost as fast as endocytosis trapped you!

This is quite enough adventure for one day. You swim to the exit point, and the control team removes you by syringe.

Testing Your Knowledge

Multiple Choice

1. The movement of molecules from an area of higher concentration to an area of lower concentration is called
   a. diffusion.
   b. endocytosis.
   c. catalysis.
   d. active transport.
   e. osmosis.

2. Which of the following is *not* true of an enzyme? An enzyme
   a. is a protein.
   b. acts as a biological catalyst.
   c. supplies energy to start a chemical reaction.
   d. is specific.
   e. lowers the energy barrier for a chemical reaction.

3. Phospholipid molecules in a membrane are arranged with their____ on the exterior and their____ on the interior.
   a. hydrophobic heads . . . hydrophilic tails
   b. hydrophilic heads . . . hydrophobic tails
   c. nonpolar heads . . . polar tails
   d. hydrophobic tails . . . hydrophilic heads
   e. hydrophilic tails . . . hydrophobic heads

4. In osmosis, water always moves toward the ____ solution, that is, toward the solution with the ____ solute concentration.
   a. isotonic . . . greater
   b. hypertonic . . . greater
   c. hypertonic . . . lesser
   d. hypotonic . . . greater
   e. hypotonic . . . lesser

5. Which of the following enables a cell to pick up and concentrate a specific kind of molecule?
   a. passive transport
   b. diffusion
   c. osmosis
   d. receptor-mediated endocytosis
   e. pinocytosis

6. A cell uses energy released by ____ reactions to drive the ____ reaction that makes ATP. Then it uses the energy released by the hydrolysis of ATP, an ____ reaction, to do various kinds of work in the cell.
   a. exergonic . . . exergonic . . . endergonic
   b. endergonic . . . exergonic . . . endergonic

   c. exergonic . . . endergonic . . . exergonic
   d. endergonic . . . endergonic . . . exergonic
   e. exergonic . . . endergonic . . . endergonic

7. Energy of activation
   a. is released when a large molecule breaks up.
   b. gets a reaction going.
   c. is released by an exergonic reaction.
   d. is stored in an endergonic reaction.
   e. is supplied by an enzyme.

8. The laws of thermodynamics state that whenever energy changes occur, ____ always increases.
   a. disorder
   b. order
   c. kinetic energy
   d. potential energy
   e. energy of activation

9. Living things transform kinetic energy into potential chemical energy in the ____ , when ____ is made.
   a. mitochondrion . . . ADP
   b. chloroplast . . . ADP
   c. chloroplast . . . an enzyme
   d. mitochondrion . . . glucose
   e. chloroplast . . . glucose

10. Why does heating interfere with the activity of an enzyme?
    a. It kills the enzyme.
    b. It changes the enzyme's shape.
    c. It decreases the energy of substrate molecules.
    d. It causes the enzyme to break up.
    e. It decreases the chance that the enzyme will meet a substrate molecule.

11. An enzyme is specific. This means
    a. it has a certain amino acid sequence.
    b. it is found only in a certain place.
    c. it functions only under certain environmental conditions.
    d. it speeds up a particular chemical reaction.
    e. it occurs in only one type of cell.

12. Diffusion of water across a selectively permeable membrane is called
    a. active transport.
    b. osmosis.
    c. exocytosis.
    d. passive transport.
    e. facilitated diffusion.

Essay

1. Describe the kinds of molecules that cannot easily diffuse through cell membranes. How do proteins facilitate diffusion of these substances?

2. Make a sketch showing why an enzyme acts only on a specific substrate.

3. Most enzyme-catalyzed chemical reactions in humans occur most readily around body temperature, 37°C. Why do these reactions slow down at lower temperatures? Why do they slow down at higher temperatures?

4. Which contains more potential energy, a large, complex molecule like a protein, or the smaller amino acid subunits of which it is composed? Is the joining of amino acids to form a protein an exergonic or endergonic reaction? Why must this be the case? Where does the cell obtain energy to carry out such reactions?

5. Describe the circumstances under which plant and animal cells gain and lose water by osmosis. Which of the following is the least serious problem: water gain by a plant cell, water loss by a plant cell, water gain by an animal cell, or water loss by an animal cell? Why?

Applying Your Knowledge

Multiple Choice

1. If a cell is like a factory, then enzymes are like
   a. the plans for the factory.
   b. the machines in the factory.
   c. the power plant for the factory.
   d. the raw materials used by the factory.
   e. the walls of the factory.

2. A molecule that has the same shape as the substrate of an enzyme would tend to
   a. speed metabolism by guiding the enzyme to its substrate.
   b. speed metabolism by acting as a cofactor for the enzyme.
   c. speed metabolism because it would also be a catalyst.

   d. save the cell energy by substituting for the substrate.
   e. slow metabolism by blocking the enzyme's active site.

3. A plant cell is placed in a solution whose solute concentration is twice as great as the concentration of the cell cytoplasm. The cell membrane is selectively permeable, allowing water but not the solutes to pass through. What will happen to the cell?
   a. No change will occur because it is a plant cell.
   b. The cell will shrivel because of osmosis.
   c. The cell will swell because of osmosis.
   d. The cell will shrivel because of active transport of water.
   e. The cell will swell because of active transport of water.

4. A white blood cell is capable of producing and releasing thousands of antibody molecules every second. Antibodies are large, complex protein molecules. How would you expect them to leave the cell?
   a. active transport
   b. exocytosis
   c. receptor-mediated endocytosis
   d. passive transport
   e. pinocytosis

5. Which of the following would be *least* likely to diffuse through a cell membrane without the help of a transport protein?
   a. a large polar molecule
   b. a large nonpolar molecule
   c. a small polar molecule
   d. a small nonpolar molecule
   e. Any of the above would easily diffuse through the membrane.

6. Red blood cells shrivel when placed in a 10% sucrose solution. When first placed in the solution, the solute concentration of the cells is ____ the concentration of the sucrose solution. After the cells shrivel, their solute concentration is ____ the concentration of the sucrose solution.
   a. less than . . . greater than
   b. greater than . . . less than
   c. equal to . . . equal to
   d. less than . . . equal to
   e. greater than . . . equal to

7. A nursing infant is able to obtain disease-fighting antibodies, which are large protein molecules, from its mother's milk. These molecules probably enter the cells lining the baby's digestive tract via
    a. osmosis.
    b. passive transport.
    c. exocytosis.
    d. active transport.
    e. endocytosis.

8. Some enzymes involved in the hydrolysis of ATP cannot function without the help of sodium ions. Sodium in this case functions as
    a. a substrate.
    b. a cofactor.
    c. an active site.
    d. a noncompetitive inhibitor.
    e. a vitamin.

9. The relationship between an enzyme's active site and its substrate is most like which of the following?
    a. a battery and a flashlight
    b. a car and a driver
    c. a key and a lock
    d. a glove and a hand
    e. a hammer and a nail

10. In which of the following do both examples illustrate kinetic energy?
    a. positions of electrons in an atom . . . a ball rolling down a hill
    b. heat . . . arrangement of atoms in a molecule
    c. a rock resting on the edge of a cliff . . . heat
    d. a ball rolling down a hill . . . heat
    e. light . . . arrangement of atoms in a molecule

11. Which of the following is a difference between active transport (AT) and facilitated diffusion (FD)?
    a. AT involves transport proteins and FD does not.
    b. FD can move solutes against a concentration gradient.
    c. FD requires energy from ATP and AT does not.
    d. FD involves transport proteins and AT does not.
    e. AT requires energy from ATP and FD does not.

12. An enzyme and a membrane receptor molecule are similar in that they
    a. are always attached to membranes.
    b. act as catalysts.
    c. require ATP to function.
    d. supply energy for the cell.
    e. bind to molecules of a particular shape.

Essay

1. The burning of glucose molecules in paper is an exergonic reaction, which releases heat and light. If this reaction is exergonic, why doesn't the book in your hands spontaneously burst into flame? You could start the reaction if you touched this page with a burning match. What is the role of the energy supplied by the match?

2. Seawater is hypertonic in comparison to body tissues. Explain what would happen to his stomach cells if a shipwrecked sailor drank seawater.

3. The laws of thermodynamics have imaginatively been described as the house rules of a cosmic energy card game: "You can't win, you can't break even (and to stay alive, you can't get out of the game)." State the law that says living things can't win the energy game. State the law that says they can't break even.

4. A farm worker accidentally was splashed with a powerful insecticide. A few minutes later he went into convulsions, stopped breathing, and died. The insecticide acted as a competitive inhibitor of an enzyme important in the function of the nervous system. Describe the structural relationship between the enzyme, its substrate, and the insecticide.

5. Lecithin is a substance used in foods such as mayonnaise as an emulsifier, which means that it helps oil and water mix. Lecithin is a phospholipid; a lecithin molecule has a polar "head" and a nonpolar "tail." How might the structure of lecithin allow water to surround fat droplets? Sketch a microscopic view of some fat droplets in mayonnaise, and show how you think the fat, surrounding water, and lecithin molecules might be arranged.

Using Your Knowledge

1. Soap bubbles have many of the physical and chemical properties of cell membranes. Several books are available in bookstores and libraries that describe the physics and chemistry of bubbles. You might want to blow some bubbles, observe their behavior, and learn the scientific principles that shape them.

2. Are you reading this under an electric light? Recall that energy cannot be created or destroyed, but only changed from one form into another. In what form was the light energy coming from the lamp before the light bulb changed it into light? Where did it come from? Trace the steps in the transformation of this energy as far back as you can. What happens to the light energy after it shines on you and the book?

3. Various household products contain enzymes—cleaners, laundry detergent, meat tenderizer, and so on. See if you can find products containing enzymes. Where are the enzymes in these products obtained? What are their substrates?

How Cells Harvest Chemical Energy   **6**

You need to eat and breathe because your cells need food and oxygen for energy. In every cell in your body, organic molecules and oxygen interact in a complex process called cellular respiration. In this process, food molecules such as glucose are broken down and the energy contained in their chemical bonds is used to make ATP. The ATP made in cellular respiration is then used to drive cellular activities. Right now, ATP produced in cellular respiration is being used to generate the nerve impulses from your eyes to your brain, to move your muscles, and to drive your heartbeat. This chapter explains how your cells harvest the energy that keeps you alive.

Organizing Your Knowledge

Exercise 1   (Introduction – Module 6.3)

Review the basic terms and concepts of cellular respiration by filling in the blanks below.

Right now, you are breathing at a steady rate of 12 to 20 breaths per minute. Breathing, or [1]_____, is necessary for life, but why? Breathing allows the body to take in [2]_____ gas and expel waste [3]_____. Your breathing is closely related to [4]_____ , the aerobic harvest of the energy in food molecules by cells. Your cells take in [5]_____ and [6]_____ from the blood. These two substances interact, the sugar is broken apart, and [7]_____ and [8]_____ are produced. In the process, some of the energy is stored in molecules of [9]_____, which provide the energy for body activities. To make enough ATP for their needs, average human beings must take in food that provides about [10]_____ kilocalories (kcal) of energy per day.

You require an [11]_____ (oxygen-containing) environment to make enough ATP to supply your body's needs. During most day-to-day activities, including [12]_____ exercise, the blood can deliver O_2 to the muscles as fast as they consume it. But if you exercise vigorously, muscle cells use up their oxygen and become [13]_____. They can continue to produce some ATP under these conditions for a while, but much less efficiently. The cells also produce [14]_____, a toxic waste product, which makes muscles [15]_____ and [16]_____.

Some organisms, such as the [17]_____ described in the introduction to this chapter, can live in both aerobic and anaerobic environments. When first added to bread dough or a wine vat, yeast cells are in an [18]_____ environment. They consume [19]_____ and [20]_____. The [21]_____ they produce causes the dough to rise. In a sealed wine vat, oxygen is quickly used up, and the yeasts then must live under [22]_____ conditions. They still consume [23]_____, but under anaerobic conditions they produce [24]_____ and [25]_____ as waste products.

Exercise 2   (Modules 6.4 – 6.7)

How does the cell capture the energy of organic molecules in ATP? Review the basic concepts by completing this crossword puzzle.

Across

1. The cell transfers energy by shuttling ____ from molecule to molecule.

6. Energy released in the electron transport chain moves ____ (H⁺) across a membrane.

7. ATP is made from ____ and inorganic phosphate.

10. Substrate-level ____ is a simpler way to make ATP than chemiosmotic phosphorylation.

11. A sequence of electron carriers forms the electron ____ chain.

14. ____ is loss of an electron.

16. A molecule such as glucose is oxidized when it ____ an electron.

18. ____ is short for "oxidation-reduction."

19. Oxygen is ____ in cellular respiration.

Down

2. ____ is gain of an electron.

3. A coenzyme called ____ is used to carry hydrogen atoms in redox reactions.

4. A molecule is reduced when it ____ an electron.

5. Most cells make most of their ATP via a process called ____.

8. A ____ enzyme strips hydrogen atoms from organic molecules.

9. ATP ____ use energy from a gradient of ion concentration to make ATP.

12. In substrate-level phosphorylation an enzyme transfers a ____ group to ATP.

13. NADH delivers electrons to an electron ____.

15. Glucose is ____ in cellular respiration.

17. NAD⁺ picks up electrons and hydrogen, forming ____.

Exercise 3   (Module 6.8)

The illustration in Module 6.8 introduces the three stages of cellular respiration. After studying it, see if you can label the diagram below without referring to the text. Include **electron transport chain and chemiosmosis, pyruvic acid, mitochondrion, CO_2, high-energy electrons carried by NADH, Krebs cycle, glycolysis, cytoplasmic fluid, ATP, glucose**, and **NADH and $FADH_2$.**

Exercise 4   (Module 6.9)

Glycolysis is the first of three steps in cellular respiration. Review glycolysis by matching each phrase on the right with a term on the left. Some terms are used twice.

A. NADH
B. Pyruvic acid
C. ATP
D. NAD^+
E. Glucose
F. Glycolysis
G. ADP and P
H. Oxygen
 I. Intermediate

_____ 1. Compound formed between glucose and pyruvic acid
_____ 2. Not involved in glycolysis
_____ 3. Fuel molecule broken down in glycolysis
_____ 4. Produced by substrate-level phosphorylation
_____ 5. Invested to energize glucose molecule at start of process
_____ 6. Reduced as glucose is oxidized
_____ 7. Glucose converted to two molecules of this
_____ 8. Assembled to make ATP
_____ 9. "Splitting of sugar"
_____ 10. Carries hydrogen and electrons from oxidation of glucose

Exercise 5 (Modules 6.10 – 6.11)

Pyruvic acid from glycolysis is chemically altered and then enters the Krebs cycle, a series of steps that completes the oxidation of glucose. To review these processes, fill in the blanks in the diagram below. (Try to do as many as you can without referring to the text.) Include the following: **NAD⁺, pyruvic acid, CO_2, FADH₂, NADH, coenzyme A, ATP, acetyl CoA**

Exercise 6 (Module 6.12)

Circle the correct words or phrases in parentheses to complete each sentence.

The [1] (*final, second*) stage of cellular respiration is the electron transport chain and synthesis of [2] (*glucose, ATP*) by [3] (*chemiosmosis, active transport*). The electron transport chain is a sequence of [4] (*electron, proton*) carriers built into the [5] (*outer, inner*) membrane of the mitochondrion. Molecules of [6] (*ADP, NADH*) bring hydrogens and electrons to the chain from glycolysis and [7] (*the Krebs cycle, chemiosmosis*). The electrons move along the chain from carrier to carrier in a series of redox reactions, finally joining with [8] (H_2O, CO_2, O_2) and H⁺ from the surrounding solution to form [9] (H_2O, CO_2, O_2). Energy released by the electrons is used to move protons—[10] (*H⁺ ions, ADP molecules*)—by [11] (*active transport, passive transport*) into the space between the inner and outer mitochondrial membranes.

The buildup of protons in the intermembrane space—a proton gradient—constitutes [12] (*kinetic, potential*) energy that the cell can tap to make [13] (*ATP, glucose*). The concentration of protons tends to drive them back through the membrane into the [14] (*inner compartment of the mitochondrion, cytoplasm of the cell*), but protons can cross the membrane only by passing through special protein complexes, called [15] (*coenzyme A's, ATP syn-*

thases). As these complexes allow protons back through the membrane, their enzymes harness the energy of the moving protons to phosphorylate [16] (*NAD, ADP*) and make [17] (*NADH, ATP*). Thus, electron transport and chemiosmosis transform [18] (*most, some, a small portion*) of the energy extracted from glucose into the phosphate bonds of ATP.

Exercise 7   (Modules 6.12 – 6.13)

These diagrams will help you review electron transport, chemiosmosis, and how poisons disrupt them. In the first diagram, show how the processes work normally. Trace movement of an electron with a red arrow, movement of H⁺ ions (active transport and chemiosmosis) with black arrows, and formation of ATP with a yellow arrow.

In the second diagram, draw arrows showing the movement of electrons and H⁺ and the formation of ATP, as in the first diagram. Then draw a blue line where each poison acts, to show how each of the poisons short-circuits the normal processes. Label the poisons **rotenone, cyanide, carbon monoxide, DNP,** and **oligomycin**.

1. Normal electron transport and chemiosmosis:

2. Effects of poisons:

Exercise 8   (Modules 6.9 – 6.14)

Check your overall understanding of cellular respiration by matching each of the phrases
below with one of the three stages of the process. Use G for Glycolysis, K for Krebs cycle,
E for Electron transport and chemiosmosis.

_____ 1. Generates most of the ATP formed by cellular respiration
_____ 2. Begins the oxidation of glucose
_____ 3. Occurs outside the mitochondrion
_____ 4. Produces 4 ATP/glucose by substrate-level phosphorylation, but 2 ATP/glucose
used to get it started
_____ 5. Oxidizes NADH and $FADH_2$, producing NAD^+ and FAD
_____ 6. Carried out by enzyme in matrix (fluid) of the mitochondrion
_____ 7. Here electrons and hydrogen combine with O_2 to form H_2O
_____ 8. Occurs along inner mitochondrial membrane
_____ 9. Generates most of the CO_2 produced by cellular respiration
_____10. $FADH_2$ and NADH deliver hydrogen ions and electrons to this stage
_____11. ATP synthase makes ATP

Exercise 9   (Module 6.15)

Review this module by filling in the blanks below.

1. _____ anaerobes can make their ATP by fermentation or aerobic respiration.
2. _____ is an organism that normally uses aerobic respiration to produce ATP, but it can
generate ATP without oxygen, via alcoholic fermentation.
3. Fermenters replenish their supply of NAD^+ by using NADH to oxidize _____ acid.
4. When oxygen is scarce, human _____ cells can make ATP by lactic acid fermentation.
5. Fermentation enables cells to make ATP in the absence of _____ .
6. For every molecule of glucose consumed, glycolysis produces two molecules of pyruvic acid, two molecules of ATP, and two molecules of _____ .
7. The waste products of alcoholic fermentation are _____ and carbon dioxide.
8. _____ acid fermentation is used to make cheese and yogurt.
9. Fermentation generates two _____ molecules for every molecule of glucose consumed.
10. A cell can use _____ to generate a small amount of ATP, but it must somehow recycle its
supply of NAD^+.
11. Like aerobic respiration, alcoholic fermentation produces _____ gas as a waste product.
12. Strict _____ require anaerobic conditions and are poisoned by oxygen.

Review the molecules that can be used as fuel for cellular respiration by writing their names in the blanks in this diagram. Include **glucose, amino acids, fats, fatty acids, proteins, sugars, polysaccharides**, and **glycerol**.

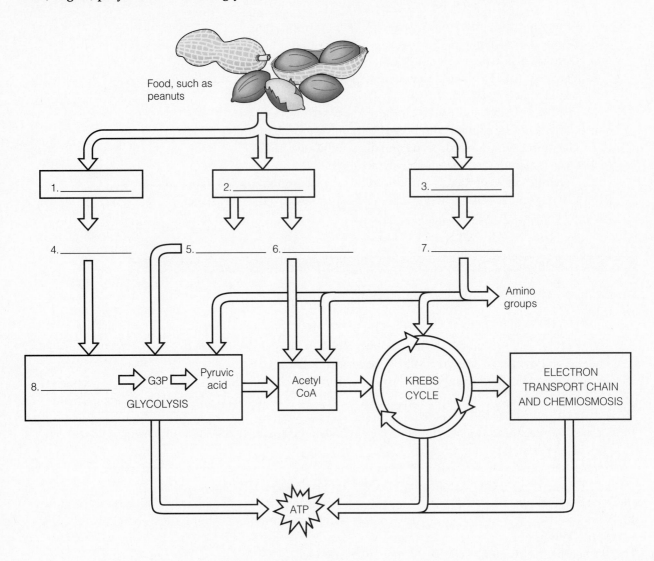

Exercise 11   (Module 6.17)

Show how a cell obtains organic molecules for biosynthesis of proteins, polysaccharides, and fats by drawing the missing arrows on this diagram.

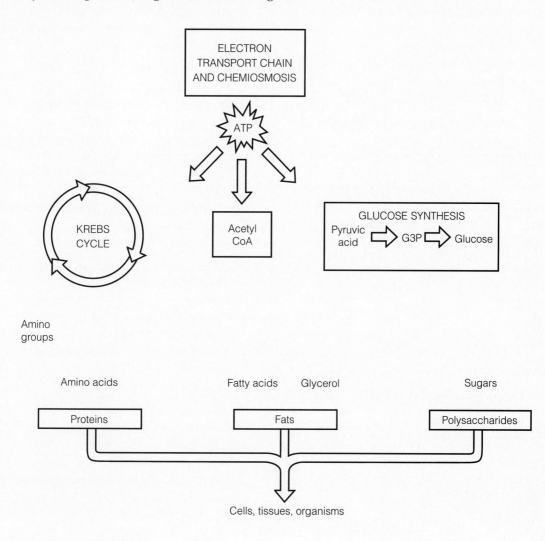

Testing Your Knowledge

Multiple Choice

1. A molecule is oxidized when it
   a. changes shape.
   b. gains a hydrogen (H^+) ion.
   c. loses a hydrogen (H^+) ion.
   d. gains an electron.
   e. loses an electron.

2. The main function of cellular respiration is
   a. breaking down toxic molecules.
   b. making ATP that powers cell activities.
   c. making food.
   d. producing chemical "building blocks" for cell structures.
   e. breaking down ATP, so that ADP and P can be reused.

3. In cellular respiration, _____ is oxidized and _____ is reduced.
   a. O_2 . . . ATP
   b. ATP . . . O_2
   c. glucose . . . O_2
   d. CO_2 . . . H_2O
   e. glucose . . . ATP

4. Most of the ATP produced in cellular respiration comes from
   a. glycolysis.
   b. chemiosmosis.
   c. lactic acid fermentation.
   d. biosynthesis.
   e. the Krebs cycle.

5. _____ is used and _____ is produced in the overall process of cellular respiration.
   a. CO_2 . . . H_2O
   b. O_2 . . . glucose
   c. H_2O . . . ATP
   d. glucose . . . CO_2
   e. ATP . . . O_2

6. The energy given up by electrons as they move through the electron transport chain is used to
   a. break down glucose.
   b. make NADH and $FADH_2$.
   c. pump H^+ through a membrane.
   d. oxidize water.
   e. manufacture glucose.

7. Fermentation is essentially glycolysis plus an extra step in which pyruvic acid is reduced to form lactic acid or alcohol and CO_2. This last step

   a. removes poisonous oxygen from the environment.
   b. extracts a bit more energy from glucose.
   c. enables the cell to recycle NAD^+.
   d. inactivates toxic pyruvic acid.
   e. enables the cell to make pyruvic acid into substances it can use.

8. A small amount of ATP is made in glycolysis and the Krebs cycle
   a. by transfer of a phosphate group from a fragment of glucose to ADP.
   b. using energy from the sun to perform the process of photosynthesis.
   c. by transport of electrons through a series of carriers.
   d. when electrons and hydrogen atoms are transferred to NAD^+.
   e. as a product of chemiosmosis.

9. The ATP synthase in a human cell gets energy for making ATP directly from
   a. sunlight.
   b. flow of H^+ through a membrane.
   c. oxidation of glucose.
   d. movement of electrons through a series of carriers.
   e. reduction of oxygen.

10. Which of the following describes glycolysis?
    a. It begins the oxidation of glucose.
    b. It produces a small amount of ATP.
    c. It generates NADH.
    d. It splits glucose to form two molecules of pyruvic acid.
    e. All of the above.

11. Most of the NADH that delivers high-energy electrons to the electron transport chain comes from
    a. chemiosmosis.
    b. the cytoplasm.
    c. glycolysis.
    d. biosynthesis.
    e. the Krebs cycle.

12. When protein molecules are used as fuel for cellular respiration, _____ are produced as waste.
    a. amino groups
    b. fatty acids
    c. sugar molecules
    d. molecules of lactic acid
    e. ethanol and CO_2

Essay

1. Describe the relationship between breathing and cellular respiration.

2. Compare the advantages and disadvantages of aerobic cellular respiration with the advantages and disadvantages of fermentation as methods of making ATP for cellular activities.

3. Compare the two mechanisms that generate ATP in cellular respiration—chemiosmotic phosphorylation and substrate-level phosphorylation. In what stage(s) of cellular respiration does each occur? Where does each get the energy for making ATP? Which produces the most ATP under aerobic conditions? Under anaerobic conditions?

4. Describe three ways in which poisons can interfere with cellular respiration.

5. Explain the roles of glycolysis and the Krebs cycle in the biosynthesis of organic molecules.

Applying Your Knowledge

Multiple Choice

1. Which of the following illustrates oxidation?
    a. $CO_2 + H_2O \longrightarrow C_6H_{12}O_6$
    b. $C_6H_{12}O_6 \longrightarrow CO_2 + H_2O$
    c. $ADP + P \longrightarrow ATP$
    d. $ATP \longrightarrow ADP + P$
    e. $H_2O \longrightarrow O_2 + H$

2. In an experiment, mice were fed glucose containing a small amount of radioactive oxygen, $C_6H_{12}O_6$. The mice were closely monitored, and in a few minutes radioactive oxygen atoms showed up in
    a. CO_2.
    b. NADH.
    c. H_2O.
    d. ATP.
    e. O_2.

3. In a second experiment, mice were allowed to breathe oxygen gas laced with radioactive oxygen, O_2. In this experiment, the radioactive oxygen atoms quickly showed up in
    a. CO_2.
    b. NADH.
    c. H_2O.
    d. ATP.
    e. glucose, $C_6H_{12}O_6$.

4. A chemist has discovered a drug that blocks glucose phosphate isomerase, an enzyme that catalyzes the second reaction in glycolysis. He wants to use the drug to kill bacteria in people with infections. But he can't do this because
    a. bacteria are facultative anaerobes; they usually don't need to do glycolysis.
    b. glycolysis produces so little ATP that the drug will have little effect.
    c. human cells also do glycolysis; the drug might also poison them.
    d. bacteria do not perform glycolysis.
    e. glycolysis can occur without the action of enzymes.

5. A glucose molecule is completely broken down in glycolysis and the Krebs cycle, but these two processes yield only a few ATPs. Where is the rest of the energy the cell obtains from the glucose molecule?
    a. in FAD and NAD^+
    b. in the oxygen used in the electron transport chain
    c. lost as heat
    d. in NADH and $FADH_2$
    e. in the CO_2 molecules released by the processes

6. Which of the following contains energy that a cell could use to make ATP?
    a. O_2
    b. CO_2
    c. NAD^+
    d. NADH
    e. H_2O

7. NADH is sometimes used by the cell in biosynthesis of needed organic molecules. Based on what you know about NADH, which of the following might be its function in biosynthesis?
    a. oxidizing organic molecules
    b. aiding in direct phosphorylation
    c. reducing organic molecules
    d. producing NAD^+
    e. breaking down ATP

8. Gram for gram, sugars are not as good as fats as a source of energy for cellular respiration, because sugars
   a. produce toxic amino groups when broken down.
   b. contain more hydrogen.
   c. usually bypass glycolysis and the Krebs cycle.
   d. contain fewer hydrogen atoms and electrons.
   e. are not as easily reduced.

9. A microbiologist discovered a new antibiotic that slowed the growth of bacteria by interfering with cellular respiration. She found that bacteria treated with the antibiotic produced about 15 ATP molecules for every glucose molecule they consumed. Which of the following hypotheses could explain the antibiotic's effect? The treated bacteria
   a. cannot perform glycolysis.
   b. have partially crippled electron transport chains.
   c. cannot produce NADH.
   d. have to rely at least partially on biosynthesis for their ATP.
   e. are forced to rely on fermentation for ATP.

10. A drug that blocks dehydrogenase enzymes would cause cells to
    a. run out of NADH.
    b. suffocate.
    c. deplete their supply of ADP.
    d. rely totally on electron transport and chemiosmosis for energy.
    e. be poisoned by lactic acid.

Essay

1. Fermentation is a much less efficient way to make ATP than aerobic cellular respiration. This being the case, why do you think the fermenters have not been driven to extinction by competition with aerobes?

2. Without oxygen, cellular respiration grinds to a standstill, although glycolysis can continue to make some ATP anaerobically for a short time. When oxygen runs out, why does electron transport stop? Why do you think the Krebs cycle stops?

3. FAD and NAD^+ are made from the B vitamins riboflavin and niacin. Why do you think these substances are required in such tiny amounts in your diet? How would a deficiency in one of these vitamins interfere with cell function?

4. After a biochemical analysis of the victim's tissues, brilliant biologist/criminal investigator J. C. Mickleberry announced his findings: "Contrary to the conclusions of the police, the victim did not suffocate. The electron carriers in his mitochondria were all in the oxidized state. We will need to perform a second autopsy to determine the actual cause of death." Explain how the data led Mickleberry to his conclusion.

5. A microbiologist placed a test tube full of yeast in a flask of sugar water and periodically took samples from the flask. At first, the amount of sugar in the flask decreased gradually. Then there was a sharp drop in sugar, accompanied by the appearance of ethanol in the flask. Explain these results.

Using Your Knowledge

1. Look around your home and find as many products as possible that either contain yeast or were made by yeast. Why were yeasts used to make these products?

2. Health and fitness experts recommend 20 minutes of aerobic exercise at least three times a week for peak cardiovascular conditioning and weight control. What kinds of exercise do you do? Do you exercise enough?

Photosynthesis: Using Light To Make Food **7**

Seen from space, Earth is a blue-green jewel—blue light reflected from the oceans and green light reflected from the mantle of vegetation that covers much of the land. From orbit, astronauts can see the deep green that cloaks the equator and watch the advance and retreat of green as the seasons unfold in the temperate regions. Green is the color of chlorophyll, the most important chemical on Earth. In the process of photosynthesis, chlorophyll captures the energy of the sun. It enables plants and other producers to store solar energy in the food used by all living things. This chapter is about photosynthesis, the process that starts with sunlight shining on chlorophyll.

Organizing Your Knowledge

Exercise 1   (Module 7.1)

Autotrophs are able to produce their own organic molecules. All organisms that use light energy to make food are called photosynthetic autotrophs. Circle the organisms below that are photosynthetic autotrophs.

mushroom	pine tree	squirrel	green bacterium
rosebush	seaweed	moss	bread mold
parasitic bacterium	alga	sponge	grass

Exercise 2   (Modules 7.1 – 7.7)

Review some of the basic terminology of photosynthesis by completing this crossword puzzle.

Across

5. ____ is oxidized in the process of photosynthesis.

6. The ____ is the cell organelle where photosynthesis takes place.

10. ____ are the light-catching membranes in a chloroplast.

11. Stacks of thylakoids in a chloroplast are called ____.

13. ____ energy travels through space as rhythmic waves.

14. When chlorophyll absorbs a photon, an ____ is excited.

16. Sugar is actually made in the ___ cycle.

18. The Calvin cycle occurs in the ____—the fluid of the chloroplast.

19. ____ is the green pigment in a leaf.

20. ____ are pores in a leaf.

Down

1. A ____ is an antennalike energy-harvesting unit in a thylakoid.

2. A ____ is a fixed quantity of light energy.

3. ____ is the process by which plants make food from carbon dioxide and water.

4. The color of light is related to its ____.

7. ____ is the source of energy for photosynthesis.

8. Carbon ____ is the incorporation of carbon dioxide into organic compounds.

9. The reaction ____ is the chlorophyll molecule that donates excited electrons.

12. ____ are yellow-orange pigments in a chloroplast.

15. ____ is the green tissue in the interior of a leaf.

17. A photosynthetic ____ is an organism that uses light to make food.

Exercise 3   (Modules 7.3 – 7.4)

Write the overall equation for photosynthesis in the boxes below. Show the substances used on the left and those produced on the right. Use different colors for carbon, hydrogen, oxygen in carbon dioxide, and oxygen in water, and then use your color code to show where atoms of C, H, and O on the left end up in the products on the right. On the lines under the substances used, state which is oxidized and which is reduced.

Exercise 4   (Modules 7.3 – 7.5)

Label this diagram summarizing the light-dependent and light-independent reactions of photosynthesis. Include **outer membrane of chloroplast, thylakoids, granum, stroma, light reactions, Calvin cycle, light, H_2O, O_2, electrons, NADPH, ATP, CO_2, sugar, ADP + P,** and **NADP$^+$.**

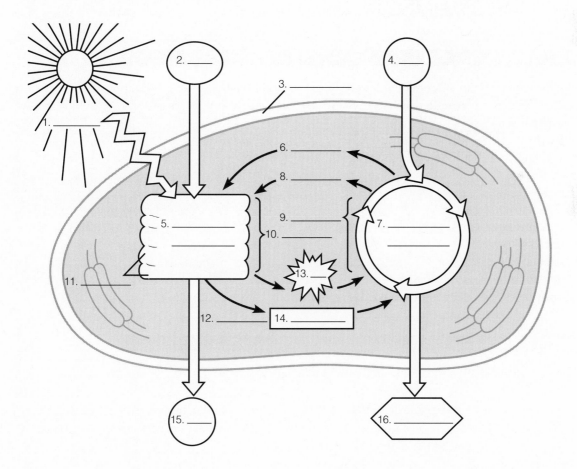

Exercise 5   (Modules 7.3 – 7.5)

Refer to the equations and diagrams in the modules to match each of the phrases on the left
with one of the ingredients or products of photosynthesis listed on the right.

_____ 1. Oxidized in light reactions

_____ 2. Reduced in Calvin cycle

_____ 3. Carries H and electrons from light reactions to Calvin cycle

_____ 4. Food produced by photosynthesis

_____ 5. Source of H and electrons in glucose

_____ 6. Source of O atoms in glucose

_____ 7. Where O atoms from water end up

_____ 8. Oxidized in Calvin cycle

_____ 9. Reduced in light reactions

_____10. Supplies energy to Calvin cycle

_____11. Where C and O atoms in carbon dioxide end up

_____12. Recycled from Calvin cycle to make ATP

_____13. Supplies energy to light reactions

_____14. Gas produced by reactions in the thylakoids

_____15. Gas consumed by reactions in the stroma

_____16. Source of carbon for carbon fixation

_____17. Source of H for the Calvin cycle

_____18. Picks up energized electrons from reactions in the thylakoids

A. Carbon dioxide, CO_2

B. Water, H_2O

C. Glucose, $C_6H_{12}O_6$

D. Oxygen, O_2

E. ADP + P

F. ATP

G. $NADP^+$

H. NADPH

I. Light

Exercise 6   (Modules 7.6 – 7.7)

Order the following forms of electromagnetic energy from the shortest wavelength (1) to
the longest (9). At which wavelength do photons have the most energy? Which are used by
plants in photosynthesis?

_____ a. Green light

_____ b. Radio waves

_____ c. X rays

_____ d. Red light

_____ e. Ultraviolet light

_____ f. Infrared

_____ g. Microwaves

_____ h. Blue light

_____ i. Gamma rays

Exercise 7   (Modules 7.7 – 7.11)

To review photosynthesis, fill in the blanks in the following story.

        Next time you eat an apple, reflect on the apple tree's ability to make the sugars
it contains, using sunlight to assemble simple substances from air and soil. This process is
called [1]_____, and it takes place in structures called [2]_____
in cells of tissues called the [3]_____ inside the leaves of the apple tree.
Photosynthesis actually consists of two processes: In the [4]_____ reactions,
[5]_____ molecules in membranes called [6]_____ in the chloro-

plast capture light energy. In the [7]_____ cycle, which takes place in the [8]_____ surrounding the thylakoids, this energy is used to make sugar, a process called [9]_____ fixation.

Chlorophyll molecules absorb [10]_____, packets of light energy. Chlorophyll absorbs only certain wavelengths, or colors, of light, mainly in the [11]_____ and [12]_____ parts of the spectrum. It reflects [13]_____ light. Other pigments, such as [14]_____, can absorb colors that chlorophyll cannot use directly, and transfer this energy to chlorophyll. Chlorophyll and other pigments are clustered on the thylakoid membranes in groups called photosystems. All the pigment molecules in a photosystem pass their energy along to a single chlorophyll molecule, called the [15]_____, in the middle of the photosystem. There are two kinds of photosystems, photosystem I and photosystem II, which absorb slightly different colors of light.

Most bacteria and all plants power the production of food through a process called [16]_____ electron flow. This process starts when a pair of photons energize a chlorophyll molecule in photosystem I. This raises two of the chlorophyll's [17]_____ to an excited state. The excited electrons are passed to the chlorophyll at the reaction center, which passes them to a protein called the [18]_____. From here, the electrons travel along an [19]_____ chain and end up as high-energy electrons in a molecule called [20]_____. (There is more to the story in plants such as our apple tree, and we will return to it. However, in certain bacteria, this process alone is used to generate ATP. As the electrons pass through the electron transport chain, some of their energy is used to generate energy by chemiosmosis. Rather than passing to NADPH, the electrons simply return to photosystem I. For this reason, this bacterial process—which also assists plants—is called [21]_____ electron flow.)

In the cells in the leaves of an apple tree, photosystem II replaces the electrons lost by photosystem I. A chlorophyll molecule of photosystem II absorbs two [22]_____ and ejects two [23]_____ . These pass to the primary electron acceptor and on to an electron transport chain. The electrons pass down the chain and eventually end up replacing the two electrons lost by the chlorophyll of [24]_____ On their way down the electron transport chain, the electrons perform important work. One of the electron carriers in the chain uses the energy released by the electrons to transport [25]_____ ions from the [26]_____ into the space inside the [27]_____. This creates a buildup of H^+, a concentration [28]_____ of H^+ across the membrane. The H^+ ions then diffuse through the membrane via a protein complex called [29]_____, which captures their energy to make [30]_____. In photosynthesis, this chemiosmotic production of ATP is called [31]_____. How does photosystem II replace its lost electrons? It gets them by splitting [32]_____. When the electrons of photosystem II are jarred loose, the reaction center develops a strong attraction for electrons. It obtains them by breaking apart a molecule of [33]_____. This leaves two H^+ ions (which pass into the thylakoid space) and an [34]_____ atom. This atom combines with another from another water molecule to form a molecule of [35]_____ gas, which diffuses out of the leaf—a product of photosynthesis important to us and other animals.

At this point, the cells of the apple leaf have captured the energy of the sun in molecules of NADPH and ATP, but so far they have made no sugar. The NADP and ATP are used, and sugar is made, in the [36]_____ cycle, the second portion of [37]_____ that takes place in the [38]_____ of the chloroplast, around the thylakoids. Using carbon from [39]_____ obtained from the air, energy from [40]_____, and hydrogen and high-energy electrons carried by [41]_____, the enzymes of the Calvin cycle construct [42]_____, a high-energy sugar molecule. In a series of steps, these molecules are combined to form the important six-carbon sugar [43]_____ and other organic compounds, in the leaves and in other parts of the plant.

The cellulose that gives an apple its crunch and the sugar that gives it its sweet taste are made from the glucose made in photosynthesis. In your intestine, the sugars enter your blood and are transported to your body cells. There the chemical pathways of cellular [44]_____ release the energy in the sugar molecules and use it to build [45]_____, which is in turn used to power cellular work. Energy from the sun, captured by the apple and passed on to you, enables you to see, to move, and to think about this amazing story.

Exercise 8   (Module 7.12)

Plants employ a variety of ways of fixing CO_2 and saving water. State whether each of the following statements relates to C_3 plants, C_4 plants, or **CAM** plants.

_____ 1. May waste energy on photorespiration on a hot day
_____ 2. Trap carbon in four-carbon compound, which donates it to Calvin cycle
_____ 3. Corn and sugarcane
_____ 4. Open stomata and trap CO_2 at night
_____ 5. Most plants
_____ 6. Soybeans, oats, wheat, rice
_____ 7. Can grow in hot, dry climates
_____ 8. Also can grow in hot, dry climates
_____ 9. Pineapple and many cacti
_____10. Calvin cycle uses CO_2 directly from the air

Exercise 9   (Module 7.13)

Test your understanding of this module by stating whether you think each of the following would be likely to **warm** or **cool** the Earth through alteration of the greenhouse effect.

_____ 1. Increased burning of coal to produce electricity
_____ 2. Increased rate of growth of algae in the oceans
_____ 3. Increased cloud cover, which would reflect more sunlight
_____ 4. Better gas mileage, so cars would go farther on a tank of gasoline
_____ 5. Increased cutting of tropical rain forests
_____ 6. Using nuclear power instead of coal and oil to make electricity
_____ 7. Reforesting deforested and overgrazed land
_____ 8. Increased rate of decomposition of organic matter

Testing Your Knowledge

Multiple Choice

1. The ultimate source of energy in the sugar molecules produced by photosynthesis is
   a. sugar.
   b. the sun.
   c. oxygen.
   d. ATP.
   e. chlorophyll.

2. Which of the following is produced by the light reactions of photosynthesis and consumed by the Calvin cycle?
   a. NADPH
   b. O_2
   c. H_2O
   d. sugar
   e. ADP + P

3. Which of these wavelengths is least useful for photosynthesis?
   a. green
   b. yellow
   c. blue
   d. orange
   e. red

4. When chloroplast pigments absorb light,
   a. they become reduced.
   b. they lose potential energy.
   c. their electrons become excited.
   d. the Calvin cycle is triggered.
   e. their photons become excited.

5. The process of noncyclic electron flow generates high-energy electrons, which end up in ____ . It also produces ____ and ____ .
   a. ATP . . . NADPH . . . O_2
   b. O_2 . . . sugar . . . ATP
   c. chlorophyll . . . ATP . . . NADPH
   d. water . . . sugar . . . O_2
   e. NADPH . . . ATP . . . O_2

6. The overall function of the Calvin cycle is
   a. capturing sunlight.
   b. making sugar.
   c. producing CO_2.
   d. splitting water.
   e. oxidizing glucose.

7. Which of the following correctly matches each of the inputs of the Calvin cycle with its role in the cycle?
   a. CO_2: high-energy electrons, ATP: energy, NADPH: oxidation
   b. CO_2: carbon, ATP: energy, NADPH: high-energy electrons
   c. CO_2: high-energy electrons, ATP: carbon, NADPH: energy
   d. CO_2: energy, ATP: carbon, NADPH: high-energy electrons
   e. CO_2: hydrogen, ATP: carbon, NADPH: energy

8. The main photosynthetic autotrophs in aquatic environments are
   a. plants and animals.
   b. plants and fungi.
   c. animals and algae.
   d. algae and bacteria.
   e. plants and bacteria.

9. CAM plants, such as pineapples, can photosynthesize in hot, dry environments because they
   a. store CO_2 during the day for use at night.
   b. store water at night for use during the day.
   c. store light energy during the day for use at night.
   d. use no water to perform photosynthesis.
   e. store CO_2 at night for use during the day.

10. Which of the following is *not* a product of the light reactions of photosynthesis?
    a. O_2
    b. sugar
    c. high-energy electrons
    d. ATP
    e. NADPH

11. Which of the following is oxidized in photosynthesis?
    a. O_2
    b. CO_2
    c. $C_6H_{12}O_6$
    d. ATP
    e. H_2O

12. In photosynthesis, plants use carbon from ____ to make sugar and other organic molecules.
    a. water
    b. the air
    c. chlorophyll
    d. the sun
    e. soil

Essay

1. Photosynthesis uses water and carbon dioxide to produce sugar and oxygen gas. Scientists long wondered whether the oxygen atoms in the oxygen gas produced in photosynthesis were obtained from carbon dioxide or water. Describe the experiments that enabled them to find out, and the results of these experiments.

2. Draw two squares, one labeled to represent the **light reactions** and the other to represent the **Calvin cycle**. Using arrows, show the inputs and outputs of each process. Include the following: **NADPH, ADP + P, O_2, light, CO_2, sugar, H_2O, ATP, $NADP^+$, electrons**.

3. Photosynthesis has been called "the most important chemical process on Earth." Explain why.

4. How is photosynthesis in C_3 plants slowed under hot, dry conditions? Describe how modifications of photosynthesis in C_4 and CAM plants enable them to overcome this problem.

5. State two activities of humans that tend to intensify the greenhouse effect. Why are people concerned about this? State two actions we could take that would reduce our contribution to the greenhouse effect.

Applying Your Knowledge

Multiple Choice

1. A photon of which of these colors would carry the most energy?
   a. green
   b. yellow
   c. blue
   d. orange
   e. red

2. A plant is placed in a sealed greenhouse with a fixed supply of water, soil, and air. After a year, the plant weighs 5 kg more than at the start of the experiment, and the ____ weighs almost 5 kg less.
   a. soil in the pot
   b. water left in the room
   c. organic matter in the soil
   d. air in the room
   e. soil in the pot together with the water in the soil

3. In a rosebush, chlorophyll is located in
   a. chloroplasts, which are in mesophyll cells in the thylakoids of a leaf.
   b. mesophyll cells, which are in the thylakoids in chloroplasts in a leaf.
   c. thylakoids, which are in mesophyll cells in the chloroplasts in a leaf.
   d. chloroplasts, which are in thylakoids in the mesophyll cells of a leaf.
   e. thylakoids, which are in chloroplasts in the mesophyll cells of a leaf.

4. In an experiment, a plant was given ____ containing radioactive O-18, and the radioactive oxygen atoms were used to make sugar.
   a. water
   b. chlorophyll
   c. oxygen gas
   d. ATP
   e. carbon dioxide

5. The *photo* part of the word *photosynthesis* refers to _____ , whereas *synthesis* refers to _____.
   a. the reactions that occur in the thylakoids . . . carbon fixation
   b. the reactions in the stroma . . . the reactions in the thylakoids
   c. the Calvin cycle . . . carbon fixation
   d. the Calvin cycle . . . the reactions in the stroma
   e. the light reactions . . . reactions in the thylakoids

6. The energy used to produce ATP in the light reactions of photosynthesis comes from
   a. the "burning" of sugar molecules.
   b. splitting water.
   c. movement of H^+ through a membrane.
   d. carbon fixation.
   e. fluorescence.

7. The following (*P* through *U*) are the main steps of chemiosmotic ATP synthesis in the light-dependent reactions of photosynthesis. Which answer places them in the correct order?

   P. H^+ concentration gradient established.

   Q. H^+ diffuses through ATP synthase.

   R. Carriers use energy from electrons to move H^+ across membrane.

   S. Electrons from photosystem II pass along electron transport chain.

   T. Light excites electrons in photosystem II.

U. Energy of H$^+$ flow used by ATP synthase to make ATP.
   a. PQTSRU
   b. STPQRU
   c. TSRPQU
   d. TSRUQP
   e. PQUSTR

8. The way ATP synthase captures the energy of H$^+$ ions in the light reactions appears to be similar to the way
   a. a turbine in a dam harnesses running water.
   b. an automobile engine uses hydrocarbons for fuel.
   c. heating of water forms steam in a nuclear reactor.
   d. electricity makes a light bulb glow.
   e. a rotating wheel stores mechanical energy.

9. An oceanographer has suggested slowing the rate of greenhouse warming by fertilizing the ocean to increase the growth of algae. How would this reduce the greenhouse effect?
   a. It would produce oxygen, which reflects sunlight from the atmosphere.
   b. It would "repair" the Earth's ozone layer.
   c. It would use up CO_2, which traps heat in the atmosphere.
   d. It would change the color of the ocean, reflecting the sun's heat.
   e. It would trap sunlight that would otherwise warm the Earth.

10. Which of the following would not be capable of performing photosynthesis?
   a. bacterium
   b. pine tree
   c. mushroom
   d. seaweed
   e. algae

Essay

1. In an experiment, plants were grown under colored filters that allowed equal amounts of light of different colors to strike different plants. Under which filter do you think plants grew the slowest? Why?

2. Carotenoids are yellow and orange pigments involved in photosynthesis. What colors of light must carotenoids absorb? Reflect? How is this useful to the plant?

3. Remember that the greater the concentration of hydrogen ions (H$^+$), the lower the pH of a solution. How do you suppose the pH of the solution in the thylakoid space compares with the pH of the solution in the stroma? What is responsible for the difference?

4. Compare chemiosmotic ATP production in photosynthesis (photophosphorylation) with ATP production in cellular respiration (oxidative phosphorylation): What is the source of high-energy electrons for each? What harnesses the energy of the electrons, and what is the energy used for? Where does each process take place? Where do the electrons end up? What actually manufactures the ATP? What is the immediate source of energy for ATP synthesis?

5. ATP is used to power the movement of your muscles as you turn the pages of this book. Where did the energy in the ATP come from? Trace the energy in the ATP molecules back to the sun.

Using Your Knowledge

1. What are some products you use or activities in which you participate that contribute to greenhouse warming? How might you modify your lifestyle to reduce your contribution to greenhouse warming?

2. Because photosynthesis uses carbon dioxide, deforestation increases greenhouse warming and planting trees reduces it. You can do your part to slow greenhouse warming by planting trees and encouraging others in your community to do so. Some groups that offer advice and information about planting trees are:

Children of the Green Earth, PO Box 95219, Seattle, WA 98145; phone 206-781-0852

Greenhouse Crisis Foundation, 1130 17th St. NW, Suite 630, Washington, DC 20036; phone 202-466-2823

International Society of Arboriculture, 303 W. University Ave., Urbana, IL 61801; phone 217-328-2032

National Arbor Day Foundation, 100 Arbor Ave., Nebraska City, NE 68410; phone 402-474-5655

TreePeople, 12601 Mulholland Dr., Beverly Hills, CA 90210, phone 818-753-4600

The Cellular Basis of Reproduction and Inheritance

8

In laboratories all over the world, biologists are studying cell division to try to understand why cells divide, what determines when they divide, what happens during cell division, and why some cells divide and others do not. The process of cell division is fundamental to inheritance, reproduction, growth, and development. Much of the inquiry into cell division is basic research; but this research has immense practical importance because understanding cell division will enable us to understand cancer, birth defects, heredity, and inherited diseases. This chapter concerns the connections among cell division, reproduction, and inheritance.

Organizing Your Knowledge

Exercise 1 (Introduction – Module 8.3)

Review the concepts introduced in these modules by filling in the blanks.

"Like begets [1]_____." This old saying means offspring look like their parents. Technically, only offspring produced by [2]_____ reproduction look exactly like their parents, because they inherit all their [3]_____ from a single parent. For example, when an amoeba divides, its [4]_____ is duplicated, and identical sets of [5]_____ (the structures that contain most of the amoeba's DNA) are allocated to opposite sides of the cell. The parent amoeba splits, and the two daughter amoebas that are formed are genetically [6]_____ to each other and to the [7]_____ cell.

Bacteria also reproduce asexually, via a type of cell division called [8]_____ _____. Most genes in bacteria are carried on a single [9]_____ DNA molecule, which is much [10]_____ and [11]_____ in structure than the multiple chromosomes of eukaryotes. The bacterium replicates its DNA and attaches the copies to different points on the plasma [12]_____. As the cell grows, the chromosomes become separated. Finally, the plasma membrane and cell [13]_____ grow inward, separating the chromosomes and dividing the cell in two. Like the reproduction of an amoeba, binary fission produces daughter cells identical to the parent cell.

The offspring produced by sexual reproduction resemble their parents, but they are not identical to their parents or to each other. Sexual reproduction begins with the production of an [14]_____ and a [15]_____, specialized cells that join to produce an offspring. The egg and sperm fuse, and the [16]_____ egg inherits a unique combination of genes from both parents. Through repeated cell divisions, the fertilized egg develops into an organism with a unique combination of traits—for example, a cat with long, gray fur or a human with blue eyes and freckles. Thus, sexual reproduction produces [17]_____ among offspring. Thus, through sexual reproduction "like begets like," but not exactly.

Exercise 2   (Modules 8.4 – 8.5)

Read these modules, and then test your knowledge of chromosomes by completing the crossword puzzle.

Across

4. ____ cells have more chromosomes and genes than bacteria do.

5. A typical ____ has about 3000 genes.

6. Chromosome duplication produces two sister ____.

8. A bacterium has a ____ simple chromosome.

9. A chromosome contains one long ____ molecule.

10. Chromosomes are named for their affinity to ____ used in microscopy.

12. Eukaryotic chromosomes involve more ____ molecules than those of bacteria.

14. When a chromosome divides, one sister chromatid goes to each ____ cell.

15. Proteins help determine chromosome shape and control ____ activity.

Down

1. Sister chromatids are joined at the ____.

2. ____ cells have characteristic numbers of chromosomes.

3. When a cell is not dividing, chromosomes form long, thin fibers called ____.

7. Genes in eukaryotic cells are grouped into ____ chromosomes.

9. Chromosomes are clearly visible only when a cell is ____.

11. Egg and ____ cells contain half the chromosomes of somatic cells.

13. ____ cells carry about 100,000 genes in 46 chromosomes.

Review the cell cycle: First identify the parts of the cycle and place them in order by writing the name of each phase or process on the diagram. Choose from **S, interphase, mitosis, G_1, mitotic phase, M, cytokinesis, G_2**. Then add a brief description of what is happening during that portion of the cycle. Choose from **cell division, DNA synthesis, cell growth, division of cytoplasm, activity between divisions, division of nucleus and chromosomes, activity between DNA synthesis and division**, and **mitosis plus cytokinesis**.

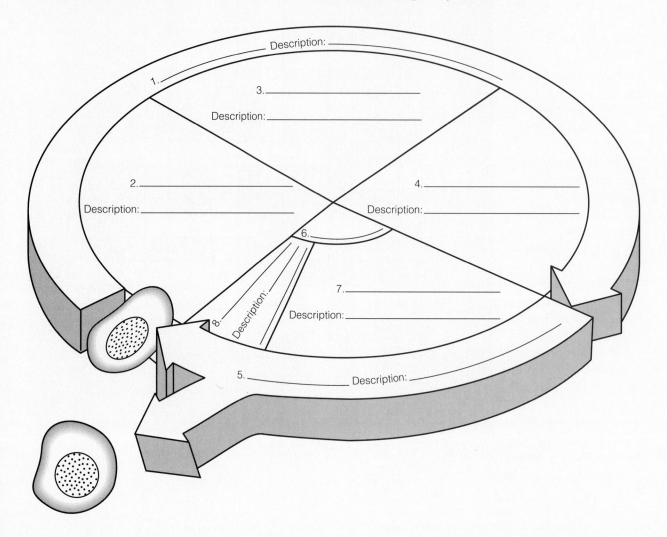

Exercise 4   (Module 8.7)

Summarize cell division. Briefly describe the appearance and activities of each of these cell parts during interphase and the four stages of mitosis. Include a simple sketch for each phase.

	Interphase	Prophase	Metaphase	Anaphase	Telophase
Nucleus and nuclear envelope					
Mitotic spindle					
Chromosomes					
Cell size and shape					
Sketch					

Exercise 5   (Module 8.7)

This is a photograph of cells in an onion root tip, an area of rapid cell division. In which stage of mitosis (or interphase) is each of the numbered cells?

1. _____

2. _____

3. _____

4. _____

5. _____

Exercise 6   (Module 8.8)

Match the word or phrase on the right with the correct role in mitosis in an animal cell on the left.

A. Where spindle microtubules attach to chromosomes
B. Move chromosomes
C. Pulled apart by spindle microtubules
D. Material around centrioles from which mitotic spindle grows
E. Chromosomes come to rest here during metaphase

_____ 1. Metaphase plate
_____ 2. Kinetechores
_____ 3. Sister chromatids
_____ 4. Spindle microtubules
_____ 5. Microtubule organizing center

Exercise 7   (Module 8.9)

Read this module and then write a statement containing exactly 30 words (no more, no less!) comparing cytokinesis in plant and animal cells. (Writing *exactly* 30 words will force you to think about the processes and choose your words carefully. It's fun to try it.)

Exercise 8   (Modules 8.10 – 8.12)

Review the functions of cell division by filling in the blanks below.

Mitotic cell division has several important functions. Some animals rely on cell division for [1]_____ reproduction. *Hydra*, for example, produces buds that detach from the parent and take up life on their own. Cell division is also responsible for [2]_____, as seen in human embryos and plant roots. In an adult human, some cells, such as [3]_____ and [4]_____ cells, cease to divide. Others, such as cells of the [5]_____, divide only if the organ is damaged. Some cells, such as those of the surface of the [6]_____ and the lining of the [7]_____, are constantly being abraded and lost. These cells are [8]_____ by cell division. In each of these cases, the new cells have exactly the same [9]_____ and [10]_____ of chromosomes as the parent cells, because of the way duplicated chromosomes divide in the process of [11]_____.

Growth, cell replacement, and reproduction require control of the rate and timing of cell division. Much has been learned by studying cells grown in laboratory [12]_____. Cells growing in a laboratory dish will divide only when in contact with a solid [13]_____ and will multiply only until they touch one another, a phenomenon known as [14]_____. Apparently, cells rely on chemicals called [15]_____ for division, and they will stop dividing when these substances are depleted.

Whether a cell will divide seems to be determined during the [16]_____ phase of the cell cycle, when a cell either does or does not go through a critical stage called the [17]_____. If the cell passes this point, it will divide. If not, it will switch to a nondividing state.

Sometimes cells escape these control mechanisms, divide uncontrollably, and invade other body tissues. These [18]_____ cells can kill the organism. In cell culture, they can grow without being attached to a solid surface, are unaffected by density-dependent inhibition, and are less affected than normal cells by growth factors. Cancer cells can go on dividing indefinitely (unlike normal cells, which can divide in culture for only [19]_____ to [20]_____ generations). Cancer treatments, such as [21]_____ and [22]_____, slow cancer by interfering with [23]_____. The anticancer drugs vinblastin and taxol prevent cell division by disrupting the [24]_____of the mitotic spindle.

Exercise 9   (Modules 8.13 – 8.14)

Describe the relationship between the terms or items in each of the following pairs.

1. Sex chromosomes and autosomes

2. The two chromosomes of a homologous pair

3. The two sister chromatids of a single chromosome

4. A diploid cell and a haploid cell

5. A somatic cell and a gamete

6. An egg and a zygote

7. Fertilization and meiosis

8. Mitosis and meiosis

Exercise 10 (Module 8.15)

Review meiosis by drawing in the chromosomes to complete this sequence of diagrams. Some have been done for you. Label **meiosis I, meiosis II**, the **phases** of meiosis I and II, a pair of **homologous chromosomes**, two **sister chromatids**, and an example of **crossing over**. To make you think carefully about meiosis, the diploid number is 6 in this example. (It is 4 in Module 8.15.)

1. _____

10. _____

2. _____ 3. _____ 4. _____ 5. _____
 and _____

11. _____

6. _____ 7. _____ 8. _____ 9. _____
 and _____

Exercise 11   (Module 8.16)

Compare mitosis and meiosis by completing this chart.

Mitosis	Meiosis
1.	Produces haploid daughter cells unlike parent cell
Involves one cell division	2.
Produces two daughter cells	3.
4.	Homologous chromosomes pair and then separate
Individual chromosomes line up at metaphase plate	5.
No crossing over occurs	6.
7.	Produces cells for sexual reproduction

Exercise 12   (Modules 8.17 – 8.19)

These modules discuss how orientation of chromosomes, random fertilization, and crossing over can lead to varied offspring. The diagram below shows the two homologous pairs of chromosomes in a cell with a diploid number of 4. Three different genes are also shown. On separate paper, complete the sketches described in questions 1 through 3.

1. Show how two different orientations of the chromosomes during metaphase I of meiosis could lead to the four different combinations of genes in gametes (assuming crossing over does not occur). (You don't need to show meiosis step by step—just the outcome.)

2. Show how crossing over could recombine genes on the larger pair of chromosomes, producing different gametes.

3. How many different combinations of genes in gametes are possible if these two processes happen simultaneously? Try to sketch all of them.

Testing Your Knowledge

Multiple Choice

1. There are a number of differences between fission of a bacterium and human cell division. Which of the following is *not* one of them?
   a. A bacterium has only one chromosome.
   b. Duplicated bacterial chromosomes attach to the plasma membrane.
   c. Bacteria are smaller and simpler than human cells.
   d. Bacteria have to duplicate their DNA before dividing.
   e. Human chromosomes are larger and more complex.

2. You would be unlikely to see which of the following human cells dividing?
   a. nerve cell
   b. skin cell
   c. cancer cell
   d. cell from an embryo
   e. intestinal lining cell

3. Which of the following correctly matches a phase of the cell cycle with its description?
   a. M—duplication of DNA
   b. S—immediately precedes cell division
   c. G_2—cell division
   d. G_1—immediately follows cell division
   e. All of the above are correctly matched.

4. Which of the following is *not* true of human somatic cells?
   a. They arise by mitotic cell division.
   b. They are haploid.
   c. They are body cells other than eggs and sperm.
   d. They are larger and more complex than bacterial cells.
   e. They contain 46 chromosomes.

5. In telophase of mitosis, the mitotic spindle breaks down and nuclear membranes form. This is essentially the opposite of what happens in
   a. prophase.
   b. interphase.
   c. metaphase.
   d. S phase.
   e. anaphase.

6. Sister chromatids
   a. cross over during prophase I of meiosis.
   b. separate during the first meiotic division.
   c. are produced during S phase between cell divisions.
   d. cross over during prophase II of meiosis.
   e. are also called homologous chromosomes.

7. Which of the following is *not* a function of mitotic cell division in animals?
   a. asexual reproduction
   b. growth
   c. repair of damaged organs
   d. production of gametes
   e. cell replacement

8. Meiosis
   a. is responsible for body growth and repair.
   b. halves the number of chromosomes in cells.
   c. is the process by which the body produces diploid cells.
   d. follows mitosis and splits the cytoplasm in two.
   e. is important in asexual reproduction.

9. Crossing over is
   a. important in genetic recombination.
   b. what makes a cell become cancerous.
   c. a key process that occurs during mitosis.
   d. an important mechanism of chromosome repair.
   e. what prevents cells from multiplying indefinitely in cell culture.

10. Human _____ are diploid, and human _____ are haploid.
    a. sex chromosomes . . . autosomes
    b. autosomes . . . sex chromosomes
    c. somatic cells . . . gametes
    d. gametes . . . somatic cells
    e. chromosomes . . . chromatids

11. Which of the following does *not* lead to genetic variability?
    a. random fertilization
    b. crossing over during meiosis
    c. division of chromosomes during anaphase of mitosis
    d. orientation of chromosomes during metaphase I of meiosis
    e. mutation

Essay

1. Explain why, strictly speaking, the phrase "Like begets like" applies only to asexual reproduction.

2. Briefly describe mitosis and cytokinesis and state their functions.

3. Describe how cancer cells differ from normal body cells.

4. Compare mitosis and meiosis. What are their functions? Which produces haploid cells? Diploid cells? What kinds of cells undergo mitosis and meiosis? What kinds of cells are produced by each? How many cells are produced?

5. How might the genes on the sister chromatids of a certain chromosome compare with each other and with the genes on the sister chromatids of the homologous chromosome?

6. Describe three aspects of sexual reproduction that lead to the production of varied offspring.

Applying Your Knowledge

Multiple Choice

1. In some organisms such as certain fungi and algae, cells undergo mitosis repeatedly without subsequently undergoing cytokinesis. What would result from this?
   a. a decrease in chromosome number
   b. inability to duplicate DNA
   c. division of the organism into many cells, most lacking nuclei
   d. large cells containing many nuclei
   e. a rapid rate of sexual reproduction

2. A human bone marrow cell, in prophase of mitosis, contains 46 chromosomes. How many chromatids does it contain altogether?
   a. 46
   b. 92
   c. 23
   d. 23 or 46, depending when during prophase you look
   e. 46 or 92, depending when during prophase you look

3. Which of the following is the most significant difference between mitosis and meiosis?
   a. Chromosomes are duplicated before mitosis.
   b. Meiosis is not followed by cytokinesis.
   c. Homologous pairs of chromosomes are split up in meiosis.
   d. A spindle formed of microtubules moves the chromosomes in mitosis.
   e. Crossing over occurs in mitosis.

4. If there are 22 chromosomes in the nucleus of a toad skin cell, a toad egg would contain ____ chromosomes.
   a. 22
   b. 44
   c. 11
   d. 33
   e. 88

5. Which of the following carry the same genetic information?
   a. sister chromatids
   b. X and Y chromosomes
   c. all autosomes
   d. homologous chromosomes
   e. all haploid cells

6. A cell biologist carefully measured the quantity of DNA in grasshopper cells growing in cell culture. Cells examined during the G_2 phase of the cell cycle contained 200 units of DNA. What would be the amount of DNA in one of the grasshopper daughter cells seen in telophase of mitosis?
   a. 50 units
   b. 100 units
   c. between 50 and 100 units
   d. 200 units
   e. 400 units

7. What would be the quantity of DNA in one of the grasshopper cells produced by telophase II of meiosis?
   a. 50 units
   b. 100 units
   c. between 50 and 100 units
   d. 200 units
   e. 400 units

8. The two chromosomes of a homologous pair
   a. carry identical genetic information at corresponding locations.
   b. carry information for the same traits at different locations.
   c. carry identical genetic information at different locations.
   d. carry information for the same traits at corresponding locations.
   e. Any of the above is possible.

9. A picture of a dividing pigeon cell taken through a microscope shows that the cell contains 7 chromosomes, each consisting of 2 chromatids. This picture must have been taken during
   a. metaphase of mitosis.
   b. prophase I of meiosis.
   c. telophase II of meiosis.
   d. prophase II of meiosis.
   e. telophase of mitosis.

10. A culture of mouse cells is treated with a chemical that interferes with the activity of microfilaments. Which of the following will probably be affected the most?
    a. mitosis
    b. chromosome duplication
    c. pairing of homologous chromosomes
    d. cytokinesis
    e. joining of sister chromatids at the centromere

11. A zoologist examined an intestine cell from a crayfish and counted 200 chromosomes, each consisting of 2 chromatids, at prophase I of mitosis. What would he expect to see in each of the four cells at telophase II of meiosis if he looked in the crayfish ovary?
    a. 50 chromosomes, each consisting of 2 chromatids
    b. 50 chromosomes, each consisting of 1 chromatid
    c. 100 chromosomes, each consisting of 2 chromatids
    d. 100 chromosomes, each consisting of 1 chromatid
    e. 200 chromosomes, each consisting of 1 chromatid

12. One chromosome of a homologous pair carries the genes *J* and *K*. The other chromosome of the pair carries the genes *j* and *k* at corresponding loci. Crossing over results in exchange of chromosome segments and production of gametes with new combinations of genes. A "recombinant" type gamete resulting from this crossover might contain
    a. genes *J* and *K*.
    b. genes *j* and *K*.
    c. genes *J* and *j*.
    d. genes *j* and *k*.
    e. genes *K* and *k*.

Essay

1. Recall how chemotherapy slows the growth of cancerous tumors. Common side effects of chemotherapy are hair loss, nausea, and loss of appetite. Why do you think that chemotherapy has its strongest side effects on the skin and lining of the digestive tract?

2. Most of the grasshopper cells examined by the cell biologist in multiple choice questions 6 and 7 above contained either 100 or 200 units of DNA. Some of the interphase cells, however, contained between 100 and 200 units. Explain what was happening in these cells.

3. A slide of dividing cells in an onion root tip (Figure 8.12A in the text) is a "snapshot" in time. Each cell is stopped at the particular point in its cell cycle when the slide was made. A biology student examined such a slide under a microscope. Out of 100 cells she caught in the act of dividing, 52 were in prophase, 8 in metaphase, 11 in anaphase, and 29 in telophase. Assuming that the cells are growing and dividing independently, what do these data tell you about the phases of mitosis in onion cells?

4. A white blood cell from a female golden retriever was found to contain a total of 78 chromosomes. How many different kinds (sizes and shapes) of chromosomes would you expect to find in the cell? Why?

5. The somatic cells of a mosquito contain three pairs of chromosomes—two large ones, two medium-sized ones, and two small ones. One large chromosome bears the *A* gene; its homologue bears the *a* gene. One medium-sized chromosome bears the *B* gene; its homologue bears the *b* gene. One small chromosome bears the *C* gene; its homologue bears the *c* gene. Sketch cells at metaphase I of meiosis in the ovary of the mosquito, showing the different alignments of chromosomes that are possible. Then show how these lead to different combinations of genes in the gametes. How many different combinations are possible in the eggs?

In the sperm of a male mosquito with the same genes, how many different gene combinations are possible? If the male and female mate and her eggs are fertilized with his sperm, how many different combinations are possible in the zygotes?

6. One chromosome of a homologous pair carries genes Q and R. Its homologue carries genes q and r. Show how crossing over during meiosis could produce gametes with new combinations of genes. What combinations of genes occur in parental-type gametes? In recombinant-type gametes?

Using Your Knowledge

1. What are some features that you (and your siblings, if you have any) appear to have inherited from your parents? Why don't you look exactly like your parents (or siblings)?

2. Do you know the warning signs of cancer? Cancer is more common in older people, but even college students in their teens and twenties are not immune to its deadly effects. Next time you see a Cancer Society brochure or advertisement, it would be a good idea to read it.

Patterns of Inheritance

At some point in your life you must have wondered about your inherited physical characteristics. Why is your hair dark or light, curly or straight? Do your eyes look like your mother's, your father's, or like neither of them? How do your eyes and hair, your height, and the color of your skin compare to those of your parents and your siblings? What made you female or male? People have asked these kinds of questions for thousands of years, but we only began to get some answers a little over a century ago, when Gregor Mendel started to experiment with inheritance in plants in his abbey garden. This chapter concerns the principles and patterns of inheritance.

Organizing Your Knowledge

Exercise 1   (Modules 9.1 – 9.4)

These modules discuss the basic principles of heredity and introduce the vocabulary of genetics. Read the modules carefully, and then practice using the vocabulary by matching each phrase on the right with a word or phrase on the left.

A. Allele
B. Homozygous
C. Hybrid
D. Genotype
E. Segregation
F. F_2 generation
G. True-breeding
H. Heterozygous
I. Self-fertilization
J. Dominant
K. P generation
L. Monohybrid cross
M. Trait
N. Phenotype
O. Cross
P. F_1 generation
Q. Recessive
R. Homologous chromosomes
S. Gene
T. Phenotypic ratio
U. Pangenesis
V. Cross-fertilization
W. Punnett square
X. Blending

_____ 1. A unit that determines heritable traits
_____ 2. Organisms that always produce offspring identical to parents
_____ 3. The offspring of two different varieties
_____ 4. When two alleles of a pair are different, the one that is masked
_____ 5. An incorrect idea that acquired characteristics are passed on
_____ 6. Parent organisms that are mated
_____ 7. A diagram that shows possible combinations of gametes
_____ 8. A breeding experiment that uses parents different in one trait
_____ 9. One of the alternative forms of a gene for a trait
_____ 10. Relative numbers of organisms with various traits
_____ 11. An organism that has two different alleles for a characteristic
_____ 12. Old idea that hereditary materials from parents mix in offspring
_____ 13. An organism's genetic makeup
_____ 14. Separation of allele pairs that occurs during gamete formation
_____ 15. Fertilization of a plant by pollen from a different plant
_____ 16. An organism that has two identical alleles for a characteristic
_____ 17. Offspring of the P generation
_____ 18. A characteristic
_____ 19. What an organism looks like; its expressed traits
_____ 20. Offspring of the F_1 generation
_____ 21. When pollen fertilizes eggs from the same flower
_____ 22. A hybridization
_____ 23. When two alleles of a pair are different, the one expressed
_____ 24. Where alleles for a certain trait are located

Exercise 2   (Modules 9.3 – 9.4)

Test your knowledge of Mendel's principles by answering the following questions. You may want to test your ideas on scratch paper.

1. A pea plant with green pods is crossed with a plant with yellow pods. All their offspring have green pods.

    a. Which allele is dominant? Which allele is recessive?

    b. Using letters, what is the genotype of the green parent? The yellow parent?

    c. What are the genotypes of the offspring?

2. F_1 pea plants from the above cross are crossed. Use a Punnett square to figure out the genotypic and phenotypic ratios in the F_2 generation.

    a. Genotypic ratios:

    b. Phenotypic ratios:

3. Two black mice mate. Six of their offspring are black and two are white.

    a. What are the genotypes of the parents?

    b. For which offspring are you sure of the genotypes?

Exercise 3 (Module 9.5)

Mendel studied the inheritance of two traits at once and found that each pair of alleles segregates independently during the formation of gametes. In other words, if a tall pea plant with purple flowers is crossed with a short plant with white flowers, some of their descendants can be tall with white flowers. The tall and purple alleles do not have to stick together—they are independent.

So far, the textbook has discussed inheritance in peas and budgies. Just to be different, let's look at a genetic cross involving rabbits. In rabbits, the allele for brown coat is dominant, the allele for white coat recessive. The allele for short fur is dominant, the allele for long fur recessive. Imagine mating a true-breeding brown, short-haired rabbit with a white, long-haired rabbit. Using Module 9.5 as a model, write the genotypes of rabbits and gametes in the P, F₁, and F₂ generations in the blanks below. You may want to modify the drawings to show the phenotypes of the rabbits in the F₁ and F₂ generations. Use the Punnett square to figure out the phenotypic ratios in the F₂ generation—the proportion of rabbits that you can expect to be brown and short-haired, brown and long-haired, white and short-haired, and white and long-haired.

Exercise 4   (Module 9.6)

After reading this module on testcrosses, test your understanding by answering the following questions.

1. Recall that brown coat color in rabbits is dominant and white color is recessive. Suppose you have a group of rabbits—some brown and some white.
   a. For which phenotype(s) do you know the genotype(s)?

   b. For which phenotype(s) are you unsure of the genotype(s)?

2. Using B and b to symbolize the brown and white alleles:
   a. What are the possible genotypes of a white rabbit in your group?

   b. What are the possible genotypes of a brown rabbit?

3. Suppose you wanted to find out the genotype of a brown rabbit. What color rabbit would you mate it with?

4. A brown buck (male) is mated with a white doe (female). In their litter of 11 young, six are white and five are brown. Using a Punnett square to check your answer, what is the genotype of the buck?

5. Use a Punnett square to figure out the ratio of brown and white offspring that would have been produced by the above mating if the brown buck had been homozygous.

6. If half the offspring from a testcross are of the dominant phenotype and half are of the recessive phenotype, is the parent of the dominant phenotype (but unknown genotype) homozygous or heterozygous?

7. If all the offspring from a testcross are of the dominant phenotype, is the parent with the dominant phenotype (but unknown genotype) homozygous or heterozygous?

Exercise 5   (Module 9.7)

The rules of probability can be used to predict the flip of a coin, the drawing of a card from a deck, or the roll of a pair of dice. They also govern segregation and recombination of genes. Read Module 9.7 carefully, and then fill in the blanks below.

     The probability scale ranges from [1]_____ (an event that is certain not to occur) to [2]_____ (an event that is certain to occur). The probabilities of all possible outcomes for an event must add up to [3]_____. Imagine rolling a pair of dice, one die at a time. Each of the six faces of a die has a different number of dots, from one to six. If you roll a die, the probability of rolling a one is [4]_____. The probability of rolling any number other than one is [5]_____. The outcome of a given roll is unaffected by what has happened on previous rolls. In other words, each roll is an [6]_____ event.

     If you roll two dice simultaneously, what is the probability of "snake eyes" (both ones)? The roll of each die is an independent event. The probability of such a compound event (both dice coming up ones) is the [7]_____ of the separate probabilities of the independent events. Therefore, the probability of rolling two ones is [8]____ × [9]_____ = [10]_____. This is called the rule of [11]_____.

This rule also governs the combination of genes in genetic crosses. The probability that a heterozygous *(Pp)* individual will produce an egg containing a *p* allele is [12]_____. The probability of producing a *P* egg is also [13]_____. If two heterozygous individuals are mated, what is the probability of a particular offspring being [14]_____ recessive *(pp)*? The probability of producing a *p* egg is 1/2. The probability of producing a *p* sperm is also 1/2. The production of egg and sperm are independent events, so to calculate their combined probability we use the rule of [15]_____. Thus the chance that two *p* alleles will come together at fertilization to produce a *pp* offspring is [16]____ × [17]____ = [18]____.

Back to the dice for a moment. What is the probability that a roll of two dice will produce a three and a four? There are two different ways this can occur. One die can come up a three and the other a four, or one can come up a four and the other a three. The probability of the first combination is 1/6 × 1/6 = 1/36. The probability of the second is also 1/6 × 1/6 = 1/36. According to the rule of [19]_____, the probability of an event that can occur in two or more alternative ways is the [20]_____ of the separate probabilities of the different ways. The probability of rolling a three and a four is therefore [21]_____ + [22]_____ = [23]_____.

Similarly, what is the probability that a particular offspring of two heterozygous parents will itself be heterozygous? The probability of the mother producing a *P* egg is [24]_____. The probability of the father producing a *p* sperm is also [25]_____. Therefore, the probability of a *P* egg and a *p* sperm joining at fertilization is [26]_____ × [27]_____ = [28]_____. Or a *p* egg and a *P* sperm could join. The probability of this occurring is also [29]_____. According to the rule of addition, the probability of an event that can occur in two alternative ways is the sum of the separate probabilities. Therefore, the probability of heterozygous parents producing a heterozygous offspring is [30]_____ + [31]_____ = [32]_____.

Exercise 6   (Module 9.8)

After you read this module, use the information in the illustration to solve the following problems. You will probably want to work out Punnett squares on scratch paper.

1. A man and woman, both without freckles, have four children. How many of the children would you expect to have freckles?

2. Both Fred and Wilma have widow's peaks. Their daughter Shirley has a straight hairline. What are Fred and Wilma's genotypes?

3. Arthur has mid-digital hair, but his wife, Susan, and their son, James, do not. If they have another child, what is the probability that he or she will have mid-digital hair?

4. A man and woman both suffer from astigmatism, but their daughter does not. What is the probability that their next child will suffer from astigmatism?

Exercise 7   (Modules 9.9 – 9.13)

These modules discuss examples of inheritance that are a bit more complex than the simple patterns of heredity observed by Mendel. After reading the modules, see if you can match each description with a pattern of inheritance. Choose from:

A. incomplete dominance
B. multiple alleles
C. codominance
D. pleiotropy
E. polygenic inheritance

_____ 1. There are three different alleles for blood group—I^A, I^B, and i, but an individual has only two at a time.

_____ 2. Crosses between two cremello (off-white) horses always produce cremello offspring. Crosses between chestnut (brown) horses always result in chestnut offspring. A cross between chestnut and cremello horses produces palomino (a golden-yellow color somewhat intermediate between chestnut and cremello) offspring. If two palominos are mated, their offspring are produced in the ratio of 1 chestnut : 2 palominos : 1 cremello.

_____ 3. The sickle-cell allele, s, is responsible for a variety of phenotypic effects, from pain and fever to damage to the heart, lungs, joints, brain, or kidneys.

_____ 4. In rabbits, an allele for full color (C) is dominant over an allele for chinchilla (c') color. Both full color and chinchilla are dominant over the white allele (c). A rabbit can be CC, Cc', Cc, $c'c'$, $c'c$, or cc.

_____ 5. In addition to the A and B molecules found on the surface of red blood cells, humans also have M and N molecules. The genotype L^ML^M produces the M phenotype. The genotype L^NL^N gives the N phenotype. Individuals of genotype L^ML^N have both kinds of molecules on their red blood cells, and their phenotype is MN.

_____ 6. If a red shorthorn cow is mated with a white bull, all their offspring are roan, a phenotype that has a mixture of red and white hairs.

_____ 7. Independent genes at four different loci are responsible for determining an individual's HLA tissue type, important in organ transplants and certain diseases.

_____ 8. A recessive allele causes a human genetic disorder called phenylketonuria. Homozygous recessive individuals are unable to break down the amino acid phenylalanine. As a consequence, they have high levels of this substance in their blood and urine, reduced skin pigmentation, lighter hair than their normal brothers and sisters, and often some degree of mental impairment.

_____ 9. When graphed, the number of individuals of various heights forms a bell-shaped curve.

_____10. Chickens homozygous for the black allele are black, and chickens homozygous for the white allele are white. Heterozygous chickens are gray.

Genes are located on chromosomes. Genes undergo segregation and independent assortment because the chromosomes that carry them undergo segregation and independent assortment during meiosis. The illustration below is similar to that in Module 9.14 and shows how alleles and chromosomes are arranged in an F_1 rabbit and how meiosis sorts the alleles into their gametes. The diagram below shows only the chromosomes. Put a letter in each of the numbered boxes to show how segregation and independent assortment of chromosomes cause segregation and independent assortment of alleles.

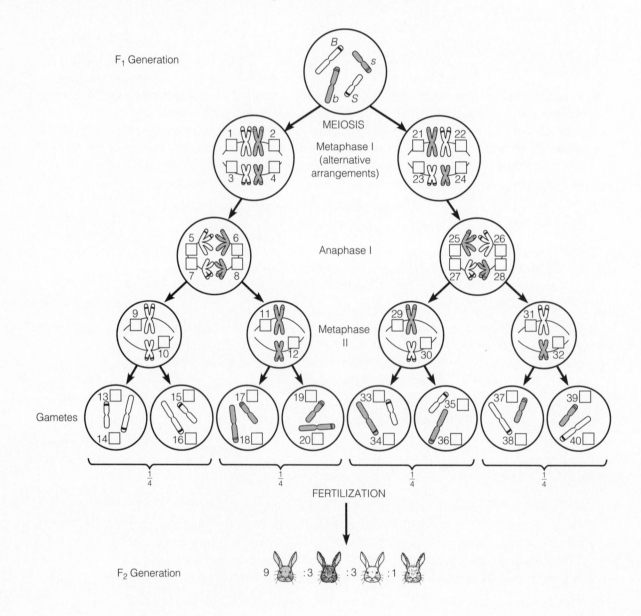

Exercise 9   (Modules 9.15 – 9.17)

These three modules discuss the inheritance of linked genes—genes on the same chromosome. Their pattern of inheritance is inconsistent with Mendel's "rules," but they illustrate important principles of chromosome structure and behavior. After reading the modules, match each of the observations below with the statement that explains the observation. Take your time; this exercise is not easy.

Observations

____ 1. When two heterozygous round yellow peas are crossed, their offspring are produced in a 9:3:3:1 ratio (9 round yellow : 3 round green : 3 wrinkled yellow : 1 wrinkled green).

____ 2. When two peas heterozygous for purple flowers and long pollen are crossed, the expected 9:3:3:1 ratio is not seen in the offspring. Very few purple round and red long plants are produced. The ratio is close to 3 purple long : 1 red round. Similarly, when a fruit fly with red eyes and long wings *(SsCc)* is crossed with a fly with scarlet eyes and curled wings *(sscc)*, offspring are not produced in the expected 1:1:1:1 ratio. Most offspring are red long and scarlet curled.

____ 3. When two heterozygous purple long peas are crossed, most of their offspring are purple and long or red and round. But a very small number of offspring are purple and round or red and long. Similarly, when the *SsCc* and *sscc* fruit flies are mated, nearly all their offspring are *SsCc* and *sscc*. However, a small number of offspring (about 6% of the total) are *Sscc* and *ssCc*.

____ 4. When a fruit fly with red eyes and long wings *(SsCc)* is crossed with a fly with scarlet eyes and curled wings *(sscc)*, 94% of their offspring are *SsCc* and *sscc*, and 6% are *Sscc* and *ssCc*. In other words, the recombination frequency between the *s* and *c* alleles is 6%. When a fly with red eyes and pale body *(SsEe)* is crossed with a fly with scarlet eyes and ebony body, 27% of their offspring are *Ssee* and *ssEe*. The recombination frequency between alleles *s* and *e* is 27%.

____ 5. When a fly with long wings and pale body *(CcEe)* is crossed with a fly with curled wings and ebony body *(ccee)*, 21% of their offspring are *Ccee* and *ccEe*. The recombination frequency between alleles *c* and *e* is 21%.

Explanations

A. The greater the distance between two genes, the greater the opportunity for crossing over to occur between them. If crossing over is more likely, more recombinant offspring will result. If two genes are farther apart, the recombination frequency will be greater between them.

B. Pairs of alleles on different chromosomes segregate independently during gamete formation. They follow Mendel's principle of independent assortment. In other words, genes for different traits on different chromosomes do not tend to "stick together" when passed on to offspring.

C. If two genes are on the same chromosome, or linked, they tend to be inherited together. Alleles on the same chromosome do not segregate independently. They tend to "stick together," violating Mendel's principle of independent assortment.

D. Recombination frequencies can tell you how far apart genes are on a chromosome. If you know the distance from *a* to *b*, the distance from *a* to *c*, and the distance from *b* to *c*, you can map the sequence of genes on the chromosome.

E. Homologous chromosomes cross over during meiosis and exchange segments. This recombines linked genes into assortments not seen in the parents.

Exercise 10   (Module 9.18)

What determines an individual's sex? Sex is generally determined by genes and chromosomes, but the process of sex determination works differently in different species. Match each group of organisms below with their system of sex determination.

A. Most plants, including peas, corn
B. Humans, fruit flies, some plants
C. Ants, bees
D. Earthworms, snails
E. Some butterflies, birds, fishes
F. Grasshoppers, crickets, roaches

____ 1. Females are *ZW*, males *ZZ*.
____ 2. Females are diploid, males haploid.
____ 3. Females are *XX*, males *XO* (one *X*).
____ 4. Hermaphroditic; all produce eggs and sperm.
____ 5. Females are *XX*, males *XY*.
____ 6. Monecious; all produce eggs and sperm.

Exercise 11   (Module 9.19)

Genes located on the sex chromosomes—called sex-linked genes—determine many traits unrelated to maleness or femaleness. Red-green color blindness is a recessive sex-linked trait in humans. After reading Module 9.19, see if you can describe the inheritance of color blindness by filling in the blanks below.

The genes for normal color vision and red-green color blindness, like most human sex-linked traits, are carried on the [1]_____ chromosome. A capital letter *C* represents the [2]_____ allele for normal vision; a small *c* represents the color-blindness allele. A male with normal color vision has the genotype [3]_____. (Because these genes are carried on the *X* chromosome, their symbols are shown as superscripts on the letter *X*.) A color-blind male has the genotype [4]_____.

A color-blind male will transmit the allele for color blindness to all his [5]_____ but none of his [6]_____. This is because only his daughters inherit his [7]_____ chromosome, and only his [8]_____ chromosome is passed to all his sons. All the children of a color-blind male and a homozygous-dominant female will have normal color vision. Their sons will inherit only the normal vision allele, but their daughters will be [9]_____ of the color-blindness allele, thus possessing the genotype [10]_____.

A heterozygous female carrier transmits the color-blindness allele to [11]_____ of her offspring. If she and a male with normal vision have children, [12]_____ of their sons will be normal and [13]_____ will be color blind. [14]_____ of their daughters will be normal, because they inherit at least one dominant allele from their [15]_____. But half these daughters will be [16]_____ of the color-blindness trait, because they inherit the color-blindness allele from their mother.

Color blindness is much more common in men than in women. If a man inherits a single color-blindness allele from his [17]_____, the gene will be expressed and he will be color blind. Because a man has only one [18]_____ chromosome, whatever genes it carries are seen in the man's phenotype. If a woman inherits just one color-blindness allele, she has normal vision, because the dominant normal allele on her other *X* chromosome masks the effects of the color-blindness allele. For a woman to be color blind, she would have to inherit [19]_____ alleles from both her mother and her father, which is much less likely.

Testing Your Knowledge

Multiple Choice

1. How did Mendel's studies in genetics differ from earlier studies of breeding and inheritance?
   a. Mendel worked with plants; earlier studies used animals.
   b. Mendel was able to explain the "blending" hypothesis.
   c. Mendel's work was more quantitative.
   d. Mendel worked with wild species, not domesticated ones.
   e. Mendel found that offspring inherit traits from both parents.

2. A true-breeding fruit fly would be _____ for a certain trait.
   a. homozygous dominant
   b. homozygous recessive
   c. heterozygous
   d. Any of the above can be true-breeding.
   e. a or b

3. When looking at the inheritance of a single trait, Mendel found that a cross between two true-breeding peas (between purple and white, for example) always yielded a _____ in the F_2 generation.
   a. 1:1 phenotypic ratio
   b. 3:1 genotypic ratio
   c. 1:2:1 phenotypic ratio
   d. 3:1 phenotypic ratio
   e. 1:1 genotypic ratio

4. Alternative forms of genes for a particular trait are called
   a. homologous chromosomes.
   b. alleles.
   c. linked genes.
   d. genotypes.
   e. phenotypes.

5. A fruit fly has two genes for eye color, but each of its sperm cells has only one. This illustrates
   a. independent assortment.
   b. linked genes.
   c. pleiotropy.
   d. polygenic inheritance.
   e. segregation.

6. Mendel made some crosses where he looked at two or three traits at once—round yellow peas crossed with wrinkled green peas, for example. He did this because he wanted to find out
   a. how new characteristics originated.
   b. whether different traits were inherited together or separately.
   c. how plants and animals adapt to their environments.
   d. whether the traits influence each other—whether color affects degree of roundness, for example.
   e. Actually, Mendel never had a clear purpose in mind.

7. A pea plant with purple flowers is heterozygous for flower color. Its genotype is Pp. The P and p alleles in the pea plant's cells are located
   a. next to each other on the same chromosome.
   b. at the same location on homologous chromosomes.
   c. on the X and Y chromosomes.
   d. some distance apart on the same chromosome.
   e. at different locations on homologous chromosomes.

8. When an individual has both I^A and I^B blood group alleles, both genes are expressed and the individual has group AB blood. This is an example of
   a. codominance.
   b. a dihybrid.
   c. pleiotropy.
   d. incomplete dominance.
   e. linked genes.

9. How many genes are there on one chromosome?
   a. one
   b. two
   c. hundreds
   d. thousands
   e. millions

10. Which of the following is *not* true of linked genes?
    a. They tend to be inherited together.
    b. They violate Mendel's principle of independent assortment.
    c. They are on the same chromosome.
    d. They can form new combinations via crossing over.
    e. They are relatively few in number; most genes are unlinked.

11. Morgan and his students were able to map the relative positions of genes on fruit-fly chromosomes by
    a. coloring chromosomes with dyes and observing under a microscope.
    b. scrambling the chromosomes and observing how the flies changed.
    c. crossing various flies and looking at the proportions of offspring.
    d. transplanting chromosomes from one fly to another.
    e. looking at crosses that showed independent assortment.

12. The sex chromosomes of a human female are ____. The sex chromosomes of a human male are ____.
    a. *XX . . . XY*
    b. *YY . . . XX*
    c. *XX . . . YY*
    d. *XY . . . XX*
    e. *YY . . . XY*

13. Most sex-linked traits in humans are carried on the ____ chromosome, and the recessive phenotypes are seen most often in ____.
    a. *X . . .* women
    b. *X . . .* men
    c. *Y . . .* women
    d. *Y . . .* men

Essay

1. Explain why Gregor Mendel was able to figure out the principles of heredity, while many other investigators before (and some after) Mendel failed to do so.

2. If you flip two coins, the probability that you will get two heads is 1/4, but the probability that you will get one head and one tail is 1/2. Explain why.

3. Why are organisms such as peas and fruit flies better subjects for genetics studies than human beings?

4. What determines a human's sex? Describe two other systems of sex determination in different organisms.

Applying Your Knowledge

Multiple Choice

1. A brown mouse is mated with a white mouse. All of their offspring are brown. If two of these brown offspring are mated, what fraction of their offspring will be white?
   a. all
   b. none
   c. 1/4
   d. 1/2
   e. 3/4

2. Suppose you wanted to know the genotype of one of the brown F_2 mice in question 1. The easiest way to do it would be to
   a. keep careful records of the parent mice.
   b. mate it with a brown mouse.
   c. mate it with a mouse of its own genotype.
   d. mate it with a white mouse.
   e. It can't be done.

3. Some dogs bark while trailing; others are silent. The barker gene is dominant, the silent gene recessive. The gene for normal tail is dominant over the gene for screw (curly) tail. A barker dog with a normal tail who is heterozygous for both traits is mated to another dog of the same genotype. What fraction of their offspring will be barkers with screw tails?
   a. 3/4
   b. 9/16
   c. 3/16
   d. 1/4
   e. 1/16

4. Two heterozygous tall pea plants with purple flowers are crossed. The probability that one of their offspring will have white flowers is 1/4. The probability that one of their offspring will be short is 1/4. What is the probability that one of their offspring will be short with white flowers?
   a. 0
   b. 1/16
   c. 1/8
   d. 1/4
   e. 1/2

5. A young unmarried woman had a baby and wished to collect child support from the father. Her blood group is AB. The baby's blood group is A. There are two possible fathers: Jim is group A, and Michael is group O. Which man could be the father?
   a. either
   b. Jim
   c. Michael
   d. neither
   e. impossible to tell given this evidence

6. Which of the following illustrates pleiotropy?
   a. In fruit flies, the genes for scarlet eyes and hairy body are located on the same chromosome.
   b. Matings between earless sheep and long-eared sheep always result in short-eared offspring.
   c. Wheat kernels can range from white to red in color, a trait controlled by several genes.
   d. The human cystic fibrosis gene causes many symptoms, from respiratory distress to digestive problems.
   e. An individual with both I^A and I^B alleles has blood group AB.

7. When two gray-bodied fruit flies are mated, their offspring total 86 gray-bodied males, 81 yellow-bodied males, and 165 gray-bodied females. The allele for yellow body is
   a. sex-linked and dominant.
   b. not sex-linked and dominant.
   c. sex-linked and recessive.
   d. not sex-linked and recessive.
   e. impossible to say on the basis of this information.

8. In fruit flies, the allele for red eyes is dominant, and the allele for purple eyes is recessive. Normal gray body is dominant, and black body is recessive. A geneticist mated a heterozygous red-gray male with a purple-black female. She predicted that there would be four phenotypes of offspring in equal numbers, but she was wrong. Instead, 48% of the offspring were red-gray, 46% were purple-black, 3% were red-black, and 3% were purple-gray. She concluded that in the male, the red and gray genes were linked, and in the female the purple and black genes were linked. If this is the case, how would you account for the red-black and purple-gray offspring?
   a. This is an example of pleiotropy.
   b. Body color and eye color are quantitative traits.

c. Crossing over during meiosis recombined the genes.
   d. This cross shows incomplete dominance at work.
   e. Segregation of alleles occurred during meiosis.

9. Red-green color blindness is a human recessive sex-linked trait. A man and a woman with normal vision have a color-blind son. What is the probability that their next child will also be a color-blind son?
   a. 0
   b. 1/8
   c. 1/4
   d. 1/2
   e. 3/4

Essay

1. Two apparently normal parents have a daughter who suffers from agammaglobulinemia, an inherited defect of the immune system, which is supposed to protect the body from infection. Use a Punnett square to show how two normal parents could have a child afflicted with an inherited disease. What are the parents' genotypes? The daughter's genotype? What is the probability that their second child will also have agammaglobulinemia?

2. A pea plant with purple flowers and green pods is crossed with a plant that has white flowers and yellow pods. All the offspring have purple flowers and green pods. If two of these F_1 peas are crossed, what phenotypes will be seen in the F_2 generation, and in what proportions?

3. Freckles is dominant, no freckles recessive. A man with freckles and a woman with no freckles have three children with freckles and one with no freckles. What are the genotypes of the parents and children?

4. The inheritance of flower color in snapdragons illustrates incomplete dominance: When a red snapdragon is crossed with the white one, all their offspring are pink. What offspring would be produced, in what proportions, if two of these pink snapdragons were crossed? What offspring would be produced, in what proportions, if a pink snapdragon was crossed with a white one?

5. A man whose blood group is A and a woman whose blood group is B have a son who is type O. What are their genotypes? What is the probability of the couple's next child being type B?

6. Recall that some traits, such as skin color, appear to be controlled by several genes. This creates a continuum of variation. If this polygenic explanation for the inheritance of human skin pigmentation is correct, how do the skin colors of the following four individuals compare? Which of the couples could have children with the widest range of skin colors? Why? Couple 1: *aaBbCC* and *aaBbCC*. Couple 2: *AaBbCc* and *AaBbCc*.

7. In fruit flies, the allele for red eyes is dominant over the allele for pink eyes. Straight wings is dominant over curled wings. Imagine that a red-eyed, straight-winged fly that is heterozygous for both traits is mated with a fly with pink eyes and curled wings. Predict the offspring that would be produced by this cross (genotypes, phenotypes, and fraction of each) if these two genes were on different chromosomes.

   When a geneticist actually carried out this mating, the offspring were as follows: 49% red eyes and straight wings, 49% pink eyes and curled wings, 1% red eyes and curled wings, and 1% pink eyes and straight wings. Does this agree with your prediction? How would you explain these results?

8. Numerous fruit-fly matings showed that the *h* allele for hairy body, the *b* allele for spineless bristles, and the *s* allele for striped body were all located on the same chromosome. The recombination frequency between alleles *h* and *b* was 4%. The recombination frequency between alleles *s* and *b* was 15%. Why are the recombination frequencies between *h* and *b* and between *s* and *b* different? The recombination frequency between alleles *h* and *s* was 10%. What is the order of these three genes on the chromosome?

Using Your Knowledge

1. Gregor Mendel and Thomas Hunt Morgan were two of the more colorful figures in the history of biology. You may want to look them up in an encyclopedia or seek out their biographies in the library.

2. If you live near your family, try to figure out the patterns of inheritance of some of the traits mentioned in Module 9.8 among your parents and siblings.

Molecular Biology of the Gene

10

Most biologists would agree that the most significant biological discovery of the twentieth century is the discovery of the structure of the gene. At the beginning of the century Mendel's rules were rediscovered, and genes were traced to the chromosomes. Soon it was possible to map the locations of genes, and scientists started wondering what exactly genes were made of and how they shaped an organism. By mid-century, it was clear that DNA is the genetic material and that genes act by directing the synthesis of proteins. Soon researchers discovered the double helix structure of DNA and deciphered the genetic code by which DNA shapes the body. Now scientists have learned how to make genes and move them from one organism to another. In the future, will this century be remembered as the space age? the atomic age? the computer age? Maybe, but our accelerating knowledge of genetics suggests that future generations might remember this era as the gene age. This chapter describes the molecular biology of the gene and the discoveries that made possible our new understanding of genes.

Organizing Your Knowledge

Exercise 1   (Modules 10.1 – 10.3)

Review the discovery that DNA is the genetic material and the structures of DNA and RNA. Then match each phrase on the right with the correct term(s) on the left. Note that some answers are used more than once, and some questions have multiple answers.

A. Adenine (A)	_____ 1. The basic chemical unit of a nucleic acid
B. Base	_____ 2. The "transforming factor" that alters pneumonia bacteria
C. Cytosine (C)	_____ 3. The two kinds of nucleic acids
D. DNA	_____ 4. The three parts of every nucleotide
E. *E. coli*	_____ 5. A pair of these forms a "rung" in the DNA ladder
F. Double helix	_____ 6. Used to "label" DNA and protein in experiments
G. Guanine (G)	_____ 7. The component of a bacteriophage that enters the host cell
H. Hydrogen bond	_____ 8. Two alternating parts that form the nucleic acid "backbone"
I. Radioactive isotope	_____ 9. The four bases in DNA
J. Covalent bond	_____ 10. The DNA base complementary to T
K. Bacteriophage	_____ 11. A virus that attacks bacteria
L. Protein	_____ 12. The substance a phage leaves outside its host cell
M. Nucleic acid	_____ 13. Ribose in RNA and deoxyribose in DNA
N. Nucleotide	_____ 14. Watson and Crick deduced the structure of this molecule
O. Centrifuge	_____ 15. The four bases in RNA
P. Phosphate	_____ 16. The DNA base complementary to G
Q. Polynucleotide	_____ 17. A bacterium attacked by T2 and T4 phages
R. RNA	_____ 18. The sequence of these encodes DNA information
S. Sugar	_____ 19. Eukaryotic chromosomes consist of this and DNA
T. Thymine (T)	_____ 20. The overall shape of a DNA molecule
U. Uracil (U)	_____ 21. Links adjacent nucleotides in a polynucleotide chain
	_____ 22. Machine used to separate particles of different sizes
	_____ 23. Links a complementary pair of bases together
	_____ 24. A polymer of nucleotides
	_____ 25. RNA base that is not in DNA

Exercise 2   (Modules 10.2 – 10.3)

Review the structure of DNA by labeling these diagrams. Include **nucleotide, polynu-cleotide, sugar (deoxyribose), phosphate group, sugar-phosphate backbone, pyrimi-dine bases, purine bases, thymine (T), adenine (A), guanine (G), cytosine (C), hydrogen bond, complementary base pair, double helix.**

Exercise 3   (Module 10.4)

Reproduction and inheritance involve copying DNA instructions, so that they can be passed to the next generation. This process is carried out by DNA polymerases, enzymes which use each strand of the DNA helix as a template on which to build a complementary strand. Review DNA replication by completing the diagrams below. The first diagram shows the parent DNA molecule; label the nucleotides in the right-hand strand. Complete and label the second diagram, so that it shows the parent strands separating and being used as templates. Label the third diagram, so that it shows two completed daughter molecules of DNA. Color the original DNA strands blue and the new strands red.

Exercise 4   (Modules 10.5 – 10.6)

In a cell, the genotype—genetic information in DNA—is expressed as phenotype in the form of proteins—structural proteins that shape the organism and enzymes that carry out metabolism. Review the relationship between genotype and phenotype by completing this crossword puzzle.

Across

3. A gene consists of hundreds or ____ of nucleotide bases.

6. The information in DNA specifies the synthesis of ____.

8. Garrod noted the gene-protein link in "inborn errors of ____."

9. Genetic instructions are written in 3-base "words" called ____.

10. An organism's expressed traits (what it looks like) make up its ____.

12. To make a protein, DNA information is first transcribed into ____.

13. The DNA language consists of a linear sequence of nucleotide ____.

14. The ____ is an organism's genetic makeup.

15. Phenotype is expressed via structural proteins and ____.

17. Making a polypeptide according to an RNA message is called ____.

Down

1. The base sequence of RNA is ____ to the DNA from which it is transcribed.

2. Genotype is the inheritable information encoded in ____.

4. Each codon in DNA and RNA specifies a certain ____ acid in a polypeptide.

5. Translation is conversion of an RNA message into a ____.

7. Transfer of information from DNA into an RNA molecule is called ____.

11. One ____ specifies how to build one polypeptide.

16. Using bread ____, Beadle and Tatum showed that a gene codes for an enzyme.

Exercise 5   (Modules 10.6 – 10.13)

These modules explain how the information in a gene is used to build a protein. Review the processes of transcription and translation by filling in the blanks below.

The first step in making a protein is transcription of the gene. This occurs in the [1]_____ of a eukaryotic cell. An enzyme called [2]_____ carries out the process of transcribing RNA from the DNA. It starts at a specific nucleotide sequence called a [3]_____, next to the gene. RNA polymerase attaches, and the two DNA strands separate. RNA polymerase moves along one strand, and as it does, RNA [4]_____ take their places one at a time along the DNA template. They hydrogen bond with complementary bases, following the same pairing rules as in DNA—C with G, and U (replacing T in RNA) with A. As the RNA molecule elongates, it peels away from the DNA. Finally, the enzyme reaches the [5]_____, a base sequence that signals the end of the gene, and the polymerase molecule lets go of the gene and the RNA molecule. In a prokaryote, the RNA transcribed from a gene, called [6]_____ (mRNA), can be used immediately in polypeptide synthesis. In a eukaryotic cell, mRNA leaves the nucleus and enters the [7]_____, where translation takes place.

Translation of the "words" of the mRNA message into the [8]_____ sequence of a protein requires an interpreter—[9]_____ (tRNA)—which links the appropriate [10]_____ with each [11]_____ in the mRNA message. A tRNA molecule is a folded strand of RNA. At one end, a special [12]_____ attaches a specific amino acid. The other end of the tRNA molecule bears three bases called the [13]_____, which is complementary to a particular mRNA codon. During the translation process, the tRNA matches its amino acid with an mRNA codon.

[14]_____ are the "factories" where the information in mRNA is translated and polypeptide chains are constructed. A ribosome consists of protein and [15]_____ (rRNA). Each ribosome has a groove that serves as a binding site for mRNA. There are two binding sites for tRNA: The P site holds the tRNA carrying the growing [16]_____, while the A site holds a tRNA bearing the next amino acid.

Translation begins with initiation. An mRNA and a special [17]_____ tRNA bind to the ribosome and a specific mRNA codon, the [18]_____, where translation begins. The initiator tRNA generally carries the amino acid methionine (Met). Its anticodon UAC binds to the start codon, AUG. The initiator tRNA fits into the P site on the ribosome.

The next step in [19]_____ synthesis is elongation—adding amino acids to the growing chain. The anticodon of an incoming tRNA, carrying its amino acid, pairs with the mRNA codon at the open A site. With help from the ribosome, the polypeptide separates from its tRNA and forms a peptide bond with the [20]_____ attached to the tRNA in the A site. Then the "empty" tRNA in the P site leaves the ribosome, and the tRNA in the A site, with the polypeptide chain, is shifted to the P site. The mRNA and tRNA move as a unit, allowing the next codon to enter the A site. Another tRNA, with a complementary anticodon, brings its amino acid to the A site. Its amino acid is added to the chain, the tRNA leaves, and the complex shifts again. In this way, [21]_____ are added to the chain, one at a time.

Finally, a [22]_____ reaches the A site of the [23]_____, terminating the polypeptide. A stop codon causes the polypeptide to separate from the last tRNA and the [24]_____. The polypeptide folds up, and it may join with other polypeptides to form a larger [25]_____ molecule.

Exercise 6   (Module 10.13)

This module summarizes the key steps in the flow of genetic information from DNA to protein. Study the diagrams carefully, then label the numbered parts and processes.

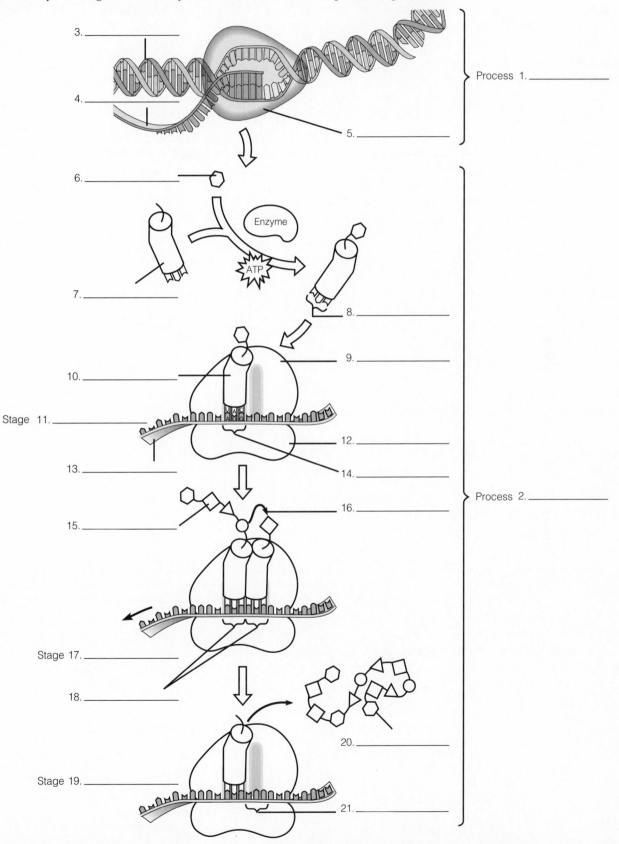

Exercise 7 (Modules 10.14 – 10.15)

These modules describe the genetic code, how biologists cracked the code, and how mutations change the meaning of the coded genetic message. Use the genetic code chart (Figure 10.14A in the textbook) to translate the following mRNAs into amino acid sequences and answer the questions.

mRNA nucleotide sequence:
(mRNA 1)

1. Amino acid sequence:

Mutation in mRNA:
(mRNA 2)

2. Amino acid sequence:
3. Number of bases changed in mRNA:
4. Type of mutation:
5. Number of amino acids changed:

Mutation in mRNA:
(mRNA 3; compare to 1)

6. Amino acid sequence:
7. Number of bases changed in mRNA (look carefully!):
8. Type of mutation:
9. Number of amino acids changed (compared to mRNA 1):
10. Which mutation had the greatest effect and why?

Exercise 8   (Modules 10.16 – 10.19)

These modules describe the structures and life cycles of viruses. Match each phrase on the right with a term from the left. Some answers are used more than once.

A. RNA viruses
B. Prophage
C. AIDS
D. Glycoprotein spikes
E. Virus
F. DNA
G. Lytic cycle
H. Vaccine
I. Nucleus
J. Membranous envelope
K. Bacteriophage
L. Provirus
M. HIV
N. Reverse transcriptase
O. Lysogenic cycle
P. Retrovirus
Q. White blood cell
R. Prophage genes
S. Tobacco mosaic
T. Plasma membrane
U. RNA
V. Cytoplasm

_____ 1. Consists of nucleic acid packaged in protein
_____ 2. Leads quickly to breaking open of host cell
_____ 3. Phage DNA inserted into bacterial chromosome
_____ 4. When virus "hides" as part of bacterial chromosome
_____ 5. Responsible for toxins of diphtheria, botulism
_____ 6. Rod-shaped plant virus
_____ 7. This or DNA may be virus genetic material
_____ 8. Cause of flu, colds, polio, mumps, AIDS
_____ 9. Helps flu or mumps virus enter and leave host cell
_____ 10. Used by mumps virus or HIV to attach to host receptors
_____ 11. Mumps virus reproduces here
_____ 12. Mumps virus makes this and protein from RNA template
_____ 13. Mumps virus gets envelope from this part of host cell
_____ 14. Herpesvirus reproduces here
_____ 15. Genetic material of herpesvirus
_____ 16. DNA of herpesvirus inserted into host cell DNA
_____ 17. Can be used to prevent a viral disease
_____ 18. Virus that causes AIDS
_____ 19. Genetic material of HIV
_____ 20. RNA virus that reproduces by means of DNA
_____ 21. Enzyme that can make DNA from RNA template
_____ 22. Form in which HIV "hides" in host cell
_____ 23. Acquired immune deficiency syndrome
_____ 24. Kind of cell infected by HIV
_____ 25. Viruses like T2 and T4 that infect bacteria

Testing Your Knowledge

Multiple Choice

1. In an important experiment, bacteriophages were allowed to infect bacteria. In the first trial, the phages used contained radioactive DNA, and radioactivity was detected in the bacteria. Next, other phages containing radioactive protein were allowed to infect bacteria, and no radioactivity was detected in the bacteria. When the experimenters compared the results of these two trials, they concluded that
   a. genes are made of DNA.
   b. bacteriophages can infect bacteria.
   c. DNA is made of nucleotides.
   d. genes carry information for making proteins.
   e. genes are on chromosomes.

2. An RNA or DNA molecule is a polymer made of subunits called
   a. bases.
   b. amino acids.
   c. nucleotides.
   d. nucleic acids.
   e. pyrimidines.

3. The information carried by a DNA molecule is in
   a. its amino acid sequence.
   b. the sugars and phosphates forming its backbone.
   c. the order of the bases in the molecule.
   d. the total number of nucleotides it contains.
   e. the RNA units that make up the molecule.

4. A gene is
   a. the same thing as a chromosome.
   b. the information for making a polypeptide.
   c. made of RNA.
   d. made by a ribosome.
   e. made of protein.

5. DNA replication occurs
   a. whenever a cell makes protein.
   b. to repair gene damage caused by mutation.
   c. before a cell divides.
   d. whenever a cell needs RNA.
   e. in the cytoplasm of a eukaryotic cell.

6. The flow of information in a cell proceeds
   a. from RNA to DNA to protein.
   b. from protein to RNA to DNA.
   c. from DNA to protein to RNA.
   d. from RNA to protein to DNA.
   e. from DNA to RNA to protein.

7. Which of the following is *not* needed for DNA replication?
   a. ribosomes
   b. DNA
   c. nucleotides
   d. enzymes
   e. All of the above are needed.

8. Which of the following processes occur in the cytoplasm of a eukaryotic cell?
   a. DNA replication
   b. translation
   c. transcription
   d. DNA replication and translation
   e. translation and transcription

9. Beadle and Tatum showed that each kind of mutant bread mold lacked a specific enzyme. This experiment demonstrated that
   a. genes carry information for making proteins.
   b. mutations are changes in genetic information.
   c. genes are made of DNA.
   d. enzymes are required to repair damaged DNA information.
   e. cells need specific enzymes in order to function.

10. During the process of translation (polypeptide synthesis), ____ matches a nucleic acid codon with the proper amino acid.
    a. a ribosome
    b. DNA polymerase
    c. ATP
    d. transfer RNA
    e. messenger RNA

11. How does RNA polymerase "know" where to start transcribing a gene into mRNA?
    a. It starts at one end of the chromosome.
    b. Transfer RNA acts to translate the message to RNA polymerase.
    c. It starts at a certain nucleotide sequence called a promoter.
    d. The ribosome directs it to the correct portion of the DNA molecule.
    e. It looks for the AUG start codon.

12. When RNA is being made, the RNA base ___ always pairs with the base ___ in DNA.
    a. U . . . T
    b. T . . . G
    c. U . . . A
    d. A . . . U
    e. T . . . A

13. A mutagen is
    a. a gene that has been altered by a mutation.
    b. something that causes a mutation.
    c. an organism that has been changed by a mutation.
    d. the portion of a chromosome altered by a mutation.
    e. any change in the nucleotide sequence of DNA.

14. How do retroviruses, such as HIV, differ from other viruses?
    a. They are much simpler than other viruses.
    b. They contain DNA that is used as a template to make RNA.
    c. They can reproduce only inside of living cells.
    d. They contain nucleic acids that code for the making of proteins.
    e. They contain RNA that is used as a template to make DNA.

Essay

1. Sketch a short piece of a DNA molecule, five base pairs long. Use simple shapes to represent bases, sugars, and phosphates. Show proper base pairing, and label a nucleotide, a base, a phosphate group, a sugar, A, C, T, G, the double helix, and hydrogen bonds.

2. Explain why, in DNA, T pairs only with A and not with C or G.

3. Why does it take a group of three DNA nucleotides to specify one amino acid in a protein? Wouldn't it be simpler to have a one-to-one code, where one nucleotide specified one amino acid?

4. What is a mutation? What causes mutations? Why are most mutations harmful? Why aren't all mutations harmful?

5. Which type of mutation—a base substitution or a base deletion—is likely to have the greatest effect on the organism? Why?

6. Describe step by step, but in simple terms, the roles of mRNA, tRNA, ribosomes, and amino acids in making a polypeptide.

Applying Your Knowledge

Multiple Choice

1. Which of the following are arranged in the correct order by size, from largest to smallest?
   a. chromosome-gene-codon-nucleotide
   b. nucleotide-chromosome-gene-codon
   c. codon-chromosome-gene-nucleotide
   d. gene-chromosome-codon-nucleotide
   e. chromosome-gene-nucleotide-codon

2. A geneticist raised a crop of T2 bacteriophages in a medium containing radioactive phosphorus, so that the DNA of the bacteriophages was labeled with radioactivity. The labeled phages were then allowed to infect nonradioactive bacteria. In a few hours, these bacteria burst open, releasing many bacteriophages. Some of these phages contained labeled
   a. DNA.
   b. RNA.
   c. protein.
   d. all of the above.
   e. DNA and protein only.

3. A messenger RNA molecule for making a protein is made in the nucleus and sent out to a ribosome. The ribosome reads the mRNA message and makes a protein containing 120 amino acids. The mRNA consisted of at least how many codons?
   a. 30
   b. 40
   c. 120
   d. 360
   e. 480

4. The nucleotide sequence of a DNA codon is ACT. A messenger RNA molecule with a complementary codon is transcribed from the DNA. In the process of protein synthesis, a transfer RNA pairs with the mRNA codon. What is the nucleotide sequence of the tRNA anticodon? (Careful—This one is harder than it appears.)
   a. TGA
   b. UGA
   c. ACT
   d. TGU
   e. ACU

5. Imagine an error occurring during DNA replication in a cell, so that where there is supposed to be a T in one of the genes there is instead a G. What effect will this probably have on the cell?
   a. Each of its kinds of proteins will contain an incorrect amino acid.
   b. An amino acid will be missing from each of its kinds of proteins.
   c. One of its kinds of proteins might contain an incorrect amino acid.
   d. An amino acid will be missing from one of its kinds of proteins.
   e. The amino acid sequence of one of its kinds of proteins will be completely changed.

6. A cell is grown in a solution containing radioactive nucleotides, so that its DNA is labeled with radioactivity. It is removed from the radioactive solution and grown in a normal medium, so that any new DNA strands it makes will not be radioactive. In the normal medium, the cell replicates its DNA and divides. The two daughter cells also replicate their DNA and divide, producing a total of four cells. If a dotted line represents a radioactive DNA strand and a solid line represents a nonradioactive DNA strand, which of the following depicts the DNA of the four cells?

a.

b.

c.

d.

e.

7. A particular ____ carry the information for making a particular polypeptide, but ____ can be used to make any polypeptide.
   a. gene and ribosome . . . a tRNA and an mRNA
   b. gene and mRNA . . . a ribosome and a tRNA
   c. ribosome and mRNA . . . a gene and a tRNA
   d. gene and tRNA . . . a ribosome and an mRNA
   e. tRNA and ribosome . . . a gene and an mRNA

8. A sequence of pictures of polypeptide synthesis shows a ribosome holding two transfer RNAs. One tRNA has a polypeptide chain attached to it; the other tRNA has a single amino acid attached to it. What does the next picture show?
   a. The polypeptide chain moves over and bonds to the single amino acid.
   b. The tRNA with the amino acid leaves the ribosome.
   c. The amino acid moves over and bonds to the polypeptide chain.
   d. The tRNA with the polypeptide chain leaves the ribosome.

e. A third tRNA with an amino acid joins the pair on the ribosome.

9. A microbiologist analyzed chemicals obtained from an enveloped RNA virus (similar to a mumps virus) that infects monkeys. He found that the virus envelope contained a protein characteristic of monkey cells. Which of the following is the most likely explanation for this?
   a. The virus gets its envelope when it leaves its host cell.
   b. The virus forced the monkey cell to make proteins for its envelope.
   c. The virus has a lysogenic life cycle.
   d. The virus gets its envelope when it enters its host cell.
   e. The virus fools its host cell by mimicking its proteins.

10. At one point as a cell carried out its day-to-day activities, the nucleotides G A T were paired with the nucleotides C U A. This pairing occurred
    a. in a double-stranded DNA molecule.
    b. during translation.
    c. during transcription.
    d. when an RNA codon paired with a tRNA anticodon.
    e. impossible to say, given this information.

Essay

1. *E. coli* bacteria are used in many genetic studies. Type A *E. coli* can live on a simple nutrient medium, because they have all the genes necessary to produce the chemicals they need. Type V *E. coli* can live only on a nutrient medium to which a certain vitamin has been added, because they lack a gene that enables them to make this vitamin for themselves. It has been found that bacteria can absorb genes from other dead, groundup bacteria. Describe an experiment using type A and type V *E. coli* to determine whether genes are made of protein or DNA.

2. It is possible to extract DNA from cells and analyze it to determine the relative amounts of the four DNA bases. The DNA of a goldfish contains more T and less G than human DNA, but in both goldfish and human DNA the amount of T is equal to the amount of A. Explain why.

3. Eric said to Renee, "The amino acid sequence of the proteins in your hair determines how curly or straight your hair will be." Renee replied, "I don't think that's right. Your genes determine whether your hair is curly or straight. That's why it's inherited." Who is right? Explain.

4. The DNA base sequence for a short gene is:

   TATGATACCTTGATAGCTATCTGATTG.

   What is the amino acid sequence of the polypeptide produced according to this DNA information? Use the genetic code chart (Figure 10.14A in the text) and your knowledge of transcription and translation to figure out the message.

5. A biochemist found that a bacterium produced an mRNA molecule consisting of 852 nucleotides and translated this mRNA into a polypeptide containing 233 amino acids. How many nucleotides in the mRNA message would actually be needed to carry the message for the polypeptide, and how many were "extras"? How would the bacterium know *which* nucleotides made up the message?

6. The virus that causes chickenpox can disappear for years and then reappear in a line of painful sores where a nerve cell passes through the skin. How can viruses go away and then reappear like this? Where are the viruses during the intervening period of time?

Using Your Knowledge

1. One of the most important things you can learn in your study of biology is how to keep from becoming infected with HIV, the AIDS virus. Information is available from college health centers and the National AIDS Hotline, 1-800-342-AIDS.

2. In recent years, there have been several major outbreaks of "childhood" viral diseases such as measles on college campuses. Most young children are vaccinated, so unvaccinated individuals are not likely to contract these diseases "naturally." Unfortunately, these viruses can have serious effects, especially on adults. Do you know whether you have been vaccinated for measles, mumps, chickenpox, and so on? Did you have any of these diseases when you were younger? Your family or family doctor may have information about this. If you are not immune to these diseases, you may want to ask your doctor or college health center about vaccination.

Control of Gene Expression

One of the wonders of biology is how a fertilized egg—a single, relatively generalized cell—can develop into an adult human being—a multicellular organism consisting of hundreds of kinds of specialized cells. The zygote contains all the genes needed to construct and direct the activities of an entire human being. As the zygote divides, different groups of cells differentiate, taking on different shapes and different functions, finally forming the cooperating tissues and organs of the adult body. Muscle cells, bone cells, nerve cells—experiments suggest that these different kinds of cells all contain identical sets of genes. What makes the cells different is that different subsets of genes are expressed in each kind of cell. A sophisticated regulatory system switches genes on and off, responding to environmental changes and shaping the differentiation of various cell lineages. Occasionally, genes escape from their normal controls, producing birth defects or cancer. This chapter considers gene regulation, a subject of much mystery and intensive research.

Organizing Your Knowledge

Exercise 1   (Modules 11.1 – 11.3)

Genes can be turned on and off. In multicellular organisms, this is important in cellular differentiation. Only certain genes in each kind of cell are activated, shaping the structure and function of the cell for a certain role. Genes also turn on and off as a bacterium responds to changes in its environment. In bacteria, genes are grouped, with control sequences called operators and promoters, into clusters called operons. The *lac* and *trp* operons are two such gene clusters in the bacterium *E. coli*. Study the diagrams in Modules 11.2 and 11.3, and then match each of the components of the *lac* and *trp* operon systems with its function.

lac operon:

_____ 1. Regulatory gene
_____ 2. Repressor protein + lactose
_____ 3. Repressor protein without lactose
_____ 4. RNA polymerase
_____ 5. Promoter
_____ 6. Operator
_____ 7. Operon genes
_____ 8. Enzymes

A. Keeps RNA polymerase from attaching to promoter and transcribing genes
B. Transcribes genes into RNA for protein synthesis
C. Repressor protein attaches here
D. Use lactose
E. Information for making repressor protein
F. Where RNA polymerase starts transcribing genes
G. Allows RNA polymerase to transcribe genes
H. Information for making enzymes that use lactose

trp operon:

_____ 1. Regulatory gene
_____ 2. Repressor protein + tryptophan
_____ 3. Repressor protein without tryptophan
_____ 4. RNA polymerase
_____ 5. Promoter
_____ 6. Operator
_____ 7. Operon genes
_____ 8. Enzymes

A. Keeps RNA polymerase from attaching to promoter and transcribing genes
B. Transcribes genes into RNA for protein synthesis
C. Repressor protein attaches here
D. Make tryptophan
E. Information for making repressor protein
F. Where RNA polymerase starts transcribing genes
G. Allows RNA polymerase to transcribe genes
H. Information for making enzymes that make tryptophan

Exercise 2   (Modules 11.4 – 11.6)

In eukaryotic organisms, cells become specialized—differentiated—for different functions. It has been demonstrated with frogs and plants that differentiated cells retain a complete set of genes for making all the specialized cells in the whole organism. Each differentiated cell *has* all the genes, but *different* genes are *active* in different kinds of cells. Check your understanding of this concept by coloring in the boxes below to show which genes you think would be active in each of the cell types.

	Stomach Gland Cell	*Hair Follicle Cell*	*Stem Cell* (Becomes Red Blood Cell)
Hemoglobin gene	☐	☐	☐
Keratin (hair protein) gene	☐	☐	☐
Krebs cycle enzyme gene	☐	☐	☐
Digestive enzyme gene	☐	☐	☐
Insulin gene	☐	☐	☐

Exercise 3 (Modules 11.7 – 11.8)

Complete this crossword puzzle to review the roles of DNA packing and protein activators in gene expression.

Across

4. Scientists think most eukaryotic regulatory proteins act as ____.

6. One X chromosome in each of a woman's cells is ____.

12. The DNA-histone beaded fiber is further wrapped into a tight ___ fiber.

14. Nucleosomes may control gene ____ by limiting access to DNA.

15. The DNA supercoil is further wrapped and folded into a ____.

17. Insect molting hormone affects genes by binding to a ____ protein.

19. A cell may have 1000 ____ proteins that may help regulate genes.

20. DNA fits into a cell because of a system of folding, or ____.

21. In eukaryotes, structural and regulatory genes are ____.

22. The twisted DNA further coils into a ____ with a diameter of 200 nm.

Down

1. The color pattern of a ____ cat reflects the influence of the Barr body.
2. The DNA-histone complex looks like "____ on a string."
3. DNA is wound around small proteins called ____.
5. Changes in fruit-fly chromosome puffs are brought about by a ____.
7. A ____ is a complex of DNA wrapped around eight histone molecules.
8. DNA packing seems to control gene expression at the ____ stage.
9. A ____ body is a highly compacted X chromosome.
10. Fruit-fly larvae have ____ chromosomes in their salivory gland cells.
11. ____ on giant chromosomes show patterns of tighter and looser DNA packing.
13. Activators trigger RNA ____ to begin transcription.
16. An activator may help position RNA polymerase on a gene's ____.
18. In most eukaryotic cells, most ____ are not expressed.
20. The pattern of chromosome ____ changes as a fruit-fly larva develops.

In eukaryotes, gene expression is also regulated after transcription of genes into mRNA and during and after translation of mRNA into protein. Review these processes by matching each of the processes on the right (listed in order of occurrence) with a description on the left.

A. Altering a protein to form an active final product
B. Retaining or breaking down mRNA molecules, controlling the amount to which they are translated
C. Action of inhibitors that may block synthesis of protein from mRNA message
D. Addition of nucleotides to beginning and end of mRNA; may protect mRNA and enhance translation
E. Removal of noncoding introns from mRNA and joining of exons that will be expressed
F. Retaining or breaking down proteins, depending on cell's needs
G. Altering the sequence of coding nucleotides in mRNA, producing a "final draft"

_____ 1. RNA splicing
_____ 2. Addition of cap and tail to RNA
_____ 3. RNA editing
_____ 4. Selective degradation of mRNA
_____ 5. Inhibition of translation
_____ 6. Activation of finished protein
_____ 7. Selective degradation of proteins

This module summarizes the mechanisms that regulate gene expression in eukaryotes. After reviewing the module, match each of the mechanisms of regulation with the stage of gene expression where it acts. Choose from **mRNA degradation, DNA packing, cleavage/modification/activation, transport, protein degradation, addition of cap and tail, transcription, splicing/editing, translation**

Exercise 6   (Module 11.13)

To review this module, write a single sentence describing the functions and relationships of homeotic genes, homeotic proteins, and homeoboxes.

Exercise 7   (Modules 11.14 – 11.17)

Review the causes and mechanisms of cancer by filling in the blanks in the following story.

Lung cancer kills about 400,000 people per year. In the U.S., more than 150,000 people will develop lung cancer this year. Long the most common kind of cancer in men, lung cancer recently passed breast cancer to become the most frequent cancer in women as well.

Cancer is uncontrolled multiplication of cells. In most cells, control mechanisms regulate [1]_____ and [2]_____. Sometimes cells escape from these controls and multiply wildly. The cells form abnormal masses called [3]_____, which displace nearby normal tissues and can spread through the body. Localized tumors can be treated by [4]_____, but if a tumor spreads, it is often necessary to treat the resulting cancer with [5]_____or [6]_____ that inhibit actively dividing cells.

There are hundreds of different kinds of cancers in humans, named for the tissue or organ where they occur and grouped into four categories. Lung cancers are usually [7]_____, cancers that originate in the body coverings—in this case, the cells lining the lung.

Because the growing tumors block breathing passages, the first symptoms of lung cancer are usually coughing and difficulty breathing. The tumorous masses show up on chest X rays, and usually a small sample of lung tissue is taken to examine the tumor cells.

What causes lung cancer? Cancer-causing agents are called [8]_____. [9]_____, such as X rays and UV light, are known to cause some cancers, but most are caused by chemicals. Carcinogens in tobacco smoke appear to be the major cause of lung cancer. An increase in cigarette smoking since 1900 has been paralleled by a rise in lung cancer rates. Tobacco has also been linked to other forms of cancer, such as cancer of the throat and stomach.

Scientists have discovered much about the cellular mechanisms of cancer by studying cancers caused by viruses in humans and other animals. They were surprised to find that cancer-causing viruses carry cancer-causing genes, called [10]_____, as part of their genome. When the viruses insert their genes into the chromosomes of a host cell, the cancer-causing genes are inserted as well. Even more surprising, researchers found that oncogenes are simply altered versions of genes normally found in all cells. These genes, called [11]_____, usually code for proteins called [12]_____ factors, which normally stimulate cell [13]_____. What appears to cause most cancers is a [14]_____ in a proto-oncogene in a body (somatic) cell that changes it into an oncogene.

If lung cancer works like other cancers, the carcinogens in tobacco smoke probably cause a mutation in a proto-oncogene. The mutation changes the gene so that it now codes for a hyperactive protein—one that stimulates cell division more than normal—or excess production of protein. This causes the cell to divide without restraint.

Changes in genes whose products *inhibit* cell division—so-called [15]_____ genes—can also contribute to the development of cancer. A mutation can prevent the tumor-suppressor protein from being made and allow uncontrolled cell division. In most cases studied so far, changes in both proto-oncogenes and tumor-suppressing genes seem to be necessary for a full-fledged tumor to develop.

Exercise 8   (Module 11.17)

After reading this module, match each of the following human cancers with the carcinogen implicated as its major cause. (Answers may be used more than once.)

A. Ultraviolet light
B. Testosterone
C. Tobacco
D. Estrogen
E. Alcohol; hepatitis viruses
F. Animal fat; low dietary fiber
G. X rays

_____ 1. Lung
_____ 2. Colon
_____ 3. Breast
_____ 4. Prostate
_____ 5. Bladder
_____ 6. Uterus
_____ 7. Mouth and throat
_____ 8. Leukemia
_____ 9. Melanoma (skin)
_____ 10. Liver

Exercise 9   (Module 11.18)

There are two general hypotheses that have been proposed to explain the changes that accompany aging. Perhaps cells just wear out; they accumulate mutations and other damage as they get older, and this damage impairs their function. Alternatively, cells could be genetically programmed to change as they age; the decline that occurs with aging could be in the genes. State whether you think each of the statements below supports the "wear-and-tear" hypothesis (W), the "genetic program" hypothesis (G), or both (B). (Not all statements are clear-cut; there may not be a single right answer for every statement. The idea is to think about aging and its possible causes.)

_____ 1. The concentration of toxic wastes is greater in older cells.
_____ 2. Cells in cell culture divide about 50 times and then die.
_____ 3. Mutations are more numerous in older cells.
_____ 4. Different genes are expressed in the cells of a 70-year-old than in the cells of a 20-year-old.
_____ 5. DNA repair is more efficient in younger cells.
_____ 6. At the age of 10, a child with a rare disorder called progeria can show signs of aging that are normally seen in 70-year-olds.
_____ 7. DNA replication slows down in older cells.
_____ 8. Smoking and exposure to sunlight can cause skin to age prematurely.
_____ 9. There is some evidence that the same vitamins that prevent cancer (C and E) can also slow some chemical changes associated with aging.

Testing Your Knowledge

Multiple Choice

1. Your muscle and bone cells are different because
    a. they contain different sets of genes.
    b. they are differentiated.
    c. they contain different operons.
    d. different genes are switched on and off in each.
    e. their oncogenes are expressed differently.

2. Operons enable bacteria to
    a. function in frequently changing environments.
    b. resist attack by predators.
    c. correct mutations that might interfere with their genetic instructions.
    d. differentiate.
    e. mutate and evolve more rapidly.

3. If the nucleus of a frog egg is destroyed and replaced with the nucleus of an intestine cell from a tadpole, the egg can develop into a normal tadpole. This demonstrates that
    a. intestine cells are fully differentiated.
    b. there is little difference between an egg cell and an intestine cell.
    c. an intestine cell possesses a full set of genes.
    d. intestine cells are not differentiated.
    e. frogs can regenerate lost parts.

4. DNA packing—the way DNA is folded into chromosomes—affects gene expression by
    a. controlling access to DNA.
    b. positioning related genes near each other.
    c. protecting DNA from mutations.
    d. enhancing recombination of genes.
    e. allowing "unpacked" genes to be eliminated from the genome.

5. Histones are
    a. master genes that affect development.
    b. groups of genes that respond to environmental changes.
    c. proteins around which DNA is coiled.
    d. cancer-causing genes.
    e. portions of genes that are eliminated by DNA splicing.

6. In most eukaryotic cells most genes are not expressed. This suggests that most eukaryotic regulatory proteins act as
    a. exons.
    b. repressors.
    c. introns.
    d. oncogenes.
    e. activators.

7. After an mRNA molecule is transcribed from a eukaryotic gene, portions called ____ are removed and the remaining ____ are spliced together to produce an mRNA molecule with a continuous coding sequence.
    a. proto-oncogenes . . . oncogenes
    b. exons . . . introns
    c. oncogenes . . . proto-oncogenes
    d. introns . . . exons
    e. promoters . . . operators

8. Which of the following mechanisms of gene regulation operates after mRNA transcription but before translation of mRNA into protein?
    a. mRNA splicing and editing
    b. DNA packing
    c. repressors and activators
    d. protein degradation
    e. all of the above

9. Homeobox polypeptide segments
    a. serve as histones, facilitating DNA packing.
    b. bind to DNA and activate or repress gene transcription.
    c. are responsible for cancer if their normal controls are altered.
    d. act as enzymes, carrying out important chemical reactions.
    e. carry out mRNA splicing in the cell nucleus.

10. The genes that malfunction in cancer normally
    a. control RNA transcription.
    b. are responsible for organizing DNA packing.
    c. code for enzymes that repair damaged DNA.
    d. are not present in most body cells unless inserted by a virus.
    e. regulate cell division.

11. Which of the following cancers is much less common than the others? Cancer of the
    a. prostate.
    b. liver.
    c. lung.
    d. breast.
    e. colon.

12. Which of the following is a known or likely carcinogen?
    a. ultraviolet light
    b. viruses
    c. hormones
    d. X rays
    e. all of the above

Essay

1. In the proper growth medium, a single cell from a Boston fern can be stimulated to grow into an entire plant. What does this signify with regard to cellular differentiation in plants?

2. What are introns and exons? Discuss three possible biological functions of introns.

3. What is a homeotic gene? Why does a mutation in a homeotic gene have a much more drastic effect on the organism than a mutation in other genes?

4. Very similar homeoboxes control gene transcription in the DNA of frogs, flies, and humans. Why do biologists consider this significant?

5. How do cancer cells differ from normal body cells? Describe the changes in a cell that can make the cell become cancerous.

6. If a person wishes to avoid cancer, what factors in the environment should he or she try to avoid? What dietary and health habits would you recommend?

7. Compare two general hypotheses that have been proposed to explain aging.

Applying Your Knowledge

Multiple Choice

1. When a certain bacterium encounters the antibiotic tetracycline, the antibiotic molecule enters the cell and attaches to a repressor protein. This keeps the repressor from binding to the bacterial chromosome, allowing a set of genes to be transcribed. These genes code for enzymes that break down the antibiotic. This arrangement of genes is best described as
   a. oncogenes.
   b. regulatory genes.
   c. an operon.
   d. carcinogens.
   e. a nucleosome.

2. A genetic defect in humans results in absence of sweat glands in the skin. Some men have this defect all over their bodies, but in women it usually is expressed in a peculiar way. A woman with the defect typically has small patches of skin with sweat glands and other patches where sweat glands are lacking. This pattern suggests the phenotypic effect of
   a. a carcinogen.
   b. the Barr body.
   c. RNA splicing.
   d. an oncogene.
   e. the homeobox.

3. A bacterium makes the amino acid glycine or absorbs it from its surroundings. A biochemist found that glycine binds to a repressor protein and causes the repressor to bind to the bacterial chromosome, "turning off" an operon. If this is like other operons, the genes of this operon probably code for enzymes that
   a. control bacterial cell division.
   b. break down glycine.
   c. produce glycine.
   d. cause the bacterium to differentiate.
   e. manufacture the repressor protein.

4. In humans, the hormone testosterone enters cells and binds to specific proteins, which in turn bind to specific sites on the cells' DNA. These proteins probably act to
   a. help RNA polymerase transcribe certain genes.
   b. alter the pattern of DNA splicing.
   c. turn proto-oncogenes into oncogenes.
   d. unwind the DNA so that its genes can be transcribed.
   e. cause mutations in the DNA.

5. A geneticist studying chromosome puffs found that ____ is more intense around the puffs than elsewhere.
   a. conversion of proto-oncogenes to oncogenes
   b. activity of RNA polymerase
   c. protein degradation
   d. DNA polymerase activity
   e. repression of homeotic genes

6. It is possible for a cell to make proteins that last for months; hemoglobin in red blood cells is a good example. However, many proteins are not this long-lasting. They may be degraded in days or even hours. Why do cells make proteins with such short lifetimes if it is possible to make them last longer?
   a. Most proteins are used only once.
   b. Most cells in the body live only a few days.
   c. Cells lack the raw materials to make most of the proteins they need.
   d. Only cancer cells, which can keep dividing, contain long-lasting proteins.
   e. This enables cells to control the amount of protein present.

7. Researchers who wish to grow cells in laboratory nutrient solutions sometimes fuse their cells with cancer cells to create hybrids. Why might they want to do this?
   a. This reduces the possibility of mutation; cancer cells have already mutated.
   b. Cancer cells are more resistant to attack by viruses.
   c. Cancer cells continue to divide indefinitely.
   d. The genes of cancer cells are better understood than those of normal cells.
   e. Cancer cells are more highly differentiated than other cells.

8. Which of the following would be most likely to lead to cancer?
   a. multiplication of a proto-oncogene and inactivation of a tumor-suppressor gene
   b. hyperactivity of a proto-oncogene and activation of a tumor-suppressor gene
   c. inactivation of a proto-oncogene and multiplication of a tumor-suppressor gene
   d. inactivation of both a proto-oncogene and a tumor-suppressor gene
   e. hyperactivity of both a proto-oncogene and a tumor-suppressor gene

9. A cell biologist found that two different proteins with largely different structures were translated from two different mRNAs. These mRNAs, however, were transcribed from the same gene in the cell nucleus. What mechanism below could best account for this?
   a. Different systems of DNA packing could result in two different mRNAs.
   b. A mutation might have altered the gene.
   c. Exons from the same gene could be spliced in different ways to make different mRNAs.
   d. The two mRNAs could be transcribed from different chromosome puffs.
   e. One mRNA probably came from a proto-oncogene, the other from the corresponding oncogene.

10. Researchers have found homeotic genes in humans, but they are not yet certain how these genes shape the human phenotype. Considering the functions of homeotic genes in other animals, which of the following is most likely to be their function in humans?
    a. determining skin and hair color
    b. regulating cellular metabolic rate
    c. determining head and tail, back and front
    d. determining whether an individual is male or female

e. regulating the rate and timing of cell division

Essay

1. Mutations sometimes affect operons. Imagine a mutation in the regulatory gene that produces the repressor of the *lac* operon in *E. coli*. The altered repressor is no longer able to bind to the operator. What effect will this have on the bacterium?

2. Describe how three different types of cells in your body are specialized for different functions. How do their differences reflect differences in gene expression? Suggest a gene that might be active in each of the cells but none of the others. Suggest a gene that might be active in all the cells. Suggest a gene that is probably not active in any of the cells.

3. A biochemist was studying a membrane-transport protein consisting of 258 amino acids. She found that the gene coding for the transport protein consisted of 3561 nucleotides. The mRNA molecule from which the transport protein was transcribed contained 1455 nucleotides. What is the minimum number of nucleotides needed to code for the protein? How can the protein be transcribed from an mRNA that is larger than necessary? How can this mRNA be made from a gene that is so much larger?

4. Explain how, in a eukaryotic cell, a gene on one chromosome might affect the expression of a gene on a different chromosome. How might a gene in a certain cell affect expression of a gene in a different cell?

5. A certain kind of leukemia can be caused by a virus, a chemical, or radiation. Explain how these different factors can all trigger identical forms of cancer in the same kind of tissue.

Using Your Knowledge

1. What kinds of changes in your lifestyle and diet might reduce your chances of developing cancer?

2. If you smoke, quitting is the single most effective change you can make for a longer, healthier life. Period.

Recombinant DNA Technology

12

In recent years—within the lifetimes of most students reading this textbook—biologists have learned to construct genes and use them to alter organisms. A new toolkit consisting of enzymes, radioactive isotopes, cell culture techniques, and "gene machines" enables us to find genes, read them, compare them, insert them into new organisms, and harvest their products. Genetic engineering has revolutionized our understanding of genes and promises to transform our lives in many ways, from medicine to agriculture. But our new understanding of genes and our ability to manipulate them are accompanied by some dangers and numerous social and ethical questions. This chapter concerns recombinant DNA technology—its methods, promise, and potential pitfalls.

Organizing Your Knowledge

Exercise 1   (Module 12.1)

Bacteria can transfer DNA via conjugation, transformation, and transduction. Match the following statements with one of the methods of bacterial DNA transfer. (Some statements are true of all methods of DNA transfer.)

A. Conjugation
B. Transformation
C. Transduction
D. All three above

_____ 1. What happened in Griffith's experiment with Pneumococcus
_____ 2. DNA may be integrated into chromosome of recipient
_____ 3. Taking up of DNA from the fluid surrounding a cell
_____ 4. Alters genes of recipient cell
_____ 5. Figure 5 at the left
_____ 6. Male and female cells joined by sex pili
_____ 7. Figure 7 at the left
_____ 8. Bacterial "mating"
_____ 9. Figure 9 at the left
_____ 10. Creates a recombinant cell
_____ 11. Transfer of genes by a bacteriophage

Exercise 2   (Modules 12.2 – 12.4)

Bacteria use plasmids as carriers, or vectors, for gene transfer. We have used this ability to transfer genes, to engineer bacteria that can perform particular jobs and produce useful products. Sometimes genes called transposons move on their own, within the genome. Review concepts related to plasmids and transposons by matching each phrase on the right with a term on the left. (Some answers are used more than once.)

A. *E. coli*
B. Plasmid
C. F factor
D. Corn
E. Bacterial chromosome
F. Transposon
G. Gonorrhea bacterium
H. Biotechnology
I. Vector
J. R plasmid

_____ 1. Plasmid that carries genes that destroy antibiotics
_____ 2. Use of living organisms to do practical tasks
_____ 3. Bacterium made resistant to antibiotics by plasmid genes
_____ 4. Bacterium used in most recombinant DNA work
_____ 5. Organism in which "jumping genes" were first discovered
_____ 6. Piece of DNA with genes for bacterial conjugation
_____ 7. Contains most bacterial genes
_____ 8. General name for a carrier of genes
_____ 9. Can convert a recipient cell to a male
_____ 10. F factor can integrate its genes here
_____ 11. Can carry a desired gene from a plant or animal into a bacterial cell
_____ 12. Role of plasmid in recombinant DNA technology
_____ 13. Mobile gene that can move to and disrupt another part of the genome
_____ 14. Small circular DNA molecule

Exercise 3   (Modules 12.4 – 12.10)

Plasmids can be used to engineer bacteria to produce desired genes or proteins. Review techniques used to splice and clone genes by filling in the blanks.

Gene engineers use plasmids as [1]_____ to insert genes into bacteria or eukaryotic cells. Imagine that you wanted to build a bacterium capable of making large quantities of human growth hormone (HGH), which is a protein. Your first step would be to obtain the [2]_____ that codes for HGH. One way to do this is to use a [3]_____ enzyme to cut up all the DNA in a human cell. The enzyme recognizes short nucleotide [4]_____ within DNA molecules and cut the DNA at specific points in these [5]_____ sequences. Restriction enzymes cut the two DNA strands unevenly, leaving single-stranded ends that can hydrogen bond with complementary single-stranded [6]"_____ends." A restriction enzyme can chop up a cell's DNA into thousands of pieces, each consisting of a few genes.

The next step in making human growth hormone is isolating a supply of [7]_____ to use as vectors, for carrying the DNA fragments into bacteria. These are treated with the same restriction enzyme that was used to cut up the human DNA, producing plasmids with sticky ends that are [8]_____ to sticky ends of the human DNA fragments.

Now the human DNA fragments are mixed with plasmids. The sticky ends on the fragments base pair with the sticky ends on the plasmids, but these connections are weak and temporary. An enzyme called DNA [9]_____, which normally functions in DNA [10]_____, is used to catalyze the formation of covalent

bonds between adjacent nucleotides in the DNA fragments and plasmids. This forms
[11]_____ DNA, a DNA molecule with a new, human-made combination of
genes.

In the next step, each recombinant plasmid is added to a bacterium. Under specific conditions, a bacterium will take up the plasmid DNA from solution by the process of [12]_____. The bacterium, with its recombinant plasmid, is allowed to grow and reproduce on a nutrient medium. Each bacterium replicates its own DNA and the plasmid DNA and then divides repeatedly. Each bacterium grows into a colony of identical cells, all containing the recombinant DNA. This production of multiple copies of the genes is called gene [13]_____. Cloning all the different DNA fragments obtained from the human cell produces a genomic [14]_____ of DNA segments. Because this procedure does not target a particular gene (at least so far), it is called the [15]"_____ approach" to gene cloning. (DNA fragments can also be spliced into phages, [16]_____ that infect bacteria. The phages reproduce in bacteria to produce libraries of cloned DNA pieces.)

Eukaryotic genes are often too large and unwieldy to clone easily because they contain noncoding [17]_____. In this situation, the enzyme [18]_____ transcriptase can be used to make an intron-free gene. If you wanted to make a human growth hormone gene, the place to start would be a cell from the pituitary gland, where HGH is made. In the cell, the HGH gene (and others) is transcribed into RNA. Enzymes then remove the introns from the RNA and splice the remaining [19]_____ together to make mRNA. The mRNA is then extracted from the cell and reverse transcriptase (obtained from a [20]_____) is added. The reverse transcriptase transcribes a strand of DNA along the mRNA molecule. The RNA and DNA then separate, and a second DNA strand is synthesized, producing double-stranded DNA. This artificial gene lacks introns, so it is more manageable than the original gene. It also can be transcribed and translated by [21]_____, which lack the ability to splice RNA. Artificial genes produced in this fashion are cut and pasted into plasmids, using restriction enzymes and ligase, and then cloned in bacteria. There are many mRNA molecules in a pituitary gland cell, so this method also produces a [22]_____ library; however, this library is smaller than a library produced by cutting up the entire genome, because it is limited only to the genes actually [23]_____ in a pituitary gland cell.

At this point the gene engineer has isolated and cloned the HGH gene, but where is it? A genomic library can consist of thousands of bacterial colonies. The bacteria of one of the colonies contains the HGH gene, but which one? One way to look for the gene is to look for HGH, its [24]_____ product. But usually you look for the gene itself, a search that is made easier by using a molecular [25]_____. If you know part of the amino acid sequence of HGH, you can work backward to figure out the probable nucleotide sequence of part of the HGH gene—AAGTGTAG, for example. Now you can produce an artificial RNA (or DNA) molecule with a complementary base sequence—[26]_____, in this case. This complementary molecule is labeled with a [27]_____ isotope and is called a probe because it can be used to find the gene. To find the bacterial clone that holds the gene, single-stranded DNA are obtained from each colony of bacteria. The probe is then mixed with the DNAs, and it hydrogen bonds only with the recombinant DNA with a complementary base sequence—the HGH gene. Once you have figured out which bacterial colony in the library contains the HGH gene, you can grow these bacteria in larger amounts.

If you know the nucleotide sequence of the desired gene, you can bypass much of the trouble of cloning and searching for the gene by simply synthesizing it in the laboratory. There are now gene machines that can put together artificial genes several hundred [28]_____ in length. There are also automatic DNA sequencers capable of determining the sequences of large genes in a few days. The genes are cut into fragments by restriction enzymes and machine-analyzed. The sequences are then fed into computer data banks for interpretation.

The final step in engineering bacteria to produce human growth hormone is to grow the bacteria in large quantities (usually done in large vats) and extract the protein. The bacteria will manufacture the protein on command if you have spliced the proper control sequences into your recombinant plasmids. Now it is only necessary to collect and purify the protein (and get approval from the Food and Drug Administration!) to start treating patients with recombinant DNA HGH.

Exercise 4   (Modules 12.4 – 12.10)

Review recombinant DNA techniques by matching each of the diagrams (or parts of diagrams) below with one of the following processes: **isolating plasmid from *E. coli*; extracting DNA from a eukaryotic cell; obtaining copies of gene and protein from cloned bacteria; cutting DNA with restriction enzyme; joining plasmid and DNA fragment using DNA ligase; cloning recombinant DNA; using reverse transcriptase to make an artificial gene; using a molecular probe to find a gene; inserting a plasmid into a bacterium via transformation; mixing plasmids and DNA fragments with sticky ends.**

1. _____

2. _____

3. _____

4. _____

5. _____

6. _____

7. _____

8. _____

9. _____

10. _____

Exercise 5 **(Modules 12.11 – 12.16)**

DNA technology promises great benefits, but it also presents some risks and raises ethical questions. Review some of the uses and dilemmas of DNA technology by completing this crossword puzzle.

Down

2. Some vaccines use a harmless variant of the ____ virus.

4. It may be possible to genetically engineer plants to fix ____.

6. Yeast is used to prduce interferons and ____ B vaccine.

7. A ____ organism contains genes from another species.

8. A vaccine can contain a ____ from the surface of a bacterium or virus.

10. ____ is the eukaryote most commonly used for genetic engineering.

11. Mice with human genes are susceptible to ____.

12. It is possible that recombinant organisms could transfer ____ to other species.

14. It is easier to manipulate genes in single cells than in ____ organisms.

16. Yeast and bacteria contain ____ that can be used as gene vectors.

17. ____ are often the best organisms for making large amounts of gene product.

19. Microbes used in recombinant DNA work are genetically ____ to stop their spread.

20. A ____ is used to stimulate the immune system to fight disease.

25. ____ is used to treat dwarfism.

Across

1. Cystic ____ has been treated with gene therapy.
3. Some vaccines contain harmless artificial ____ of disease agents.
5. Gene ____ is alteration of human genes to remedy a genetic disease.
8. A plasmid vector from a soil bacterium is used to get genes into ____ cells.
9. Safety measures keep engineered microbes from leaving the ____.
13. Monoclonal ____ are used in diagnosis and cancer treatment.
15. ____ is the bacterium most commonly used in recombinant DNA work.
18. Recombinant DNA technology has had a major impact on the ____ industry.
21. DNA sequencing will make it possible to "read" the entire human ____.
22. ____ plasmids carry genes into plant cells.
23. Human ____ and growth hormone were the first recombinant pharmaceuticals.
24. Recombinant DNA technology raises many important questions.
25. Factor VIII is used to treat ____, a blood-clotting disease.
26. DNA is injected into animal eggs through a fine ____.
27. Bacteria can be engineered to ____ proteins into the surrounding medium.

Testing Your Knowledge

Multiple Choice

1. The primary difference between bacterial sex and sexual reproduction in plants and animals is that
   a. bacterial sex involves more than two individuals.
   b. bacterial sex does not involve genetic recombination.
   c. bacteria exchange RNA, not DNA.
   d. bacterial sex does not produce offspring.
   e. eggs and sperm are different, but bacterial gametes are all alike.

2. Sometimes a bacteriophage transfers a gene from one bacterium to another. This process is called
   a. transduction.
   b. conjugation.
   c. cloning.
   d. DNA splicing.
   e. transformation.

3. R plasmids are important because they
   a. control bacterial reproduction.
   b. are used as vectors to transfer genes between plants.
   c. make bacteria resistant to antibiotics.
   d. can be used to produce transgenic animals.
   e. protect bacteria against mutations.

4. The ability of a male *E. coli* to carry out conjugation is usually due to a piece of DNA called
   a. a probe.
   b. an R plasmid.
   c. recombinant DNA.
   d. an F factor.
   e. a Ti plasmid.

5. Which of the following would *not* normally be used in recombinant DNA experiments?
   a. DNA ligase
   b. restriction enzyme
   c. plasmid
   d. transposon
   e. *E. coli*

6. In recombinant DNA experiments, ____ is used to cut pieces of DNA, and ____ joins these segments to form recombinant DNA.
   a. a restriction enzyme . . . DNA ligase
   b. a transposon . . . a restriction enzyme
   c. a plasmid . . . DNA ligase

   d. DNA ligase . . . a restriction enzyme
   e. a transposon . . . a plasmid

7. A genomic library is
   a. where you look to find out how to make recombinant DNA.
   b. a listing of the known nucleotide sequences for a particular species.
   c. all the genes contained in one kind of cell.
   d. a collection of cloned DNA pieces from an organism's genome.
   e. a place where one can obtain DNA samples from various species.

8. A molecular probe might be used to
   a. insert genes into a host cell.
   b. make DNA for gene cloning.
   c. splice pieces of DNA.
   d. cut pieces of DNA down to manageable size.
   e. find a nucleotide sequence.

9. It is sometimes necessary to genetically engineer mammalian cells to produce proteins because they
   a. can produce larger quantities of protein than bacteria.
   b. can read eukaryotic genes and bacteria cannot.
   c. can add sugars to make glycoproteins and bacteria cannot.
   d. are easier to grow than bacteria.
   e. can be induced to secrete proteins into their environment.

10. Scientists have produced a smallpox virus that contains genes from several other disease-causing microorganisms. They hope to use the virus
    a. in a vaccine against several diseases.
    b. as a compact genomic library.
    c. to perfect a germ-warfare weapon with no antidote.
    d. as a gene vector for human gene therapy.
    e. as a tool for diagnosing various infectious diseases.

11. So far, genetic engineering of plants has not resulted in significant increases in food production because
    a. new plant diseases have evolved faster than expected.
    b. genetically engineered plants are very expensive.
    c. the vectors used do not work with many food plants.

d. most crop plants already grow at the maximum theoretical rate.

e. agricultural scientists have not pursued it seriously.

Essay

1. How might transposons harm an organism? How might they contribute to the evolution of a species?

2. Explain how DNA segments can be cut and spliced together to produce recombinant DNA. How do the segments "find" each other and stick together? How is recombinant DNA then cloned to produce multiple copies of the gene?

3. Explain why bacteria and yeast are often used as "factories" for gene products.

4. Describe some uses of recombinant DNA technology in agriculture.

5. Describe some uses of recombinant DNA technology in medicine.

6. What are some potential risks of genetically engineered organisms being accidentally or purposefully released into the environment? What kinds of safety measures guard against accidental release? Are you concerned about these risks? Why or why not?

7. Discuss some of the ethical questions raised by recombinant DNA technology. Which of these issues do you think is the most important? Why?

Applying Your Knowledge

Multiple Choice

1. A geneticist found that a particular nucleotide sequence was found on different chromosomes in different mouse skin cells. This suggested that
    a. transformation was occurring in some skin cells.
    b. transposons were moving around.
    c. the cells were engaging in conjugation.
    d. plasmids were transferring genes from cell to cell.
    e. the mice were transgenic animals.

2. A microbiologist analyzed the DNA of *E. coli*
    before and after conjugation. She found that
    a. both cells lost some genes and gained others.
    b. both cells gained genes but lost none of their original genes.
    c. one cell lost genes and the other gained genes.
    d. one cell gained genes and the genes of the other were unchanged.
    e. the genes of both cells remained unchanged.

3. Which of the following do "sticky ends" and molecular probes have in common?
    a. They both are used as gene vectors in genetic engineering.
    b. They both involve complementary base pairing.
    c. They both are parts of RNA molecules.
    d. They both are produced by the action of restriction enzymes.
    e. They both are important aspects of bacterial sex.

4. Because eukaryotic genes contain introns, they cannot be translated by bacteria, which lack RNA splicing machinery. If you want to engineer a bacterium to produce a eukaryotic protein, you can synthesize a gene without introns (if you know the nucleotide sequence) or
    a. alter the bacteria used so that they can splice RNA.
    b. use a molecular probe to find a gene without introns.
    c. work backward from mRNA to a piece of DNA without introns.
    d. use a phage to insert the desired gene into a bacterium.
    e. use a restriction enzyme to remove introns from the gene.

5. When recombinant plasmids are added to *E. coli*, not all the bacteria take them in. This makes it difficult to spot the bacteria that actually contain the plasmids. For this reason, R plasmids are often used as gene vectors in recombinant DNA experiments. Using R plasmids would allow experimenters to find the bacteria with recombinant DNA easily by
    a. using a molecular probe.
    b. killing off the bacteria without the R plasmids.
    c. feeding them a special diet.
    d. exposing the bacteria to restriction enzymes.
    e. watching what happens when the bacteria reproduce.

6. Scientists wished to create an organism capable of breaking down several kinds of toxic wastes, so they combined the genes of several bacteria to create a single "superbacterium." They probably did not need to use which of the following in creating the superbacterium?
   a. molecular probes
   b. F factors
   c. plasmids
   d. restriction enzymes
   e. DNA ligase

7. A crop scientist spliced genes for disease resistance into Ti plasmids and then treated tomato plants with the plasmids. Some parts of some plants resisted the disease, but most of the plants eventually died. The researcher could increase his chances for success by
   a. treating single cells and cloning whole plants from the cells.
   b. using molecular probes to figure out where to put the genes.
   c. using R plasmids rather than Ti plasmids to introduce the genes.
   d. inserting the genes into the cells of the tomato plants with a needle.
   e. employing bacteriophages as vectors to get the genes into the plants.

8. A molecular biologist used a virus to introduce a gene coding for a certain enzyme into mouse cells. Most of the mouse cells were able to make the enzyme, but most of them lost the ability to make some other protein (different ones in different cells), and many died. Which of the following best explains these results?
   a. The viruses caused the mouse cells to become diseased.
   b. The viruses transferred genes from one mouse cell to another.
   c. The viruses inserted the enzyme gene into mouse cell genes.
   d. The viruses activated transposons, which disrupted other genes.
   e. The enzyme acted as a restriction enzyme, cutting up mouse DNA.

Essay

1. A mutant strain of *E. coli* bacteria will not grow unless they are supplied with the amino acid lysine. Another strain will not grow without a different amino acid, proline. When *E. coli* of the two strains are mixed, a few bacteria appear in the culture that are able to grow without either of the amino acids. Name and briefly describe three possible mechanisms that might account for this change.

2. Describe how you would go about genetically engineering a bacterium to produce human epidermal growth factor (EGF), a protein used in treating burns. Use the following: DNA ligase, *E. coli*, plasmids, genetic code chart, restriction enzyme, machine for synthesizing a gene, description of the amino acid sequence of EGF, and glassware and equipment for growing and handling bacteria and extracting protein.

3. In a "shotgun" experiment, all the DNA in a cell is cut up and cloned, producing a genomic library of DNA fragments, each fragment in a different bacterial colony. Why do you think some researchers question the safety of the shotgun approach?

4. A researcher is searching for the bacterial clone containing a particular cloned gene. She knows that part of the nucleotide sequence of the gene is ATGGCTATC. Explain how she might find the bacteria that contain the gene.

5. A microbiologist developed a strain of *E. coli* that were easily killed by sunlight and whose diet required two unusual amino acids not normally found outside the laboratory. Why would such a bacterium be useful in recombinant DNA work?

Using Your Knowledge

1. Have you used any products produced by means of DNA technology? Think about drugs (HGH, insulin), home pregnancy tests, hospital diagnostic tests. As this book goes to print, transgenic fruits and vegetables (tomatoes) are about to appear in supermarkets.

2. Hardly a day goes by without a story about DNA technology appearing in the newspaper. For the next few days, look for examples of DNA technology in the news. You may see articles about new drugs, cancer treatment, genetic diseases, transgenic crop plants, mapping of the human genome, clarification of evolutionary relationships, and even crimes solved using recombinant DNA techniques. Do you understand these articles better after studying recombinant DNA technology? What did the authors omit that you would like to know?

The Human Genome

13

In laboratories all over the world, biologists are mapping human chromosomes and recording the language of our genes. We have learned more about human genes in the past decade than in all of previous human history. This unprecedented gain in knowledge will enable us to cure or prevent many genetic defects. It will enable us to better understand the development of a baby and the origin of the human species. As we increase our understanding of our genes, we are also beginning to understand that our knowledge can be misused and that knowing about our genes carries with it the responsibility to use our knowledge wisely. This chapter outlines what we know and what we are learning about the human genome.

Organizing Your Knowledge

Exercise 1   (Modules 13.1 – 13.2)

This module gives a number of facts and figures that will help put the human genome into perspective. It is not important to memorize the figures, but it is important for you to get an idea of how big and how small some of these things are. Choose the correct number to complete each of the statements. Choose from **2, 3, 10, 23, 46, 80, 1000, 2000, 100,000, 300,000, 3 billion.**

1. Number of chromosomes in a diploid human cell: _____
2. Number of chromosomes in a haploid set: _____
3. Number of nucleotide pairs in a haploid set of human chromosomes: _____
4. Total length of the DNA in a human cell, in meters: _____
5. Length of DNA in a cell divided by the cell's diameter: _____
6. Thickness of a DNA molecule, in micrometers: _____
7. Estimated number of genes in a human cell: _____
8. Number of genes in *E. coli*: _____
9. Amount of DNA in a human cell divided by amount of DNA in *E. coli*: _____
10. Percentage of DNA in a human cell that is thought to be noncoding DNA: _____
11. Amount of human DNA in introns divided by amount in exons: _____

Exercise 2   (Module 13.2)

Review karyotyping by matching each statement on the right with a word on the left.

A. Metaphase

B. Mitosis

C. Microscope

D. Banding pattern

E. Karyotype

F. Centrifuge

G. White blood cell

H. Stain

____ 1. A picture of the chromosomes in a cell

____ 2. Cell often used to prepare karyotype

____ 3. Cells are treated to stimulate this process

____ 4. Cell division is stopped at this stage

____ 5. Used to sediment and separate cells

____ 6. Used to bring out chromosome banding pattern

____ 7. Used to examine and photograph chromosomes

____ 8. Helps identify chromosomes and their parts

Exercise 3   (Modules 13.3 – 13.5)

These modules discuss human abnormalities resulting from extra or missing chromosomes. Review them by completing this crossword puzzle.

Across

1. *XXX* females have limited ____ and may be mentally retarded.

5. The *Y* chromosome carries ____ genes than the *X* chromosome.

7. The incidence of Down syndrome increases with the ____ of the mother.

8. Down syndrome individuals are prone to ____ and Alzheimer's disease.

9. A single *Y* chromosome is enough to produce "____."

12. Nondisjunction results in extra or missing chromosomes in ____.

15. ____ is when members of a pair of chromosomes fail to separate during meiosis.

18. Most human offspring with abnormal numbers of chromosomes are spontaneously ____.

21. Down syndrome is the most serious common ____ defect in the U.S.

23. Pregnant women over 35 years of age are candidates for ____ testing.

25. In meiosis II, ____ chromatids may fail to separate.

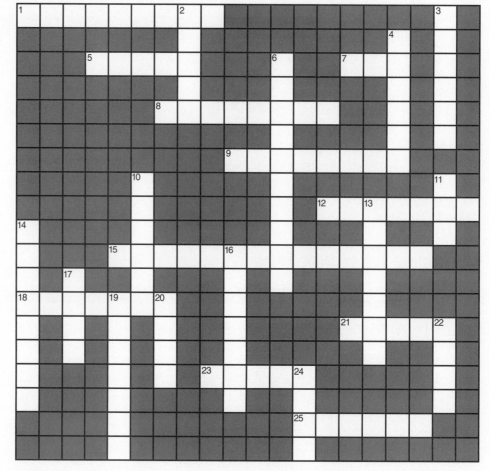

Down

2. A person with Down syndrome has ____ number-21 chromosomes.

3. A woman with only one *X* chromosome is said to have ____ syndrome.

4. *XXY* or *XXYY* males have small ____ and feminine body contours.

6. In meiosis I, a pair of ____ chromosomes may fail to separate.

10. If a sperm fertilizes an egg with an extra chromosome, the ____ will have an extra chromosome.

11. Nondisjunction can alter the number of ____ chromosomes, as well as autosomes.

13. Chromosomal defects usually result from errors in ____.

14. Individuals with Down syndrome are usually mentally ____.

16. People with Down syndrome live ____-than-normal life spans.

17. Meiosis begins before a woman is ____ and is completed years later.

19. An extra chromosome 21 is called ____ 21.

20. A person with an extra number-21 chromosome has ____ syndrome.

22. ____ of the eggs of a Down syndrome woman will have an extra chromosome.

24. An unusual number of sex chromosomes has ____ effect than an unusual number of autosomes.

Exercise 4 (Module 13.6)

Chromosomes sometimes break, their parts can become scrambled, and abnormalities can result. Match each of the diagrams of chromosome alterations with its name and a description of its effects.

Diagram   *Name*   *Effects*

1. —— ——

2. —— ——

3. —— ——

4. —— ——

Names: A. Deletion
   B. Duplication
   C. Inversion
   D. Translocation

Effects: W. May cause chronic myelogenous leukemia in somatic cells
   X. Least likely to have serious effects, because genes are still present in normal numbers
   Y. Likely to have the most serious effects, as in cri du chat syndrome
   Z. A chromosome fragment breaks off and joins a homologous chromosome

Exercise 5 (Modules 13.7 – 13.8)

Family trees called pedigrees are used to trace the inheritance of human genes. The two pedigrees below show the inheritance of sickle-cell anemia (described in Table 13.8), which is caused by an autosomal recessive allele. In the first pedigree, the square and circle symbols are colored, as far as genotypes are known. Fill in the genotypes—*SS*, *Ss*, or *ss*—below the symbols. Use question marks to denote unknown genotypes. Complete the second pedigree by coloring in the symbols, following the rules described in Module 13.7. Again denote unknowns with question marks.

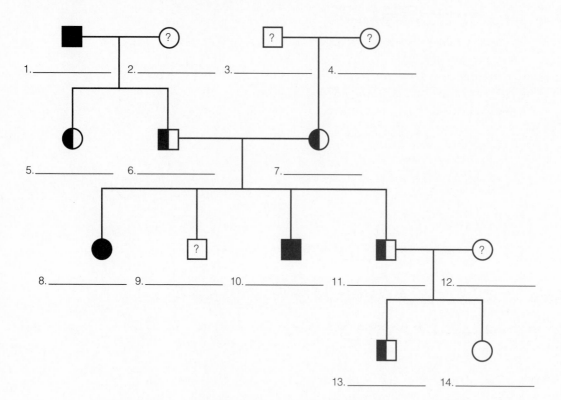

1._____ 2._____ 3._____ 4._____

5._____ 6._____ 7._____

8._____ 9._____ 10._____ 11._____ 12._____

13._____ 14._____

Exercise 6   (Modules 13.8 – 13.9)

These modules discuss common human genetic diseases and their inheritance. Indicate whether each of the statements below is true or false, and change false statements to make them true.

_____ 1. Cystic fibrosis is the most common lethal genetic disease in the United States.
_____ 2. A genetic disorder is expected in half the children of two carriers of a recessive allele.
_____ 3. About 50 human genetic disorders are known to be inherited as Mendelian traits.
_____ 4. Most people afflicted with genetic disorders are born to afflicted parents.
_____ 5. Most serious human genetic disorders are recessive.
_____ 6. Queen Victoria's descendants spread phenylketonuria through the royalty of Europe.
_____ 7. Most genetic diseases are evenly distributed among ethnic groups.
_____ 8. Most known sex-linked disorders result from alleles in the *Y* chromosome.
_____ 9. Hemophilia is a blood-clotting disorder caused by a sex-linked recessive allele.
_____10. Dominant alleles are always more common in the population than recessive alleles.
_____11. Most color-blind individuals are male.
_____12. The gene for Duchenne muscular dystrophy has been located on the *X* chromosome.
_____13. Sex-linked genetic disorders are more common in females than in males.
_____14. Geneticists agree that inbreeding always increases the risk of inherited diseases.
_____15. Huntington's disease is lethal, but it does not cause death until after the age of 40.
_____16. Fathers pass *X*-linked alleles to all of their sons but to none of their daughters.
_____17. Extra fingers and toes is a dominant trait.
_____18. Sickle-cell anemia is common among African-Americans.

Exercise 7   (Modules 13.10 – 13.13)

These modules describe how the inheritance of a genetic disorder can be traced and how the likelihood of its occurrence can be predicted. Review by filling in the blanks.

Greg and Amy were happy that she was pregnant, but their joy was mixed with anxiety. Recently, the couple had received some tragic news: Greg's father had been diagnosed with Huntington's disease (HD) at the age of 46. This was not unexpected; Greg's grandmother had died of the disease. Because Huntington's disease is caused by a [1]_____ allele, Greg's father had a [2]_____ chance of inheriting it from his mother. Greg and Amy were at a clinic to consult with a genetic [3]_____ about Greg's and the unborn child's chances of inheriting the disease.

Sharon, the counselor, reviewed Amy's and Greg's family histories. She had constructed a family tree, called a [4]_____, for each of them. Sharon said, "Our first priority is to figure out the chances of Greg and the baby inheriting the Huntington's gene. Greg, until your father was diagnosed, you stood a [5]_____ chance of inheriting the HD allele from your grandmother. Now, your odds of developing Huntington's are fifty-fifty. Fortunately, we know a lot more about Huntington's disease now than we used to, thanks to Nancy Wexler's work in Venezuela."

"Why Huntington's? And why Venezuela?"

Sharon continued, "Dr. Wexler herself is at risk of developing HD. She went to Venezuela because HD is especially common there. Dr. Wexler has put together a giant pedigree that links over 10,000 people. She and her team were able to use the pedigree and thousands of blood tests and karyotypes to trace the inheritance of HD. They have

found that the HD gene is on chromosome [6]_____. They used
[7]_____ DNA techniques, or genetic engineering, to locate the gene on the
chromosome and find a [8]_____—a bit of DNA that lies close to the gene—
that is inherited by people who have the disease."

Amy interjected, "So what does this mean for us and for our baby?"

"It means that Greg and the baby can be tested to find out whether they might
have the HD allele and eventually develop the disease."

There was a brief silence, and Greg whispered, "I want to know that I don't have
the disease, but I don't want to know that I do." Greg knew that there is no
[9]_____ for HD. Many people at risk for HD have decided not to take the
test until there is a treatment for the disease. After discussing the test's implications with
the counselor, Greg said to Amy, "If it were just you and me, I wouldn't want to know.
But this doesn't involve just you and me."

A technician withdrew blood for the test, and a week later Amy and Greg were
back in the genetic counselor's office. She breezed through the door and said with a
smile, "You're in the clear."

Greg whooped and hugged his wife, then the counselor. After the celebrating
subsided, Sharon's expression became a bit more serious. "Besides testing for the HD
marker, we did a routine screen for several other disorders, including
[10]_____ disease—even though you are not Jewish—and PKU. It turns out
you are a carrier for PKU. Don't be too concerned though. Because PKU is caused by a
[11]_____ allele, you both have to be carriers for your baby to inherit it, and
even then the probability that the baby will have the disease is only [12]_____."

"What exactly is PKU? Is it dangerous?"

Sharon explained that PKU, short for [13]_____, is an inherited inability to break down an [14]_____called phenylalanine. "The phenylalanine
can build up in the blood and cause mental retardation. It is very unlikely that you both
carry the gene, but if you do, we can test your fetus. If PKU is detected early, retardation
can be prevented by putting the child on a special [15]_____, low in phenylalanine. To test the fetus we'll have to perform [16]_____—taking a sample
of the [17]_____fluid. We can check for PKU by testing for certain chemicals
in the fluid itself. While we're at it we will culture some of the fetal cells from the fluid
and take a [18]_____—a picture of the chromosomes—to check for
[19]_____ syndrome. Or we could get faster results by using a newer technique called [20]_____ sampling. The placenta cells we sample with this
technique are [21]_____ more rapidly, so . . . "

Amy interrupted: "Wait a minute. Do you have to get samples? Can't you just do
[22]_____ imaging to look at the baby?"

"We really can't check chromosomes or PKU by just looking at the fetus, either
with ultrasound or with [23]_____, where the doctor uses a viewing scope to
look at the fetus. Amy, because you are over [24]_____ years old, I think it
is important to get a sample of amniotic fluid so that we can check for Down syndrome.
Plus, the karyotype will answer another question I'm sure you are eager to know the answer to—whether your baby is a boy or a girl."

Exercise 8   (Modules 13.14 – 13.16)

Powerful molecular biology techniques now allow us to amplify, analyze, and compare genes. Review these methods by matching each phrase on the right with a term on the left.

A. RFLP analysis

B. Paternity

C. Gel

D. Nucleotide sequence

E. Positive pole

F. Restriction fragment

G. DNA polymerase

H. RFLP

I. Genome

J. White blood cell

K. Band

L. Restriction enzyme

M. Genetic marker

N. Blood

O. DNA fingerprint

P. Polymerase chain reaction

Q. Electrophoresis

R. Contamination

S. Restriction site

_____ 1. Similar to members of your family but not to strangers

_____ 2. Cuts different people's DNA at different points

_____ 3. Piece of DNA cut up by restriction enzymes

_____ 4. Place where enzyme cleaves DNA; depends on person

_____ 5. Restriction fragment length polymorphism

_____ 6. Type of cell often used in RFLP analysis

_____ 7. Separates DNA fragments by size and electrical charge

_____ 8. Restriction fragments move through this

_____ 9. Restriction fragments are attracted to this

_____ 10. Where specific restriction fragment collects in gel

_____ 11. Nucleotide sequence that indicates a specific allele

_____ 12. Used to locate a particular genetic marker

_____ 13. Method for making many copies of a DNA molecule

_____ 14. Used to replicate DNA in a test tube for PCR method

_____ 15. DNA testing can be used to link this with a crime

_____ 16. An individual's electrophoresis band pattern

_____ 17. DNA fingerprinting is also used to determine this

_____ 18. Only a small part of this is used in forensic DNA tests

_____ 19. A possible problem in using PCR in forensic DNA tests

Exercise 9   (Modules 13.17 – 13.19)

Current advances in genetics hold much promise, but they also raise important ethical questions. Summarize the methods and issues outlined in these three modules by filling in the blanks below.

Is it possible to map the entire human genetic code? If we can find defective genes, can we also fix them? Pioneering researchers of the human genome project and gene therapy are poised to answer these important questions. Human gene therapy—[1]_____ genetic disorders—employs the genetic engineering methods used to alter other organisms. Cells in the [2]_____ and lymph nodes, which produce most cells of the [3]_____ and immune system, are the targets of the first gene therapy tests. Imagine that a patient's bone marrow cells are unable to make an important [4]_____ because they contain a defective [5]_____. The normal gene can be cloned using recombinant DNA methods, then inserted into a [6]_____, which "infects" the patient's bone marrow cells with the normal gene. The "repaired" cells are then injected back into the patient.

This procedure has already been successfully used to treat severe combined immunodeficiency syndrome, a potentially lethal inability to fight [7]_____. Clinicians have also targeted other genetic disorders, such as [8]_____ and muscular dystrophy.

There are some technical problems with gene therapy. It is difficult to make the transferred genes [9]_____ in target cells. Thus far, gene therapy can be used only on cells that continue to [10]_____ throughout life, so it cannot be used on cells like [11]_____ cells that cease to proliferate at some point.

A serious ethical problem is whether we should treat only somatic (body) cells or also germ cells—cells of the ¹²_____ and ¹³_____—that develop into gametes and carry genes that are passed on to offspring. It is one thing to treat a patient and quite another to alter a gene that might conceivably be useful to future generations. Do we really want to eliminate such a gene from the human gene pool? In part because of the atrocities of Nazi Germany, our society rejects ¹⁴_____, the idea that "undesirable" traits should be eliminated from the human population. On the other hand, if we know how to correct a genetic defect, is it ethical to withhold treatment?

These questions will only become more insistent as we learn more about the human genome. Advances in DNA technology have made it possible to map the entire human DNA ¹⁵_____ sequence. Scientists expect to soon have the entire human genetic "encyclopedia" deciphered (or at least written down). They call this multi-million-dollar undertaking the human ¹⁶_____ project.

The project involves four lines of attack: First, researchers want to produce a genetic (linkage) ¹⁷_____ of the entire human genome. They will use data from genetic crosses to study the ¹⁸_____ of linked genes on chromosomes and the ¹⁹_____ between the genes. They hope to use pedigrees and ²⁰_____ analysis to find several thousand genetic ²¹_____ that can be used as "signposts" on the chromosomes.

Second, researchers plan to produce a ²²_____map of each chromosome, by using ²³_____ enzymes to fragment the chromosomes and then determining the order of the fragments. Next, they want to ²⁴_____ the entire genome—determine the order of base pairs in each chromosome. Even with the help of automatic sequencers and computers, this will take a long time, because a human cell contains about ²⁵_____ billion nucleotide pairs. Finally, the researchers want to perform similar analyses of other organisms—²⁶_____, yeast, a plant, fruit flies, and mice. This will provide useful points of comparison for the human data.

Potential benefits of the human genome project are varied and immense. Basic science will be enriched by new insights into such areas as ²⁷_____ development and evolution. New information about human genes will yield new methods for diagnosing and treating genetic disorders and more common ailments, from heart disease to allergies.

Like gene therapy, the human genome project raises many ²⁸_____ questions. Who will decide whether a gene is "good" or "bad"? Who should have access to genetic data? Might some people become victims of genetic discrimination, excluded from jobs or insurance because of their genotype? The frontier of genetics offers many opportunities, but only if we approach this frontier responsibly.

Exercise 10   (Modules 13.16 – 13.19)

These modules raise a number of ethical questions and issues that we face as we begin to use our understanding of molecular biology to shape our own genetic destiny. Imagine that you are a journalist specializing in biological issues. On a separate piece of paper write six headlines for articles outlining ethical questions in human genetics.

Testing Your Knowledge

Multiple Choice

1. There are thought to be about ____ genes in a human cell.
   a. 23
   b. 46
   c. 1000
   d. 100,000
   e. 3 billion

2. There is about 1000 times as much DNA in a human cell as in an *E. coli* cell but only about 50 times as many genes. Why?
   a. A human cell has much more noncoding DNA.
   b. The DNA is much more tightly coiled in a human cell.
   c. Most of the genes in a human cell are turned off.
   d. *E. coli* genes are less efficient than human genes.
   e. Human cells are much smaller than *E. coli* cells.

3. Each cell in an individual with Down syndrome contains ____ chromosomes.
   a. 3
   b. 22
   c. 24
   d. 45
   e. 47

4. Geneticists suspect that the extra chromosome seen in Down syndrome usually comes from the egg, rather than the sperm, because
   a. eggs are produced so rapidly that there is more chance for error.
   b. Down syndrome is due to a dominant gene in women, a recessive gene in men.
   c. most women inherit Down syndrome from their mothers.
   d. eggs are produced in much larger numbers than sperm.
   e. meiosis takes longer in the ovary, increasing the likelihood of error.

5. Disorders involving unusual numbers of sex chromosomes show that "maleness" is caused by the
   a. presence of an *X* chromosome.
   b. presence of a *Y* chromosome.
   c. absence of an *X* chromosome.
   d. absence of a *Y* chromosome.

   e. absence of an *X* chromosome and presence of a *Y* chromosome.

6. Which of the following chromosomal alterations would you expect to have the most drastic consequences?
   a. inversion
   b. duplication
   c. translocation
   d. deletion
   e. a and b are equally the most serious

7. The most common lethal genetic disease in the United States is
   a. sickle-cell anemia.
   b. cystic fibrosis.
   c. Huntington's disease.
   d. hemophilia.
   e. PKU.

8. Which of the following human genetic disorders is sex linked?
   a. hemophilia
   b. PKU
   c. cystic fibrosis
   d. sickle-cell anemia
   e. all of the above

9. There are various procedures that can be used to detect genetic disorders before birth. Among the tests discussed in this chapter, ____ is the least invasive, while ____ carries the highest risk.
   a. chorionic villi sampling . . . amniocentesis
   b. ultrasound imaging . . . fetoscopy
   c. fetoscopy . . . chorionic villi sampling
   d. fetoscopy . . . amniocentesis
   e. ultrasound imaging . . . chorionic villi sampling

10. Electrophoresis is used to
    a. separate fragments of DNA.
    b. clone genes.
    c. cut DNA into fragments.
    d. match a gene with its function.
    e. amplify small DNA samples to obtain enough for analysis.

11. A genetic marker is
    a. a place where a restriction enzyme cuts DNA.
    b. a chart that traces the family history of a genetic trait.
    c. a nucleotide sequence near a particular gene.
    d. a radioactive probe used to find a gene.
    e. an enzyme used to cut DNA.

12. Curing a genetic disorder by replacing the defective gene with a normal one is called
    a. amniocentesis.
    b. electrophoresis.
    c. gene cloning.
    d. karyotyping.
    e. gene therapy.

Essay

1. Explain how an error in meiosis can cause a baby to be born with an extra or missing chromosome.

2. About one in 700 babies born in the United States possesses an extra chromosome 21, resulting in Down syndrome. Why are few individuals seen who have extra copies of other chromosomes?

3. John's father died of Huntington's disease. What is the probability that John has inherited the HD gene? There is now a test that can detect the presence of the HD gene before the disease develops, but John is reluctant to be tested. Why do many at risk of developing Huntington's disease decline to be tested?

4. Describe how RFLP analysis is done: How is DNA cut into fragments? How are the fragments separated? What does the DNA fingerprint look like, and why? Why do different people have different DNA fingerprints?

5. What is the most difficult ethical question concerning human gene therapy?

6. Describe the four main objectives of the human genome project. What are some of the benefits expected from this project?

7. As we learn more about the human genome, how might individual DNA data be misused?

Applying Your Knowledge

Multiple Choice

1. A karyotype would be least likely to show which of the following?
   a. an extra chromosome
   b. part of a chromosome duplicated
   c. a missing chromosome
   d. part of a chromosome turned around
   e. a translocation

2. Duchenne muscular dystrophy is caused by a sex-linked recessive allele. Its victims are almost invariably boys, who usually die before the age of 20. Why is this disorder almost never seen in girls?
   a. Sex-linked traits are never seen in girls.
   b. The allele is carried on the Y chromosome.
   c. Nondisjunction occurs in males but not in females.
   d. Males carrying the allele don't live long enough to be fathers.
   e. A sex-linked allele cannot be passed on from mother to daughter.

3. Humans have 23 pairs of chromosomes, while our closest relatives, chimpanzees, have 24. Chromosome studies indicate that at some point early in human evolution, two chromosomes simultaneously broke into a large portion and a small portion. The large parts combined to form a large chromosome, and the small parts combined to form a much smaller chromosome (which was subsequently lost). This important chromosomal change could best be described as
   a. nondisjunction followed by deletion.
   b. translocation followed by deletion.
   c. duplication followed by deletion.
   d. translocation followed by inversion.
   e. nondisjunction followed by inversion.

4. Tay-Sachs disease runs in Rebecca's family. On a family pedigree, she saw a half-darkened circle. This represented
   a. a male with Tay-Sachs.
   b. a female with Tay-Sachs.
   c. a carrier male.
   d. a carrier female.
   e. a male of unknown genotype.

5. On a pedigree tracing the inheritance of PKU, a horizontal line joins a black square and a half-black circle. What fraction of this couple's children would you expect to suffer from PKU?
   a. none
   b. 1/4
   c. 1/2
   d. 3/4
   e. all

6. Michelle and Keith are apparently normal, but their daughter was born with alkaptonuria, an inherited metabolic disorder. If alkaptonuria is like most human hereditary disorders, the

probability of their next child being born with alkaptonuria is

a. 0.

b. 1/4.

c. 1/2.

d. 2/3.

e. 3/4.

7. Several inherited disorders are much more common in close-knit religious communities, such as the Amish, than in the general population. This is at least partly due to the fact that

a. people in such communities are more likely to marry relatives.

b. shared environmental conditions such as diet can increase mutation rate.

c. modern medical care is not widely available in such communities.

d. community members care for each other and disorders are passed on.

e. individuals in such communities are more likely to have large families.

8. Tom's brother suffers from PKU. What are Tom's chances of being a carrier of PKU?

a. 0

b. 1/4

c. 1/2

d. 2/3

e. 3/4

9. If the Huntington's allele is dominant, why is it not more common in the population?

a. It is dominant but seldom passed on by those who have it.

b. It is recessive, like all human disorders.

c. Dominance has nothing to do with how common a gene is.

d. It is lethal, so it is unlikely to be passed on.

e. It is actually extremely common but only rarely harmful.

10. Which of the following would be most useful for preventing a particular genetic disorder?

a. knowing how the allele causes its phenotypic effects

b. being able to identify carriers

c. a test that can determine whether a fetus suffers from the disorder

d. knowing which chromosome bears the allele that causes the disorder

e. tracing the trait back through parents and grandparents

11. DNA fingerprints were used to determine whether Sam could be the father of Becky's baby. Which of the following would show that Sam is not the father? If ____ genetic fingerprint showed some bands not in ____ genetic fingerprint.

a. Sam's . . . the baby's

b. Becky's . . . the baby's

c. the baby's . . . Sam's

d. the baby's . . . Becky's

e. the baby's . . . Sam's or Becky's

12. Archaeologists unearthed a human skull with a small dried fragment of the scalp still attached. They extracted a tiny amount of DNA from the scalp tissue. How could they obtain sufficient DNA for an analysis of the ancient man's genes?

a. subject the DNA to electrophoresis

b. use a molecular probe

c. subject the specimen to amniocentesis

d. use the polymerase chain reaction

e. subject the DNA to restriction enzymes

Essay

1. Which of the following abnormalities in sex chromosomes would result in a predominantly male phenotype? A predominantly female phenotype? Explain why. *XXXX, XXY, XYYY, X, XXYY*

2. How likely is it that you carry an allele for a serious genetic disorder? Why is it much less likely that you will have a child with such a disorder?

3. In humans, the presence of a fissure (gap) in the iris of the eye (called "coloboma iridis") is due to a sex-linked recessive gene. Show how an apparently normal couple could have a child with this condition. Is the affected child more likely to be a boy or a girl?

4. In most human genetic diseases, the normal allele codes for an important protein, usually an enzyme. The allele that causes the disorder codes for a nonfunctional enzyme. Explain why the allele for such a disorder is recessive. (Hint: Why are homozygous dominant and heterozygous individuals of normal phenotype, and why do only homozygous recessive individuals suffer from the disorder?)

5. Imagine that you are a genetic counselor, and a couple who are planning to have children come to you for advice. Diane's brother has hemophilia. There is no history of hemophilia in Craig's family. What is the probability that their child will have hemophilia? (Recall that hemophilia is caused by a sex-linked recessive allele.)

6. Manuel, a political activist, "disappeared" during the reign of a dictator. After the dictator was overthrown, human remains thought to be those of Manuel were found buried in a prison compound. A sample of DNA was extracted from the remains and subjected to restriction fragment analysis, along with DNA samples from Manuel's parents. The patterns of restriction fragments are shown below. Could the remains be those of Manuel, the missing activist? Explain your answer.

Human remains

Manuel's father

Manuel's mother

7. A certain genetic disorder results from the lack of a blood enzyme that is secreted by bone marrow cells. A second disease occurs when nerve cells are unable to produce a particular enzyme. Which of these disorders would be a better candidate for gene therapy and why?

Using Your Knowledge

1. Is there any record of a genetic disorder in your family? Which one? How is it inherited? If you are able to collect data from your family, you might find it interesting to construct a pedigree for this genetic disorder. Alternatively, you could construct a pedigree for one of the simple human genetic traits illustrated in Chapter 9.

How Populations Evolve

Charles Darwin made sense out of biology. He looked at what many others had seen and had an idea that no one had ever thought of—evolution through the mechanism of natural selection. This idea is as important to biology as the atom is to chemistry. Natural selection explains both the diversity of life and its adaptation to the environment. Copernicus and Galileo discovered where we fit in the universe. Einstein showed us our place in space and time. Darwin revealed our relationship to the living world. It all started on an ocean voyage, when Darwin was only 22 years old. This chapter starts with Darwin's voyage of discovery and explains where that voyage has since taken us in our understanding of life.

Organizing Your Knowledge

Exercise 1   (Module 14.1)

Charles Darwin was not the first person to ponder the origin of species. Match each of the following with his place in unraveling the history of life. Don't focus on names and dates but rather on how ideas about the origin and history of life have changed over the centuries.

A. Darwin
B. Lyell
C. Wallace
D. Anaximander
E. Aristotle
F. Lamarck
G. Buffon

_____ 1. Ancient Greek who believed living things have changed
_____ 2. Greek philosopher who believed species to be fixed
_____ 3. Fossils led this Frenchman to believe the Earth is old
_____ 4. Proposed that acquired characteristics are inherited
_____ 5. Wrote *On the Origin of Species*
_____ 6. Believed in gradual geological change, not catastrophes
_____ 7. Conceived a theory of evolution identical to Darwin's

Exercise 2   (Module 14.2)

This module identifies five categories of evidence for evolution. Name the category of evidence to which each of the following examples belongs.

Category

_____

_____

_____

_____

_____

_____

Example
1. Fertilized eggs of earthworms, insects, and snails all go through the same pattern of cell division.
2. The DNA of humans and chimpanzees is about 98% identical.
3. Remains of upright-walking but small-brained apes have been found in Africa.
4. All animals with backbones have 12 pairs of nerves extending from the brain.
5. A protein called albumin is very similar in dogs and wolves, less similar in dogs and cats.
6. The farther an island is from the mainland, the more different its plants and animals are.
7. Animals called trilobites were common in the oceans 300 million years ago, but they have been extinct for millions of years.

Exercise 3   (Modules 14.3 – 14.6)

Read these modules, and then review selection and population genetics by completing the following story.

     If you think that the more you mow your lawn, the meaner the weeds get, you may be right. Researchers have found that in lawns that are mown regularly, the dandelions fight back! Of course, dandelions don't "know" what they are doing. But the dandelions in a regularly mown lawn reproduce faster than their ancestors in more "natural" environments.

     ¹_____, the English scientist who first devised the theory of ²_____, would have explained it this way: Not all dandelions are alike; they ³_____ in color, size, and rate of maturation. Many of these characteristics are ⁴_____, or passed on to offspring. Every dandelion flower is capable of producing hundreds, perhaps thousands, of white-tufted seeds in a season. This constitutes an ⁵_____ of offspring, because ⁶_____ are limited; not every dandelion seed will find just the right environment in which to grow. Darwin speculated that those individuals whose inherited characteristics ⁷_____ them best to their environment would be more likely to ⁸_____ and ⁹_____ than less ¹⁰_____ individuals. Their type would become more common in the next generation. Darwin called this phenomenon ¹¹_____. Examples of this are seen among ¹²_____ and moths in Britain. Those whose coloration protects them best from predation leave more descendants. Favored traits are represented more and more and unfavorable ones less and less in succeeding generations. Reproduction is central to natural selection; in fact, natural selection can be defined as ¹³_____ reproduction.

     Natural selection is at work when you mow the lawn. It might be helpful to discuss the dandelions in terms of ¹⁴_____ genetics, the study of how genes affect population changes. The dandelions in your lawn make up a ¹⁵_____, a group of individuals of the same species. The species is all dandelions, the group of populations whose individuals can interbreed. Depending on where you live, your lawn may be more or less ¹⁶_____ from other dandelion populations. Through some quirk of nature, dandelions only reproduce asexually, even though they continue to produce nectar and pollen. So unlike most other flowers, dandelions in your lawn cannot cross-pollinate with other populations, although seeds may blow in from elsewhere, and seeds from your lawn might blow across town.

     The characteristics of dandelions—color, height, shape of root system, and so on—are dictated by their genes. All the genes in the dandelion population make up the ¹⁷_____. Apparently, the height of dandelions when they mature and produce seed and how long it takes to do this are controlled by genes, and these traits can vary from dandelion to dandelion. If you mow the lawn often enough, the slower-growing, taller dandelions get lopped off before they can produce any seeds. In terms of reproductive success, these dandelions are the less ¹⁸_____ individuals in the population. Since they don't produce many offspring, this slow type will not be as numerous in the next generation, and their ¹⁹_____ will make up a smaller proportion of the gene pool. On the other hand, dandelions that don't have to grow as tall or take as long to produce seed can reproduce between mowings. Their genes, and their fast-growing traits, will be better represented in the next generation. Such a small change in the frequencies of alleles in the gene pool is called [20]_____.

This story illustrates a familiar example of natural selection. Over time, the shorter, faster-growing dandelions will predominate in the lawn. Note that natural selection involves differences between individuals, but individual dandelions do not evolve. An individual does not change its growth rate. But because there is variation in survival and reproduction of individuals with different characteristics, the [21]_____ does evolve.

Exercise 4   (Modules 14.7 – 14.8)

Imagine a population of 100 annual wildflowers, some red and some yellow. The red allele, *R*, is dominant; the yellow allele, *r*, is recessive. There are 36 *RR* plants in the population, 48 *Rr* plants, and 16 *rr* (yellow) plants. If the Hardy-Weinberg principle holds in this case, what will be the frequencies of the various genotypes and the frequencies of the two alleles, *R* and *r*, in the next generation? Follow the example in Module 14.7 as a guide, and fill in the blanks below to figure out the frequencies for this example.

First, figure out genotype frequencies for the current generation:

		Red	Red	Yellow
A.	Phenotypes	Red	Red	Yellow
B.	Genotypes	____	____	____
C.	Number of plants (total 100)	____	____	____
D.	Genotype frequencies (number of genotypes/100)	____	____	____

Next, figure out the frequencies of *R* and *r* alleles in the gene pool:

E.	Number of *R* alleles in gene pool	____	____	____
F.	Number of *r* alleles in gene pool	____	____	____
G.	Allele frequencies (number of *R* alleles/200 or number of *r* alleles/200)	Frequency of *R:* p = ____		Frequency of *r:* q = ____

Now you know the frequency of *R* and *r* gametes these plants will produce:

H. Gamete frequencies
   (= allele frequencies)   Frequency of *R:* ____   Frequency of *r:* ____

Now you can use the rule of multiplication to calculate the frequencies of the three possible genotypes of plants in the second generation:

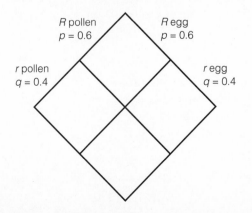

I. Phenotype   Red   Red   Yellow
J. Genotype   ____   ____   ____
K. Genotype frequencies   ____   ____   ____

Now you can figure out the frequencies of *R* and *r* alleles in the gene pool for the second generation (assuming the population stays at 100 individuals):

L. Number of *R* alleles in gene pool ____   ____   ____
M. Number of *r* alleles in gene pool ____   ____   ____
N. Allele frequencies
    (number of *R* alleles/200
    or number of *r* alleles/200)   Frequency of *R*: $p =$ ____   Frequency of *r*: $q =$ ____
O. What happened to the genotype and allele frequencies in the second generation?

    What would you predict for the next generation? Why?

Exercise 5   (Modules 14.9 – 14.11)

The Hardy-Weinberg equilibrium is an idealized model. Equilibrium is maintained only if five conditions are met. This happens only in the fertile imaginations of biologists, not in real populations. Real populations always deviate from one or more of the conditions, and their gene pools change over time. The Hardy-Weinberg equilibrium is nevertheless a useful standard to compare real populations whose gene pools are changing.

Let's continue to look at the wildflower population introduced in Exercise 4. If it is like other real populations, its gene pool is changing. For each of the scenarios below, state which of the Hardy-Weinberg conditions the population deviates from, and explain what agent of microevolution causes the gene pool to change. Also state which of these deviations would cause the flowers to adapt to their environment.

1. A windstorm blows in hundreds of seeds from a nearby meadow, where nearly all the flowers are yellow.

2. A cosmic ray hits one of the red flowers just as it is replicating its DNA. Quite by chance, a red allele is transformed into a yellow allele.

3. The flowers tend to grow in red or yellow patches. A landslide buries and kills a huge patch of red flowers.

4. The red pigment in the petals of the red flowers is poisonous and tends to protect them from beetles that eat the developing seeds. The yellow flowers are not protected in this way.

5. The bees that pollinate the flowers tend to develop a "search image." Once they start visiting flowers of a certain color, they stick to that color. So pollen from red flowers is more likely to be delivered to other red flowers, and pollen from yellow flowers is more likely to fertilize other yellow flowers.

Exercise 6   (Modules 14.12 – 14.17)

Review the various terms and concepts relating to variation by completing this crossword puzzle.

Across

2. Mutation and sexual ____ generate variation.
3. Some traits controlled by several genes vary ____.
6. Cheetahs passed through a ____, which left them with reduced variation.
9. A population is ____ if two or more morphs are noticeably present.
11. ____ can create new alleles.
14. There are often many hidden recessive alleles in ____ organisms.
15. Only the ____ component of variation leads to adaptation.
16. ____, which multiply rapidly, can evolve by means of mutation alone.

Down

1. Meiosis and ____ contribute to sexual recombination.
3. A ____ mutation is a large-scale change in a chromosome.
4. ____ variation occurs when the environment differs from place to place.
5. Endangered species such as cheetahs show extreme genetic ____.
7. A ____ mutation is a change in a single nucleotide in a gene.
8. Heterozygote ____ means that heterozygotes do better than homozygotes.
10. One form of a phenotypic trait is called a ____.
12. A ____ is a gradual change in a trait along a geographical continuum.
13. Natural selection tends to ____ variation.

Returning one last time to the wildflowers discussed in Exercises 4 and 5, complete the following scenario regarding natural selection.

An early writer on evolution described natural selection as "nature red in tooth and claw." This may be true for lions and zebras on the Serengeti, but natural selection is usually more subtle. All living things are engaged in what Charles Darwin called a [1]"_____ for existence." He called natural selection "survival of the [2]_____," but this involves more than brute strength. Biologists define [3]_____ as the relative contribution that an individual makes to the gene pool of the next generation. It has more to do with reproduction than strength or cunning.

For the wildflowers in our previous example, the struggle for existence involves physical traits such as color and shape of leaves and metabolic characteristics like efficiency in capturing sunlight and resistance to cold. The [4]_____ of the plant is a composite of all its characteristics. Only the [5]_____, not the [6]_____, is exposed to the environment. A red flower may be protected from predation, while a yellow flower is eaten. Genes for red color increase fitness, and these genes are more likely to be passed on to the next generation. The frequency of red genes in the [7]_____ increases, as does the frequency of red individuals in the [8]_____. Of course, the fitness of a genotype for a particular characteristic depends on its entire [9]_____(all its genes). A red flower with a weak stem that snaps easily in the wind might be less fit than a yellow flower with a stronger stem.

The [10]_____ also shapes phenotype. Soil conditions, for example, can alter flower color. Genes may set limits on the height of a plant, but weather may determine whether genetic potential is realized.

There are three different ways in which natural selection can affect a population. Many traits are not simple "either-or" alternatives like red and yellow flower color. Traits like height vary continuously and can be described by a [11]_____-shaped curve: There may be a few very short plants in the population, a majority of plants of medium height, and a few very tall plants. Imagine our wildflowers growing in a cold, windy environment. Very tall plants might freeze before their seeds mature. Shorter plants would stay warmer, but very short plants might have trouble dispersing their seeds to favorable environments. In this kind of situation, [12]_____ selection favors the intermediate variants, not too tall and not too short. Next imagine a situation where the environment is gradually becoming drier. In this case, [13]_____ selection might favor those individuals genetically programmed to grow the deepest roots. This kind of natural selection is most common during periods of environmental [14]_____. Finally, [15]_____ selection occurs when two different sets of environmental conditions favor the extreme phenotypes and act against [16]_____ types. For example, plants with shallow, spreading roots might be at an advantage in dry rocky soil, where water tends to penetrate quickly. At the same time, a deep taproot system might be favored in richer soil that holds water longer. Intermediate root systems would be at a disadvantage in both environments.

Testing Your Knowledge

Multiple Choice

1. During his voyage around the world, Charles Darwin was inspired to think about evolution by
   a. books that he read.
   b. fossils he collected.
   c. studying adaptations of organisms to their environments.
   d. unique organisms he saw in the Galapagos Islands.
   e. all of the above.

2. _____ and _____ generate variation, while _____ results in adaptation to the environment.
   a. genetic drift . . . natural selection . . . mutation
   b. mutation . . . sexual recombination . . . natural selection
   c. overproduction of offspring . . . mutation . . . sexual recombination
   d. natural selection . . . mutation . . . sexual recombination
   e. sexual recombination . . . natural selection . . . mutation

3. Breeding of plants and animals by humans is called
   a. natural selection.
   b. sexual recombination.
   c. founder effect.
   d. artificial selection.
   e. neutral variation.

4. Bacteria can adapt to changes in the environment by means of mutation alone because
   a. they are so small in size.
   b. their populations are very isolated from one another.
   c. a bacterium is much more likely to mutate than a larger organism.
   d. they multiply so rapidly.
   e. their populations are so large.

5. The smallest unit that can evolve is a
   a. species.
   b. genotype.
   c. gene.
   d. population.
   e. morph.

6. "Differential reproduction" is just another way of saying
   a. natural selection.
   b. mutation.
   c. variation.
   d. recombination.
   e. genetic drift.

7. Which of the following changes in the gene pool results in adaptation to the environment?
   a. nonrandom mating
   b. genetic drift
   c. natural selection
   d. gene flow
   e. mutation

8. The ultimate source of all genetic variation is
   a. natural selection.
   b. genetic drift.
   c. sexual recombination.
   d. the environment.
   e. mutation.

9. In evolutionary terms, an organism's fitness is measured by its
   a. health.
   b. contribution to the gene pool of the next generation.
   c. mutation rate.
   d. genetic variability.
   e. stability in the face of environmental change.

10. Animals that possess homologous structures probably
   a. are headed for extinction.
   b. evolved from the same ancestor.
   c. have increased genetic diversity.
   d. by chance had similar mutations in the past.
   e. are not related.

11. Sexual recombination occurs when chromosomes are shuffled in _____ and fertilization.
   a. mitosis
   b. genetic drift
   c. natural selection
   d. mutation
   e. meiosis

12. Darwin
    a. was the first person to realize that organisms can evolve.
    b. believed that organisms could pass on acquired changes to offspring.
    c. was eager to publish his theory so that he could get all the credit.
    d. worked out the mechanism of evolution—natural selection.
    e. was the first to realize that fossils are remains of ancient organisms.

Essay

1. Explain how heritable variations, overproduction of offspring, and limited natural resources cause a species to adapt to its environment.

2. Briefly describe five categories of evidence for evolution.

3. What is the difference between a population and a species?

4. Horses look a lot like zebras. How would Darwin have explained this?

5. Sometimes a harmful allele may be present in the gene pool at a relatively high frequency. How might this be explained in terms of genetic drift? How might it relate to the fact that most organisms are diploid? What might it have to do with heterozygote advantage?

6. Describe how each of the following might alter the gene pool: genetic drift, nonrandom mating, gene flow, mutation, natural selection.

Applying Your Knowledge

Multiple Choice

1. In a population of bears, which would be considered the fittest?
   a. the biggest bear
   b. the bear having the largest number of mutations
   c. the bear that blends in with its environment the best
   d. the strongest, fiercest bear
   e. the bear that leaves the most descendants

2. A geneticist mixed together many different kinds of fruit flies—some with long wings, some with short wings, some with red eyes, some with brown eyes, and so on. He allowed the flies to feed, mate randomly, and reproduce by the thousands. After many generations, most of the flies in the population had medium wings and red eyes, and most of the extreme types had disappeared. This experiment appears to demonstrate
   a. stabilizing selection.
   b. geographical variation.
   c. diversifying selection.
   d. genetic drift.
   e. fitness.

3. Rabbits living farther north tend to have smaller ears. This is an example of
   a. a cline.
   b. polymorphism.
   c. artificial selection.
   d. heterozygote advantage.
   e. genetic drift.

4. Blue poppies native to China are grown at a plant-breeding center in California, where those with the thickest leaves survive and reproduce best in the drier climate. This evolutionary adaptation of the poppies to their new environment is due to
   a. genetic drift.
   b. stabilizing selection.
   c. directional selection.
   d. neutral variation.
   e. diversifying selection.

5. Biologists have noticed that most human beings enjoy sex. How would they explain this, in evolutionary terms?
   a. If sex were not enjoyable, the human species would have died out.
   b. Early humans who enjoyed sex most had the most babies.
   c. Only body structures evolve, not behavior, so enjoyment cannot evolve.
   d. This was due to a random mutation, so it did not affect evolution.
   e. Biologists are baffled by the phenomenon of sex.

6. Which of the following would result in evolutionary adaptation of a mouse population to its environment?
   a. Half the mice are killed by an avalanche.
   b. A mutation for spotted fur occurs.
   c. Several mice leave the area and mate with individuals elsewhere.
   d. Mice with thicker fur best survive a cold winter.
   e. Mice are most likely to mate with close neighbors.

7. The relationship of genome to organism is the same as ___ to population.
   a. species
   b. gene
   c. gene pool
   d. mutation
   e. variation

8. Critics like to point out that the theory of evolution is flawed because it is based on random changes—mutations. They say that a random change in an organism (or a car or a TV set) is likely to harm it, not make it function better. What logical statement could a defender of evolution make in reply to this criticism?
   a. Fossils prove without a doubt that mutations drive evolution.
   b. Mutation is random, but natural selection is not.
   c. Mutation has little to do with evolution.
   d. This is a weak spot in the theory that remains to be worked out.
   e. Mutations are not actually random.

9. A zoologist found that in the population of frogs in MacGregor's pond, half the genes for skin color in the gene pool were alleles for green spots, and half the genes were alleles for brown spots. Which of the following could cause these proportions to change?
   a. A drought shrinks the pond so that only five frogs remain.
   b. Brown-spotted females prefer to mate with brown-spotted males.
   c. Green-spotted frogs can hide more easily among the pond weeds.
   d. Filling a nearby pond causes those frogs to move to MacGregor's pond.
   e. Any of the above could cause the proportions to change.

10. Each of us is part of the ongoing evolution of the human species. Which of the following occurrences would have the greatest impact on the future biological evolution of the human population?
    a. You work out every day so that you stay physically fit and healthy.
    b. A mutation occurs in one of your skin cells.
    c. You move to Hawaii, the state with the longest life expectancy.
    d. A mutation occurs in one of your sperm or egg cells.
    e. You encourage your children to develop their intellectual abilities.

Essay

1. A butterfly has a long tubelike proboscis that it uses to suck the nectar from a certain kind of deep, tubular flower. Its close relatives have much shorter proboscises. How might Lamarck have explained the existence of this long proboscis? How would Darwin have explained it?

2. A long section of *On the Origin of Species* describes the breeding of pigeons and how pigeon breeders have produced many shapes, sizes, and colors of pigeons, all starting from a common wild pigeon. Why did Darwin think this was important?

3. Jeff and Trina collected some wildflower seeds from a meadow and scattered them along the roadside near their home. As the weather got hotter, some of the seedlings began to wither. Jeff said, "Hey, those scrawny plants are going to have to adapt or they are going to die." Trina replied, "*They* may not adapt, but the wildflower *population* will." What did she mean?

4. A population of 100 fish in an aquarium consists of 49 homozygous dominant green individuals, 42 heterozygous green individuals, and 9 white individuals. What are the frequencies of the three genotypes in the population? What are the frequencies of the green and white alleles in the gene pool? If the population is at Hardy-Weinberg equilibrium, what will the genotype and gene frequencies be in the next generation of fish?

5. Describe five ways in which the fish population in question 4 could be caused to deviate from Hardy-Weinberg equilibrium. Is this likely? Which of these deviations might cause the fish to become better adapted to their environment?

6. An ecologist studying predators and their prey in White Sand Dunes State Park found that nocturnal pocket mice exist in dark and light morphs. In this area, 42% of the mice are dark and 58% are light. Because owls swallow their prey whole and then cough up "pellets" consisting of bones and fur, the ecologist was able to discover that 61% of the mice caught by owls were dark, and 39% were light. What is likely to happen to the pocket mouse population in the future, and why?

Using Your Knowledge

1. There is an ongoing controversy in the United States about the teaching of evolution in public schools. "Scientific creationists" demand equal time for explanations other than evolution by natural selection. Has this been an issue in your community or state? What evidence and arguments are presented by each side? How does each side answer the other's criticisms?

2. Take a walk around your neighborhood, preferably in a natural area or park where living things are abundant. For each organism that you see, note an adaptation to its environment. How did the evolutionary processes discussed in this chapter shape the organism to fit its environment?

The Origin of Species

15

Waialeale, a mountain on the Hawaiian island of Kauai, is one of the rainiest places on Earth. Here 15 meters of rain a year are not unusual. The Alakai swamp lies at an elevation of 1300 meters on the flanks of the mountain. This boggy forest is one of the best places to see honeycreepers, birds unique to the Hawaiian Islands. Walking on a trail through the forest, you might see an apapane, a common red honeycreeper that uses its curved bill to eat nectar and caterpillars. The bright red iiwi laps up nectar with its long ivory-colored beak. You would be unlikely to spot the greenish akialoa, which has a needle-sharp beak half as long as its body. It was last seen in 1973 and may be extinct. Other honey-creepers have stout hooked bills used for crushing twigs to get insects, or short heavy beaks for cracking seeds. There are twenty-odd species of honeycreepers on the various islands, and several extinct species are known. All are thought to be descended from a single ancestral species—perhaps a finch—that reached the islands perhaps 5 million years ago. Speciation—how new species such as the diverse forms of honeycreepers arise—is the subject of this chapter.

Organizing Your Knowledge

Exercise 1   (Modules 15.1 – 15.2)

Briefly explain why it might be difficult to apply the biological species concept in each of the following situations.

1. Fossils of "Java Man" and "Peking Man" are both thought to represent a single species—*Homo erectus*.

2. A tiger and a lion can interbreed in a zoo and produce a hybrid offspring called a tiglon.

3. Dogs come in many shapes and sizes, from Chihuahuas to St. Bernards.

4. There are many strains and species of *Streptococcus* bacteria, which reproduce asexually.

5. Among *Clarkia* wildflowers in California, flowers of population A can interbreed and produce fertile offspring when crossed with flowers from population B. Similarly, B can interbreed with C. But A and C cannot successfully interbreed.

6. One bird guide calls flycatchers of the genus *Empidonax* "the bane of birdwatchers." Several species look so much alike that birders can distinguish them only by their songs.

7. A song sparrow population in Baja, California, is separated from other song sparrows by over a hundred miles of desert.

Exercise 2   (Module 15.3)

Review the reproductive barriers that separate species by categorizing the following examples. State whether each barrier is prezygotic (Pre) or postzygotic (Post) and then name the specific kind of barrier (such as temporal isolation or hybrid inviability) it exemplifies. The chart in Module 15.3 is a helpful summary.

Pre or Post   *Kind of Barrier*   *Example*

1. _____   _____   The salamanders *Ambystoma tigrinum* and *A. maculatum* breed in the same areas. *A. tigrinum* mates from late February through March. *A. maculatum* does not start mating until late March or early April.

2. _____   _____   Two species of mice are mated in the lab and produce fertile hybrid offspring, but offspring of the hybrids are sterile.

3. _____   _____   When fruit flies of two particular species are crossed in the lab, their offspring are unable to produce eggs and sperm.

4. _____   _____   A zoologist observed two land snails of different species that were trying to mate with little success because they apparently did not "fit" each other.

5. _____   _____   Male fiddler crabs (genus *Uca*) wave their large claws to attract the attention of females. Each species has a slightly different wave.

6. _____   _____   When different species of tobacco plants are crossed in a greenhouse, the pollen tube usually bursts before the eggs are fertilized.

7. _____   _____   Blackjack oak (*Quercus marilandica*) grows in dry woodlands, and scrub oak (*Q. ilicifolia*) grows in dry, rocky, open areas. Pollen of one species seldom pollinates the other.

8. _____   _____   The tiglon offspring of a lion and tiger are often weak and unhealthy.

Exercise 3   (Module 15.4)

Formation of new species often begins with geographical isolation. For each of the organisms listed below, name two geographical barriers that might lead to allopatric speciation.

1. Daisy

2. Mouse

3. Trout

4. Oak tree

5. Sparrow

6. Sea star

Exercise 4   (Module 15.5)

Speciation and adaptive radiation often occur on islands like the Galapagos, Hawaiian Islands, and Madagascar (Chapter 38). But for geographical isolation to occur, an island does not have to be a bit of land in the midst of an ocean. An evolutionary island could be, for example, a patch of grassland surrounded by forest. List three other examples and the kinds of species that could be isolated and evolve in each.

1.

2.

3.

Exercise 5   (Modules 15.6 – 15.7)

New species can result if a change in chromosome number produces a reproductive barrier that isolates organisms from their parent populations. Study the examples given in Modules 15.6 and 15.7. Then state whether you think each of the organisms in italics below would be able to reproduce (yes or no), whether it could represent a new species (yes or no), and how many chromosomes it would have.

Repro- duction	New Species	No. of Chrom.	Example
1. _____	_____	_____	A flower has 18 chromosomes ($2n = 18$). Nondisjunction occurs, and diploid gametes are formed. The *zygotes formed* are tetraploids, capable of self-fertilization when mature.
2. _____	_____	_____	A tree has 22 chromosomes ($2n = 22$). Nondisjunction produces a diploid pollen grain, which fertilizes a normal haploid egg. The resulting polyploid zygote develops into a *full-grown tree.*
3. _____	_____	_____	Antelope of two different species mate in a zoo. Species A has 36 chromosomes; species B has 34 chromosomes. They produce a *hybrid offspring.*
4. _____	_____	_____	A lily of species A has 16 chromosomes. Species B has 20 chromosomes. A hybrid with 18 chromosomes undergoes nondisjunction, producing a *lily* that is capable of self-pollination.
5. _____	_____	_____	A botanist treats some tissue from a strawberry plant ($2n = 14$) with a chemical that causes nondisjunction. This doubles the number of chromosomes in a cell. This cell is then cultured on a special medium, until it eventually develops into a *strawberry plant* that can self-pollinate.
6. _____	_____	_____	Pollen from the cell-cultured strawberry in question 5 is placed on the flower of a plant of the parent species, producing a triploid *hybrid zygote.*

Contrast the gradualist model of speciation with the punctuated equilibrium model by completing the chart below.

	Gradualism	*Punctuated Equilibrium*
1. Long-term tempo of evolution smooth or jumpy?		
2. Same processes as microevolution, or different process?		
3. Kind of evidence in fossil record?		
4. Speciation fast or slow?		
5. Continuous change through life of species, or quick change and stability?		

Exercise 7   (Summary)

Use the concepts and terms from this chapter to complete the following story about (imaginary) butterflies and asters.

The yellowspot butterfly is found over hundreds of square miles of land in the delta of an African river. Its primary food source is a species of purple aster—a flower related to daisies and dandelions. Patches of asters are scattered in sunny meadows in the delta, some several miles apart. The butterflies do not usually venture far for food. Each patch of asters supports a separate [1]_____ of yellowspots, but because the butterflies sometimes wander and mate with butterflies in other areas, all the butterflies have been classified as members of the same [2]_____.

Insect taxonomists noted that one population of yellowspots was centered across the main river channel from the other populations. They suspected that the river might act as a [3]_____ to the butterflies, since they do not usually fly far over water. The researchers examined the butterflies from the isolated population and found that the butterflies from across the river were a bit smaller than most yellowspots and more orange in color, so the researchers nicknamed them "orangespots." The biologists found that the differences in appearance were inherited. They suspected these could reflect [4]_____ due to chance differences in the butterflies that founded the orangespot population. The researchers also noted that the environment was slightly warmer and drier on the far side of the river, so [5]_____ may have caused the change in the butterfly population there.

The scientists suspected that the two populations could represent different species. To test this hunch, they had to find out whether butterflies from the far side of the river could [6]_____ with individuals from the main population. The biologists captured some butterflies from both areas and placed them in a cage. Surprisingly, the orangespots and yellowspots ignored each other. The researchers found that the female butterflies rest on leaves and flash their wing spots to attract the males. Yellowspot females flash their wings at a much faster rate than orangespot females. Apparently the wing-flashing display acts as a [7]_____-zygotic reproductive barrier, specifically [8]_____. It keeps the two populations of butterflies from interbreeding. In one instance, the eggs of an orangespot female were fertilized by a yellowspot male, but the embryos soon died. Apparently, there is also a [9]_____ -zygotic reproductive barrier between the butterflies. The particular type of barrier is [10]_____ in this case. The researchers realized that the yellowspot and orangespot butterflies are separate [11]_____, incapable of interbreeding.

A study of river sediments showed that the channel separating the two butterfly populations formed about a thousand years ago, when the river shifted course. It has indeed acted as a geographical barrier, [12]_____ the two populations from one another and eventually leading to [13]_____ speciation. Biologists were particularly interested in how quickly (in geological terms) the new butterfly species must have evolved. This rapidity would seem to support the [14]_____ model of species evolution.

While studying the habits of the butterflies, the biologists found a group of asters with oval-shaped leaves and slightly larger flowers than all the others. A study of their chromosomes showed that the unusual plants were [15]_____, having more than the usual two sets of chromosomes. All the other purple asters in the area had two sets of chromosomes and a $2n$ chromosome number of 26. The unusual plants were [16]_____, having four sets of chromosomes, for a total of 52. [17]_____ must have occurred during meiosis in one of the diploid asters, creating diploid gametes. Because the plant could self-pollinate, zygotes were formed with four sets of chromosomes. The plants that developed from these zygotes could interbreed with one another, but they were reproductively [18]_____ from the parent species. They represented a new [19]_____, produced in one generation through the process of [20]_____ speciation.

Testing Your Knowledge

Multiple Choice

1. Two animals are considered different species if they
   a. look different.
   b. cannot interbreed.
   c. live in different habitats.
   d. are members of different populations.
   e. are geographically isolated.

2. Which of the following is the first step in allopatric speciation?
   a. genetic drift
   b. geographical isolation
   c. polyploidy
   d. hybridization
   e. formation of reproductive barrier

3. The science of naming and classifying organisms is called
   a. biology.
   b. polyploidy.
   c. genetics.
   d. taxonomy.
   e. gradualism.

4. Most of the time, species are identified by their appearance. Why?
   a. If two organisms look alike, they must be the same species.
   b. This is the criterion used to define a biological species.
   c. If two organisms look different, they must be different species.
   d. This is the most convenient way of identifying species.
   e. Most organisms reproduce asexually.

5. A new species can arise in a single generation
   a. through geographical isolation.
   b. in a very large population that is spread over a large area.
   c. if a change in chromosome number creates a reproductive barrier.
   d. if allopatric speciation occurs.
   e. according to the gradualist model of speciation.

6. The evolution of numerous species, such as Darwin's finches, from a single ancestor is called
   a. adaptive radiation.
   b. sympatric speciation.
   c. gradualism.
   d. nondisjunction.
   e. geographical isolation.

7. According to the _____ model, evolution occurs in spurts; species evolve relatively rapidly, then remain unchanged for long periods.
   a. nondisjunction
   b. gradualist
   c. adaptive radiation
   d. punctuated equilibrium
   e. geographical isolation

8. Sympatric speciation is
   a. the appearance of a new species in the same area as the parent population.
   b. the process by which most animal species have evolved.
   c. initiated by the appearance of a geographical barrier.
   d. the emergence of many species from a single ancestor.
   e. especially important in the evolution of island species.

9. Individuals of different species living in the same area may be prevented from interbreeding by responding to different mating dances. This is called
   a. ecological isolation.
   b. hybrid breakdown.
   c. mechanical isolation.
   d. temporal isolation.
   e. behavioral isolation.

10. It is unlikely that the human population will give rise to a new species because
    a. the human population is too large.
    b. geographical isolation is unlikely to occur.
    c. a change in chromosome number would be fatal.
    d. the human population is too diverse.
    e. natural selection cannot affect humans.

Essay

1. Give three situations in which it might be difficult to use the test of interbreeding to determine whether two organisms are of the same or different species.

2. Describe step by step how geographical isolation could lead to speciation.

3. Describe what has to happen for two species with different numbers of chromosomes to interbreed and produce a fertile hybrid. Will this hybrid be able to interbreed with either of its parents? Why? How common is this in nature?

4. State whether a large, widely distributed population or a small, isolated population is more likely to undergo speciation, and explain why.

5. Compare the gradualist and punctuationist models of species evolution. What would the fossil record of speciation look like if it supported the gradualist model? What would it look like if punctuated equilibrium was the case? Which model does the fossil record seem to support most often? Why?

Applying Your Knowledge

Multiple Choice

1. Three species of frogs—*Rana pipiens, Rana clamitans,* and *Rana sylvatica*—all mate in the same ponds, but they pair off correctly because they have different calls. This is an example of
   a. prezygotic barrier . . . behavioral isolation.
   b. postzygotic barrier . . . hybrid breakdown.
   c. prezygotic barrier . . . temporal isolation.
   d. postzygotic barrier . . . behavioral isolation.
   e. prezygotic barrier . . . gametic isolation.

2. Which of the following reproductive barriers actually prevents individuals of different species from mating with each other?
   a. hybrid inviability
   b. hybrid sterility
   c. gametic isolation
   d. hybrid breakdown
   e. behavioral isolation

3. Bullock's oriole and the Baltimore oriole are closely related, but are they the same species? To find out, you could see whether they
   a. sing similar songs.
   b. look alike.
   c. live in the same areas.
   d. have the same number of chromosomes.
   e. successfully interbreed.

4. Sometimes two quite different populations interbreed to a limited extent, so that it is difficult to say whether they are clearly separate species. This does not worry biologists much because it
   a. is quite rare.
   b. is true for almost every species.
   c. supports the theory of punctuated equilibrium.
   d. may illustrate the formation of new species in progress.
   e. happens only among plants, not among animals.

5. Two species of water lilies in the same pond do not interbreed because one blooms at night and the other during the day. The reproductive barrier between them is an example of
   a. temporal isolation.
   b. gametic isolation.
   c. mechanical isolation.
   d. hybrid breakdown.
   e. ecological isolation.

6. Comparison of fossils with living humans seems to show that there have been no significant physical changes in *Homo sapiens* in 30 thousand to 50 thousand years. What might an advocate of punctuated equilibrium say about this?
   a. It is about time for humans to undergo a burst of change.
   b. That is about how long we have been reproductively isolated.
   c. It is impossible to see major internal changes by looking at fossils.
   d. You would expect lots of changes in the skeleton in that time period.
   e. Lack of change is consistent with the punctuated equilibrium model.

7. Which of the following is an example of a postzygotic reproductive barrier?

a. One species of frog mates in April; another mates in May.
b. Two fruit flies of different species produce offspring that are sterile.
c. The sperm of a marine worm only penetrate eggs of the same species.
d. A species of flower grows in forested areas, another in meadows.
e. Two pheasant species perform different courtship dances.

8. Lake Malawi, in the African Rift Valley, is home to over a hundred species of cichlid fishes, each with a slightly different diet and habits. All these fishes probably evolved from one ancestor, an example of
   a. sympatric speciation.
   b. hybrid breakdown.
   c. adaptive radiation.
   d. gradualism.
   e. punctuated equilibrium.

9. A botanist found that a kind of white daisy had a diploid chromosome number of 16. In the same area, he found a yellowish daisy. Its cells contained 24 chromosomes. He found that the yellowish daisy was a polyploid descendant of the white daisies. Which of the following would describe this unusual plant? P: tetraploid, Q: triploid, R: probably sterile, S: a new species.
   a. P R S
   b. Q R
   c. Q R S
   d. P S
   e. Q S

10. A fossil expert finds an impression of an ancient marine creature called a trilobite in a layer of rock. In the adjacent layer is another species of trilobite, clearly related to the first but quite different in form. If the expert is a gradualist, how might he or she interpret this?
    a. This kind of change is exactly what gradualism would predict.
    b. Sympatric speciation must have occurred.
    c. Intermediate forms could have existed but were not fossilized.
    d. Internally, the creatures were identical; only the outer shell changed.
    e. This kind of abrupt transition is rare in the fossil record.

Essay

1. There are dozens of species of small rodents, such as rats and mice, in western North America, but relatively few species, ranging over much larger areas, in the east. Suggest a hypothesis related to speciation that might explain this.

2. The mating of a horse and a donkey produces a mule, which is strong and hard working but sterile. A horse cell contains 64 chromosomes, a donkey cell 62 chromosomes. How many chromosomes would you expect to find in a cell from a mule? Why? Explain why a mule is sterile.

3. A dog (*Canis familiaris*) and a coyote (*Canis latrans*) will readily mate in captivity, and their offspring are healthy and fully fertile. Are we justified in saying they are distinct species? What might this tell us about the history of dogs and coyotes?

4. In terms of speciation, how might freshwater streams in a desert be like islands in the ocean?

5. Various sections of the Hawaiian Islands were formed by lava flows at different times. Kenneth Kaneshiro compared older populations of fruit flies living in older habitats with presumably younger populations living in more recently formed habitats. He found that male fruit flies in both areas are generally eager to mate, but females of older populations are much more selective about their partners than females of newer populations. Suggest a hypothesis to explain why this might be the case.

6. In the Hawaiian Islands, there are several native species of moths in the genus *Hedylepta* that feed exclusively on bananas. Other related species specialize in palms, grasses, or legumes. Bananas were brought to Hawaii by the native Polynesians about a thousand years ago. Why might the banana-eating species be of particular interest to biologists working out the details of punctuated equilibrium theory?

7. Critics of evolution often say things like, "Sure, moths and bacteria can adapt to a changing environment. But peppered moths are still peppered moths, and *Streptococcus* is still *Streptococcus*. Nobody has ever seen a new *species* evolve." Is this criticism valid? How would you respond to this comment?

Using Your Knowledge

1. The Nature Conservancy is an environmental organization that buys land to protect rare and endangered species of plants and animals. Are there any Nature Conservancy preserves near your home or college? What species are protected? How did these species come to be located there? Why are they rare?

2. The desire to name and categorize things seems to be a basic human characteristic. This may be why many people like to collect things. Some biologists believe that we have gotten carried away with this desire to classify when it comes to the idea of species. They think that the concept of a "species" is just that—only a concept. They believe that a species is not a real biological entity, only a convenient idea. There are just too many flaws and exceptions in the species concept for it to be valid. Populations, however, are real. Based on what you know and have read in this chapter, do you think species are real or not? What kinds of evidence or information might support each side of this discussion?

Tracing Evolutionary History

Imagine walking down a dusty trail along a riverbank. It is a hot summer day; insects circle and buzz lazily in the shade of overhanging trees. Here and there, the trail skirts high bluffs of gray and brown rock. The rock is laid down in layers—different colors, different thicknesses, and different textures. In places, the river has undercut the bank, and sheets of rock have peeled away and tumbled downhill. You round a bend and see a large slab upended beside the trail. There is a crack running all the way through the slab, so it is easy to pull the layers apart. You have to jump back as the surface layer falls away and shatters at your feet. The newly exposed surface is solid with the impressions of hundreds of fossil seashells. It is like finding a buried treasure. Nobody has ever seen these fossils before; they have been hidden for millions of years. Where did the shells come from? How did they get here? How old are they? What can they tell us about the history of life? These are some of the questions explored in this chapter.

Organizing Your Knowledge

Exercise 1   (Modules 16.1 – 16.2)

Review fossils by completing this crossword puzzle.

Across

1. Hard parts, such as bones, teeth, and ____, fossilize most easily.
4. Particles of sand may be compressed to form ____.
5. ____ forms from mud.
6. Sometimes a whole organism is preserved, if ____ does not occur.
10. Insects are often preserved in ____, fossilized tree resin.
12. An impression left by a seashell may fill with minerals to form a ____.
13. A ____ is a scientist who studies fossils.

Down

1. Layers of sedimentary rock are called ____.
2. ____ occurs when remains of dead organisms are turned to stone.
3. Fossils are usually found in ____ rocks.
7. The ____ fossils are found in the deepest strata.
8. A ____ is a preserved imprint or remnant of an ancient organism.
9. Some fossils retain ____ materials such as chlorophyll or proteins.
11. A seashell might make an imprint in mud, which forms a ____.

Exercise 2   (Module 16.3)

Review the geological time line by numbering each of the following events in order and naming the geological era when each occurred.

Order	Era	Event
_____	_____	A. Cone-bearing plants and dinosaurs dominant
_____	_____	B. Humans appear
_____	_____	C. Origin of first animals
_____	_____	D. First vertebrates
_____	_____	E. Increase in mammals, birds, insects, and flowers
_____	_____	F. Extinction of dinosaurs
_____	_____	G. Invasion of land by plants and arthropods
_____	_____	H. First insects, amphibians, and reptiles
_____	_____	I. Origin of life (prokaryotes)
_____	_____	J. Mass extinction of marine life

Exercise 3   (Module 16.4)

Use the concept of half-life to answer these questions about the ages of fossils.

1. The half-life of carbon-14 is 5600 years. If a mammoth was thought to contain 16 g of carbon-14 when it was frozen in a Siberian glacier, and it now contains 4 g, how old is the mammoth? At an error level of plus or minus 10%, what are the maximum and minimum ages of the fossil?

2. The half-life of potassium-40 is 1.3 billion years. If a rock specimen contained 1 g of potassium-40 when it was formed, how much potassium-40 would be left if the rock is 3.9 billion years old?

Exercise 4   (Modules 16.5 – 16.6)

Review the concepts of continental drift by matching each of the phrases on the right with a word on the left. The illustrations in the textbook will help.

A. Pangaea	____	1. The surface of the Earth, broken into plates
B. Plate tectonics	____	2. This ocean is surrounded by violent geological events
C. Plate	____	3. Southern land mass formed when Pangaea broke up
D. Earthquake	____	4. Place where plates slide along one another
E. Himalayas	____	5. This ocean grows as North America and Eurasia split apart
F. Continental drift	____	6. Famous volcanic island
G. Gondwana	____	7. Supercontinent formed 250 million years ago
H. Mantle	____	8. Continent formed from the eastern part of Laurasia
I. Australia	____	9. Any large, moving segment of the Earth's crust
J. Pacific	____	10. Movement resulting from forces exerted at plate edges
K. Laurasia	____	11. Mountains formed by collision of Indian and Eurasian plates
L. Crust	____	12. Movement of continents over the Earth's surface
M. Fault	____	13. Hot layer that lies beneath the crust
N. Atlantic	____	14. Northern land mass formed when Pangaea broke up
O. Krakatoa	____	15. Part of the same plate as India
P. Eurasia	____	16. Geological forces that cause movements of crustal plates

Exercise 5   (Modules 16.7 – 16.10)

The following story illustrates some of the concepts and terms concerning mass extinctions and their aftermath. Fill in the blanks using words from Modules 16.7–16.10.

The disappearance of the dinosaurs at the end of the [1]_____ period, [2]_____ million years ago, is one of the great mysteries of evolution. The dinosaurs ruled the Earth for [3]_____ million years; then, in just a few million years, they vanished.

There have been several mass extinctions in the history of life; in fact, there have been [4]_____ major periods of mass extinction in the last 600 million years. At the end of the Permian period, 230 million years ago, continental drift brought all the continents together, forming the supercontinent of [5]_____. This altered climate and shorelines. More than [6]_____% of marine animal species perished, and many land-dwellers disappeared as well. The climate was [7]_____ late in the Cretaceous period, and shallow seas were disappearing. Perhaps the dinosaurs were simply victims of ongoing climate change.

Many other hypotheses have been suggested to explain the Cretaceous extinction. All of the sediments laid down at this time contain deposits of [8]_____, an element rare on the Earth but common in [9]_____. It is possible that a [10]_____ or [11]_____ struck the Earth, raising a cloud of dust that could have blocked the sun and disrupted the climate. This could have also interfered with plant growth, creating a shortage of food. Alternatively, evidence from India suggests that [12]_____ might have injected particles into the atmosphere that could have blocked the sun and played havoc with the climate and food supply.

Mammals first appeared on the scene 75 million years before the demise of the dinosaurs. The first mammals were small, scurrying creatures that played a secondary role to the great reptiles. But unlike the dinosaurs (and numerous kinds of marine life), mammals survived the Cretaceous extinction. Why did mammals survive, while the mighty dinosaurs did not? Perhaps it was due to the fact that mammals are covered with insulating [13]_____, or perhaps it was due to the way they take care of their young and feed them [14]_____. Mammals are also able to generate their own body heat, which may not have been the case for the dinosaurs. In addition, the early mammals were probably nocturnal scavengers, so they would have been well suited for finding a meal in the dark, cold aftermath of a catastrophe. Body heat, insulation, and nocturnal habits are all examples of [15]_____, characteristics that evolve in one context and later are found to be useful for another function.

An explosive increase in mammals followed the Cretaceous extinction. Freed from domination by the dinosaurs, the mammals inherited a whole world, and they adapted to every possible environment. Long-legged herbivores ran through woodlands and meadows, pursued by carnivores with powerful jaws and sharp teeth. Giant relatives of horses and rhinoceroses, such as the [16]_____, thundered across the plains. Trends such as increase in size may have resulted from unequal [17]_____, or unequal survival of [18]_____, where the environment favored larger forms. Some large mammals even returned to the sea, giving rise to modern whales and dolphins. Some mammals climbed trees, and some—the bats—even took to the air.

All of these mammalian adaptations started with changes in genes. Even a slight change can have drastic effects on development and body form if the gene that is altered is a [19]_____ gene that controls a number of [20]_____ genes. This may have happened in our primate ancestors. A subtle genetic change could have resulted in retention of the rounded skull and small face of a fetus throughout life. This retention of juvenile body features in an adult is termed [21]_____. Within the skull was room for a larger, more complex human brain. The main function of our long [22]_____ may be to program our large brains, so that we can [23]_____ from adults how to live in our complex world.

Exercise 6   (Module 16.11)

You are the first zoologist to penetrate the Timbasi Swamp and explore the Okongo Forest. Amazingly, you have identified seven species of guenon monkeys previously unknown to science. You have risked your neck to obtain blood samples from the elusive monkeys. Using similarities in blood proteins and facial markings, you have figured out the relationship of the new guenons with the monkeys on the chart. Test your understanding of the phylogenetic tree by matching each of the new monkeys listed below with one of the letters inserted into the revised phylogenetic tree.

_____   1. Ann's: More closely related to Diana than any other species
_____   2. Flat-topped: As close to Mona as Mona is to Campbell's
_____   3. Gladstone's: Closer to redtail and moustached than any other new species
_____   4. Bearded: Related to Diana but not as closely as Ann's
_____   5. Liebaert's: A ground-dweller not closely related to any of the others
_____   6. Perkins's: Related to Mona and Campbell's, but it branched off earlier
_____   7. Striped: Equally related to blue and redtail, but closer to ancestor

Exercise 7   (Modules 16.12 – 16.13)

The system of taxonomic categories used by biologists is like a set of boxes into which organisms are sorted. A cocker spaniel—*Canis familiaris*—for example, is first placed in a small box, the specific name *familiaris* that separates it from all other species. This is placed in a slightly larger box, the genus *Canis*, which also holds *Canis lupus* (the wolf) and *Canis latrans* (the coyote). This genus box is placed in a larger box, along with other genera of doglike animals, and so on, all the way up to the last box that separates animals from plants and fungi. Imagine that the nested boxes below represent the taxonomic categories, starting with species (omitting subphylum). Label the boxes to show the relationships among the categories.

7. _____ : Animalia

6. _____ : Chordata

5. _____ : Mammalia

4. _____ : Carnivora

3. _____ : Canidae

2. _____ : *Canis*

1. _____ species _____ : *familiaris*

Exercise 8   (Modules 16.13 – 16.14)

Homologous structures—similar structures derived from the same structure in a common ancestor—tell us about phylogenetic relationships among organisms. But convergent evolution can make unrelated organisms look alike; their similarities may be analogous, not homologous. Fortunately, we can dig beneath surface similarities and compare biological molecules to measure relatedness between species. DNA-DNA hybridization measures similarity indirectly, by determining how well DNA strands from different organisms bond to each other. Amino acid and DNA sequencing take molecules apart and count the similarities. Imagine that albumin, a blood protein, has been sequenced for six species of rodents, A through F. All the rodents are thought to have evolved from a common ancestor, X. The number of differences in albumin amino acid sequence is compiled in the table below, which reads like a road map mileage chart. For example, there are 4 differences between A and C, 9 differences between A and D. Use the differences in sequence to place Species A through F on the phylogenetic tree.

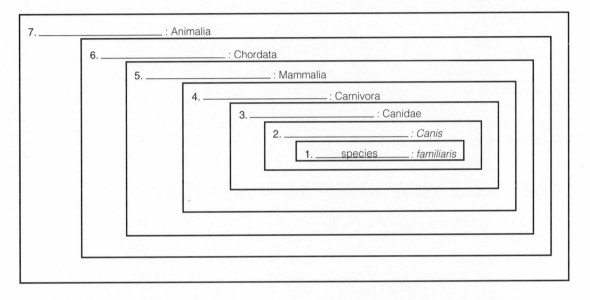

A					
B	10				
C	4	11			
D	9	5	10		
E	14	16	15	15	
F	10	2	10	6	16
	A	B	C	D	E	F

1. _____ 2. _____ 3. _____ 4. _____ 5. _____ 6. _____

Cladistic taxonomy and classical evolutionary taxonomy are different methods of interpreting phylogenetic data and classifying organisms. Read each statement below and check whether it relates to the cladistic approach, the classical approach, or both.

	Cladistic	Classical
1. Method for classifying organisms and reconstructing phylogeny	☐	☐
2. Concerned only with the order of branching of lineages	☐	☐
3. Produces cladograms	☐	☐
4. Concerned with branching and degree of divergence	☐	☐
5. Differentiates between primitive and derived characters	☐	☐
6. Puts lizards and crocodiles in one class, birds in another	☐	☐
7. Becoming more popular with researchers	☐	☐
8. Says birds are closer to crocodiles than to other reptiles	☐	☐
9. Uses anatomy and molecular biology to determine relationships	☐	☐
10. Places humans in the same family as some other apes	☐	☐
11. Places humans in their own family, separate from apes	☐	☐
12. The approach used in this textbook	☐	☐
13. Considered to be a more objective approach	☐	☐
14. Involves subjective judgments about divergence	☐	☐

Compare the five kingdoms of life by completing this chart. Some of the answers are done for you.

Characteristics	Monera	1. _____	Plantae	2. _____	3. _____
Prokaryotic or eukaryotic	4.	Eukaryotic	5.	6.	7.
Single-celled, multicellular, or both	8.	9.	10.	11.	12.
Nutrition: autotrophic, heterotrophic, or both	Both	Both	13.	14.	Heterotrophic; decomposition and absorption

Testing Your Knowledge

Multiple Choice

1. Which of the following matches an event in the history of life with the correct geological era?
   a. first mammals—Precambrian
   b. appearance of humans—Mesozoic
   c. dominance of dinosaurs—Cenozoic
   d. movement of plants and animals onto land—Paleozoic
   e. first flowering plants—Precambrian

2. Changes in regulatory genes might be very important in evolution because
   a. they are not affected by natural selection.
   b. they might cause major changes in development and body form.
   c. they speed up the rate of convergent evolution.
   d. they provide insurance against future mass extinctions.
   e. they are more readily passed on to offspring.

3. If you want to see a dinosaur, it would be best to set the controls of your time machine for the
   a. Mesozoic era.
   b. Paleozoic era.
   c. Pleistocene period.
   d. Paleozoic era.
   e. Precambrian era.

4. Which of the following taxonomic categories contains all the others?
   a. genus
   b. class
   c. family
   d. subclass
   e. order

5. Paleontologists
   a. assign names to living things.
   b. are experts on the mechanisms of continental drift.
   c. classify organisms in families, orders, classes, and phyla.
   d. study fossils.
   e. study proteins and DNA to trace evolutionary relationships.

6. Organisms that survive mass extinctions
   a. often diversify, taking advantage of new opportunities.
   b. usually are so reduced in numbers that they soon go extinct.
   c. are often "living fossils" that have existed unchanged for long periods.
   d. usually cannot cope with the new conditions that follow.
   e. usually lack preadaptations to the new environment.

7. Pangaea
   a. was a land mass that broke up to form the present-day continents.
   b. is the idea that all life on Earth is related.
   c. was an animal common in ancient seas but now extinct.
   d. is the evolutionary history of a species, family, or phylum.
   e. is the theory that crustal plates can move relative to one another.

8. The theory of continental drift helps us to explain all of the following except
   a. locations of volcanoes.
   b. formation of river systems.
   c. distribution of animals and plants.
   d. formation of mountain ranges like the Himalayas.
   e. earthquakes.

9. What evidence most strongly suggests that an impact by an asteroid or meteorite may have caused the extinction of the dinosaurs?
   a. Fossils show that dinosaurs suffered from cold and starvation.
   b. Sedimentary rocks contain a layer of mineral uncommon on Earth.
   c. There have been several near-misses in recent years.
   d. The dinosaurs disappeared rather abruptly—virtually overnight.
   e. Fossils indicate that most dinosaurs were looking up when they died.

10. The oldest fossils usually
   a. contain more radioactive isotopes than younger fossils.
   b. are found in the deepest strata.
   c. have the longest half-lives.
   d. are found above younger fossils.
   e. are found in sediments from the Cenozoic era.

Essay

1. Explain how the formation of Pangaea may have led to mass extinctions at the end of the Permian period, about 250 million years ago.

2. Place the following events in the history of life in the proper order: appearance of humans, origin of eukaryotes, dominance of dinosaurs and cone-bearing plants, origin of animals, appearance of first vertebrates, diversification of mammals, origin of flowering plants, first prokaryotes, movement of plants and animals onto land.

3. Unlike hours, days, or weeks, geological eras and periods are not of uniform lengths. The Precambrian era, for example, lasted for billions of years, the Mesozoic era only for about 180 million. Explain why the units of the geological time line are uneven.

4. Describe two different ways in which plate tectonics and continental drift cause mountains to form.

5. The following list shows the classification of a human being, *Homo sapiens*. Name the category that corresponds to each of the taxa listed.

Animalia   Primates

Chordata   Hominidae

Vertebrata   *Homo sapiens*

Mammalia

Applying Your Knowledge

Multiple Choice

1. Which of the following would cast doubt on the asteroid-impact hypothesis for the extinction of the dinosaurs?
   a. finding a crater 200 million years old
   b. finding fossil dinosaur bones beneath a layer of iridium
   c. determining that birds are closely related to dinosaurs
   d. finding fossil dinosaur bones above a layer of iridium
   e. finding that many forms of marine life disappeared at the same time

2. The wings of birds and insects have the same function, but they do not have the same evolutionary origin. Bird and insect wings are
   a. homologous.
   b. phylogenetic.
   c. analogous.
   d. binomial.
   e. taxonomic.

3. Two animals in the same family would not have to be in the same
   a. genus.
   b. kingdom.
   c. order.
   d. phylum.
   e. class.

4. We know a lot about fossil clams, snails, and corals but not very much about ancient seaweeds. Why do you suppose this is the case?
   a. There were no seaweeds in the ancient oceans.
   b. Seaweeds were probably too soft to fossilize well.
   c. Animal life was much more abundant than seaweeds in ancient times.
   d. Plants moved onto land, leaving only animals in the sea.
   e. A mass extinction wiped out seaweeds, but animals survived.

5. Fossils of an ancient reptile called *Lystrosaurus* have been found in Africa, India, and Antarctica. Which of the following best explains this distribution?
   a. They were able to move between continents before the oceans filled.
   b. Movement of India due to continental drift carried them from place to place.
   c. These areas were once next to each other and have since drifted apart.
   d. They were able to migrate over frozen seas during the Ice Ages.
   e. Changes in climate forced them to migrate from place to place.

6. Look at the monkey phylogenetic tree in Module 16.11 of the text. The species in which of the following pairs are most closely related to each other?
   a. Crowned and Mona
   b. Blue and greater spot-nosed
   c. Diana and Syke's
   d. Diana and DeBrazza's
   e. Campbell's and Mona

7. Anthropologists suggest that the grasping hands and eye-hand coordination that early humans inherited from their tree-dwelling primate ancestors enabled humans to start making and using tools. This illustrates the concept of
   a. phylogeny.
   b. homology.
   c. preadaptation.
   d. neoteny.
   e. convergent evolution.

8. Which of the following illustrates neoteny?
   a. Separate groups of mice in Africa and Asia come to look much alike.
   b. Two families of trees are descended from a common ancestor.
   c. Eyes of octopuses and fish evolved separately but look alike.
   d. Many kinds of clams appeared after the Permian mass extinction.
   e. An insect is able to reproduce while still in the immature stage.

9. A microbiologist recently discovered a new single-celled eukaryotic organism. It is green and swims, propelled by two flagella. This organism would probably be placed in Kingdom
   a. Monera.
   b. Plantae.
   c. Fungi.
   d. Protista.
   e. Animalia.

10. In a detailed phylogenetic tree showing the evolutionary relationships within a class, the trunk of the tree might represent the _____, while the ends of individual "twigs" might represent _____.
    a. phylum . . . classes
    b. class . . . species
    c. genus . . . species
    d. species . . . families
    e. class . . . phyla

Essay

1. When a tree was buried in a swamp, it contained 400 mg of carbon-14. The half-life of carbon-14 is 5600 years. How long ago was the tree buried if it now contains 50 mg of carbon-14?

2. Australia is famous for its many marsupials—mammals that are born at an early stage in development and then mature in the mother's pouch. Koalas, kangaroos, wombats, and opossums are all marsupials. Marsupials are considered offshoots of an earlier type of mammal than the more familiar placental mammals, whose young remain longer in the uterus, connected to the mother by the placenta. Marsupials are found elsewhere, but in most places most marsupials have been displaced by the more "efficient" placentals. Speculate about why the marsupials managed to hang on in Australia. Might it have had something to do with continental drift?

3. The dinosaurs perished in a mass extinction at the end of the Cretaceous period, 65 million years ago. Briefly describe alternative hypotheses that have been suggested to explain the extinction of the dinosaurs. Did other forms of life disappear during the same period? Does this cast doubt on any of the hypotheses? Why?

4. Biologists like to look at fossils to trace evolutionary relationships. But birds are very delicate and unlikely to be fossilized; only a handful of old bird fossils are known. How might the molecules in the cells of living birds be compared to determine how ducks, sparrows, and hawks are related? Briefly name the molecules, what characteristic of the molecules would be compared, and how this would indicate how closely related the birds are.

5. Molecular comparisons indicate that the vultures of Africa and America are more closely related to separate hawklike ancestors than they are to each other. If the two groups of vultures are not closely related, why do you think they look so much alike?

Using Your Knowledge

1. Have you ever collected fossils? It is a very interesting and rewarding hobby. In fact, several well-known paleontologists got started as amateurs. Are there fossil deposits in your area? If you are interested in looking for fossils, you can get information from state departments of geology or mining, geological surveys, departments of natural resources, bureaus of mines, university geology departments, or local rockhound clubs.

2. If you are interested in learning more about fossils and the history of life, many university museums have fossil collections. Some of the best dinosaur exhibits are located at the American Museum of Natural History in New York, the Smithsonian Institution in Washington, DC, the Royal Ontario Museum in Toronto, the National Museum of Canada in Ottawa, the Peabody Museum at Yale University, the Carnegie Museum in Pittsburgh, the Field Museum of Natural History in Chicago, the Denver Museum of Natural History, and the Los Angeles County Museum. Assembled dinosaur skeletons can be seen at Dinosaur National Monument in Utah, and Dinosaur Provincial Park in Alberta. You can see dinosaur tracks at Rocky Hill, Connecticut, and near Glen Rose, Texas.

The Origin and Evolution of Microbial Life: Prokaryotes and Protists

17

A drop of pond water holds a beautiful and fascinating world. Under a microscope, pond scum becomes delicate filaments of green algae, their chloroplasts spiraling inside crystal cell walls. Other algae swim and spin lazily. Golden diatoms glide across the field of view, their intricately etched silica skeletons refracting rainbows of light. Necklaces of simpler cyanobacteria wave gently to and fro. Ciliated cells shaped like funnels suck algae and even smaller bacteria into their mouths, while formless amoebas crawl slowly in search of microscopic prey. Suddenly a huge *Paramecium* darts into view, its cilia beating in rhythm as it vacuums up small food particles. These creatures remind us of a time before there were any plants or animals—before there were any multicellular organisms at all. Some take us back to the very dawn of life on Earth. This chapter concerns the origin of life and the diversity of microbial life forms.

Organizing Your Knowledge

Exercise 1   (Module 17.1)

Summarize the early history of the Earth and the beginnings of life by numbering the following events in sequence. Some of the events are described in the text and some in Figure 17.1C.

_____ A. Formation of the seas

_____ B. Loss of Earth's original hydrogen (H_2) atmosphere

_____ C. Formation of Earth

_____ D. Melting of Earth, followed by formation of core and crust

_____ E. Oldest known fossils

_____ F. The "Big Bang"—formation of the universe

_____ G. O_2 from photosynthetic bacteria produces aerobic atmosphere

_____ H. Origin of eukaryotes

_____ I. Volcanoes belch out atmosphere of H_2O, CO, CO_2, N_2, CH_4, NH_3

_____ J. Origin of life—first prokaryotes

_____ K. Oldest known animal fossils

Exercise 2 (Modules 17.2 – 17.6)

This flow chart summarizes information and theory concerning the origin of life. Fill in the boxes by choosing from the list of components. Fill in the ovals by choosing from the list of processes. Some are done for you.

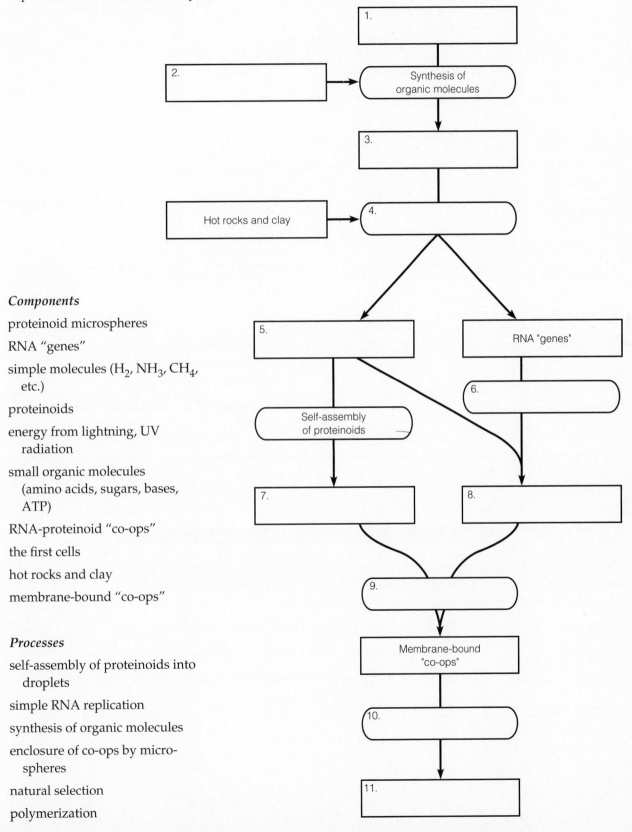

Components

proteinoid microspheres

RNA "genes"

simple molecules (H$_2$, NH$_3$, CH$_4$, etc.)

proteinoids

energy from lightning, UV radiation

small organic molecules (amino acids, sugars, bases, ATP)

RNA-proteinoid "co-ops"

the first cells

hot rocks and clay

membrane-bound "co-ops"

Processes

self-assembly of proteinoids into droplets

simple RNA replication

synthesis of organic molecules

enclosure of co-ops by microspheres

natural selection

polymerization

Compare prokaryotes and eukaryotes by completing the following chart.

Characteristic	Prokaryotes	Eukaryotes
Cell size	1.	Larger—10 to 100 μm
Nucleus	2.	Nucleus with membrane
DNA	3.	Several separate chromosomes; much protein
Reproduction	4.	Sexual reproduction
Cell division	5.	Mitosis and meiosis
Cell surface	6.	Some have cell walls and some don't; plant cell walls made of cellulose
Origin	7.	Appeared about 1.5 billion years ago

Bacteria can be categorized according to their mode of nutrition. Identify whether each of the following bacteria is an autotroph, a heterotroph, a *chemo*heterotroph (like most bacteria), or a *photosynthetic* autotroph (as are cyanobacteria).

_____ 1. Energy source—sun; carbon source—organic compounds

_____ 2. Energy source—inorganic chemicals; carbon source—CO_2

_____ 3. Energy source—organic compounds; carbon source—organic compounds

_____ 4. Energy source—sun; carbon source—CO_2

Exercise 5 (Modules 17.9 – 17.14)

These modules review some of the shapes, characteristics, and types of bacteria. Review them by completing this crossword puzzle.

Across

3. ____ are blue-green photosynthetic bacteria.

6. Archaebacteria called ____ may be similar to the earliest prokaryotes.

10. Some bacteria produce ____ that inhibit other microbes.

11. The ____ diverged from other prokaryotes early in the history of life.

12. A ____ of cyanobacteria may indicate polluted water.

13. Spherical bacteria are called ____.

14. Archaebacteria live in extreme environments such as hot ____.

17. Autotrophic bacteria get energy from ____ or inorganic chemicals.

19. ____ is thought to be the most ancient metabolic pathway.

20. Vibrios, spirilla, and spirochetes are curved or ____-shaped.

Down

1. Cocci that occur in chains are called ____.

2. Many bacteria produce dormant, resistant cells called ____.

4. The first prokaryotes probably got energy from ____ from their environment.

5. Cyanobacteria formed layered ____ in the ancient seas.

7. ____ obtain carbon from organic compounds.

8. ____ make their own organic compounds from inorganic sources.

9. All prokaryotes except archaebacteria are called ____.

12. Rod-shaped bacteria are called ____.

15. Chemoheterotrophs obtain ____ and carbon from organic compounds.

16. Many bacteria can move by means of propeller-like ____.

18. ____ enable bacteria to stick to surfaces and each other.

Exercise 6   (Modules 17.15 – 17.20)

These modules outline some of the important roles of bacteria. To review them, fill in the blanks in the story below.

"I wish I could get rid of this cold," Matt said, as he wiped his red and swollen nose. "Bacteria are pests that evolved purely to make my life miserable. I wish they'd all disappear."

Alex looked up from his Psychology textbook. "Colds are caused by viruses, not bacteria," he said. "And if bacteria disappeared, we'd be in deep trouble. We couldn't live on this planet if it weren't for bacteria. In fact, one of your ancestors was a bacterium."

"Really?"

"The first living things on Earth were [1]_____, the simple kinds of cells that we call bacteria. Bacteria eventually gave rise to creatures with more complex cells—[2]_____. That includes everything from algae to us. Incidentally, they changed the entire Earth. For example, they produced the [3]_____ in the atmosphere."

Matt sniffled. "Maybe they don't cause colds, but they do make people sick."

"Yes, there are many [4]_____, or disease-causing, bacteria," Alex replied. "About [5]_____% of all human diseases are caused by bacteria."

"So how do they make you sick?"

"A few bacteria actually destroy [6]_____, but most cause illness by producing [7]_____ . *Staphylococcus aureus*, for example, is a common skin bacterium, but it secretes substances called [8]_____ that can cause food poisoning or toxic shock in the body. Some bacteria have poisons called [9]_____ in their cell walls. A species called [10]_____ causes food poisoning that way."

Matt sneezed. "This doesn't feel like a simple cold. It could be pneumonia."

"If it is pneumonia, *then* we're talking *bacteria*. But healthy adults in developed countries don't get many bacterial infections these days. Sanitation measures, like [11]_____ and [12]_____ systems, keep many bacterial diseases from spreading, and . . . "

"You don't get pneumonia from dirty water, Mr. Know-It-All," Matt interjected.

"Right, but most forms of pneumonia, like most bacterial diseases, can be treated with [13]_____. So you don't have to worry."

"Say, since you know so much about this, maybe you can tell me how they know that a particular bacterium causes a particular disease. How do microbiologists know that a particular germ causes galloping pneumonia, or food poisoning, or whatever?"

"They use a procedure, a sort of set of rules, called Koch's [14]_____, to prove that a particular bacterium is the cause of a disease. First you have to show that the same bacterium is present in every animal that has the disease. Then you try to isolate the bacterium from an animal and grow it in a [15]_____. Then you

[16]_____ experimental hosts with the pure bacteria to see whether they get the disease. Finally, you check to see whether the suspected pathogen is present in the [17]_____. If it is, you can be pretty sure it causes the disease."

Matt changed direction. "Bacteria cause disease all right. That's why I said they should be wiped off the face of the Earth."

Alex said, "We couldn't live without them. Many bacteria are involved in the cycling and recycling of various [18]_____ between living things and the non-living environment. Like the bacteria that live in the roots of [19]_____ —plants like beans and peas—that take [20]_____ from the air and convert it into forms that plants can use."

"How does that help me?"

"It only helps you if you eat plants or animals that eat plants. I think you do, along with just about every other animal on this planet. By the way, [21]_____ that live in soil and water also trap nitrogen, along with providing some of the oxygen we breathe."

Matt brightened. "I think I understand. There is another kind of recycling that bacteria do. Bacteria decompose [22]_____ matter, like [23]_____ and [24]_____, and return chemicals to the environment. Then living things can use them again. If it weren't for bacteria, in no time we'd be up to our necks in . . . "

"Sewage. That's why bacteria are important in sewage [25]_____ plants. [26]_____ bacteria work on solid matter, called [27]_____, and [28]_____ bacteria break down liquid wastes. Some bacteria can even break down oil, and some are used to extract toxic metals, like [29]_____, from old mining sites, so that we can clean them up."

Matt sneezed. "I wish I could get rid of my toxic wastes. This is no ordinary cold. Maybe it's the flu. Maybe it's Lemon disease."

Alex sighed. "You mean Lyme disease? Been bitten by a [30]_____ lately?"

Exercise 7   (Module 17.21)

This module explains how the complex organelles of eukaryotic cells may have arisen through (1) infolding of an ancestral prokaryote's plasma membrane and/or (2) engulfing of smaller prokaryotes by an ancestral prokaryotic cell (the endosymbiotic hypothesis). Review it by answering the folowing questions.

1. Name two organelles that probably arose via endosymbiosis.

2. What is the evidence for the endosymbiotic hypothesis?

3. What is the alternative hypothesis to the endosymbiotic hypothesis?

4. Name two organelles probably derived from the host cell.

5. Is it possible that both hypotheses could be partially correct? Explain.

These modules introduce the protists, simple eukaryotes. The diagram below shows how the various groups of protists are related and the characteristics that set them apart from each other. Review them by writing the missing names or characteristics in the spaces.

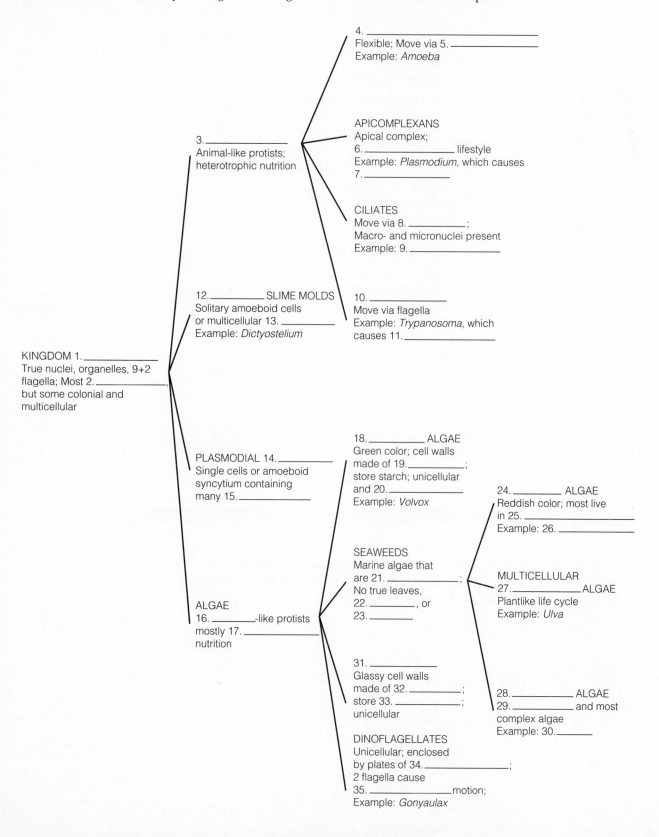

4. _____
Flexible; Move via 5. _____
Example: *Amoeba*

APICOMPLEXANS
Apical complex;
6. _____ lifestyle
Example: *Plasmodium*, which causes
7. _____

CILIATES
Move via 8. _____;
Macro- and micronuclei present
Example: 9. _____

3. _____
Animal-like protists;
heterotrophic nutrition

10. _____
Move via flagella
Example: *Trypanosoma*, which
causes 11. _____

12. _____ SLIME MOLDS
Solitary amoeboid cells
or multicellular 13. _____
Example: *Dictyostelium*

KINGDOM 1. _____
True nuclei, organelles, 9+2
flagella; Most 2. _____,
but some colonial and
multicellular

PLASMODIAL 14. _____
Single cells or amoeboid
syncytium containing
many 15. _____

18. _____ ALGAE
Green color; cell walls
made of 19. _____;
store starch; unicellular
and 20. _____
Example: *Volvox*

24. _____ ALGAE
Reddish color; most live
in 25. _____
Example: 26. _____

ALGAE
16. _____-like protists
mostly 17. _____
nutrition

SEAWEEDS
Marine algae that
are 21. _____;
No true leaves,
22. _____, or
23. _____

MULTICELLULAR
27. _____ ALGAE
Plantlike life cycle
Example: *Ulva*

31. _____
Glassy cell walls
made of 32. _____;
store 33. _____;
unicellular

28. _____ ALGAE
29. _____ and most
complex algae
Example: 30. _____

DINOFLAGELLATES
Unicellular; enclosed
by plates of 34. _____;
2 flagella cause
35. _____ motion;
Example: *Gonyaulax*

Exercise 9 (Modules 17.26 – 17.28)

Multicellular life—seaweeds, plants, animals, and fungi—probably evolved from colonial protists. Test your understanding of this process by completing each sentence on the right with a word on the left. Each answer is used only once.

A. Flagella

B. Animals

C. Protists

D. Unicellular organisms

E. Fungi

F. Colonies

G. Multicellular organisms

H. Gametes

1. Cell specialization involved separation of somatic cells and _____.
2. The links between unicellular and multicellular life were _____.
3. The first organisms on land were early plants and associated _____.
4. Among multicellular organisms, only fungi lack _____.
5. Multicellularity probably evolved several times among _____.
6. All life functions happen in one cell in all _____.
7. Different cells do different jobs in _____.
8. The oldest fossils of multicellular organisms include early _____.

Testing Your Knowledge

Multiple Choice

1. Which of the following was probably *not* present in large amounts in the atmosphere at the time life is thought to have originated?
   a. water (H_2O)
   b. methane (CH_4)
   c. ammonia (NH_3)
   d. oxygen (O_2)
   e. carbon dioxide (CO_2)

2. Which of the following appear to be most closely related to the first organisms?
   a. archaebacteria
   b. amoebas
   c. green algae
   d. eubacteria
   e. cyanobacteria

3. Biologists are interested in the role of clay in the origin of life. They think clay might have
   a. supplied the raw materials for organic compounds.
   b. catalyzed the formation of organic polymers such as proteins and RNA.
   c. formed primitive cell membranes that could grow and divide.
   d. catalyzed the formation of monomers such as amino acids and sugars.
   e. supplied the energy for metabolism in the first simple cells.

4. Which of the following is thought to have been the first step in the origin of life?
   a. cooperation among molecules
   b. formation of proteinoid microspheres
   c. formation of organic monomers
   d. replication of primitive "genes"
   e. formation of organic polymers

5. *E. coli* bacteria, which live in human intestines, are shaped like tiny, straight sausages. They are
   a. bacilli.
   b. vibrios.
   c. spirochetes.
   d. cocci.
   e. spirilla.

6. Which of the following is *not* a difference between prokaryotes and eukaryotes?
   a. Eukaryotes carry out mitosis and meiosis, and prokaryotes do not.
   b. Most of a bacterium's DNA is in a single ring; eukaryotes have multiple chromosomes.
   c. Nearly all eukaryotes have cell walls, and prokaryotes do not.
   d. Prokaryotes do not have sexual reproduction as eukaryotes do.
   e. Eukaryotic cells have nuclei and prokaryotic cells do not.

7. Most bacteria
   a. obtain energy from sunlight and carbon from organic compounds.
   b. obtain both energy and carbon from inorganic compounds.
   c. obtain energy from inorganic compounds and carbon from CO_2.
   d. obtain energy from sunlight and carbon from CO_2.
   e. obtain both energy and carbon from organic compounds.

8. Which of the following is true of cyanobacteria?
   a. They are pathogenic.
   b. They are chemoheterotrophs.
   c. They are archaebacteria.
   d. They are protists.
   e. They are autotrophs.

9. Energy metabolism, starting with glycolysis, may have evolved when early prokaryotes
   a. used up the oxygen in their environment.
   b. began producing poisonous waste products.
   c. used up the ATP in their environment.
   d. developed the ability to move through their environment via flagella.
   e. began producing significant amounts of oxygen.

10. Koch's postulates
    a. outline the probable process by which life originated.
    b. are used to demonstrate that a microorganism causes a disease.
    c. enable researchers to determine whether an organism is a bacterium.
    d. suggest how eukaryotes evolved from prokaryotes.
    e. are public health standards used to prevent the spread of disease.

11. Different groups of seaweeds can generally be distinguished on the basis of
    a. color.
    b. size.
    c. whether they are multicellular or unicellular.
    d. whether or not they have true leaves, stems, and roots.
    e. whether they are autotrophic or heterotrophic.

12. Which of the following are protozoans?
    a. diatoms, flagellates, amoebas, and ciliates
    b. apicomplexans, flagellates, amoebas, and ciliates
    c. amoebas, actinomycetes, ciliates, and flagellates
    d. flagellates, ciliates, cyanobacteria, and apicomplexans
    e. ciliates, diatoms, amoebas, and apicomplexans

13. How do algae and protozoans differ?
    a. Protozoans can move, and algae cannot.

b. Algae are free-living, and protozoans are parasitic.
   c. Protozoans are autotrophic, and algae are heterotrophic.
   d. Algae are photosynthetic, and protozoans are heterotrophic.
   e. Algae are prokaryotes, and protozoans are eukaryotes.

14. In the life cycle of a seaweed such as *Ulva*, a haploid spore develops into a(n)
    a. sporophyte.
    b. sperm.
    c. egg.
    d. gametophyte.
    e. zygote.

15. Until about 500 million years ago, all living things were
    a. asexual.
    b. autotrophic.
    c. aquatic.
    d. prokaryotic.
    e. unicellular.

Essay

1. If life could develop from nonliving chemicals 3 to 4 billion years ago, why can't the same thing happen at present?

2. Describe Stanley Miller's experiment that simulated conditions on the ancient Earth. How was the experiment carried out? What was its purpose? Its result?

3. Describe how molecules in the ancient seas may have cooperated to form the first "cells." What kinds of molecules may have been involved? How did they cooperate?

4. Name four diseases caused by bacteria. How do bacteria cause disease?

5. In what ways are bacteria useful and even vital to our well-being?

6. Primitive bacteria radically changed the Earth for all life that followed. Their effect on the environment is described as a "revolution." What did the bacteria do?

7. Explain the endosymbiotic hypothesis for the origin of eukaryotes.

8. How did multicellular eukaryotes—plants, animals, and fungi—probably arise?

Applying Your Knowledge

Multiple Choice

1. Which of the following discoveries would force scientists to revise their present theories regarding the origin of life on Earth?
   a. The Earth is found to be 6 billion years old, rather than 4.6 billion.
   b. Proteinoids can catalyze replication of small RNA molecules.
   c. There was a lot of oxygen gas in the atmosphere 3 to 4 billion years ago.
   d. Minerals in lava catalyze formation of polypeptides from amino acids.
   e. Lipids spontaneously form selectively permeable membranes.

2. The following are some major events in the early history of life.
   >   P. first heterotrophic bacteria
   >   Q. first genes
   >   R. first eukaryotes
   >   S. first autotrophic bacteria
   >   T. first animals

   Which answer below places these events in the correct order?
   a. PQSRT
   b. QSPTR
   c. QPSRT
   d. QSPRT
   e. SPQRT

3. Bacteria that live around deep-sea hot-water vents obtain energy by oxidizing inorganic hydrogen sulfide belched out by the vents. They use this energy to build organic molecules from carbon obtained from the carbon dioxide in the seawater. These bacteria might be described as
   a. photosynthetic heterotrophs.
   b. chemoautotrophs.
   c. photosynthetic autotrophs.
   d. chemoheterotrophs.
   e. none of the above.

4. The bacterium *Bacillus thuringensis* can withstand heating, dryness, and toxic chemicals that would kill most other bacteria. This indicates that it is probably able to form
   a. pseudopodia.
   b. endotoxins.
   c. endospores.
   d. pili.
   e. peptidoglycans.

5. In an experiment, a microbiologist put equal numbers of each of the following organisms into a flask of sterile broth consisting mostly of sugar and a few amino acids. She then placed the flask in the dark. Which of the organisms would be best able to survive and reproduce in this environment?
   a. chemoheterotrophic bacteria
   b. cyanobacteria
   c. diatoms
   d. thermoacidophile bacteria
   e. green algae

6. Which of the following is *not* evidence that supports the endosymbiotic theory of the origin of eukaryotes?
   a. Chloroplasts have their own DNA.
   b. The inner membrane of a chloroplast is similar to prokaryotic membranes.
   c. Mitochondria and chloroplasts are surrounded by two membranes.
   d. Mitochondria reproduce by binary fission.
   e. The DNA in the eukaryotic nucleus codes for some enzymes in mitochondria.

7. As she peered through the microscope, Paige said, "I know that this thing is supposed to be either a ciliate, a flagellate, or an amoeba, but I can't figure out which." Michelle replied, "That's easy . . . "
   a. "Watch how it moves."
   b. "How big is it?"
   c. "All you have to do is see whether it has a nucleus or not."
   d. "Watch and see what it eats."
   e. "Look at its chloroplasts."

8. Which of the following was probably *not* a direct evolutionary ancestor of a maple tree?
   a. a heterotrophic bacterium
   b. green algae
   c. an amoeba
   d. a colonial protist
   e. *All* of the above probably *were* ancestors of a maple tree.

9. Most biologists agree that seaweeds are protists, even though most other protists are microscopic unicellular creatures. Some biologists think that at least some seaweeds should be considered plants, not protists. Which of the following would support their position?
    a. Certain seaweeds have been found to be heterotrophic.
    b. Certain seaweeds contain several kinds of specialized cells.
    c. Certain seaweeds undergo sexual and asexual reproduction.
    d. Certain seaweeds are found to be prokaryotic.
    e. Certain seaweeds have cellulose cell walls and chlorophyll *a*.

10. Protozoans called choanoflagellates live in small clusters. They look very much like choanocytes, special feeding cells found in sponges, which are simple animals. Why might biologists find choanocytes of great evolutionary interest?
    a. They show how the very first organisms might have lived.
    b. They might show how the first heterotrophs lived.
    c. They might offer clues about the origin of multicellular organisms.
    d. They suggest what the first eukaryotes might have been like.
    e. They might offer clues about the first organisms to live in the sea.

Essay

1. No single simulation of conditions on the early Earth has produced all the fundamental building blocks of life. Under certain conditions, certain amino acids form. With different amounts of atmospheric chemicals, other amino acids are produced. Still other experiments generate nucleic acid bases, or sugars, or ATP. Do you think the fact that no one experiment has produced all these molecules is a problem for theories about the origin of life? Explain.

2. Scientists have found traces of amino acids on meteorites from space that have impacted on the Earth. How would you suppose they interpret this finding? What might be a source of error? Why do you suppose they think amino acids from space are significant?

3. Ultraviolet light breaks chemical bonds in large organic molecules. (That is why it is dangerous to unprotected skin.) Does this fact lend support to hypotheses about the possible role of UV radiation in the origin of life, or does it present problems for these hypotheses? Explain.

4. The diphtheria bacterium *Corynebacterium diphtheriae* grows into a mass at the back of the throat and can kill its victim by suffocation. But diphtheria victims also suffer from nervous tremors, paralysis, and heart failure. How might a bacterium that grows in the throat cause symptoms in other parts of the body?

5. Some bacteria can reproduce as often as every 20 minutes. If a hundred of these bacteria were placed in a large flask of culture medium, how many would there be after 6 hours?

Using Your Knowledge

1. Diseases such as the Black Plague, typhus, malaria, and African sleeping sickness have shaped history, politics, and geography. Can you think of other diseases that have helped to shape the modern world? There are many excellent books on this subject. If you are interested in this, you might want to explore the library.

2. When the average person hears the word "bacteria," do the beneficial effects of bacteria in ecology and human life come to mind, or does the person tend to think of harmful disease-causing organisms? You might want to conduct an informal poll.

Plants, Fungi, and the Colonization of Land **18**

Many people tend to take plants for granted. It is fun to run barefoot across a cool lawn or to kick through colorful leaves on a chilly autumn day. It is enjoyable to eat a crisp salad or a juicy red apple. In each of these instances, we appreciate plants for their textures, colors, and flavors. We seldom think of them from a biological point of view—growing, reproducing, adapting, evolving. But if plants are ignored, most of the time fungi are absolutely reviled. Fungi are all around us, but most of the time we fail to notice them, unless a piece of bread tastes a little strange, or an orange turns blue, or athlete's foot makes our toes itch. Aside from an occasional pretty mushroom, most people regard fungi with a sort of wary disgust. But fungi have their own way of living and reproducing, and our lives depend on them in several ways. This chapter examines plants and fungi in detail.

Organizing Your Knowledge

Exercise 1   (Modules 18.1 – 18.2)

Plants evolved from algae, so the two kinds of organisms share many characteristics. But plants have unique adaptations for life on land that make them different from algae. Complete the chart below by listing the differences between plants and algae. Consider such things as support, nutrition, transport, and reproduction. An example is done for you. On the next page, list the similarities between plants and algae.

Differences

Algae	*Plants*
1. Supported by water	1. Leaves and stems contain rigid supporting elements
2.	2.
3.	3.
4.	4.
5.	5.
6.	6.
7.	7.
8.	8.

Similarities

1.

2.

3.

4.

5.

Exercise 2 (Module 18.3)

This module summarizes the four major steps in plant evolution that gave rise to four major modern groups of plants. Each of the statements below describes one of those steps. Number the steps in order and fill in the names of the plant groups.

_____ A. The first plants that produced seeds arose. The modern seed plants are _____ and _____.

_____ B. Plants evolved from ancestral algae; one lineage gave rise to the _____, plants that lack vascular tissues.

_____ C. The first flowering plants (_____) appeared.

_____ D. Vascular plants evolved that had roots and strong stems supported by rigid vascular tissues, unlike the _____. Their modern-day representatives are the _____ and the seed plants.

Exercise 3 (Modules 18.4 – 18.5)

Study the diagrams in these modules to review alternation of generations in the plant life cycle. Then fill in the blanks below to complete the description of the moss life cycle.

The moss life cycle, like that of all plants, is characterized by alternation of generations. Diploid individuals called [1]_____ produce [2]_____ plants called gametophytes, which in turn produce [3]_____ sporophytes. Since it's a cycle, we could start at any point, but let's start with a spore. A haploid moss spore grows into the haploid [4]_____ plant, the leafy, cushiony growth we see on rocks or logs in a forest or bog. [5]_____ (eggs and sperm) develop in the protection of special organs called [6]_____ that are part of the gametophytes. Moss [7]_____ have [8]_____ that enable them to swim to the eggs, given a film of moisture produced by dew or raindrops. [9]_____, the fusion of egg and sperm, produces a diploid [10]_____, which remains protected in the female gametophyte. The zygote divides by mitosis and develops into the sporophyte, which consists of a [11]_____ attached to the gametophyte by a slender stalk. Within the sporangium, haploid [12]_____ are produced by the process of [13]_____. When these spores are mature, the sporangium opens and they scatter in the wind, to begin the cycle anew.

Exercise 4   (Module 18.6)

After reading the module and studying Figure 18.6, review the fern life cycle by labeling this diagram. Include the following: **haploid phase, diploid phase, sporophyte, gametophyte, zygote, sporangia, sperm, egg, meiosis, fertilization,** and **mitosis and development.** One answer is used twice. Color the haploid part of the life cycle yellow and the diploid part of the life cycle gray.

Exercise 5   (Modules 18.8 – 18.9)

Complete the following sentences with a structure or stage in the pine life cycle. Select your answers from this list: **seed, pine tree, ovule(s), sperm, egg(s), pollen grain(s), male cone(s), female cone(s), embryo, seed coat, zygote.** Some answers are used more than once.

1. A _____ is the diploid sporophyte generation of the pine life cycle.

2. The haploid gametophyte generation develops within the _____ and
   _____.

3. The small, soft _____ contain many sporangia. Meiosis occurs in the sporangia, producing many spores that develop into _____. These are male gametophytes.

4. Each scale in the larger, woody _____ bears two _____.
   Each of these develops as a sporangium covered by a tough integument.

5. The wind carries pollen grains to the female cones. Pollination occurs when a pollen grain lands on and enters an _____.

6. After pollination, meiosis occurs in the _____, producing a haploid spore that develops into the female gametophyte.

7. Over a period of months _____ are produced by the female gametophyte in the ovule. At the same time, the male gametophyte (the pollen grain) produces _____.

8. A tiny tube grows out of the _____, releasing _____ to fertilize the _____.

9. A diploid fertilized egg, or _____, develops into a sporophyte _____. The whole ovule becomes a _____.

10. The _____ consists of the _____ and a food supply made from the remains of the female gametophyte, covered by a seed coat made from the ovule's integument.

11. The seed falls on the ground. When conditions are right, the seed germinates, and the embryo, over decades, grows into a _____, the adult sporophyte. It then produces cones, and the cycle begins again.

Exercise 6   (Modules 18.11 – 18.13)

Flowers are responsible for the diversity and success of angiosperms—the flowering plants. After studying Modules 18.12 and 18.13, review the flower by matching each flower part with its function.

A. Petal

B. Anther

C. Sepal

D. Ovary

E. Stamen

F. Carpel

G. Stigma

_____ 1. Eggs develop in this chamber

_____ 2. Male reproductive structure

_____ 3. Produces pollen

_____ 4. Attracts pollinators

_____ 5. Female structure with ovary at its base

_____ 6. Protects the flower before it opens

_____ 7. Sticky tip that traps pollen

Exercise 7   (Module 18.13)

Review the life cycle of an angiosperm—a flowering plant—by labeling the diagram below. Include **haploid phase, diploid phase, sporophyte, anther, meiosis, ovary, ovule, female gametophyte, pollen grain (male gametophyte), egg, stigma, pollen tube, sperm, zygote, seed, seed coat, fertilization, embryo, fruit, food supply, pollination.** Then color the haploid part of the life cycle yellow and the diploid part gray.

Exercise 8   (Modules 18.9 – 18.15)

The life cycles of gymnosperms and angiosperms are similar, but angiosperms have added their own tricks of pollination and seed dispersal. Compare gymnosperms and angiosperms in the chart below.

Characteristic	*Gymnosperms (Conifers)*	*Angiosperms*
Characteristic reproductive structures	1.	2.
Mode of pollination (wind, insects, etc.)	3.	4.
Interval between pollination and fertilization	5.	6.
Length of life cycle (seed—adult—seed)	7.	8.
Seed protection and dispersal	9.	10.
Leaf shape	11.	12.
Thickness and strength of vascular tissues	13.	14.

Review your knowledge of the structure, life cycles, and evolution of the major plant groups. Match each statement with a group (or groups) of plants. Some statements require more than one answer.

M. Mosses
 F. Ferns
 G. Gymnosperms
 A. Angiosperms

_____ 1. Flowering plants

_____ 2. Two types of plants that produce seeds

_____ 3. These plants and their relatives formed coal deposits

_____ 4. The simplest vascular plants

_____ 5. Pines, firs, spruces, and cedars

_____ 6. A type of plant in which the gametophyte stage is dominant

_____ 7. Plants that produce fruits

_____ 8. Nonvascular plants

_____ 9. Two types of plants with flagellated swimming sperm

_____ 10. Conifers

_____ 11. Roses, apples, maples, and daisies

_____ 12. Plants with horizontal stems and leaves bearing sporangia

_____ 13. Plants with the shortest gametophyte and longest sporophyte stages

_____ 14. Two types of plants whose spores develop into pollen and ovules

_____ 15. The group that first developed good roots and rigid stems

_____ 16. Source of most lumber and wood pulp

_____ 17. The simplest plants

_____ 18. Plants that produce seeds but not fruits

_____ 19. The majority of modern plants

_____ 20. Two types of plants without seeds

_____ 21. Source of most of our food

_____ 22. Many of these plants depend on animals for pollination and seed dispersal

Exercise 10 (Modules 18.16 – 18.17 and 18.19 – 18.21)

These modules introduce fungi and describe their characteristics and importance. After reading the modules, test your knowledge of fungi by completing this crossword puzzle.

Across

1. A fungus consists of filaments called ____.
3. Fungi are responsible for the flavors of many ____.
5. ____ is caused by a human fungal parasite.
6. Lichens are very sensitive to air ____.
7. ____ are the most highly prized edible fungi.
10. ____ are unicellular fungi.
11. Yeasts are used in baking and ____.
13. Fungi cannot move, but they ____ rapidly.
14. The word ____ describes the relationship between fungus and algae in a lichen.
17. Fungi are ____ eukaryotes.
20. A parasitic fungus causes ____ elm disease.
21. A lichen consists of a fungus and green ___ or cyanobacteria.
23. Molds and ____ are typical fungi.
24. A mushroom is a fungal ____ structure.

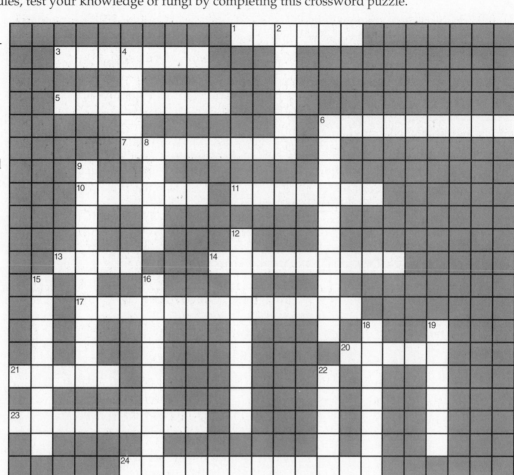

Down

2. Fungi probably moved onto land at the same time as ____.
4. People have been poisoned by ____-infested grain.
6. Fungi produce ____ and other antibiotics.
8. Parasitic fungi called smuts and ____ attack crops.
9. A root and fungus can form a ____, a mutually beneficial partnership.
12. A mycorrhizal fungus helps a plant obtain ____.
15. A network of fungal hyphae make up a ____.
16. Some fungi are parasites, some are mutualists, and some ____ organic matter.
18. Fungi digest their food ___ their bodies.
19. Fungal cell walls are made of ____.
22. Fungi are classified in their own kingdom—Kingdom ____.

Exercise 11   (Module 18.18)

Study the fungus life cycle, then complete the paragraph by filling in the blanks.

The life cycle of a fungus consists of ^{1}_____ distinct phases. Under suitable conditions, haploid ^{2}_____ germinate and produce long filaments called ^{3}_____. These strands make up a network called a ^{4}_____. The haploid mycelium grows through the substrate, secreting enzymes, digesting organic matter, and absorbing nutrient molecules. There are different kinds of mycelia, called ^{5}_____. If compatible mating types come into contact, their hyphae fuse, and the hybrid hyphae form their own mycelium. The ^{6}_____ within these hybrids do not fuse, however. This begins the ^{7}_____ phase of the fungus life cycle. Each cell in this mycelium contains two genetically distinct nuclei. Eventually, the dikaryotic mycelium forms a ^{8}_____—the familiar mushroom. Within the mushroom, haploid nuclei finally fuse, forming diploid ^{9}_____. Without undergoing mitosis, each zygote undergoes ^{10}_____, forming haploid spores. These spores, produced by the billions, are dispersed by wind, water, and animals to places where they can germinate and grow into new hyphae.

Testing Your Knowledge

Multiple Choice

1. Which of the following is *not* a difference between algae and plants?
   a. Plant cells have rigid cellulose walls, and algae cells do not.
   b. Plant gametes and zygotes are protected in moist chambers, and those of algae are not.
   c. Algae lack discrete organs—leaves, stems, roots—characteristic of plants.
   d. Plants have xylem and phloem, and algae do not.
   e. Plants have a waxy, waterproof cuticle, and algae do not.

2. Bryophytes such as mosses differ from all other plants in that they
   a. do not produce flowers.
   b. have cones but no seeds.
   c. have flagellated sperm.
   d. lack vascular tissues.
   e. produce spores.

3. The gametophyte stage of the plant life cycle is most conspicuous in
   a. ferns.
   b. mosses.
   c. angiosperms.
   d. gymnosperms.
   e. seed plants.

4. Ferns and mosses are mostly limited to moist environments because
   a. their pollen is carried by water.
   b. they lack a cuticle and stomata.
   c. they lack vascular tissues.
   d. they have flagellated sperm.
   e. their seeds do not store much water.

5. The diploid generation of the plant life cycle always
   a. produces spores.
   b. is called the gametophyte.
   c. is larger and more conspicuous than the haploid stage.
   d. develops from a spore.
   e. produces eggs and sperm.

6. During the Carboniferous period, forests consisting mainly of ____ produced vast quantities of organic matter, which was buried and later turned into coal.
   a. early angiosperms
   b. ferns and other seedless plants
   c. giant mosses
   d. gymnosperms
   e. gymnosperms and early angiosperms

7. Which of the following best describes how fertilization occurs in a conifer?
   a. A sperm cell swims through a film of moisture to fertilize the egg.
   b. A pollen grain carried by wind fertilizes the egg.
   c. A pollen grain carried by wind produces a sperm that fertilizes the egg.
   d. A sperm cell carried by wind fertilizes the egg.
   e. A pollen grain swims through a film of moisture to fertilize the egg.

8. Most species of plants are
   a. non-seed-bearing plants.
   b. angiosperms.
   c. gymnosperms.
   d. plants other than angiosperms.
   e. nonvascular plants.

9. When you look at a pine or maple tree, the plant you see is
   a. a haploid sporophyte.
   b. a diploid sporophyte.
   c. a haploid gametophyte.
   d. a diploid gametophyte.
   e. none of the above.

10. In a flowering plant, meiosis occurs in the
    _____ , producing a spore that develops into a female gametophyte.
    a. fruit
    b. seed
    c. stamen
    d. anther
    e. ovary

11. A fruit is a ripened
    a. seed.
    b. pollen grain.
    c. bud.
    d. ovary.
    e. anther.

12. Fungi cannot make their own food, and they cannot move. How do they "find" things to eat?
    a. They produce huge numbers of tiny spores.
    b. They grow rapidly.
    c. They poison anything nearby.
    d. all of the above
    e. a and b only

13. The body of a fungus consists of threadlike
    _____ , which form a network called a _____ .
    a. mycelia . . . dikaryon

b. hyphae . . . gametophyte
c. mycelia . . . hypha
d. hyphae . . . mycelium
e. sporangia . . . dikaryon

14. Where and when does fertilization occur in the mushroom life cycle?
    a. underground, as a mycelium begins to spread
    b. on the surface of the ground, when a spore germinates
    c. in a mushroom, when nuclei of a dikaryotic cell fuse
    d. underground, when hyphae of different mating types fuse
    e. in a mushroom, when eggs and sperm meet

15. Which of the following kinds of fungi would be considered the least useful or beneficial?
    a. mycorrhizal fungus
    b. yeast
    c. rust
    d. truffle
    e. decomposer

Essay

1. Compare a plant with a multicellular green alga, paying particular attention to plant adaptations to life on land.

2. Most nonbiologists consider seaweeds and fungi to be plants. Why? Why do biologists put seaweeds, fungi, and plants in separate kingdoms?

3. What are the seed plants? What adaptations have made them so successful?

4. What kinds of products do we obtain from forests? What kinds of trees supply most of our forest product needs? How can we obtain the forest products we need and still preserve our forests?

5. Sketch a flower, name its major parts, and describe their functions.

6. How do animals assist in angiosperm reproduction? How have the structures of angiosperms adapted to reflect this relationship with animals?

7. How do fungi obtain their food?

8. What two components make up a lichen? What are their roles?

Applying Your Knowledge

Multiple Choice

1. An explorer found a plant that had roots, stems, and leaves. It had no flowers but produced seeds. This plant sounds like a(n)
   a. fern.
   b. bryophyte.
   c. angiosperm.
   d. moss.
   e. gymnosperm.

2. Which of the following stages in the life of a maple tree corresponds to the leafy, spongy plant in the moss life cycle?
   a. egg and sperm
   b. adult tree
   c. flower
   d. pollen grain and ovule
   e. zygote

3. Deep in the tropical rain forest, a botanist discovered an unusual plant with vascular tissues, stomata, a cuticle, flagellated sperm, conelike reproductive structures bearing seeds, and an alternation-of-generations life cycle. He was very excited about this discovery because it would be rather unusual for a plant to have both
   a. a cuticle and flagellated sperm.
   b. vascular tissues and alternation of generations.
   c. seeds and flagellated sperm.
   d. alternation of generations and seeds.
   e. cones and vascular tissues.

4. The pinyon pine lives in near-desert areas in western North America. This habitat is a bit unusual for gymnosperms because they
   a. have a long life cycle for such a short growing season.
   b. possess flagellated sperm that must swim to the egg.
   c. produce extremely small quantities of pollen.
   d. lack vascular tissues and are unable to transport much water.
   e. produce cones rather than drought-resistant seeds.

5. Unlike most angiosperms, grasses are pollinated by wind. As a consequence, some "unnecessary" parts of grass flowers have almost disappeared. Which of the following parts would you expect to be most reduced in a grass flower?
   a. ovaries
   b. petals
   c. anthers
   d. carpels
   e. stamens

6. Fuchsia flowers are generally reddish, they hang downward, and their nectar is located deep in floral tubes. Fuchsias are typically pollinated by
   a. bees.
   b. flies.
   c. bats.
   d. butterflies.
   e. birds.

7. Some scum was found growing near the edge of a pond. Under a microscope, each of its cells were found to contain two nuclei. This means the scum must be
   a. some kind of alga.
   b. a fungus.
   c. a plant gametophyte.
   d. a liverwort.
   e. a plant sporophyte.

8. Strolling through the woods, you would be least likely to notice which of the following?
   a. a moss gametophyte
   b. a fern gametophyte
   c. an angiosperm sporophyte
   d. a fern sporophyte
   e. the dikaryotic stage of a fungus

9. The diploid phase of the life cycle is shortest in which of the following?
   a. moss
   b. angiosperm
   c. fungus
   d. fern
   e. gymnosperm

10. Which of the following is a difference between plants and fungi?
    a. Plants have diploid and haploid phases.
    b. Fungi have cell walls.
    c. Fungi are autotrophic.
    d. In fungi, zygotes undergo meiosis to produce spores.
    e. Plants undergo sexual reproduction.

Essay

1. What characteristics does a fern share with the seed plants that evolved later? In what ways are ferns similar to mosses?

2. What kinds of plants are dispersed to new habitats by spores? By seeds? What are the advantages and disadvantages of each means of dispersal?

3. When his orchard was attacked by parasitic fungi, a farmer sprayed the trees with a powerful fungicide. The next season, most of the trees were free of the parasite, but they grew poorly and produced even less fruit than they had when they were infected. What might account for this change?

4. If you have ever had athlete's foot, you are probably aware that fungal infections are rather difficult to get rid of. What do you know about the structure and lifestyle of fungi that might make them particularly persistent pests?

5. Biologists give each species a Latin name, such as *Homo sapiens* for humans and *Acer macrophyllum* for the big-leaf maple. Why do you suppose this might be difficult for lichens?

Using Your Knowledge

1. You might enjoy taking a walk through a nearby garden, park, campus, forested area, or even a backyard, and looking for examples of some of the plants and plant structures mentioned in this chapter. For example, look for mosses (gametophytes and sporophytes), mushrooms (fruiting bodies, spores, and mycelia), ferns (sporophytes, sporangia, and spores), conifers (male and female cones, pollen, seeds), and angiosperms (flowers and their parts, pollen, seeds, and fruits).

2. Are there any rare or endangered species of plants in your area? The Nature Conservancy buys property to set aside as preserves for endangered species. Your state probably has a native plant society concerned with preserving native plants and their habitats. If you would like to learn more about plant conservation or see some of these plants you can contact local or state offices of these organizations.

The Evolution of Animal Diversity

Human beings are animals, and our species evolved in an environment filled with animal life. Even now, when many humans live in cities, other species of animals affect our lives in many ways. We eat some animals, such as clams, chickens, and cattle, and make useful products from their shells, feathers, hides, and bones. Bees make honey, and their cousins the mosquitoes make us miserable. Animals still work for humans in many parts of the world, tilling fields and carrying heavy loads. Worms enrich the soil, and coral animals build reefs that protect harbors. We choose some animals to live with us as pets. We marvel at the beauty of tropical fish and rainforest butterflies and the size and strength of whales and crocodiles. Only in the last century have humans begun to appreciate the evolutionary relationships among all these fascinating animals and our own place in this story. The evolution of animal life is the subject of this chapter.

Organizing Your Knowledge

Exercise 1   (Module 19.1)

This module describes how we "draw the line" between animals and other organisms. The five kingdoms of life are illustrated below. Use colored pens or pencils to draw lines separating the kingdoms on the basis of the following characteristics.

1. Draw a black line between organisms that are prokaryotes and those that are eukaryotes. Write "prokaryotes" on one side of the line and "eukaryotes" on the other.

2. Draw a blue line between organisms that are primarily unicellular and those that are multicellular. Label these groups.

3. Separate heterotrophic and autotrophic organisms with a green line, and label each group.

4. Draw a red line between those organisms that have cell walls and those that lack cell walls, and label each. On the animal side of the line, also note that animals ingest their food, are diploid (except for eggs and sperm), and have unique embryonic stages.

Plants   Fungi   Animals

Protists

Bacteria

Exercise 2   (Module 19.2)

Using Module 19.2 as a guide, sketch and briefly describe how animals may have evolved from colonial protists.

Exercise 3   (Modules 19.3 – 19.4)

Sponges and cnidarians are the simplest animals. Complete the following description of these two phyla by filling in the blanks.

Sponges, Phylum [1]_____, and cnidarians, Phylum [2]_____, are both simple animals. Most sponges and cnidarians live in the [3]_____, but some are found in fresh water. Many sponges and all cnidarians are characterized by [4]_____ symmetry. This means their body parts are arranged in a circle around a central axis.

Sponges are by far the simpler of the two animals. A sponge is a simple tube perforated by tiny [5]_____. The body wall consists of [6]_____ layers of cells. The outer layer functions to protect the sponge. A middle layer contains a skeleton made of flexible spongin or more rigid mineral-containing particles. The sponge's inner layer consists of cells called choanocytes bearing [7]_____, which move to create a current of water that [8]_____ the sponge through the small pores and [9]_____ through a large central opening. The choanocytes trap [10]_____ from the water and then engulf them by phagocytosis. Wandering cells called amoebocytes pick up food from the choanocytes and distribute it to other cells. They also transport [11]_____ and wastes and make the skeleton.

Unlike other animals, sponges lack [12]_____ and muscles. They also lack a [13]_____ and digest their food inside their cells. During reproduction a blastula develops into a swimming larva, but sponges skip the gastrula stage. They also can reproduce asexually by [14]_____ and have great powers of [15]_____ if fragmented.

Cnidarians—animals such as [16]_____, sea anemones, and corals—are a bit more complicated. They have a [17]_____ cavity, muscles, and a [18]_____ system that enables them to respond to stimuli and coordinate muscle action. Unlike sponges, their cells are organized into [19]_____, groups of cells adapted to perform specific functions. Unlike more complex animals, most of their activities are carried out at the tissue level, not by the organs and [20]_____ of more complex creatures.

Cnidarians are radially symmetrical and come in two shapes. A ²¹_____ is a tube with tentacles radiating from one end. It is usually fixed in place. A ²²_____ is a disk with a fringe of tentacles on the edge. ²³_____ are medusas and are able to move about in the water. Some cnidarians, such as the fresh-water form called ²⁴_____, illustrated in the text, exist only in the polyp form. Others have both medusa and polyp stages in their life cycles.

A cnidarian captures small prey and pushes it into its mouth with its ²⁵_____. Special cells called ²⁶_____ on the tentacles (characteristic only of cnidarians) sting and entangle the prey. The mouth of a polyp is on top of the body, in the center of the tentacles. A jellyfish's mouth is ²⁷_____, in the center of the umbrella. The mouth leads to a digestive sac called the ²⁸_____ cavity. Food is digested here, and fluid in the cavity circulates food particles around the body. The fluid in the cavity also keeps the flimsy body "inflated" and gives the cnidarian its shape. Because the gastrovascular cavity has only one opening, ²⁹_____ are expelled through the mouth. This arrangement is called an ³⁰_____ digestive system.

Exercise 4   (Module 19.5)

The development of bilateral symmetry was a major step in animal evolution. Compare animals having bilateral symmetry with those having radial symmetry by completing the following table.

Animals with Radial Symmetry	*Animals with Bilateral Symmetry*
Examples: some sponges, cnidarians	1.
2.	Right and left sides
3.	Dorsal and ventral surfaces
No anterior or posterior end	4.
No distinct head	5.
6.	Move actively through environment

Exercise 5   (Module 19.7)

Most animals have a body cavity, a space between the digestive tract and the body wall. (There are many advantages to having a body cavity; these are reviewed in the "Testing Your Knowledge" section.) There are two kinds of body cavities: A pseudocoelom is in direct contact with the digestive tract, and a middle tissue layer lines the body covering. A coelom is a space within the middle tissue layer, which covers the digestive tract and lines the body wall. The diagrams below show the body covering and the digestive tract of three animals. Complete the diagrams by sketching in the middle tissue layer. Label the layers and color the *body covering* blue, the *digestive tract* yellow, and the *middle tissue layer* red.

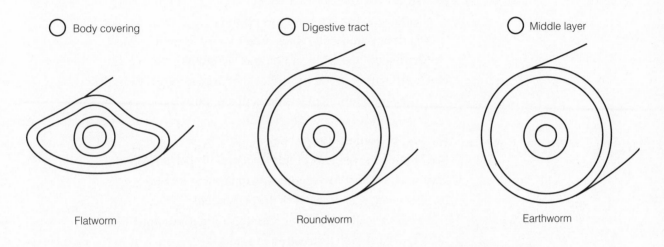

◯ Body covering   ◯ Digestive tract   ◯ Middle layer

Flatworm   Roundworm   Earthworm

Exercise 6   (Modules 19.4, 19.6, 19.8)

Review and compare the structures and lifestyles of cnidarians, flatworms, and round-worms by completing this chart.

	Cnidarians	*Flatworms*	*Roundworms*
1. Phylum name			
2. Examples			
3. Type of body symmetry			
4. Body shape(s)			
5. Body cavity			
6. Digestive tract			
7. Where they live			
8. Importance to humans			

Exercise 7   (Module 19.9)

This module discusses several of the structural and functional characteristics of mollusks. Match each of the statements on the right with a body structure on the left.

A. Coelom

B. Radula

C. Gill

D. Foot

E. Mantle

F. Circulatory system

G. Shell

_____ 1. Modified to form a lung in land snails

_____ 2. Secretes the shell

_____ 3. Used by a clam to capture food

_____ 4. Divided into hinged halves in bivalves

_____ 5. Functions in locomotion in most mollusks

_____ 6. Extracts oxygen from the water

_____ 7. Rasping organ used to scrape up food

_____ 8. Distributes nutrients, water, and oxygen around the body

_____ 9. Missing or internal in squids and octopuses

_____ 10. Outgrowth of the body surface that drapes over the animal

_____ 11. Modified to form tentacles in cephalopods

_____ 12. Small cavities around heart, kidneys, and reproductive organs

_____ 13. Long projections on the back of a sea slug

_____ 14. Rippling movements of this structure propel gastropods

_____ 15. Used by a clam for digging and anchoring in mud

_____ 16. Shoots out a jet of water to propel a squid

_____ 17. Eyes of a scallop are along the edge of this structure

_____ 18. A one-piece coiled structure in snails

Exercise 8   (Modules 19.10 – 19.11)

These two modules discuss the importance of segmentation and annelids, the segmented worms. Read the modules and review them by filling in the blanks.

The next time you dig up an earthworm, or see one wriggling on the sidewalk, pause to appreciate its beauty and complexity. Earthworms are segmented worms of the phylum [1]_____. The name, which means "ringed," refers to the repeating ringlike [2]_____ that make up the worm's body. There are three main groups of annelids. Most live in the [3]_____, but many species live in [4]_____ and moist soil.

The most distinctive external characteristic of annelids is segmentation. Internally, each segment is separated from adjacent ones by [5]_____. The [6]_____ system includes clusters of nerve cells in each segment. There are blood vessels serving each segment, and [7]_____ structures, which dispose of fluid wastes, are also repeated. The main blood vessels and the [8]_____ system are unsegmented.

What are the advantages of a segmented body? It probably is an adaptation to facilitate [9]_____. It gives the body greater [10]_____ and [11]_____. This enables an earthworm to burrow, obtaining nutrients from the soil that passes through its digestive tract. Earthworms stir up the soil, and their [12]_____ improve its texture.

The largest group of annelids are the [13]_____. Most of these worms live in the [14]_____, where they wriggle along the bottom, burrow in the mud, or construct protective [15]_____. The mobile polychaetes move by means of segmental [16]_____. In tube-dwellers, these appendages are modified for [17]_____.

The third group of annelids are the [18]_____. Some suck [19]_____, but most are free-living [20]_____ that eat small animals. Most leeches live in [21]_____, but there are some land-dwellers. Leeches have sharp [22]_____, and they secrete an anesthetic that enables them to slice painlessly through the skin and an anticoagulant that keeps blood flowing freely. The latter substance may be useful in preventing [23]_____.

Annelids are not the only segmented animals. [24]_____ are segmented; this is seen clearly in the abdomen and in the thorax, where wings and legs are repeated. Animals with backbones are also segmented. In humans, segmentation is most clearly seen in the backbone, formed from a series of bones called [25]_____, and in the abdominal [26]_____.

Exercise 9   (Modules 19.12 – 19.13)

These modules concern Phylum Arthropoda, an important group of invertebrates. Review your knowledge of arthropods by completing the crossword puzzle.

Across

1. Crabs and lobsters are ____.
4. The arthropod exoskeleton is made of ____ and protein.
7. An insect has ____ pairs of legs.
9. The ____ crab is a "living fossil" related to spiders.
10. ____ are sensory appendages on the head.
11. ____ are marine filter-feeding crustaceans.
13. Insects are the only invertebrates with ____.
15. Arthropods have ____ appendages.
17. A spider might hunt insects or catch them in a ____.
18. The ____ is an arachnid with pincers and a sting at the end of its tail.
19. ____ is shedding the old exoskeleton and growing a larger one.
20. A lobster uses its ____ for defense.
21. The ____ are the most diverse group of arthropods.
22. ____ are multilegged carnivores.

Down

2. Crabs, grasshoppers, and tarantulas are all representatives of phylum ____.
3. Scorpions, spiders, ____, and mites are all arachnids.
5. Every arthropod has a hard external skeleton called an ____.
6. An insect's body consists of head, thorax, and ____.
8. Many people are ____ to dust mites.
12. Arthropods are thought to be most closely related to ____.
14. The arthropod body consists of groups of ____.
16. An insect's wings and legs are attached to its ____.
23. ____ are wormlike plant-eaters with many short legs.

Exercise 10   (Module 19.14)

Echinoderms are a unique animal phylum. Circle the statements below that relate to echinoderms, and cross out statements that are not relevant to echinoderms.

1. Adults have radial symmetry
2. Live in salt water and fresh water
3. Larvae have radial symmetry
4. Have spines embedded under the skin
5. Most closely related to cnidarians
6. Lack segmentation
7. Bend jointed appendages to move
8. Good at regeneration
9. Have a water-vascular system

10. Move by moving tube feet
11. Adults have bilateral symmetry
12. Examples: sea urchins and sea stars
13. Live only in salt water
14. Larvae have bilateral symmetry
15. Have an endoskeleton
16. Most closely related to chordates
17. Are segmented
18. Examples: clams and snails

Exercise 11   (Module 19.15)

Like all vertebrates, we are chordates, but so are some very simple animals—the lancelets and tunicates. All chordates, from tunicates to truck drivers, share four key chordate characteristics. Label them on the drawing of a lancelet below.

Exercise 12   (Modules 19.16 – 19.22)

Vertebrates have a skull and backbone made of bone or cartilage. Most have appendages with bony supports. These modules discuss the seven major groups of vertebrates. Match each statement below with the proper group(s) of vertebrates. A few questions have more than one answer.

A. Agnathans
B. Cartilaginous fishes
C. Bony fishes
D. Amphibians
E. Reptiles
F. Birds
G. Mammals

_____ 1. The first vertebrates on land

_____ 2. Two endothermic vertebrates

_____ 3. First vertebrates with amniotic eggs

_____ 4. Have tadpole larvae

_____ 5. Gave rise to the reptiles

_____ 6. Have hair

_____ 7. Lack jaws

_____ 8. Have an operculum that helps pump water over their gills

_____ 9. Three groups that are aquatic

_____ 10. Proliferated after the dinosaurs died out

_____ 11. Trout, bass, and perch

_____ 12. The first vertebrates with legs

_____ 13. Feed their young milk

_____ 14. Two groups that evolved from reptiles

_____ 15. Sharks and rays

_____ 16. Lampreys

_____ 17. Live mostly on land but lay eggs in the water

_____ 18. Have a swim bladder that provides buoyancy

_____ 19. The young of most in this group develop inside their mothers

_____ 20. Evolved from a kind of dinosaur

_____ 21. Two groups with paired fins

_____ 22. The most primitive vertebrates

_____ 23. Placentals, marsupials, and monotremes

_____ 24. The majority of vertebrates are in this group

_____ 25. Human beings are in this group

Exercise 13   (Module 19.23 and Summary)

Module 19.23 summarizes the evolutionary relationships and characteristics of the major phyla of animals. Suppose you found an animal and wanted to know to which phylum it belonged. You could use what biologists call a "key," a series of questions that leads you to the animal's identity. Such a key is given below, in the form of a flow chart. Before you can use this key, you will need to complete it. Some questions and phylum names are given; others are missing. Using information from this chapter, complete the key. (You may then want to try it out by "keying out" an animal.)

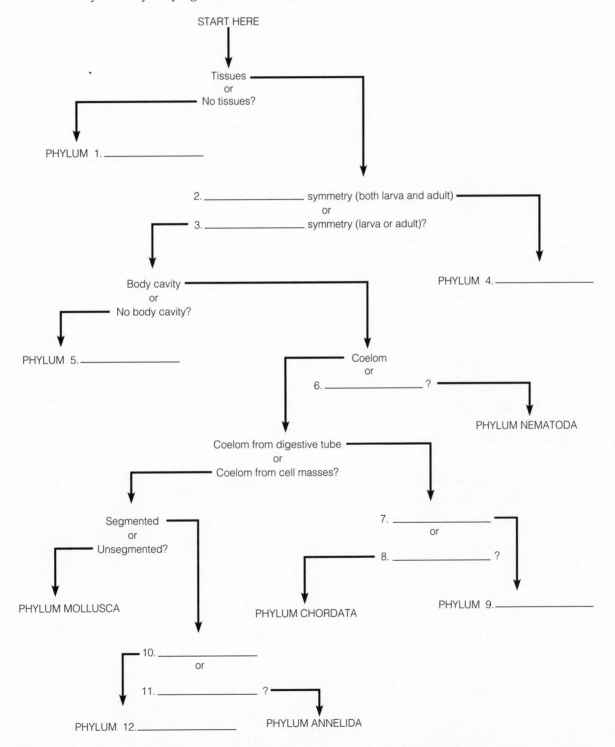

START HERE

Tissues
or
No tissues?

PHYLUM 1. _____

2. _____ symmetry (both larva and adult)
or
3. _____ symmetry (larva or adult)?

PHYLUM 4. _____

Body cavity
or
No body cavity?

PHYLUM 5. _____

Coelom
or
6. _____ ?

PHYLUM NEMATODA

Coelom from digestive tube
or
Coelom from cell masses?

Segmented
or
Unsegmented?

7. _____
or
8. _____ ?

PHYLUM MOLLUSCA

PHYLUM CHORDATA

PHYLUM 9. _____

10. _____
or
11. _____ ?

PHYLUM 12. _____

PHYLUM ANNELIDA

Testing Your Knowledge

Multiple Choice

1. You would expect to find the greatest number of phyla of animals ____ and the greatest number of species of animals ____ .
   a. on land . . . in the sea
   b. in fresh water . . . in the sea
   c. in the sea . . . on land
   d. in the sea . . . in fresh water
   e. on land . . . in fresh water

2. Which of the following is *not* a characteristic of all animals?
   a. They are multicellular.
   b. They have tissues, organs, and organ systems.
   c. They are eukaryotes.
   d. They ingest their food.
   e. They are heterotrophic.

3. Animals probably evolved from colonial protists. How do animals differ from these protist ancestors?
   a. The protists were prokaryotic.
   b. Animals have more specialized cells.
   c. The protists were heterotrophic.
   d. The protists were autotrophic.
   e. Animals are able to reproduce.

4. A ____ is the simplest animal discussed in this chapter to have ____ .
   a. sponge . . . bilateral symmetry
   b. flatworm . . . a body cavity
   c. roundworm . . . a complete digestive tract
   d. jellyfish . . . a complete digestive tract
   e. snail . . . a body cavity

5. Which of the following animals does *not* have a body cavity?
   a. flatworm
   b. ant
   c. mouse
   d. clam
   e. earthworm

6. Which of the following phyla include numerous parasites and pests?
   a. roundworms and flatworms
   b. mollusks and roundworms
   c. annelids and flatworms
   d. annelids and roundworms
   e. mollusks and flatworms

7. Which of the following animals is *not* segmented?
   a. leech
   b. snail
   c. human being
   d. lobster
   e. salmon

8. Phylum ____ includes the largest number of species.
   a. Mollusca
   b. Arthropoda
   c. Annelida
   d. Chordata
   e. Echinodermata

9. The water vascular system of a sea star functions in
   a. movement of the tube feet.
   b. circulation of nutrients around the body.
   c. pumping water for swimming movements.
   d. waste disposal.
   e. keeping all parts of the body moist at low tide.

10. A ____ is a chordate but not a vertebrate.
    a. lamprey
    b. shark
    c. lancelet
    d. sea star
    e. frog

11. How do lampreys differ from other vertebrates?
    a. They have a skeleton made of flexible cartilage.
    b. They do not have jaws.
    c. They do not have paired appendages (fins or legs).
    d. all of the above
    e. b and c only

12. The first vertebrates to live on land were
    a. agnathans.
    b. reptiles.
    c. amphibians.
    d. cartilaginous fishes.
    e. mammals.

13. There are three major groups of mammals, categorized on the basis of their
    a. size.
    b. habitat.
    c. method of locomotion.
    d. teeth and digestive system.
    e. method of reproduction.

14. Zoologists have placed chordates and echinoderms on one major branch of the animal phylogenetic tree, and mollusks, annelids, and arthropods on another major branch. Which of the following is the basis for this separation into two branches?
    a. whether or not the animals have a skeleton
    b. type of symmetry
    c. whether or not the animals have a body cavity
    d. how the body cavity is formed
    e. whether or not the animals are segmented

15. Which of the following are most numerous and successful on land?
    a. mollusks and chordates
    b. annelids and arthropods
    c. arthropods and chordates
    d. annelids and chordates
    e. mollusks and arthropods

Essay

1. Describe the characteristics that separate animals from the other kingdoms of living things.

2. Describe some of the characteristics that biologists consider important when deciding the phylum into which an animal should be classified.

3. What kinds of animals have a body cavity? What kinds lack a body cavity? Describe some of the advantages of having a body cavity.

4. Describe how the mantle, mantle cavity, and shells of snails, clams, and squids are modified for their different ways of life.

5. In terms of numbers of individuals and numbers of species, it could be argued that insects are the most successful creatures on Earth. What are some characteristics that have made them so successful?

6. The spadefoot toad of the southwestern United States is an unusual amphibian; it is capable of surviving in the desert. Few amphibians can tolerate dry desert conditions, but many reptiles—horned toads, rattlesnakes, and desert tortoises—thrive in hot, arid regions. In what ways are reptiles better adapted to life in the desert than amphibians?

7. Describe adaptations of birds for flight.

Applying Your Knowledge

Multiple Choice

1. Recent biochemical evidence indicates that cnidarians, as well as sponges, might have evolved from protists quite independently from other animals. Which of the following also suggests a separate origin?
   a. Cnidarians have no nervous tissue.
   b. Cnidarians have a gastrovascular cavity.
   c. Cnidarians are unsegmented.
   d. Cnidarians are radially symmetrical.
   e. Cnidarians have no circulatory system.

2. Which of the following includes the largest number of species?
   a. animals that are segmented
   b. animals with radial symmetry
   c. animals with a body cavity
   d. animals that are unsegmented
   e. animals with a backbone

3. Which of the following is radially symmetrical?
   a. a donut
   b. an automobile
   c. a spoon
   d. a peanut butter sandwich
   e. a wristwatch

4. A marine biologist dredged up a small animal from the bottom of the ocean. It was uniformly segmented, with short, stiff appendages and soft, flexible skin. It had a complete digestive system and a circulatory system but no skeleton. Based on this description, this animal sounds most like
   a. a lancelet.
   b. a crustacean.
   c. a mollusk.
   d. a roundworm.
   e. an annelid.

5. "Pill bugs" or "sowbugs," often found under rocks and logs in moist places, are perhaps most noticed for their ability to roll up into a ball when disturbed. Sowbugs are really crustaceans, not insects. Therefore, a sowbug does *not* have
   a. an exoskeleton.
   b. gills.
   c. 3 pairs of legs.
   d. antennae.
   e. jointed appendages.

6. Which of the following is thought to be most closely related to you?
   a. sea star
   b. snail
   c. earthworm
   d. jellyfish
   e. ant

7. There are only a few species of cartilaginous fishes, compared with the bony fishes. Cartilaginous fishes are mostly limited to a lifestyle of swimming fast in open water. Bony fishes have adapted to many different lifestyles—clinging to seaweed, hiding in crevices, even burrowing in the bottom. This could probably be attributed to the fact that bony fishes
   a. have more rigid skeletons.
   b. are smaller than cartilaginous fishes.
   c. have operculums and swim bladders.
   d. have lateral line systems and paired fins.
   e. are endothermic.

8. Which of the following is *not* thought to be in the lineage that led to human beings?
   a. an amphibian
   b. a dinosaur
   c. a jawless vertebrate
   d. a colonial protist
   e. a lobe-finned fish

9. Which of the following is *not* shared by birds and reptiles?
   a. endothermic metabolism
   b. amniotic eggs
   c. backbone of vertebrae
   d. scales made of keratin
   e. gill structures in embryo

10. Imagine that you are a paleontologist (a scientist that studies fossils of ancient life forms). In a recent dig, you unearthed bones of all of the following. Which could you have found in the oldest sediments?
    a. amphibians
    b. placental mammals
    c. dinosaurs
    d. birds
    e. marsupials

Essay

1. Sponges have no muscles and cannot move. They have no nerve cells and cannot sense the environment around them. Why are they considered animals?

2. A flattened creature called *Trichoplax*, in Phylum Parazoa (a small phylum not discussed in this chapter), is the simplest known animal. Its body consists of a simple ciliated outer layer over a core of unspecialized cells. It has no digestive tract, but it crawls over food and hunches its "back" to form a temporary hollow that serves as a digestive sac. What does this animal suggest about the early evolution of animals?

3. Imagine that you are a Peace Corps volunteer assigned to a small African village where many people are infected with pork tapeworms, which are spread from pig to pig and from pigs to people by eating infected meat. Resources are scarce; the poor villagers cannot afford expensive medicines. If these worms have life cycles like other tapeworms, suggest three ways the villagers could interrupt the worm's life cycle and prevent themselves from becoming infected.

4. Name what you consider to be a successful phylum of animals. What are your criteria for choosing these animals? What makes them successful?

5. Zoologists have found that certain marine snails and polychaete worms have similar ciliated swimming larvae. There is an animal called *Peripatus* whose segmentation, muscles, and excretory system are like those of annelids. Its respiratory system, circulatory system, and appendages are like those of arthropods. What does this evidence suggest about the evolution of annelids, mollusks, and arthropods? Is this reflected in the animal phylogenetic tree given in this chapter? Explain.

6. How do you know that a dog is a chordate? A vertebrate? Are all chordate and vertebrate characteristics seen in the adult dog? If not all are seen in the adult, what makes the dog a chordate?

7. Nearly all the land vertebrates in the Arctic and Antarctic are birds and mammals—polar bears, walruses, and penguins, for example. Why do you think there are so many birds and mammals, but virtually no reptiles or amphibians, in these regions?

Using Your Knowledge

1. There are many interesting and informative programs about animals on television, especially on PBS and the Discovery Channel. Some series that present consistently good programs are *Nature, Nova, Survival Specials, The Nature of Things, Audubon Specials,* and *National Geographic Specials.* The 12-part *Life on Earth* series traces the evolution of life and is available in some libraries and video stores.

2. Learning more about the evolution of animals makes a visit to a zoo or museum much more rewarding. Many large cities have zoos, nature centers, aquariums, and natural history museums. You might enjoy examining the invertebrates of a coral reef up close in an aquarium exhibit, comparing placentals and marsupials at the zoo and reliving the age of dinosaurs at a natural history museum.

3. If you are interested in learning about, observing, and protecting wildlife, there are several organizations you can join. Three of the largest are National Audubon Society, 950 Third Avenue, New York, NY 10022; National Wildlife Federation, 1400 16th St. NW, Washington, DC 20036; and World Wildlife Fund, 1250 24th St. NW, Washington, DC 20037.

Unifying Concepts of Animal Structure and Function

20

Take a moment to look at the skin on the back of your hand and appreciate its role in the structure and function of your body. The skin is an organ; along with hair and nails, it makes up the integumentary system. The skin's structure is well suited to its functions. Its outer layer consists of many layers of cells, dead and impermeable at the surface, living and reproducing below. The skin is tough, forming a barrier that keeps water in the body and bacteria out. It is also flexible, allowing the body to bend and move. It protects the body from changes in the external environment and at the same time actively responds to those changes. Nerve endings in the skin sense environmental variations, and sweat glands, blood vessels, and pigment cells respond to them. Other body systems interact with the skin: The digestive and respiratory systems supply it with food and oxygen, the excretory system disposes of its wastes, and the immune system protects it from infection. In exchange, the skin covers and protects these systems, playing a vital role in homeostasis—maintaining a constant internal environment suitable for continuing body functions. The skin illustrates the important connection between how the body is built and how it works. This chapter introduces the important concepts uniting animal structure and function.

Organizing Your Knowledge

Exercise 1   (Introduction and Module 20.1)

These modules discuss how animal function is correlated with structure. You are probably familiar with many other examples. In a sentence or two, state how each of the following illustrates this correspondence of structure and function. The first one is done for you.

1. A whale's tail. *The tail is flattened into broad flukes, which propel the whale through the water.*

2. A hummingbird's beak.

3. Your hand.

4. A frog's legs.

5. A mosquito's sharp, tubular mouthparts.

6. A cow's multichambered stomach.

Exercise 2   (Modules 20.2 – 20.3)

Review the hierarchy of structural levels in an animal by filling in the blanks in the following paragraphs.

The body of an animal—a cat, for example—is organized on several hierarchical levels. The smallest parts of the cat that are alive are individual [1]_____, such as the muscle fibers in the wall of the stomach. Their function is to contract and move the contents of the stomach, mixing cat food with digestive juices. Many muscle cells cooperate to form a [2]_____, the second level of body structure and function. Besides muscle, there are three other kinds of tissues that make up the cat's stomach: [3]_____ tissue, [4]_____ tissue, and [5]_____ tissue. The stomach itself, formed of these four tissues, is an [6]_____, which performs the functions of storing and digesting food. The stomach, esophagus, intestines, and digestive glands make up the digestive system, which exemplifies the [7]_____ level of structure and function. It is one of a dozen or so systems that cooperate to form the cat—an [8]_____. This is the whole animal, the highest level of the hierarchy.

To review, starting from the top down: A cat is an [9]_____. It is composed of a number of [10]_____, each of which performs specific functions such as digestion or circulation. Each system is composed of [11]_____ such as the heart or stomach, which are built from four kinds of [12]_____. At the most fundamental level of the hierarchy, a tissue is composed of individual [13]_____.

Exercise 3   (Module 20.4)

Epithelial tissue covers the body and lines body organs. Complete the following chart comparing four kinds of epithelium. Refer to the illustrations in Module 20.4 to aid you with the descriptions.

Tissue Type	Description	Body Locations	Functions
1.	2.	3.	4.
5.	6.	7.	Absorbs and secretes fluid in kidneys
8.	Single layer of thin, flattened cells	9.	10.
11.	12.	Forms mucous membranes that line digestive tract and air tubes	13.

Review the structures, functions, and locations of connective tissues by completing this crossword puzzle.

Across

2. ____ tissue contains fat.

7. Bone matrix contains fibers embedded in ____ salts.

9. ____ is a connective tissue with a fluid matrix.

10. Connective tissue is one of ____ major categories of tissues.

11. The matrix is a web of ____ in a liquid, jelly, or solid.

12. Connective tissue cells are scattered in a non-living ____.

13. Cartilage supports the nose and ears, and forms discs between the ____.

15. Adipose tissue pads and insulates the body and stores ____.

16. ____ connective tissue holds other tissues and organs in place.

Down

1. A ____ is a group of cells with a common structure and function.

3. Blood matrix is called ____.

4. Fibrous connective tissue forms ____ and ligaments.

5. ____ connective tissue has densely packed bundles of collagen fibers.

6. Some kinds of connective tissues contain rope-like ____ fibers.

8. ____ is a strong but flexible skeletal material.

14. ____ is the most rigid connective tissue.

Exercise 5   (Modules 20.6 – 20.7)

Review nervous tissue and the three types of muscle tissue by matching each phrase on the left with a tissue type on the right.

1.

_____ 1. Contractile tissue of the heart

_____ 2. Forms body communications system

_____ 3. Attached to bones by tendons

_____ 4. Branching, interconnected muscle cells

_____ 5. Cells characterized by axons and dendrites

_____ 6. Carries out voluntary body movements

_____ 7. Composed of neurons and other supporting cells

_____ 8. Cells are striated, or striped

_____ 9. Muscle cells that lack striations

_____ 10. Found in walls of digestive tract, bladder, arteries, and so on

_____ 11. Involuntary muscle of internal organs other than the heart

_____ 12. Figure 1 at the right

_____ 13. Figure 2 at the right

_____ 14. Figure 3 at the right

_____ 15. Figure 4 at the right

A. Smooth muscle
B. Cardiac muscle
C. Skeletal muscle
D. Nervous tissue

2.

3.

4.

Exercise 6   (Modules 20.8 – 20.9)

Review the functions of human organs and organ systems by filling in the blanks in the following paragraphs.

Every animal—a hamster or a human being—consists of a number of cooperating organ systems, each of which performs specific life functions. Each system consists of organs, which in turn perform specific functions. As you read these words, your brain receives nerve impulses from your eyes. It evaluates the information received and sends out responses via the spinal cord and nerves. These parts—sense organs, brain, spinal cord, and nerves—make up the [1]_____ system, one of two systems that control and coordinate body activities.

As you read, your eyes scan the page, and your hand moves to write answers in the blanks. These responses are carried out by muscles, which make up the [2]_____ system, the system responsible for body movements. To move the body, muscles pull against bones, which make up the [3]_____ system. This system also supports the body and protects delicate internal organs such as the brain.

Muscles, like all organs, require food and oxygen to function. Food is digested and absorbed by the [4]_____ system, whose parts include the stomach, intestines, and digestive glands such as the liver and pancreas. As you inhale and exhale, oxygen enters the body via the lungs, key organs of the [5]_____ system. The [6]_____ system—the heart, blood, and blood vessels—functions to transport food and oxygen to your muscles and other organs. Some of the fluid delivered to body tissues leaves the blood and is picked up by vessels of the [7]_____ system.

The fluid passes through lymph nodes, where special cells called lymphocytes attack foreign substances and microbes. The skin, or [8]_____ system, is normally able to keep most disease-causing organisms out. When they get through, they are attacked by cells of the [9]_____ system—lymphocytes, which are produced and stored in the thymus, bone marrow, and spleen, as well as lymph nodes.

As the blood delivers its cargo of food and oxygen, it also picks up waste products that must be expelled by the body. The kidneys—the key organs of the [10]_____ system—remove metabolic waste products from the blood and dispose of them via the ureters, bladder, and urethra. The kidneys also regulate the osmotic balance of blood. This activity, and many others in the body, are controlled by chemical signals called hormones. Hormones are sent out by the glands of the [11]_____ system, which acts in concert with the nervous system to coordinate the activities of all the other body systems. Endocrine glands—the pituitary, thyroid, adrenals, and others—regulate such activities as digestion, growth, metabolism, and water balance. They even help to control the process of reproduction, via their effects on the testes and ovaries, the major organs of the male and female [12]_____ systems.

Exercise 7   (Module 20.10)

Compare body imaging methods by summarizing each method on the chart.

Method	Medium Used	Used for	Disadvantages
Conventional X ray	1.	2.	3.
Computerized tomography (CT)	4.	5.	6.
Magnetic resonance imaging (MRI)	7.	8.	9.
Positron-emission tomography (PET)	10.	11.	12.

Exercise 8   (Module 20.11)

Animals must exchange materials with their environment. The cells of saclike or flattened animals (such as *Hydra* and flatworms) can exchange substances directly with the surrounding water. Larger animals, whose outer coverings are small compared with their volume, have specialized surfaces for exchanging materials with the environment. Below is a simplified diagram of an animal, similar to Figure 20.11B in the text. The circulatory system is shown. Add your own simple sketches and label the **digestive system, respiratory system, excretory system, interstitial fluid,** and **cells.** Draw and label arrows to show exchange of **food, nutrients, O$_2$, CO$_2$, metabolic wastes,** and **feces.**

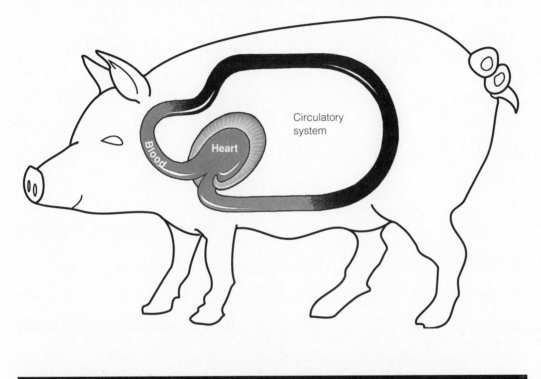

Exercise 9   (Modules 20.12 – 20.13)

To review the concepts of homeostasis and negative feedback, the most important principles of animal function, read the following paragraph and then fill in the chart.

Despite changes in the external environment, an animal can keep its internal environment remarkably constant. This maintenance of a constant internal environment is called homeostasis. The text states that animals such as the albatross maintain relatively constant salt and water balance and body temperature. Conditions within the human body are also more or less constant, fluctuating within narrow limits. For example, an organ called the pancreas monitors and regulates the amount of sugar in the blood. This maintains a constant supply of fuel for body cells, even though the body's intake of food varies widely during the day. After a meal, when blood sugar rises, the pancreas sends out a chemical signal, a hormone called insulin. Insulin causes body cells to take up and store sugar, which lowers blood sugar to the optimum range—70 to 110 mg of sugar per 100 mL of blood. This illustrates negative feedback: An increase in blood sugar triggers a response that counteracts the increase. Between meals, as cells consume sugar, the concentration of sugar in the blood starts to decrease. The pancreas responds by reducing its output of insulin and stepping up its secretion of a second hormone, glucagon. Glucagon signals certain cells to release sugar from storage, raising blood sugar to the optimum level. Thus, despite changes in the external environment (timing and content of meals), blood sugar usually fluctuates within the narrow range that is best for cells.

Compare the regulation of blood sugar described in the preceding paragraph with the control of body temperature outlined in Modules 20.12 and 20.13. Fill in the chart by identifying the components of each homeostatic control system.

	Body Temperature	*Blood Sugar*
Type of change in external environment	1.	2.
Control center	3.	4.
Stimulus	5.	6.
Kind of signal sent by control center to effector	7.	8.
Effector	9.	10.
Response	11.	12.
Set point	13.	14.

Testing Your Knowledge

Multiple Choice

1. The four major categories of tissues are
   a. bone, muscle, blood, and adipose.
   b. nervous, epithelial, connective, and muscle.
   c. muscle, epithelial, bone, and cartilage.
   d. blood, nervous, connective, and muscle.
   e. simple squamous, simple cuboidal, simple columnar, and stratified squamous.

2. Which of the following levels of structure encompasses all the others?
   a. tissue
   b. cell
   c. organ
   d. organism
   e. system

3. How many organ systems make up your body?
   a. four
   b. hundreds
   c. twelve
   d. millions
   e. It depends on the size of the person.

4. Which of the following tissues produces voluntary body movements?
   a. smooth muscle
   b. simple cuboidal epithelium
   c. cardiac muscle
   d. skeletal muscle
   e. fibrous connective tissue

5. Neurons are specialized cells characteristic of
   a. muscle tissue.
   b. nervous tissue.
   c. connective tissue.
   d. epithelial tissue.
   e. all of the above.

6. All but one of the following systems are correctly paired with one of their parts. Which pair is *incorrect*?
   a. cardiovascular system—heart
   b. respiratory system—lung
   c. endocrine system—thyroid gland
   d. integumentary system—hair
   e. excretory system—intestine

7. Which of the following do the excretory, digestive, and respiratory systems have in common?
   a. They are present only in animals with backbones.
   b. They contain specialized surfaces for exchange with the environment.
   c. They work independently, without any control by the nervous system.
   d. They enable the animal to absorb needed materials from its environment.
   e. They are isolated from the animal's internal environment.

8. A change in the body often triggers a response that counteracts the change. This kind of response is known as
   a. negative feedback.
   b. empowerment.
   c. cause and effect.
   d. positive feedback.
   e. adaptation.

9. Homeostasis is
   a. exchange of materials with the surrounding environment.
   b. the idea that all vertebrates are built in a similar way.
   c. the correlation of structure and function.
   d. maintaining a relatively constant internal environment.
   e. cooperation of body parts to form tissues, organs, and systems.

10. An animal's "internal environment" is
    a. the blood.
    b. the interior of compartments like the heart and stomach.
    c. anyplace beneath the skin.
    d. any fluid inside the body.
    e. the interstitial fluid that surrounds the cells.

11. Which of the following are listed in the correct hierarchical order?
    a. system-tissue-organ
    b. cell-tissue-organ
    c. organ-tissue-system
    d. tissue-cell-organ
    e. organism-organ-system

Essay

1. Describe how the structure of a bird wing relates to its function.

2. Name the four major types of body tissues and briefly describe the functions of each type.

3. Name the 12 organ systems of a vertebrate. Describe the function of each system in one sentence each.

4. What are the advantages of computerized tomography (CT) over conventional X rays?

5. Compare a thermostat controlling room temperature with your brain controlling body temperature. Describe the following for each system: stimulus, control center, effector, and response.

6. Some animals can "breathe" through their skins, without the aid of lungs or gills. Others can absorb food and expel wastes through the surfaces of their bodies, without specialized digestive tracts or kidneys. Why are all these animals rather small?

Applying Your Knowledge

Multiple Choice

1. Bone tissue does *not* show which of the following correlations between structure and function?
   a. It is rigid.
   b. Its cells are packed tightly together.
   c. It contains reinforcing fibers.
   d. It can grow with the animal.
   e. It contains canals for blood vessels and nerves that keep it alive.

2. When you sprain your ankle, the "straps" of tissue that hold the bones together are stretched and torn. What kind of tissue do you think is damaged in a sprain?
   a. stratified squamous epithelium
   b. visceral muscle
   c. fibrous connective tissue
   d. adipose tissue
   e. cartilage

3. Which of the following forms a thick protective barrier that keeps bacteria out of the body?
   a. skeletal muscle
   b. fibrous connective tissue
   c. stratified squamous epithelium
   d. cartilage
   e. simple columnar epithelium

4. An organ such as the heart or liver contains
   a. muscle tissue.
   b. nervous tissue.
   c. connective tissue.
   d. epithelial tissue.
   e. all of the above.

5. A new drug has been developed that impairs the movement of smooth muscle. It would affect the muscle
   a. that moves the arms and legs.
   b. of the heart.
   c. in the wall of the intestine.
   d. all of the above.
   e. b and c only.

6. Which of the following is *not* an organ?
   a. the stomach
   b. a blood vessel
   c. a neuron
   d. the heart
   e. a lung

7. A researcher wants to study the metabolic activity of various parts of exercising heart muscle. This might be accomplished by
   a. doing a CT scan.
   b. taking some X rays.
   c. doing an ultrasound scan.
   d. using MRI.
   e. doing a PET scan.

8. Which of the following best illustrates homeostasis?
   a. All the cells in the body have much the same chemical composition.
   b. Cells of the skin are constantly worn off and replaced.
   c. When blood CO_2 increases, you breathe faster and get rid of CO_2.
   d. All organs are composed of the same four kinds of tissues.
   e. The lung has a large surface for exchange of gases.

Essay

1. You have probably read books or seen nature programs on television that describe adaptations of animals to their environments. Choose an animal and briefly describe how its body shows the correlation of structure and function.

2. What might interest an anatomist about each of the following: how a fish swims, how a penguin keeps warm, how an insect defends itself from its enemies? What might interest a physiologist about each of them?

3. Read the descriptions and look at the illustrations in the text of the following tissue types, and then explain how their structure correlates with their function: bone, simple squamous epithelium, blood.

4. Briefly describe which organ systems might cooperate in delivering food and oxygen to your brain cells.

5. The parathyroid glands regulate the amount of calcium in the blood. They send out hormone signals that control how much calcium the intestine absorbs from food and how much calcium the kidneys excrete in the urine. What do you think the parathyroids cause to happen when blood calcium gets too high? What happens when blood calcium gets too low? How does this illustrate negative feedback?

Using Your Knowledge

1. Physicians often specialize in treating a particular body system. Neurologists, for example, deal primarily with the nervous system. What about dermatologists? Can you think of others? (Physicians are listed by specialty in the telephone Yellow Pages.)

2. Many everyday devices use negative feedback to maintain some constant condition. For example, there is a valve that maintains a constant water level in a toilet tank. When the toilet is flushed, the water level drops. A float drops, opening a valve, which raises the water level and the float. When the water level reaches its "set point" (when the tank is full) the rising float shuts off the water. How does this system illustrate negative feedback? Can you think of other devices that work this way?

3. Have you or someone you know ever had a CT or MRI scan? How have these devices changed health care in recent years?

Nutrition and Digestion

What was the last thing you ate? Or might you be snacking right now? What makes you feel hungry? Why do you need to eat? What is in food that your body needs? Why are some kinds of food better for you than others? What happens to food after you eat it? How is the way you eat and process food different from the eating and digestion processes of other animals? What are calories? Vitamins and minerals? Fiber? And what is the latest about sodium, saturated fat, cholesterol, and dieting? These and many other questions about nutrition and digestion are considered in this chapter.

Organizing Your Knowledge

Exercise 1   (Module 21.1)

Test your knowledge of animal diets and feeding methods by filling in this chart. Most examples are from the module; others are animals with which you are probably familiar. For diet, you may choose from omnivore, herbivore, and carnivore. There are several feeding methods: Animals can be absorptive feeders, filter feeders, suspension feeders, substrate feeders, fluid feeders, or (for want of a better name) "large piece feeders."

Animal	Diet	Feeding Method
1. Cow		
2. Earthworm		
3. Aphid		
4. Humpback whale		
5. Human being		
6. Tapeworm		
7. Shark		
8. Female mosquito		
9. Clam		
10. Grasshopper		

Exercise 2   (Modules 21.2 – 21.3)

Read about the four stages of food processing and kinds of digestive systems. Then connect the two by (1) labeling the drawings, and (2) coloring the drawings to show which digestive stages occur in each portion of these alimentary canals. On the drawings label **mouth, pharynx, esophagus, crop, gizzard, stomach, intestine,** and **anus**. Then color the alimentary canals to show where each stage in food processing occurs. Use yellow for **ingestion** (including swallowing and storage), red for **digestion**, green for **absorption**, and blue for **elimination**. (If two processes occur in the same area, mix the colors.)

Ingestion ◯   Digestion ◯   Absorption ◯   Elimination ◯

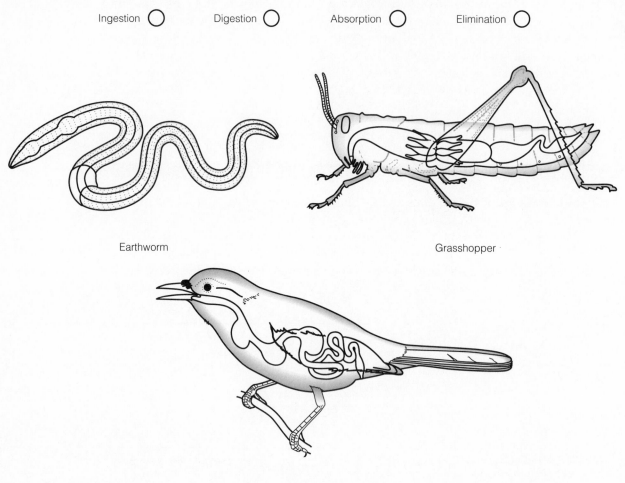

Earthworm   Grasshopper

Bird

Exercise 3   (Module 21.4)

Picture the parts of the human digestive system and their relationships to one another. Label and color these parts of the digestive system on the diagram below: **stomach, oral cavity, small intestine, esophagus, rectum, pancreas, gallbladder, mouth, large intestine, tongue, salivary glands, anus, liver, pyloric sphincter**, and **pharynx**.

Exercise 4   (Modules 21.4 – 21.7)

Identify the part of the digestive tract described in each of the following statements. Some answers occur more than once.

1. Kind of front tooth used for biting
2. Largest digestive gland
3. Muscles of esophagus that contract above a bolus
4. The opening through which wastes are eliminated
5. Where liver secretions are stored
6. A ring of muscle that controls food leaving the stomach
7. The throat
8. The digestive gland above and to the right of the stomach
9. Back teeth that grind and crush food
10. Muscles of the esophagus that contract below swallowed food
11. Where food goes after it leaves the stomach

_____ 12. Digestive gland below the stomach

_____ 13. Where taste buds are located

_____ 14. The portion of the tract that wraps around the small intestine

_____ 15. Pointed teeth that are overdeveloped in Count Dracula

_____ 16. Where the tongue and teeth are located

_____ 17. The portion of the canal between the esophagus and the small intestine

_____ 18. The technical name for a swallowed ball of food

_____ 19. Digestive glands that secrete into the oral cavity

_____ 20. Where final steps of digestion and absorption of nutrients occur

_____ 21. Where water is absorbed and feces are formed

_____ 22. A flap of cartilage that keeps food out of the trachea

_____ 23. The pyloric sphincter regulates its exit

_____ 24. The muscular tube from the pharynx to the stomach

_____ 25. The portion of the canal just before the anus

_____ 26. The mouth opens into this portion of the canal

Exercise 5 (Modules 21.7 – 21.11)

The following story will help you to visualize the movement of food through the stomach and intestine and the processes that occur there. Write the proper word in each blank.

The experimental subject has been complaining of intestinal pain, but his doctors have been unable to pinpoint its cause. Your assignment is to inspect the linings of his stomach and small intestine. You step into the Microtron and are quickly reduced to microscopic size. You enter the subject via a drink of water, and the swallowing reflex sweeps you into the [1]_____, the tube to the stomach. A wave of muscle contraction, called [2]_____, propels you forward. Ahead, the opening broadens, and you tumble into the stomach.

The subject has been on a liquid diet, so the stomach is filled with clear fluid instead of [3]_____—the normal mixture of food particles and gastric juice. Huge folds in the stomach lining look like underwater ridges and valleys. Periodic waves of peristalsis sweep across this landscape like earthquakes. As you approach the stomach wall, you note that the lining is dotted with numerous pits—the openings of tubular [4]_____, which produce [5]_____, the stomach's digestive fluid.

You enter one of the pits and swim into the gastric gland. Your instruments indicate an increase in secretion of gastric juice. The presence of food—and you—in the stomach causes cells in the stomach wall to secrete a hormone called [6]_____, which (along with nerve signals) stimulates secretion of gastric juice.

Several kinds of cells line the walls of the gastric gland. Many of the cells are undergoing [7]_____, rapidly replacing cells that are damaged in this harsh environment. Deep in the gland, large parietal cells secrete [8]_____. Nearby chief cells secrete inactive pepsinogen, which quickly changes into the enzyme [9]_____ when it comes into contact with the acid. The function of pepsin is to begin the digestion of [10]_____, preparing them for further digestion in the [11]_____. Secreting the enzyme in inactive form [superscript]12[/superscript]_____ the cells where it is produced. [superscript]13[/superscript]_____ cells secrete thick [superscript]14[/superscript]_____, which helps protect the stomach lining.

You exit the gland and continue your inspection of the stomach lining. The opening from the esophagus, where you entered the stomach, is sealed tightly. There is little evidence of backflow of acid chyme into the esophagus, which might cause [superscript]15[/superscript]_____. Nowhere do you see any evidence of a [superscript]16[/superscript]_____—a sore that develops when pepsin and acid attack the stomach lining. This can happen when there is too much pepsin or acid, or not enough [superscript]17[/superscript]_____. This can be caused by stress, infection, or substances such as alcohol or [superscript]18[/superscript]_____.

The stomach's exit is guarded by a tightly closed doughnut of muscle, the [superscript]19[/superscript]_____. As you approach, it opens to allow a squirt of liquid to leave the stomach. Soon you enter the next portion of the digestive tract, the [superscript]20[/superscript]_____.

This organ is much narrower than the cavernous stomach. It looks like a curving, twisting tunnel. The first few inches is called the duodenum. Here you see a jagged sore, tucked into a fold in the wall of the duodenum—a [superscript]21[/superscript]_____. This must be the cause of the problem. The medical team will have to deal with this.

Continuing your survey of the small intestine, you note an opening where fluid is squirting into the duodenum. This is the duct through which [superscript]22[/superscript]_____ from the gallbladder and an alkaline, enzyme-rich solution from the [superscript]23[/superscript]_____ enter the small intestine. The alkaline solution [superscript]24[/superscript]_____ acid from the stomach. Bile, produced by the [superscript]25[/superscript]_____ and stored in the gallbladder, contains bile salts that break up [superscript]26[/superscript]_____ into small droplets, a process called [superscript]27[/superscript]_____. This makes it easier for pancreatic enzymes to digest them. An enzyme called [superscript]28[/superscript]_____ breaks fat molecules down into fatty acids and glycerol. Other pancreatic enzymes, trypsin and chymotrypsin, continue the digestion of [superscript]29[/superscript]_____ that began in the stomach. They break polypeptides down to smaller polypeptides, and then peptidases break these smaller polypeptides down to [superscript]30[/superscript]_____. Pancreatic amylase hydrolyzes [superscript]31[/superscript]_____ to form maltose, a disaccharide. Other enzymes digest it and other disaccharides to form [superscript]32[/superscript]_____. Nucleases digest DNA and RNA.

Contractions in the walls of the small intestine gently propel you along. The next two portions of the small intestine are specialized for [superscript]33[/superscript]_____ of nutrients. The surface is highly folded, with numerous small projections called [superscript]34[/superscript]_____. You swim downward, and the villi surround you like the huge rubbery trunks of some inflatable forest. Their surfaces look and feel velvety, because the epithelial cells covering them bear their own tiny projections called [superscript]35[/superscript]_____. All these folds give the small intestine a huge surface area—over 600 square meters! You press your light against the surface of one of the villi. Inside, you can see the ghostly outlines of [superscript]36[/superscript]_____ vessels and a network of [superscript]37[/superscript]_____. Nutrients are pumped or [superscript]38[/superscript]_____ through the intestinal epithelium and enter these vessels. The capillaries join to form a large blood vessel that carries blood directly to the liver, where nutrients are converted to forms the body needs. For example, the liver stores excess glucose in the form of a polysaccharide called [superscript]39[/superscript]_____.

As you finish your inspection of the small intestine, you see the sphincter ahead that controls the movement of unabsorbed food material into the [superscript]40[/superscript]_____, or colon. The colon is where [superscript]41[/superscript]_____ is absorbed and undigested material is turned into [42]_____, which are stored in the rectum and expelled through the anus. That part of the digestive system is not on your itinerary! You work your way through the wall and into a blood vessel, and are soon back in the lab, munching on a sandwich.

Exercise 6   (Module 21.12)

Match each of the following animals with a description of its digestive tract.

_____ 1. Horse

_____ 2. Tiger

_____ 3. Frog

_____ 4. Cow

_____ 5. Rabbit

_____ 6. Deer

A. Digestive system gets simpler when it becomes an adult and changes to a carnivorous diet

B. Microbes in cecum digest cellulose, and some nutrients are absorbed in colon

C. Ruminant herbivore

D. Reingests fecal pellets to obtain nutrients

E. Carnivore with relatively simple digestive system

F. Rechews "cud," which is further digested by bacteria and enzymes in four-chambered stomach

Exercise 7   (Module 21.13)

This module is short, but it is an important introduction to the remainder of the chapter. See if you can summarize the module in a sentence of *exactly* 25 words.

Exercise 8   (Module 21.14)

Cellular metabolism breaks down the molecules in food and uses the energy released to build ATP, which in turn fuels body activities. The energy in food is measured in Calories (which are actually kilocalories). To get a handle on the energy contents of foods and the energy consumed by body activities, calculate how much you would need to exercise to use the calories in the food listed on the next page (similar to Table 21.14A).

First, you need to calculate your own weight in kilograms. One kilogram equals 2.2 pounds, so:

$$\text{your weight (in kg)} = \frac{\text{your weight (in lb)}}{2.2} = \underline{\quad\quad}$$

Next you need to calculate how many kcal/min you use performing various activities:

$$\text{kcal/min you use} = \text{kcal/kg/min (from table)} \times \text{your weight (in kg)}$$

Use the above formula to figure out how many kcal/min you would use jogging, swimming, and walking. Then write the figures in the table on the next page.

Finally, use the following formula to figure out how long you would have to exercise to use up the energy in a food item, and write the figures in the table:

$$\text{exercise time} = \frac{\text{energy in food}}{\text{kcal/min you use}}$$

Food-exercise energy equivalents:

	Jogging	Swimming	Walking
Speed	9 min/mile	30 min/mile	20 min/mile
kcal/kg/min	0.173	0.132	0.039
kcal/min for you:	_____	_____	_____

Number of minutes you would have to do the above exercises to consume the calories in:

Big Mac 560 kcal	_____	_____	_____
cheese pizza—slice 450 kcal	_____	_____	_____
Coca Cola Classic—10 oz 144 kcal	_____	_____	_____
whole-wheat bread—slice 86 kcal	_____	_____	_____

(You may want to look at the labels to get the energy contents of some of your favorite foods. You can record their exercise equivalents in the spaces below.)

_____ _____ _____ _____

_____ _____ _____ _____

_____ _____ _____ _____

Exercise 9   (Modules 21.16 – 21.18)

Vitamins, minerals, and certain amino acids are all essential nutrients. They are substances an animal cannot make that must be obtained in food. These and other nutrients are listed on food labels. There is a lot of detail in these modules, so try to focus on overall concepts. These concepts, the names of some nutrients, and principles of food labeling are useful in completing this crossword puzzle.

Across

4. The minerals sodium and ____ are important in acid-base and water balance.

7. Excess ____-soluble vitamins are not stored, but excreted from the body.

8. A diet low in one or more essential amino acids results in ____ deficiency.

10. The vitamin debate is mainly about ____.

13. ____ acid (vitamin C) is important in collagen synthesis.

14. Most vitamins function as ____ .

15. There are ____ essential amino acids that must be obtained from the diet.

16. Honey, corn syrup, and dextrose are all ____.

18. A ____ is an essential organic nutrient needed in very small quantities.

22. All essential amino acids can be obtained from a combination of corn and ____.

23. Packages list ____ and nutritional information.

24. ____ is a mineral important in bones, teeth, muscle, nerves, and blood clotting.

Down

1. Ingredients are listed on packages from greatest amount to ____ amount.

2. Vitamin D aids in the ____ and use of calcium and phosphorus.

3. The B ____ vitamins are important coenzymes in cellular metabolism.

5. Calcium and ____ are important in building the skeleton.

6. Most victims of protein malnutrition are ____.

9. A gram of fat contains ____ the calories of a gram of carbohydrate or protein.

11. The amino acids that body cells cannot make are called ____ amino acids.

12. ____ is a component of hemoglobin and enzymes of energy metabolism.

17. ____ are elements other than C, H, O, and N.

18. Vitamin A is important for normal ____.

19. ____-soluble vitamins (vitamins A, D, etc.) can be stored in the body.

20. Essential amino acids are most easily obtained from ____ sources.

21. Relative levels of nutrients are expressed as a percent of the U.S. ____.

Exercise 10 **(Modules 21.19 – 21.20)**

It is important to read food labels because certain ingredients and nutrients are more healthful than others. For a healthy life-style, one should try to maximize (↑) which of the following and minimize (↓) which others?

_____ 1. Dietary cholesterol

_____ 2. Vegetable oils

_____ 3. Sodium

_____ 4. Total fat intake

_____ 5. High-fiber foods

_____ 6. Exercise

_____ 7. Smoking

_____ 8. LDLs

_____ 9. HDLs

_____10. Unsaturated fats

_____11. Complex carbohydrates

_____12. Alcohol intake

_____13. Vegetables from the mustard family, such as broccoli and cabbage

_____14. Smoked and cured foods

_____15. Animal fats

Testing Your Knowledge

Multiple Choice

1. The box elder bug is an insect that sucks plant juices. It is a
   a. filter-feeding omnivore.
   b. substrate-feeding herbivore.
   c. filter-feeding herbivore.
   d. fluid-feeding herbivore.
   e. fluid-feeding carnivore.

2. Which of the following lists the four stages of food processing in order?
   a. ingestion, digestion, absorption, elimination
   b. digestion, ingestion, absorption, elimination
   c. ingestion, absorption, elimination, digestion
   d. ingestion, digestion, elimination, absorption
   e. absorption, digestion, ingestion, elimination

3. Which of the answers below would *not* work in the following sentence? "In order for the body to absorb and use _____ , they must be broken down by hydrolysis into ____."
   a. polysaccharides . . . monosaccharides
   b. amino acids . . . proteins
   c. fats . . . glycerol and fatty acids
   d. disaccharides . . . monosaccharides
   e. starch . . . monosaccharides

4. How does a gastrovascular cavity differ from an alimentary canal? The gastrovascular cavity
   a. stores food but does not digest it.
   b. is usually much larger.
   c. has only one opening.
   d. functions in digestion but not absorption.
   e. can use only small food particles.

5. In humans, most nutrient molecules are absorbed by the
   a. stomach.
   b. liver.
   c. small intestine.
   d. large intestine.
   e. pancreas.

6. The largest variety of digestive enzymes function in the
   a. large intestine.
   b. oral cavity.
   c. stomach.
   d. gallbladder.
   e. small intestine.

7. After nutrients are absorbed, the blood carries them first to the
   a. brain.
   b. pancreas.
   c. kidneys.
   d. liver.
   e. large intestine.

8. Digestion of proteins begins in the ____ , and digestion of polysaccharides begins in the ___.
   a. mouth . . . stomach
   b. stomach . . . small intestine
   c. stomach . . . mouth
   d. stomach . . . stomach
   e. small intestine . . . stomach

9. In humans, a major function of the large intestine is
   a. absorption of water.
   b. digestion of food molecules.
   c. breakdown of toxic substances.
   d. absorption of nutrients.
   e. storage of food before it is digested and absorbed.

10. The energy needed to fuel essential body processes is called
    a. essential nutrient level.
    b. metabolism.
    c. recommended daily allowance.
    d. optimum energy intake.
    e. basal metabolic rate.

11. Which of the following statements about fat is *incorrect?*
    a. People of many cultures think fat is beautiful.
    b. It is best to strive to have as little body fat as possible.
    c. Too much body fat can lead to cardiovascular disease.
    d. A moderate amount of fat contributes to a healthy immune system.
    e. Most excess calories in food are stored by the body in fat.

12. Which of the following is *not* an essential nutrient?
    a. iron
    b. glucose, a monosaccharide
    c. methionine, an amino acid
    d. sodium
    e. pantothenic acid, a vitamin

13. Which of the following is an organic molecule needed by the body in small amounts?
    a. protein
    b. zinc
    c. vitamin C
    d. monosaccharide
    e. calcium

14. ____ are needed in the diet as components of teeth and bone, regulators of acid-base and water balance, and parts of certain enzymes.
    a. Amino acids
    b. Fat
    c. Minerals
    d. Nucleic acids
    e. Vitamins

15. To maintain health, it is best to try to
    a. decrease intake of unsaturated fats.

b. decrease HDL levels in the blood.
c. decrease intake of saturated fats.
d. maintain a diet in which at least 35% of calories come from fat.
e. increase LDL levels in the blood.

Essay

1. Why do the nutrient molecules contained in the food that you eat have to be digested before your body can use them?

2. In a few sentences, compare the digestive system of a grasshopper with that of a human.

3. Describe the substances in saliva and their roles in the function of the digestive tract.

4. Plants are difficult to digest. Describe how the digestive tracts and habits of three different herbivorous mammals are suited to a plant diet.

5. Briefly describe the potential problems with fad diets. What is the best kind of diet for healthy, effective weight loss?

6. Describe the kinds of information you can get from reading the labels on food. How might this information be useful?

Applying Your Knowledge

Multiple Choice

1. Which of the following do a wolf, a hummingbird, a termite, and an elephant have in common?
   a. All are omnivores.
   b. All are substrate feeders.
   c. All are ingestive feeders.
   d. All are carnivores.
   e. All are herbivores.

2. How would you expect the digestive system of a hawk, a carnivore, to compare with that of a sparrow, a seed-eater?
   a. The hawk would have a larger gastrovascular cavity.
   b. The sparrow digestive system would be longer.
   c. The hawk would have a gizzard, but the sparrow would not.
   d. The hawk digestive system would be longer.
   e. The hawk would lack both crop and gizzard.

3. Which of the following might make the most effective anti-ulcer medication? A chemical that
   a. stimulates parietal cells of the gastric glands.
   b. inhibits secretion of the hormone gastrin.
   c. inhibits mucous cells of the gastric glands.
   d. stimulates secretion of bile.
   e. stimulates gastritis.

4. The lungs are folded into many small air sacs and blood vessels, which divide to form many small capillaries that increase the transfer of substances through their walls. The structures in the digestive system similar in function to these air sacs and capillaries are the
   a. villi.
   b. colon and rectum.
   c. gastric glands.
   d. high-density lipoproteins.
   e. sphincters.

5. Imagine that you have eaten a meal containing the following nutrients. Which would *not* have to be digested before being absorbed?
   a. protein
   b. polysaccharide
   c. disaccharide
   d. nucleic acid
   e. amino acid

6. Aunt Rose had her gallbladder removed and afterward
   a. could not eat foods containing large amounts of fat.
   b. had trouble digesting proteins.
   c. could eat monosaccharides and disaccharides but not polysaccharides.
   d. had to wash her food down with large quantities of water.
   e. needed to take an amino acid supplement.

7. Laxatives work in the large intestine to relieve constipation. Which of the following would probably *not* be an effective laxative? A substance that
   a. contains lots of fiber.
   b. promotes water absorption in the large intestine.
   c. speeds up movement of material through the large intestine.
   d. decreases water absorption in the large intestine.
   e. stimulates peristalsis.

8. It is important to get some vitamin B-1 every day, but it is all right if intake of vitamin A varies a bit. Why?
   a. Vitamin B-1 is an essential nutrient, and vitamin A is not.
   b. Vitamin A can be stored by the body, but vitamin B-1 cannot.
   c. The body needs much larger amounts of vitamin B-1 than vitamin A.
   d. The body requires vitamin B-1, but vitamin A is just an "extra."
   e. Vitamin A is water-soluble, and vitamin B-1 is fat-soluble.

9. Why does your body need grams of carbohydrates each day but only milligrams of vitamins, thousands of times less?
   a. Carbohydrates are used up, but vitamins are reusable.
   b. The body needs carbohydrates to function, but not vitamins.
   c. Vitamins contain much more energy per gram.
   d. The body makes vitamins itself but needs to get carbohydrates from food.
   e. Carbohydrates are essential nutrients, but vitamins are not.

10. A breakfast cereal listed the following ingredients (in order): whole-grain wheat, whole-grain oats, brown sugar, dates, corn syrup, sorbitol, calcium phosphate, dextrose, modified tapioca starch, BHT, vitamins. How much sugar does it contain, compared with other ingredients?
    a. less than either wheat or oats
    b. less than wheat and oats combined
    c. more than everything listed from dates to vitamins combined
    d. both b and c
    e. impossible to tell from this information

Essay

1. James and Michael are both working out to build up their muscles. Muscle is mostly protein, so they both are on high-protein diets. James is eating twice as much meat as normal. Michael is a vegetarian, so he is consuming plant foods high in protein, such as beans and peanuts. What are the pros and cons of each diet?

2. It is sometimes necessary to remove a diseased portion of the digestive system, but usually a patient can get along on a modified diet. It would be most difficult to live without which of the following, and why: stomach, large intestine, gallbladder, small intestine, salivary gland

3. The pancreas does not actually secrete the protein-digesting enzymes trypsin and chymotrypsin. It secretes substances called trypsinogen and chymotrypsinogen, which are converted to trypsin and chymotrypsin by a substance secreted by the walls of the small intestine. What do you think is the reason for this two-step process?

4. NASA calculates that an astronaut uses an average of 2500 kcal of food energy per day. How many liters of oxygen will the astronaut consume in a day? (See Module 21.14 in the text.)

5. Professor T. Herman Thimblebottom suspects that the element scandium is an essential mineral in the human diet. How might he go about demonstrating this? Why might it be difficult?

6. Look at some food labels. How many calories are contained in one serving of one of the foods? How far would you have to jog, swim, or walk to use the calories?

Using Your Knowledge

1. Do you ever read the labels on food? It can be very educational and entertaining. Look at some food labels. Which ingredients are present in the largest amounts? How much sugar does the product contain? (Remember, there are many words for sugar.) Which ingredients have value as nutrients, and which merely "improve" the taste, texture, appearance, or storage of the food? Is this food high in fat? Saturated or unsaturated fat? Calories? Sodium? How do more "natural" foods compare with more "processed" foods? How do "light" foods compare with regular ones?

2. Are you eating a healthy diet and getting enough exercise? How would you evaluate your diet, given the information in this chapter? Outline changes in your diet and lifestyle that could decrease your chances of developing cancer or cardiovascular disease.

Respiration: The Exchange of Gases

Most of us don't give respiration much thought. It just happens without thinking—a process "as natural as breathing." But when you can't breathe, you realize how important breathing really is. Have you ever been short of breath on a mountain hike or had "the wind knocked out of you" on the athletic field? Have you ever choked on a piece of food or been knocked down by a wave at the beach and come up coughing and sputtering? When breathing is interrupted, even for a few seconds, we realize how important every breath is. Breathing is part of the process of gas exchange. Not all animals breathe as we do, but they do need to obtain oxygen from their environment and dispose of carbon dioxide. This chapter is about the vital subject of gas exchange.

Organizing Your Knowledge

Exercise 1   (Module 22.1)

In most animals, there are three phases of gas exchange: **breathing, transport** of gases by the circulatory system, and **exchange** of gases with tissues. State which phase is interfered with in each of the following situations.

_____ 1. In the disease cystic fibrosis, thick mucus coats the inside of the lungs, blocking passage of gases.

_____ 2. A broken neck can paralyze the muscles of the chest.

_____ 3. Babies sometimes inhale small objects that can block the windpipe.

_____ 4. Anemia is a decrease in the oxygen-carrying protein hemoglobin.

_____ 5. During a heart attack, blockage of a blood vessel causes heart muscle cells to die from lack of oxygen.

_____ 6. An asthma attack narrows air passages into the lungs.

_____ 7. Bedridden patients sometimes get bedsores when blood vessels to the skin are pinched.

_____ 8. A mountain climber is breathing rapidly and his heart is beating strongly, but in the thin air there is not enough oxygen in the blood to diffuse into brain cells.

Exercise 2   (Module 22.2)

Match each of the following animals with a term (A–D) that describes it and a diagram (P–S) that shows its respiratory surface. Also color each respiratory surface yellow.

A. Lungs   B. Gills   C. Tracheae   D. Body surface

Animal	Term	Diagram
1. Beetle	____	____
2. Cat	____	____
3. Earthworm	____	____
4. Trout	____	____
5. Human	____	____
6. Chicken	____	____
7. Crab	____	____

P.   Q.   R.   S.

Exercise 3   (Modules 22.2 – 22.6)

Review gas-exchange mechanisms of different animals by filling in the blanks below.

Although some small animals, such as earthworms, use their ^{1}_____ as a gas-exchange organ, most animals have specialized organs that enable them to obtain ^{2}_____ and expel waste ^{3}_____. The part of an animal where gas exchange occurs is called the ^{4}_____ surface. Individual molecules of O_2 and CO_2 diffuse through a membrane only if they are dissolved in ^{5}_____, so the respiratory surface must be ^{6}_____. The surface must also be ^{7}_____ enough to take in sufficient oxygen for the body's needs and ^{8}_____ enough for gases to diffuse through it rapidly.

Most aquatic animals obtain dissolved oxygen from the surrounding water by means of ^{9}_____—featherlike extensions of the body surface. An advantage of exchanging gases in water is that the animal does not have to expend effort to keep the respiratory surface ^{10}_____. A disadvantage is that the concentration of available oxygen is much ^{11}_____ in water than in the air. The gills of a fish are efficient gas-exchange organs. Gill arches on each side of the fish's body bear numerous elongated gill filaments, and each of these bears numerous platelike ^{12}_____. The gills are red because the lamellae are filled with tiny ^{13}_____ covered by a thin layer of cells. The fish opens and closes its mouth and gill coverings to ^{14}_____ its gills, increasing contact between the water and the respiratory surface. Water flows past the lamellae in a direction ^{15}_____ to the flow of ^{16}_____ inside the lamellae. This countercurrent increases the efficiency of the gills. ^{17}_____ is transfer of a substance from a fluid moving in one direction to a fluid moving in the opposite direction. As water and blood flow past each other, ^{18}_____ diffuses from water to blood. As the blood picks up more and more oxygen, it comes into contact with water containing ^{19}_____ and ^{20}_____ available oxygen. Thus the countercurrent flow of water and blood creates a ^{21}_____ gradient that favors the diffusion of oxygen along the entire length of the lamella, greatly enhancing the efficiency of the gill.

Land animals obtain oxygen from the ^{22}_____. There are advantages to breathing air: It contains a ^{23}_____ concentration of oxygen than water, and it is ^{24}_____ to move than water. The biggest disadvantage of breathing air is that it tends to ^{25}_____ the respiratory surface. The ^{26}_____ system of an insect is an effective mechanism for gas exchange. ^{27}_____ tubes that branch throughout the body deliver gases directly to body cells without the help of the ^{28}_____ system. The tracheae branch and rebranch and end in tiny fluid-filled tubes that touch the ^{29}_____ of individual cells. ^{30}_____ dissolves in the fluid and diffuses into the cells, while ^{31}_____ diffuses out of the cells and is expelled from the insect's body.

Most land vertebrates have ^{32}_____ that function in gas exchange. The human lungs are in the chest cavity, above a thin sheet of muscle called the ^{33}_____ that functions in breathing. When you inhale, air enters through the ^{34}_____ and is warmed, humidified, and ^{35}_____ in the ^{36}_____ cavity. The air then passes into the ^{37}_____ (throat) and through the ^{38}_____, or voicebox. Here a pair of ^{39}_____

make the sounds that enable us to speak. When the vocal cords are [40]_____,
high-pitched sounds are produced. When the cords are [41]_____, they make
lower-pitched sounds. From the larynx, air passes through the [42]_____ into
a pair of [43]_____, one leading to each lung. Inside the lungs, the bronchi
branch into numerous narrow tubes called [44]_____. The surfaces of these
respiratory passageways are covered by a film of [45]_____ that traps dust
and other contaminants. The beating of numerous [46]_____ move the mucus
and trapped particles out of the respiratory tract. The bronchioles end in clusters of tiny
air sacs called [47]_____. There are [48]_____ of these tiny sacs in
each lung, so their total surface area is enormous. The lining of each alveolus is a thin
layer of epithelial cells that makes up the respiratory surface. Oxygen diffuses through a
thin layer of moisture, through the epithelium, and into a network of [49]_____
that covers the surface of the alveolus. [50]_____ diffuses out of the blood and
into the air within the alveolus. Thus the respiratory system works with the circulatory
system in the process of gas exchange.

Exercise 4   (Module 22.6)

Use this diagram to review the parts of the human respiratory system. Label and color the
parts in **bold** type. Color the **nasal cavity** purple, the **pharynx** blue, the **larynx** green, the
trachea yellow, the **bronchi** orange, and the **bronchioles** red. Also color the surrounding
lung tissue healthy pink (this woman is definitely a nonsmoker!) and the **diaphragm**
brown.

How do tobacco smoke and other air pollutants affect the respiratory system? Test your knowledge by completing this crossword puzzle.

Across

1. Cigarette smoke irritates the lining of the ____, destroying their cilia.

5. The lungs of a cigarette smoker are ____ rather than pink.

6. ____ is an air pollutant associated with respiratory disease.

7. Every cigarette probably cuts ____ minutes from the smoker's life.

10. In ____, the alveoli become brittle and the victim becomes short of breath.

12. When cilia are destroyed, only ____ can rid the lungs of pollutants.

13. ____ and cilia protect the respiratory passages.

15. ____, once used in insulation, can cause lung cancer.

16. Cigarette smoke destroys the ____ that normally sweep out pollutants.

17. Smoking increases the risk of ____ cancer by a factor of 10.

Down

2. ____ is a naturally occurring radioactive gas that may cause lung cancer.

3. Emphysema often leads to ____ disease.

4. Carbon ____ is a harmful gas in polluted air.

8. The ____ tissue lining the respiratory system is very delicate.

9. Lung ____ is the deadliest disease caused by smoking.

11. ____ dioxide is one of thousands of harmful pollutants in city air.

14. Each year, ____ kills more than 350,000 people in the United States.

Exercise 6   (Module 22.8)

State whether each of the following pertains to inhalation (I) or exhalation (E).

_____ 1. The diaphragm contracts.
_____ 2. The rib cage expands.
_____ 3. The diaphragm expands and arches upward.
_____ 4. Air enters the lungs.
_____ 5. The diaphragm moves downward.
_____ 6. Muscles between the ribs contract.
_____ 7. Muscles between the ribs relax.
_____ 8. The volume of the rib cage decreases.
_____ 9. Air pressure in the alveoli is less than that of the atmosphere.
_____10. The diaphragm relaxes.
_____11. Air pressure in the alveoli is greater than that of the atmosphere.
_____12. Air is forced out of the lungs.

Exercise 7   (Module 22.9)

Review control of breathing by the respiratory control centers: State whether each of the following changes would speed up ($\uparrow$) or slow down ($\downarrow$) your rate of breathing.

_____ 1. A rise in blood CO_2 concentration
_____ 2. Hyperventilating
_____ 3. A severe drop in blood oxygen concentration
_____ 4. An increase in blood pH
_____ 5. An increase in carbonic acid in the blood
_____ 6. A drop in blood pH
_____ 7. A decrease in blood CO_2 concentration
_____ 8. Holding your breath as long as you can, then releasing it

Exercise 8   (Modules 22.9 – 22.12)

The following story summarizes the cooperation of the respiratory and circulatory systems in gas exchange. Fill in the blanks to complete it.

Your respiratory system works together with your [1]_____ system in exchange and transport of gases. Imagine that you are riding a bicycle. Your leg muscles are working hard, consuming [2]_____ and producing [3]_____ as a waste product. Blood returning from the muscles therefore has a relatively [4]_____ concentration of oxygen and a [5]_____ concentration of carbon dioxide. One side of the heart pumps this oxygen-poor blood through the capillaries covering the [6]_____ in your lungs. You are breathing hard. A breathing control center in your [7]_____ responds to the [8]_____ in the pH of your blood caused by the increase in [9]_____. It speeds up the pace of nerve impulses sent to the [10]_____ and muscles between your [11]_____, and you breathe more rapidly and deeply. This helps expel the [12]_____ and meet the muscle's needs for more [13]_____.

The air in the lungs has a high concentration of [14]_____ and low concentration of [15]_____ relative to the blood. [16]_____ diffuses out of the blood into the air inside the alveolus, moving from a region where it is

^{17}_____ concentrated to a region where it is ^{18}_____ concentrated. ^{19}_____ similarly diffuses down its concentration gradient from the air in the alveolus into the blood.

The oxygen that enters the blood is not very ^{20}_____ in water, so little oxygen is transported in dissolved form. Most oxygen is carried by a protein called ^{21}_____, contained within ^{22}_____ blood cells. A hemoglobin molecule consists of four polypeptide chains, each of which contains a heme group with an ^{23}_____ atom at its center. Each of these atoms can carry one ^{24}_____ molecule, so each of the millions of hemoglobin molecules in a red blood cell can carry millions of O_2 molecules.

The oxygen-rich blood that leaves the lungs returns to the heart, which pumps it out to the exercising ^{25}_____ of your legs. The blood passing through the capillaries in a muscle contains a ^{26}_____ concentration of O_2 than the muscle cells where it is being used up, so O_2 diffuses out of the blood and into the cells. The cells are making CO_2 at a fast pace, so CO_2 is ^{27}_____ concentrated in the cells than in the blood. It diffuses ^{28}_____ the cells and ^{29}_____ the blood.

Some CO_2 is carried dissolved in blood plasma. Most of it enters ^{30}_____ blood cells, but most does not combine with ^{31}_____. Instead, enzymes in the blood cells cause most of it to react with ^{32}_____ molecules, forming carbonic acid (H_2CO_3). Each carbonic acid molecule then breaks apart, forming a hydrogen ion (H^+) and a ^{33}_____ ion (HCO_3^-). Hemoglobin picks up most of the H^+ ions, so it does not acidify the blood much. The bicarbonate ions diffuse out into the blood ^{34}_____, where they are part of the blood-^{35}_____ system that stabilizes the pH of the blood. If blood pH drops, the bicarbonate ions combine with H^+ ions and remove them from the plasma. If pH ^{36}_____, the bicarbonate releases these H^+ ions back into solution.

When the blood from the muscles returns (via the heart) to the lungs, the events that formed bicarbonate ions are reversed. Bicarbonate and H^+ form carbonic acid, which breaks up to form water and ^{37}_____, which in turn diffuses out of the blood into the air of the alveoli. Thus the respiratory and circulatory systems continue to work in close cooperation as you continue your bicycle ride.

Exercise 9 (Modules 22.10 – 22.12)

Number the following in order from first to last to show the path an O_2 molecule must follow through a mother to a cell in her fetus.

_____ A. The mother's heart pumps the oxygen-rich blood to her uterus.

_____ B. Oxygen diffuses out through the walls of capillaries in the uterus.

_____ C. The mother takes a deep breath of fresh air.

_____ D. Oxygen leaves the blood of the fetus and diffuses into a growing cell in the fetus's brain.

_____ E. Oxygen diffuses across the thin wall of an alveolus in the mother's lung and into a capillary.

_____ F. The mother's blood, now loaded with oxygen, returns from her lungs to the heart.

_____ G. Oxygen-rich fetal blood flows into the fetus through a vein in the umbilical cord.

_____ H. Oxygen diffuses through the wall of a capillary in the placenta and into the blood of the fetus.

_____ I. Oxygen attaches to hemoglobin in the mother's blood.

_____ J. Oxygen attaches to hemoglobin in fetal blood.

_____ K. The fetus's heart pumps the oxygen-rich blood out to its tissues.

Testing Your Knowledge

Multiple Choice

1. Which of the following has no specialized respiratory structures?
   a. crab
   b. earthworm
   c. salmon
   d. ant
   e. snake

2. The respiratory control centers are located in the
   a. heart.
   b. lungs.
   c. diaphragm and rib muscles.
   d. brain.
   e. large arteries.

3. When you exhale, the diaphragm
   a. relaxes and arches.
   b. relaxes and flattens.
   c. contracts and arches.
   d. contracts and flattens.
   e. contracts and arches, but only when you are exercising vigorously.

4. Why are bird lungs more efficient than human lungs?
   a. They use countercurrent exchange.
   b. They have more surface area than human lungs.
   c. They are able to concentrate the oxygen to much higher levels.
   d. Their alveoli are much larger.
   e. They use a one-way rather than an in-out air flow system.

5. Inhaled air passes through which of the following last?
   a. bronchiole
   b. larynx
   c. pharynx
   d. trachea
   e. bronchus

6. An advantage of gas exchange in water, compared with gas exchange in air, is that
   a. water usually contains a higher concentration of O_2 than air.
   b. water is easier to move over the respiratory surface.
   c. the respiratory surface does not dry out in water.
   d. ventilation requires less energy in water.
   e. the respiratory surface does not have to be as extensive in water.

7. In the blood, bicarbonate ions
   a. help transport oxygen.
   b. act as buffers to guard against pH changes.
   c. are transported by hemoglobin.
   d. attach to numerous CO_2 molecules, keeping them from solution.
   e. are poisonous and must constantly be removed.

8. Smoking destroys the cilia in the respiratory passageways. This
   a. makes it harder to move air in and out of the lungs.
   b. decreases the surface area for respiration.
   c. slows blood flow through lung blood vessels.
   d. makes it harder to keep the lungs clean.
   e. interferes with diffusion across the respiratory surface.

9. Most oxygen is carried by the blood ____. Most carbon dioxide is carried by the blood ____.
   a. attached to hemoglobin . . . in the form of bicarbonate ions
   b. dissolved in the plasma . . . dissolved in the plasma
   c. in the form of H^+ ions . . . in the form of bicarbonate ions
   d. attached to hemoglobin . . . attached to hemoglobin
   e. attached to hemoglobin . . . dissolved in the plasma

10. A disease called emphysema decreases the springiness of the lungs. This decreases ____ and makes it harder to breathe.
    a. the volume of each breath
    b. respiratory rate
    c. residual volume
    d. countercurrent exchange
    e. vital capacity

11. The ____ is a structure specialized for diffusion of gases and nutrients between the blood of the mother and the fetus.
    a. uterus
    b. placenta
    c. lamella
    d. alveolus
    e. umbilicus

Essay

1. Compare the advantages and disadvantages of obtaining oxygen from water and obtaining it from the air.

2. How are gills and lungs similar? How are they different?

3. Describe how the structure, number, and arrangement of alveoli are well suited to their function in gas exchange.

4. Where are your vocal cords? How do they work? How do they produce high-pitched and low-pitched sounds?

5. When you inhale, does air flow into the lungs, causing them to expand? Or do the lungs expand, causing air to flow in? Explain.

6. Explain how countercurrent exchange in a fish gill enhances absorption of oxygen from water.

Applying Your Knowledge

Multiple Choice

1. Which of the following normally contains the highest concentration of oxygen?
   a. body cells
   b. inhaled air
   c. air in the alveoli
   d. blood entering the lungs
   e. blood leaving the lungs

2. Which of the following in a human is most similar in function to the gill lamellae of a fish?
   a. vocal cords
   b. bronchioles
   c. alveoli
   d. tracheae
   e. diaphragm

3. In which of the following does oxygen pass directly from the air, through a moist surface, to individual cells, without being carried by the blood?
   a. mouse
   b. ant
   c. shark
   d. earthworm
   e. frog

4. A fish opens and closes its mouth and gill covers. A dog pants. A marine worm waves long, filmy gills in the water. All of these movements
   a. are examples of ventilation.
   b. show how circulation aids respiration.
   c. are examples of breathing.
   d. slow diffusion of CO_2.
   e. enhance countercurrent exchange.

5. ____ in CO_2 in your blood, which causes ____ in pH, would cause your breathing to speed up.
   a. An increase . . . a rise
   b. An increase . . . a drop
   c. A decrease . . . a rise
   d. A decrease . . . a drop
   e. Actually, it is rise and fall of O_2, not CO_2, that controls breathing.

6. Which of the following would have the same O_2 content?
   a. blood entering the lungs—blood leaving the lungs
   b. blood entering the right side of the heart—blood entering the left side of the heart
   c. blood entering the tissue capillaries—blood leaving the tissue capillaries
   d. blood entering the right side of the heart—blood leaving the right side of the heart
   e. blood leaving the tissue capillaries—blood leaving the lungs

7. Patients with chronic lung disease and difficulty breathing often adapt to the high concentration of CO_2 in their blood. The breathing centers stop responding to CO_2 level. If such a patient has difficulty breathing, medical personnel are reluctant to give the patient pure oxygen. Based on what you know about control of breathing, why do you think this is the case?
   a. The patient's body would use the oxygen to make even more CO_2.
   b. The oxygen would increase concentration of bicarbonate, altering pH.
   c. Increased oxygen in the blood might slow or stop breathing.
   d. The body is not used to the oxygen, and the patient would overdose.
   e. The patient would breathe too fast and become tired out.

8. In an old science fiction movie, the hero tried to drown a giant ant by holding its head under water. Would this work? Why?
   a. Yes. Ants use lungs to breathe much as we do.
   b. Yes. The skin surface, covered with water, could not get O_2 from the air.
   c. No. Ants use gills for respiration, like crabs do.
   d. No. Ants breathe through holes in the sides of their bodies.
   e. No. The ant could get oxygen by diffusion from the water.

9. A zoologist compared the respiratory efficiency and swimming speed of different fish. He found that less efficient fish tended to have
   a. greater ventilation.
   b. a thicker respiratory surface.
   c. more hemoglobin.
   d. a faster heart rate.
   e. a more extensive respiratory surface.

10. A biochemist mixed 10 drops of acid with 100 mL of water, and the pH dropped from 7.4 to 5.0. She then mixed 10 drops of acid with 100 mL of blood. The pH dropped from 7.4 to 7.2. What is the reason for this difference?
    a. Blood is thicker than water.
    b. Blood is already very acidic, so the acid has less effect.
    c. Blood is saturated with oxygen; there is little room for acid.
    d. Blood contains buffers that reduce pH change.
    e. Water is already more acidic than blood; there is little room for more.

Essay

1. Trace the path of an oxygen molecule from the air to one of your brain cells, naming all the places and structures it passes through on its way.

2. You are on the team to design a robot that will patrol the devastated terrain around the Chernobyl nuclear power plant. The robot will function like a living organism, gathering organic debris for "food" and obtaining oxygen from the surrounding air. What features would you want to include in your design of its respiratory surface?

3. Carbon monoxide molecules in cigarette smoke and automobile exhaust attach to hemoglobin molecules where oxygen normally attaches, and they hold on more strongly than oxygen. What effect would this have on the body?

4. A man smokes a pack (20) of cigarettes a day for 40 years. If each cigarette shortens his life by 5 minutes, how much "before his time" will he die?

5. In a submarine, the oxygen supply was accidentally interrupted, causing the oxygen content of the air to drop. A machine that removes carbon dioxide continued to function, so there was no corresponding buildup of CO_2. None of the sailors felt short of breath or noticed anything wrong until several individuals fainted. Why do you think they did not feel short of breath?

Using Your Knowledge

Do you smoke, or are you close to a person who smokes? Quitting the smoking habit is one of the most important and effective changes you can make to improve your health and your quality of life. Quitting is not easy; the nicotine in tobacco is highly addictive. But there is help available. College health centers and wellness programs, lung and heart associations, and state and county health departments usually have information, classes, and programs to help people stop smoking. Thousands have kicked the habit, and you can too! Good luck!

Circulation

If you are of average size, your heart pumps about 5 liters of blood every minute. This is about equal to the total volume of blood in your body. The blood rushes under pressure through thick-walled arteries to smaller arterioles and finally through a network of millions of tiny capillaries. Here the blood slows down and materials are exchanged between it and the extracellular fluid that bathes all body cells. A muscle cell, brain cell, or bone cell may be far from the external environment, but it is never more than a fraction of a millimeter from the blood in one of these capillaries. Via the capillary net, every cell obtains food and oxygen and disposes of carbon dioxide and other wastes. Chapter 23 is the story of blood, the vessels that carry it, and the heart that pumps it.

Organizing Your Knowledge

Exercise 1   (Modules 23.1 – 23.2)

These modules give an overview of circulatory systems—their basic functions and parts. Review the information by matching the following words and phrases. Each answer is used only once.

A. Ventricle	_____ 1. Network of small blood vessels
B. Open system	_____ 2. Circulatory system of vertebrates
C. Capillary	_____ 3. Inadequate for transport over long distance
D. Atrium	_____ 4. Vessel that carries blood away from heart
E. Blood	_____ 5. Solution in spaces between cells
F. Closed system	_____ 6. Heart chamber that pumps blood out via artery
G. Capillary bed	_____ 7. Carries out circulatory functions in a jellyfish
H. Diffusion	_____ 8. Circulatory system of insects, spiders, and so on
I. Interstitial fluid	_____ 9. Heart chamber that receives blood from veins
J. Artery	_____10. Vessel that conveys blood from arteries to veins
K. Gastrovascular cavity	_____11. Vessel that returns blood to the heart
L. Vein	_____12. The circulatory fluid

Exercise 2   (Module 23.3)

The fish heart pumps blood through two sets of capillaries. As the blood of a fish passes through the tiny gill capillaries, it loses pressure. Therefore, once it has picked up oxygen, it delivers this oxygen to the capillaries in body tissues rather half-heartedly. The mammal heart is like two fish hearts side by side. Each side pumps blood only through one set of capillaries. The right heart pumps blood only to the lungs. The left heart then raises the pressure and pushes this blood on its way to body tissues. On the next page, compare fish and mammal circulatory systems. Trace the flow of blood in each animal by numbering the parts blood passes through in order.

A. Fish

_1_ a. Ventricle

_____ b. Systemic capillaries

_____ c. Atrium

_____ d. Gill capillaries

B. Mammal

_1_ a. Right ventricle

_____ b. Left atrium

_____ c. Lung capillaries

_____ d. Right atrium

_____ e. Systemic capillaries

_____ f. Left ventricle

Exercise 3   (Module 23.4)

To review the mammalian cardiovascular system, start by labeling the parts indicated in this diagram. Then color the vessels that carry oxygen-rich blood red and those that carry oxygen-poor blood blue. Finally, trace the path of blood flow by numbering the circles (1–12).

Exercise 4 (Module 23.4)

One way to learn about circulation and the parts of the heart is to trace the circulation of a drop of blood from one location in the body to another, naming all the structures that the drop of blood passes on its way. Use Figures 23.4B and 23.4C in the text. (A tip: Remember that blood always has to circulate from an artery to capillaries to a vein. There aren't any shortcuts. For example, to get from the capillaries of the big toe to capillaries of the little toe, blood must go back to the heart, to the lungs, back to the heart, and then to the little toe.)

Imagine a drop of blood starting in a brain capillary, circulating to the foot, and then circulating to the hand. The blood flows from the capillary, into a vein that runs down the neck, and empties into the [1]_____, the large vein that serves the head and arms. From there, the blood enters the [2]_____ of the heart. This chamber pushes the blood into the [3]_____, which pumps it out through the [4]_____ to the lungs. In the capillaries of the lungs, the blood picks up oxygen. Then it returns to the heart via the [5]_____. It enters the [6]_____, which pumps the blood to the [7]_____. This chamber pumps blood out through the [8]_____, the largest blood vessel in the body. This vessel branches and rebranches, and finally the blood is delivered to [9]_____ in the foot, where nutrients and oxygen are dropped off at the tissues. The drop of blood travels back to the heart via leg veins, which join the [10]_____, which empties into the heart. The [11]_____ atrium and ventricle again pump blood through the [12]_____ circuit to the lungs. The [13]_____ atrium and ventricle then pump the blood out through the aorta into the [14]_____ circuit. This time the drop of blood flows down a large [15]_____ to the hand, where the blood again passes through a network of capillaries.

Try going on, tracing the flow of blood from the hand, to the intestine to pick up food molecules, or to the kidneys for filtration, and then back to the brain. Once you have done this a couple of times, try it without the diagram, or make your own sketch.

Exercise 5 (Modules 23.4 – 23.5)

Complete the following chart, comparing the structure of capillaries, arteries, and veins.

	Capillaries	*Arteries*	*Veins*
Carry blood from	1.	2.	3.
Carry blood to	4.	5.	6.
Thickness of walls (thick, thin, or in between)	7.	8.	9.
Layers in walls (names)	10.	11.	12.
Valves? (yes or no)	13.	14.	15.

Exercise 6 (Module 23.6)

The following exercise relates to the function of heart valves and cardiac output. Indicate which of the following statements refer to the atrioventricular (AV) valves and which refer to the semilunar (SL) valves. Write the name of the valve next to each phrase.

_____ A. Between atria and ventricles
_____ B. Open during systole
_____ C. At exits from ventricles
_____ D. Prevent backflow from ventricles to atria
_____ E. Prevent backflow from aorta and pulmonary artery to heart
_____ F. Close during systole

Exercise 7 (Module 23.6)

Cardiac output is the amount of blood that the left ventricle pumps into the aorta per minute. It is equal to the amount of blood pumped per beat times the number of beats per minute.

1. What is your cardiac output right now? An average heart pumps about 75 mL of blood per beat. To calculate cardiac output, take your pulse, then multiply 75 mL times the number of beats per minute.

   Cardiac output = 75 mL × _____ beats per minute = _____ mL of blood per minute

2. Now run in place for a minute, take your pulse, and make the same calculation.

   Cardiac output = 75 mL × _____ beats per minute = _____ mL of blood per minute

3. How much did output change with exercise? _____ mL of blood per minute

Exercise 8 (Modules 23.7 – 23.8)

Find a word or phrase in Module 23.7 or 23.8 that goes with each of the following. Each answer is used only once.

_____ 1. Locations of heart attack pain

_____ 2. Region that sets heart rate

_____ 3. Death of cardiac muscle cells due to heart vessel blockage

_____ 4. Relays signal to contract from atria to ventricles

_____ 5. Vessel blocked in heart attack

_____ 6. Buildup of lipids on the inside of blood vessels

_____ 7. Most common surgery for blocked arteries

_____ 8. Location of cardiovascular control centers

Exercise 9   (Modules 23.9 – 23.10)

Pay special attention to Figure 23.9A; it contains a lot of information. Included are sizes, arrangement, and names of the blood vessels and the changes that occur in pressure and velocity as blood passes through them. After reading the modules and studying the figures, match each of the following statements with one of the blood vessels listed on the left.

A. Aorta
B. Arteries
C. Arterioles
D. Capillaries
E. Venules
F. Veins
G. Vena cava

_____ 1. Pressure is lowest here.
_____ 2. Pressure and speed drop the most in these vessels.
_____ 3. Pressure is usually measured in this kind of vessel.
_____ 4. Blood moves fastest here.
_____ 5. These vessels have the strongest pulse.
_____ 6. Blood moves most slowly here.
_____ 7. These vessels are the narrowest.
_____ 8. Diastolic pressure here might be 80 mm.
_____ 9. Pressure here might be 20 mm.
_____ 10. Pressure is highest here.
_____ 11. Systolic pressure here might be 120.
_____ 12. Velocity of blood increases sharply as blood flows through these vessels.
_____ 13. Reduced flexibility and clogging of these vessels may cause hypertension.
_____ 14. Blood flows rapidly here, but there is no pulse.
_____ 15. Muscles and breathing help propel blood through these vessels.

Exercise 10   (Module 23.11)

The two mechanisms that control blood distribution, discussed in Module 23.11, sound very much alike. In the first example (Figure 23.11A), smooth muscle in arterioles *leading to* a capillary bed relax or contract to allow blood into the capillaries or divert it away. In the second example (Figure 23.11B), muscle at the beginnings of the *capillaries themselves* control blood flow. Changes in the tissues surrounding capillaries often influence increase or decrease in vessel diameter. Can you imagine two physical or chemical changes that might occur in exercising leg muscles that might trigger an increase in blood flow through the capillaries in the muscles? Write those changes below.

Exercise 11 (Module 23.12)

Through the thin walls of capillaries, materials are exchanged between the blood and the interstitial fluid that surrounds body cells. Complete this diagram showing the movement of blood through a capillary and exchange of materials with the interstitial fluid. Draw a red arrow to show blood moving from an arteriole into the capillary. Draw a blue arrow to show blood moving out of the capillary into a venule. (Why use two different colors?) Two forces are responsible for exchange of fluid between the blood and the interstitial fluid: Blood pressure tends to push fluid out of a capillary, and osmotic pressure tends to push water in. Draw large and small arrows at the arterial end of the capillary to show the relative strengths of blood pressure and osmotic pressure. Do the same at the venous end. Now draw arrows to show the net movement of fluid at the venous and arterial ends. (Do your arrows show net movement of fluid into or out of the capillary at the arterial end? Why? Is there net movement of fluid into or out of the capillary at the venous end? Why?)

Exercise 12 (Modules 23.13 – 23.15)

Most of the information on blood is summarized in the figure in Module 23.13. Study the composition of blood, and then compare the three blood cell types by filling in the blanks in the following table.

Cell Type	Relative Size	Relative Numbers	Function
Platelets	Smallest	1.	2.
3.	4.	5.	Oxygen transport
Leukocytes	6.	Least numerous	7.

Exercise 13 (Module 23.16 and Summary)

All the components necessary for blood clotting are present in blood all the time. Tissue damage activates them so that clotting occurs, in a sort of "chain reaction." Try to visualize blood flow and blood clotting by filling in the missing words in the following story.

You step into the Microtron, and you are quickly reduced to a size slightly smaller than a red blood cell. The support team injects you into a small artery in the arm. Blood pressure is fairly [1]_____ here. You feel a boom of pressure on your eardrums about once per second; this is simply the [2]_____, and it will gradually disappear as the blood in the artery flows into the narrower [3]_____ that lead to the capillary beds. Bright lights on the subject's arm enable you to see what is around you: Most of the cells around you are [4]_____, flexible disks carrying [5]_____ to the body's cells. There are also a few larger, irregular [6]_____, important in body defense. They slowly crawl along the blood vessel walls. Some even move against the current. It is best to avoid them, because some are [7]_____, capable of eating bacteria and debris. All around you are tiny "blobs." These must be [8]_____, which are involved in maintaining osmotic balance, defense, and blood clotting. There are also swarms of small fragments of cell cytoplasm, called [9]_____, that assist in the clotting process.

As you enter a [10]_____, blood slows almost to a stop and the scene brightens. The walls of these vessels are a single layer of [11]_____, only one cell thick. You can even see gaps between cells, where fluid in the capillary is exchanged with the [12]_____ fluid. There isn't much room here though. The capillaries are so narrow that the red blood cells have to line up single file in some places.

You are just under the skin of the fingertip. The team pricks the subject's skin with a pin. You are moving directly toward the wound, so you use your gripper to hang onto the vessel wall. The clotting process is already underway. The damaged lining of the vessel exposes [13]_____ to the blood. [14]_____ stick to the exposed tissue and release a cloud of chemicals. These chemicals cause even more platelets to adhere. But in this case, the damage is too serious for a platelet plug to stop the leak. Chemicals released from the platelets and damaged cells in the vessel wall trigger [15]_____ in the blood to change into an enzyme called thrombin. This enzyme then causes the small blobs of fibrinogen floating in the blood to change shape and form sticky strands of [16]_____. These strands stretch like a tangle of cords across the hole in the vessel, trapping red blood cells. The blood strains against the fibrin clot, but finally it holds and leakage stops.

A phagocyte has caught your leg! You break free, but in the process you loosen a big chunk of the clot. It could travel to the heart, lodge in one of the [17]_____, and cause a [18]_____! You let the flow of blood carry you along. As blood leaves the capillary bed and enters a [19]_____, blood flow speeds up. The vessel walls thicken, and it gets darker again. You enter an even larger [20]_____, and the blood slows down even more. Ahead you can dimly make out the flaps of a [21]_____, which keeps the blood moving toward the heart. Fortunately, the clot is briefly caught in an eddy downstream from the valve. You break it into fragments small enough to pose no threat to the subject. This is a good time to make your exit, and you are soon back in the lab discussing your adventure.

Exercise 14   (Summary)

This chapter introduces a lot of vocabulary connected with circulation. Circle the term that does not fit with the others in each of these groups, and briefly explain what the other terms in each group have in common.

1. venule   artery   atrium   capillary

2. fibrin   leukocyte   platelet   red blood cell

3. sphygmomanometer   systolic pressure   fibrinogen   hypertension

4. pulmonary veins   right ventricle   lungs   aorta

5. fibrin   leukocyte   platelet   hemophilia

6. semilunar valve   AV node   SA node   pacemaker

7. diastole   systole   pulse   leukocyte

8. oxygen   red blood cell   hemoglobin   epithelium

9. pulmonary artery   systemic circuit   aorta   left ventricle

Testing Your Knowledge

Multiple Choice

1. Rhythmic stretching of the arteries caused by heart contractions is called
   a. hypertension.
   b. heart murmur.
   c. hemophilia.
   d. pulse.
   e. diastole.

2. Which of the following animals has an open circulatory system?
   a. fish
   b. human
   c. frog
   d. *Hydra*
   e. spider

3. Which of the following *cannot* move freely in and out of a capillary?
   a. sugar
   b. oxygen
   c. carbon dioxide
   d. water
   e. plasma protein

4. Heart valves function to
   a. keep blood moving forward through the heart.
   b. mix blood thoroughly as it passes through the heart.

   c. control the amount of blood pumped by the heart.
   d. slow blood down as it passes through the heart.
   e. propel blood as it passes through the heart.

5. Which of the following correctly traces the electrical impulses that trigger each heartbeat?
   a. ventricles—pacemaker—AV node—atria
   b. pacemaker—AV node—atria—ventricles
   c. atria—pacemaker—AV node—ventricles
   d. pacemaker—atria—AV node—ventricles
   e. pacemaker—AV node—atria—ventricles

6. A heart attack occurs when
   a. a heart valve malfunctions.
   b. a coronary artery is blocked.
   c. the heart is weakened by overwork.
   d. the aorta is blocked.
   e. the pulmonary artery is blocked.

7. The cells responsible for defense against infections are
   a. red blood cells.
   b. white blood cells.
   c. epithelial cells.
   d. platelets.
   e. pacemaker cells.

8. If cells are not receiving enough oxygen, a hormone signals the bone marrow to produce more
   a. leukocytes.
   b. fibrin.
   c. plasma.
   d. platelets.
   e. erythrocytes.

9. The primary sealants that plug leaks in blood vessels are
   a. platelets and fibrin.
   b. red blood cells and albumin.
   c. fibrin and white blood cells.
   d. white blood cells and platelets.
   e. hemoglobin and platelets.

10. Which of the following best describes an artery?
    a. carries blood away from the heart
    b. carries oxygenated blood
    c. contains valves
    d. has thin walls
    e. carries blood away from capillaries

11. Blood moves most slowly in
    a. capillaries.
    b. the aorta.
    c. veins.
    d. arterioles.
    e. venules.

12. In a fish, blood circulates through ____ , while in a mammal, it circulates through ____ .
    a. two circuits . . . four circuits
    b. one circuit . . . two circuits
    c. four circuits . . . two circuits
    d. one circuit . . . four circuits
    e. two circuits . . . one circuit

13. The amount of blood flowing to skeletal muscles is greatly increased during exercise. This redirection of blood into muscles is accomplished by
    a. contraction of muscle in the walls of arteries.
    b. relaxation of muscle in walls of arterioles.
    c. opening of valves in veins.
    d. opening of valves in arteries.
    e. relaxation of muscle in the walls of veins.

Essay

1. Explain how the circulatory system changed to accommodate lung breathing and greater activity as land vertebrates evolved.

2. Cardiac output is the amount of blood the left ventricle of the heart pumps per minute. Cardiac output can be as much as four times greater when you are exercising than when you are at rest. What two things could the heart increase to increase cardiac output when you are exercising?

3. How does high blood pressure contribute to cardiovascular disease?

4. Briefly describe and explain changes in blood pressure as blood flows from arteries, to capillaries, to veins.

5. Describe the structural characteristics of capillaries that make them well suited for exchange of materials between blood and tissues.

Applying Your Knowledge

Multiple Choice

1. Just after blood leaves the left ventricle of the human heart, it passes through the
   a. pulmonary artery.
   b. left atrium.
   c. aorta.
   d. superior vena cava.
   e. right ventricle.

2. In which of the following animals are blood and interstitial fluid the same?
   a. grasshopper
   b. jellyfish
   c. fish
   d. dog
   e. sparrow

3. The heart specialist listened to Paul's heart through a stethoscope. Instead of the normal "lub-dupp, lub-dupp" heart sounds, he heard "siss-dupp, siss-dupp." The doctor said, "Hmm . . . I'm not sure it is anything to worry about, but I think there is something wrong with
   a. one of your coronary arteries."
   b. the pacemaker."
   c. an atrioventricular valve."
   d. your aorta."
   e. a semilunar valve."

4. Which of the following terms would be *least* useful in describing the circulatory system of a fish?

a. capillary bed
b. pulmonary artery
c. ventricle
d. atrium
e. cardiovascular system

5. In circulating by the shortest route from the lungs to the foot, how many times would a drop of blood pass through the left ventricle?
a. 0
b. 1
c. 2
d. 3
e. 4

6. A recording of the electrical activity of a patient's heart shows that the atria are contracting regularly and normally, but every few beats the ventricles fail to contract. Which of the following is probably not functioning properly?
a. AV node
b. semilunar valve
c. coronary artery
d. pacemaker
e. AV valve

7. Which of the following functions most like a valve in a vein?
a. a kitchen faucet
b. a revolving door
c. the volume control on a radio
d. a turnstile
e. a sliding patio door

8. In circulating from the brain to the arm, a drop of blood would *not* have to pass through which of the following?
a. left atrium
b. aorta
c. superior vena cava
d. pulmonary vein
e. inferior vena cava

9. A patient with which of the following conditions would probably *not* be helped by injections of purified stem cells?
a. impaired resistance to infection
b. leukemia
c. AIDS
d. hypertension
e. sickle-cell anemia

10. Emphysema damages the tissues of the lungs and slows pulmonary blood flow. This causes blood to back up, stretching and weakening the walls of the heart and blood vessels. Which

of the following do you think would be most affected by this backup of blood from the lungs?
a. aorta
b. right atrium
c. left atrium
d. right ventricle
e. left ventricle

Essay

1. A runner's heart rate is 160 beats per minute, and 90 mL of blood is pumped by the left ventricle with each beat. What is the runner's cardiac output?

2. The figures for blood pressure in an artery are usually given like this: 130/80 mm Hg. But it takes only one figure to specify blood pressure in a vein: 5 mm Hg. Why the difference?

3. Recall the forces that cause fluid to leave and reenter a capillary. How do you think high blood pressure would affect this balance of forces? How does this help explain that one of the symptoms of high blood pressure is swelling of the tissues with fluid?

4. Sometimes a baby is born with its large blood vessels reversed: The right ventricle pumps blood out through the aorta, and the left ventricle is connected to the pulmonary artery. The system is otherwise normal. How would this alter blood flow? Why would this be disastrous if not quickly corrected by surgery?

Using Your Knowledge

1. What kinds of changes in the life-styles of many Americans have led to the recent downturn in the incidence of cardiovascular disease? Have you made any changes in your health habits to avoid future heart or circulatory problems? Are there any changes you would like to make?

2. Have you had your blood pressure checked lately? Do you remember what your blood pressure was? How high does pressure have to be to indicate high blood pressure? Even young people can have high blood pressure. If you haven't had yours checked lately, it might be a good idea. Many pharmacies have simple devices you can use to check your blood pressure yourself.

The Immune System

If it were not for your immune system, you would have died long ago from measles, chickenpox, pneumonia, influenza, or any of hundreds of other diseases. Many times during your life, dangerous bacteria or viruses broke through the skin or other nonspecific defenses. Each time, your immune system protected you, and each time it was 100% successful. Microorganisms can overwhelm the body so quickly that 99% effectiveness won't do. The cells and antibodies of the immune system must specifically identify and destroy every invader and at the same time spare the body's own cells. Your immune system's remarkable ability to recognize and protect your "self" from "nonself" is the topic of this chapter.

Organizing Your Knowledge

Exercise 1   (Module 24.1)

Nonspecific obstacles are the body's first line of defense. Match each of the following nonspecific defenses with a phrase from the left.

A. Protein that cuts holes in bacteria   _____   1. Natural killer cell
B. Filters inhaled air   _____   2. Skin
C. Protective enzyme in sweat and tears   _____   3. Macrophage
D. Barrier that is first line of defense   _____   4. Interferon
E. Large phagocytic cell of interstitial fluid   _____   5. Stomach acid
F. Digests most microorganisms in food   _____   6. Complement
G. White blood cell that engulfs bacteria   _____   7. Lysozyme
H. Attacks cancer cells and cells infected by viruses   _____   8. Hair
 I. Helps cells resist viruses   _____   9. Neutrophil

Exercise 2   (Module 24.2)

Study this module on the inflammatory response. Then, from memory, fill in the missing words in this diagram.

Pin — Skin surface

A. _____

Chemical signals

Blood vessel

C. _____

Phagocytes and D. _____ move into area

Phagocytes

(1) Tissue injury; release of chemical signals such as B. _____

(2) Dilation and increased E. _____ of local blood vessels; migration of phagocytes to the area

(3) Phagocytes (macrophages and neutrophils) consume F. _____ and G. _____; tissue heals

Exercise 3   (Modules 24.3 – 24.5)

These modules introduce some basic concepts and terminology related to the immune and lymphatic systems. To test your knowledge of this material, state the difference between the terms in each of the following pairs.

1. antigen–antibody

2. T cell–B cell

3. cell-mediated immunity–humoral immunity

4. lymphatic system–circulatory system

5. passive immunity–active immunity

6. blood capillary–lymphatic capillary

7. specific defenses–nonspecific defenses

What is immunity? Why do you sometimes have to get sick to become immune to a disease? To answer these questions, fill in the blanks in the following story, using words and phrases from Modules 24.5 – 24.8.

Brian stopped by one afternoon to visit his friends Tom and Alicia and their 2-year-old daughter Samantha. Little did any of them suspect that Samantha was coming down with the measles. She had not yet been immunized, and she had caught the virus from a little boy in her day-care group.

Brian had never had the measles either. Once he had been exposed, and his
[1]_____ defenses had been breached, it was too late to do anything. It was up to the specific defenses of the [2]_____ system to fight the invading viruses.

Both B and T lymphocytes were involved in the battle, but we will concentrate on the B cells. The viruses's protein coats contained protein molecules foreign to Brian's body; these [3]_____ are what triggered his immune response. Throughout his body were many different types of B lymphocytes, each capable of responding to a different antigen. On each of these B cells were [4]_____ that acted as receptors for various potential foreign antigens. Only a specific B-cell type possessed antibodies whose [5]_____ were complementary to the shape of the measles virus antigens. Eventually the measles viruses encountered some of these B cells in a [6]_____ node. Only these cells were "chosen" to be activated to fight the invading viruses. This process is called clonal [7]_____.

The stimulated B cells began to multiply, forming a [8]_____, a population of genetically identical [9]_____ cells. These cells secreted [10]_____ capable of locking on to the viral antigens and inactivating the measles viruses.

The initial phase of immunity described so far is called the [11]_____ immune response. Unfortunately, Brian had to suffer through the measles, because this response is too slow to wipe out the invaders before they cause harm. It usually takes several [12]_____ for lymphocytes to become activated and form effector cell clones, and by that time Brian was already sick.

After he recovered, however, Brian was [13]_____ to measles. When his college roommate came down with measles just before final exams, Brian was safe. A second exposure to the same antigen triggers the [14]_____ immune response, which is much quicker and stronger than the primary response and also lasts longer. The explanation for this is that the first exposure actually triggers the formation of two cell clones, the effector cells that fought the original infection and also a clone of [15]_____ cells, which are held in reserve. Whereas effector cells may live only a few days, these other cells may last for [16]_____. They are capable of mounting a quick and powerful secondary response.

Although Brian was protected from the measles, he still caught a cold the day before his math exam. But that is another story entirely.

Exercise 5   (Module 24.9)

This module details the role of B cells in immunity. Review the action of B cells by labeling and coloring the following diagram. First identify and label the **primary immune response** and the **secondary immune response**. Label **antigens** and color them red. Label **B-cells** and color them light blue. Draw green antigen receptors on the B-cell types, and also draw green **antibodies** produced by plasma cells. Be sure to make antibodies "fit" antigens. Label **plasma cells** and color them dark blue. Label **memory B cells** and color them purple. When you are done, explain the diagram.

1. _____

2. _____

3. _____

4. _____

5. _____

6. _____

7. _____

8. _____

9. _____

10. _____

Exercise 6   (Modules 24.10 – 24.11)

The variable regions at the tips of an antibody's "arms" recognize antigens, and the constant region in the base of the Y helps destroy and eliminate antigens. As shown in Figure 24.11 in the text, there are a number of ways in which antibodies assist in destruction of antigens. These processes have been scrambled in the diagrams below. Identify each process and label the diagrams accordingly. Choose from **phagocytosis, precipitation of dissolved antigens, neutralization, cell lysis, activation of complement,** and **agglutination of cells.**

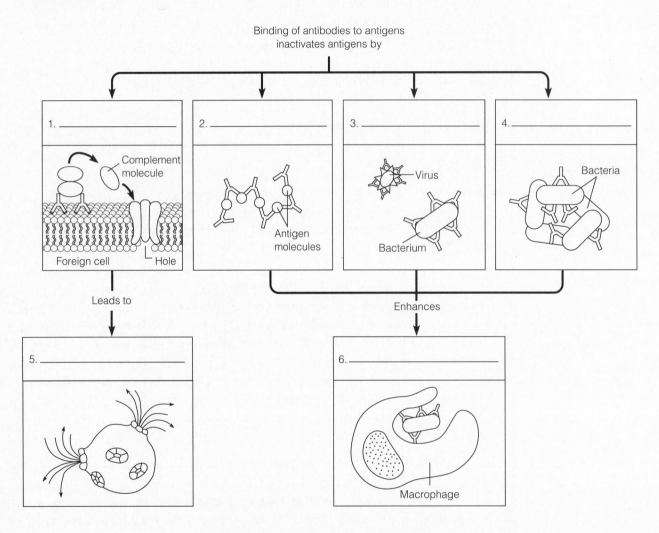

Exercise 7   (Module 24.12)

The usefulness of monoclonal antibodies stems from their ability to recognize and attach to specific molecules, which otherwise might be difficult to single out. See if you can think of several areas not mentioned in the text where this ability might be useful, and write your ideas on a separate piece of paper. Try not to limit your speculations to medicine. For example, how might monoclonal antibodies be useful in criminology, wildlife management, or archaeology?

Exercise 8 (Module 24.13)

T cells are responsible for cell-mediated immunity. Contrast this arm of the immune system with humoral immunity, described in Module 24.9. Then match each of the following components with its role in cell-mediated immunity.

A. Protein that recognizes antigens
B. Chemical that attacks infected cell
C. Signal that activates T and B cells
D. Macrophage that displays antigen
E. Identity marker on body cells
F. Cell that activates T and B cells
G. Cell that attacks infected cells
H. Stops immune activities when done

_____ 1. Suppressor T cell
_____ 2. Antigen-presenting cell
_____ 3. Helper T cell
_____ 4. T-cell receptor
_____ 5. Self molecule
_____ 6. Perforin
_____ 7. Interleukin-2
_____ 8. Cytotoxic T cell

Exercise 9 (Modules 24.14 – 24.19)

Sometimes the immune system fails to respond to invaders. Sometimes it goes overboard, responding to harmless substances in the environment, or even attacking the body's own tissues. Briefly describe how the normal function of the immune system goes awry in each of the following situations.

1. An immunodeficiency disease, like severe combined immunodeficiency

2. An allergy to cat hair

3. A cancerous brain tumor

4. Rejection of a heart transplant

5. Rheumatoid arthritis, an autoimmune disease

6. AIDS

7. Catching a cold after a stressful period, like final exams

Understanding the immune system is important and useful. Understanding AIDS could save your life. Choose the correct italicized words or phrases to complete this story.

Lauren was worried and anxious because last week, her friend Robert told her that he had tested positive to antibodies against (1) *HIV, SCID*, the virus that causes AIDS. Now Lauren was waiting for the results of a blood test to determine whether she had been infected by Robert.

Lauren and Robert had had a sexual relationship that had ended about 6 months before. Because she was using birth control pills, they had never used (2) *an IUD, condoms, withdrawal,* which might have given them some protection. Besides, Lauren thought AIDS was primarily a disease of drug addicts and homosexuals. When he called, Robert told her that he was very sorry; he had no idea that he could have AIDS. He said that he had probably been infected by a woman with whom he had had an earlier relationship. She had in fact experimented with drugs and had probably been infected by blood on a needle shared with a person infected with AIDS. Robert was probably infected via contact with the woman's (3) *blood, skin, saliva* during intercourse. Lauren may have contracted AIDS from Robert's blood or semen.

As Robert and Lauren now knew, AIDS is increasing in the heterosexual, non-drug-using population. In the United States, over (4) *200,000, 10,000* individuals have been diagnosed with the disease. Worldwide, more than (5) *1 million, 10 million* people may be infected.

Lauren had learned about AIDS in her biology class. The virus attacks the body's immune system, specifically (6) *B lymphocytes, helper T cells, cytotoxic T cells.* The virus contains (7) *RNA, DNA.* When it enters a cell, this molecule directs the synthesis of (8) *DNA, RNA* that becomes part of the cell's chromosomes. The viral instructions may remain dormant for some time but eventually direct the cell to make more viruses. The viruses then destroy the host cell and leave to infect other cells. Because HIV attacks cells of the immune system, the body of the AIDS victim is wracked by infection. Most AIDS victims die of another disease or (9) *autoimmunity, cancer, allergies,* usually within (10) *1 year, 3 years.* There is no vaccine against AIDS. New drugs such as (11) *HIV, AZT* offer a longer life to those infected by AIDS, but there is no cure for the disease.

Lauren had had a bad cold for the previous 2 weeks. At first she had thought nothing of it, but now she wondered if her immune system were beginning to falter. The telephone rang. She took a deep breath, sat down at the desk, and picked up the phone. The voice at the other end was professional. All Lauren could get out was, "Negative? You mean I'm, I'm . . . I don't have AIDS?"

Testing Your Knowledge

Multiple Choice

1. An antigen is
   a. a protein molecule that helps defend the body against disease.
   b. a type of white blood cell.
   c. an invading virus or bacterium.
   d. a foreign molecule that evokes an immune response.
   e. a body cell attacked by an invading microorganism.

2. Your lymphatic system fights infection and
   a. delivers food and water to tissues.
   b. carries glandular secretions.
   c. maintains high blood pressure.
   d. allows red blood cells to approach cells more closely.
   e. drains fluid from tissues.

3. How do memory cells differ from effector cells?
   a. Memory cells are more numerous.
   b. Memory cells are responsible for the primary immune response.
   c. Memory cells attack invaders and effector cells do not.
   d. Memory cells live longer.
   e. Memory cells are capable of producing antibodies.

4. Which of the following triggers tissue inflammation?
   a. accumulation of phagocytes in an injured area
   b. release of interferon by infected cells
   c. increased blood flow in an infected or injured area
   d. fever
   e. release of chemicals such as histamine by damaged cells

5. A cell capable of producing monoclonal antibodies is produced by fusing a lymphocyte with a
   a. cancer cell.
   b. red blood cell.
   c. bone marrow cell.
   d. T cell.
   e. macrophage.

6. A clone of lymphocytes
   a. produces different antibodies.
   b. lives in the same area of the body.
   c. consists of immature cells, incapable of carrying out an immune response.
   d. makes antibodies against the same antigens.
   e. consists of helper, suppressor, and cytotoxic T cells.

7. Most individuals infected with HIV
   a. can live for 10 to 15 years.
   b. never develop AIDS.
   c. die from autoimmune reactions.
   d. suffer from increased sensitivity to foreign antigens.
   e. die from other infections or cancer.

8. Tissues are "typed" before an organ transplant to make sure that the _____ of donor and recipient match as closely as possible.
   a. T cells
   b. antibodies
   c. "self" molecules
   d. histamines
   e. B cells

9. A vaccine contains
   a. white blood cells that fight infection.
   b. antibodies that recognize invading microbes.
   c. inactivated disease-causing microbes.
   d. a hormone that boosts immunity.
   e. lymphocyte antigens.

10. When you are immune to a disease
    a. antibodies against the disease are constantly circulating in your blood.
    b. certain lymphocytes are able to make the proper antibodies quickly.
    c. your nonspecific defenses are strengthened.
    d. B cells are stimulated to quickly engulf invaders.
    e. antigens are altered so invaders can no longer attack your tissues.

11. An antibody is a
    a. protein that attaches to an antigen.
    b. foreign substance or organism.
    c. white blood cell that attacks invading bacteria or viruses.
    d. molecule on a body cell that identifies the cell as "self."
    e. large carbohydrate molecule that helps defend the body.

12. B lymphocytes
    a. attack cells that have been infected by viruses.
    b. engulf and destroy bacteria and viruses.
    c. multiply and make antibodies that circulate in blood and lymph.
    d. are responsible for cell-mediated immunity.
    e. all of the above

13. The AIDS virus infects mostly
    a. cells of the nervous system.
    b. B lymphocytes.
    c. cytotoxic T cells.
    d. suppressor T cells.
    e. helper T cells.

14. The biggest difference between cell-mediated immunity and humoral immunity is
    a. how long their protection lasts.
    b. whether a subsequent secondary immune response can occur.
    c. whether clonal selection occurs.
    d. how they respond to and dispose of invaders.
    e. how fast they can respond to an invader.

15. Viruses and bacteria in body fluids are attacked by
    a. antibodies from B cells.
    b. cytotoxic T cells.
    c. complement proteins.
    d. helper T cells.
    e. suppressor T cells.

16. What do the antibodies secreted by plasma cells (the effector cells of humoral immunity) do to attack their targets?
    a. activate complement to punch holes in them
    b. clump cells together so that phagocytes can ingest them
    c. cause antigen molecules to settle out of solution
    d. attach to antigens and detoxify them
    e. all of the above

Essay

1. Explain how vaccination allows you to develop immunity to a disease without becoming ill. Include in your explanation the primary and secondary immune responses.

2. Compare humoral and cell-mediated immunity. In your comparison, discuss types of lym-phocytes involved, roles of antibodies in the immune response, where invaders are attacked, and methods used to destroy invaders.

3. When biologists first started to work out the mechanisms of immunity, they found that the body produced antibodies that matched the shape of invading antigens. The researchers suspected that they would find that the immune system somehow analyzed the antigens and custom-built antibodies to fit. Did they find this to be the case? Explain.

4. Joan was telling her friend Jim why she had not been able to make it to biology class for several days: "My throat has been so sore I could hardly swallow," she croaked. "And the glands in my neck are really sore and swollen." Jim said, "We have been talking about this in class. They are not really 'glands,' you know, and the reason they are sore is . . ." Complete Jim's explanation.

5. Describe the inflammatory response and how it helps the body deal with injury or infection.

Applying Your Knowledge

1. Which of the following is not present until after the primary immune response occurs?
   a. memory cells
   b. macrophages
   c. helper T cells
   d. complement proteins
   e. antigens

2. The relationship between an antigen and an antibody is most like
   a. a battery and a flashlight.
   b. a hand and a glove.
   c. a hammer and a nail.
   d. a left foot and a right foot.
   e. a recipe and a cake.

3. A group of researchers have tested many chemicals and found several that have potential for use in treating malfunctions of the immune system. Which of the following would seem to have the most promise as a drug for inhibiting autoimmune diseases?
   a. Compound A13: acts like histamine
   b. Compound Q6: stimulates cytotoxic T cells
   c. Compound N98: a potent allergen

d. Compound B55: stimulates suppressor T cells

e. Compound M31: stimulates helper T cells

4. The body produces antibodies complementary to foreign antigens. The process by which the body comes up with the correct antibodies to a given disease is most like
   a. going to a tailor and having a suit made to fit you.
   b. ordering the lunch special at a restaurant without looking at the menu.
   c. going to a shoe store and trying on shoes until you find a pair that fits.
   d. picking out a video that you haven't seen yet.
   e. selecting a lottery prize winner by means of a random drawing.

5. Rhonda has been diagnosed as suffering from an immunodeficiency disease. Her doctor suspected Rhonda might have an immunodeficiency because
   a. Rhonda strongly rejected an organ transplant.
   b. Rhonda suffered from numerous allergies.
   c. Rhonda's blood showed high levels of numerous antibodies.
   d. Rhonda seemed to be immune to her own "self" molecules.
   e. Rhonda suffered from repeated, prolonged infections.

6. The idea behind vaccination is to induce _____ without the vaccinated individual having to get sick.
   a. passive immunity
   b. the primary immune response
   c. anaphylactic shock
   d. nonspecific defenses
   e. inflammation

7. Researchers found that when laboratory rats were already infected with a virus, they were better able to resist infection by a second completely different virus. The first infection apparently caused _____ , which protected the rats from the second infection.
   a. increased stress
   b. secretion of interferons
   c. production of antibodies
   d. passive immunity
   e. cell agglutination

8. Which of the following would be effective in eliminating bacteria but ineffective against viruses? (Hint: Viruses are not cells.)
   a. activation of complement proteins
   b. secretion of interferon by infected cells
   c. neutralization by antibodies
   d. agglutination by antibodies
   e. perforin secretion by cytotoxic T cell

9. An allergen acts like
   a. an antigen.
   b. histamine.
   c. interferon.
   d. an antibody.
   e. complement.

10. In a series of immune system experiments, the thymus glands were removed from baby mice. Which of the following would you predict as a likely result?
    a. The mice suffered from numerous allergies.
    b. The mice never developed cancerous tumors.
    c. The mice suffered from autoimmune diseases.
    d. The mice readily accepted tissue transplants.
    e. The mice were unable to produce an inflammatory response.

Essay

1. More people die each year from bee stings than from rattlesnake bites. The individual who is stung dies of anaphylactic shock, a massive, life-threatening allergic response. Anaphylactic shock usually occurs the second time an individual is stung by a bee. Why does this occur the second time and not the first time? Explain in terms of the function of the immune system.

2. Researchers have not yet come up with a cure for the common cold, but they have made some interesting observations, among them: (1) There are more than 50 different known kinds of rhinoviruses, the viruses that cause colds; and (2) an average 2-year-old might catch three or four colds per year, but an 80-year-old catches a cold only once every 3 or 4 years. How do these findings relate to each other and the function of the immune system?

3. Before traveling to Africa, John got a "gamma-globulin" injection containing antibodies against hepatitis. Two years later, John planned another trip to Africa. His doctor recommended another gamma-globulin injection because John was no longer immune to hepatitis. What kind of immunity did John acquire from the first injection? Why didn't the immunity last?

4. After the flu season, blood samples were obtained from flu patients, and antibodies in the blood were analyzed. Researchers found that the antibodies produced by different patients in response to the same virus were often quite different. In some cases, the antigen-binding sites of the antibodies were completely different shapes. Explain how this could occur.

5. Antibodies are Y-shaped, with antigen-binding sites at the tip of each "arm" of the Y. How does the fact that an antibody has two antigen-binding sites help the antibody inactivate invaders? (Think what would happen if they had only one antigen-binding site.)

6. Organ transplant recipients must be very careful to avoid infections, especially viral infections. Explain why.

Using Your Knowledge

1. AIDS activists charge that the U.S. government was slow to respond to the AIDS crisis because the disease was perceived to affect mostly homosexuals and drug users, and not the "mainstream." Do you think the government has been slow to respond? Do you think enough is being done to confront AIDS? If not, what else could be done?

2. Recently, several college campuses were disrupted by major measles epidemics. Are you immune to measles? Have you been immunized, or did you have measles when you were younger? How about other diseases? Do you know where to look or whom to ask to find out whether you are immune to specific diseases?

Control of the Internal Environment

25

A tiny sparrow hops across snow drifts on a subfreezing day, searching for food. How is it able to remain alive, even active, in such a hostile environment? Despite the cold, the sparrow's internal environment, the fluid that bathes its cells, remains remarkably constant. Numerous mechanisms keep it that way despite external fluctuations in temperature. Rapid cell metabolism allows the bird to generate its own body heat. Its feathers, circulatory system, and behavior control heat loss. It needs to keep eating to keep its temperature up, but metabolism of food alters body chemistry. To compensate, the sparrow's liver makes minute-to-minute adjustments in blood nutrient content and sweeps the blood for toxins. The kidneys actively sort the blood's chemical constituents, helping to maintain optimum solute content, adjust pH, get rid of waste products, and save water at the same time. This chapter discusses three processes that help maintain a constant internal environment—thermoregulation, osmoregulation, and excretion of wastes.

Organizing Your Knowledge

Exercise 1   (Introduction)

The great white shark is no "cold fish," and this module tells why. It also introduces the terms "ectotherm" and "endotherm" and explains why biologists prefer these terms to "cold-blooded" and "warm-blooded." After reading the introduction, choose answers to the questions below from the following list: **insect (invertebrate), bird, great white shark, polar bear, frog (amphibian), snake (reptile), most fish.**

1. Which of the animals above are popularly regarded as "cold-blooded"?

2. Which of the animals above are popularly regarded as "warm-blooded"?

3. Which of the animals above would be cold in a cold environment?

4. Which of the animals above would be warm in a warm environment?

5. Which of the animals above would be warmer than their surroundings in a cold environment?

6. Which of the animals above are endothermic?

7. Which of the animals above are ectothermic?

Exercise 2   (Modules 25.1 – 25.4)

Does the room temperature where you are right now feel comfortable? Do you feel a bit chilly? Or are you sweating because it is hot and humid? These modules discuss thermoregulation, and they provide numerous examples of methods animals use to regulate their internal temperatures. Animals regulate temperatures two ways: (1) by changing rate of heat production and (2) by adjusting rate of heat gain or loss. State whether each of the following is

A. a method of warming or cooling the body by regulating heat production, or
B. a method of warming the body by reducing heat loss, or
C. a method of warming the body by increasing heat gain, or
D. a method of cooling the body by increasing heat loss

_____ 1.   Moisture evaporates from a lizard's nostrils.
_____ 2.   A robin fluffs up its feathers to trap more air near the skin.
_____ 3.   A rabbit grows a thicker coat in the winter.
_____ 4.   A lizard comes out of its burrow and turns broadside to the sun.
_____ 5.   An elephant sprays itself with cold water.
_____ 6.   Bees cluster together and shiver.
_____ 7.   Blood vessels dilate in a jackrabbit's ears.
_____ 8.   You shiver.
_____ 9.   A kangaroo rat presses itself against the cool wall of its burrow.
_____ 10.   A countercurrent heat exchanger cools the blood flowing to a duck's feet.
_____ 11.   Hormones increase a mouse's metabolic rate.
_____ 12.   A polar bear's transparent hairs beam sunlight to its black skin.
_____ 13.   A cat licks itself, and saliva evaporates from its skin.
_____ 14.   A bee seeks flowers that focus sunlight on its body.

Exercise 3   (Modules 25.5 – 25.7)

Animals must maintain a correct balance of water and solutes in body fluids to stay alive. The concentration of seawater is good enough for many marine animals, so they are osmoconformers, simply matching the osmotic concentration of their environment. Other marine creatures, as well as freshwater and land animals, are osmoregulators. They actively move water and solutes in and out of their cells to maintain body fluid compositions different from their environments. Land animals have particular problems with loss of water and solutes. Use the information in the modules to complete this chart comparing osmoconformers and various osmoregulators.

	Marine Worm	Freshwater Fish	Saltwater Fish	Land Animal
Osmoregulator or osmoconformer?	1.	2.	3.	4.
Tends to gain or lose water?	5.	6.	7.	8.
Tends to gain or lose solutes (ions)?	9.	10.	11.	12.
Method of compensating for gain or loss?	13.	14.	15.	16.

Exercise 4   (Module 25.8)

Breakdown of proteins and nucleic acids produces nitrogen-containing waste products. Different animals dispose of nitrogen in different ways. Summarize the kinds of animals that excrete each of the following nitrogenous wastes, and discuss the advantages and disadvantages of each.

	Ammonia	*Urea*	*Uric Acid*
Animals excreting this compound	1.	2.	3.
Advantages of excreting this compound	4.	5.	6.
Disadvantages of excreting this compound	7.	8.	9.

Exercise 5   (Module 25.9)

The human excretory system performs important functions in fluid homeostasis. This exercise and the next will help you to become familiar with the kidneys and nephrons, their functional units. First, complete this diagram of the excretory system. Label the **renal artery** and color it red. Label the **renal vein** and color it blue. Label the **kidneys** and color them brown. Then label and color the **ureters** (light brown), the **bladder** (yellow), and the **urethra** (green).

Exercise 6   (Module 25.9)

Note in Figure 25.9 in the text how the millions of nephrons are arranged in each kidney. Label and color the following diagram of a nephron. Color arterioles and arterial capillaries red; color venous capillaries and the renal vein blue; and color the entire renal tubule, from Bowman's capsule through the collecting duct, yellow. Label **arteriole to glomerulus, glomerulus, arteriole from glomerulus, capillaries, Bowman's capsule, proximal tubule, loop of Henle, distal tubule**, and **collecting duct.**

Exercise 7   (Modules 25.9 – 25.11)

Nephrons regulate the water and solute content of blood in a four-step process: filtration, reabsorption, secretion, and excretion. The next two exercises review nephron function, which is the heart of this chapter. After studying the modules, match each of the following parts of a nephron with its function.

A. Mainly functions in water reabsorption
B. "Refine" filtrate by reabsorption and secretion
C. Delivers blood to glomerulus
D. Carries urine to central "pelvis" of kidney; also functions in water reabsorption
E. Porous ball of capillaries where filtration occurs
F. Collects filtrate from glomerulus

_____ 1. Loop of Henle

_____ 2. Bowman's capsule

_____ 3. Proximal and distal tubules

_____ 4. Glomerulus

_____ 5. Collecting duct

_____ 6. Renal artery

Exercise 8   (Module 25.11)

After studying kidney function, match each of the following components of blood with what happens to it in a nephron. Hint: Ask yourself, "Is this something the body wants to keep or something it wants to get rid of?" (Answers may be used more than once.)

A. Not filtered
B. Filtered and mostly reabsorbed
C. Filtered, then mostly *not* reabsorbed, and finally excreted in urine
D. Filtered, mostly *not* reabsorbed, *also* secreted, and finally excreted in urine

_____ 1. Glucose
_____ 2. Water
_____ 3. Urea
_____ 4. H^+
_____ 5. Blood protein
_____ 6. Amino acid
_____ 7. Red blood cell
_____ 8. Nicotine (drug)
_____ 9. Salt (NaCl)

Exercise 9   (Module 25.12)

Kidney dialysis has saved many lives, often as a stopgap until a kidney transplant can be arranged.

1. What are some advantages and disadvantages of dialysis, compared with a transplanted kidney?

2. What do you think would be some advantages and disadvantages of a kidney transplant, compared with dialysis?

Exercise 10   (Module 25.13 and Summary)

The liver complements the kidneys in regulating body fluid composition. Unlike the kidneys, the liver can chemically alter substances in the blood and store useful substances for later use. The liver releases its waste products into the blood, and they are excreted by the kidneys. Fill in the blanks in the following scenario summarizing the roles of the liver and kidneys in dealing with some substances in food.

Tom finished his breakfast with a cup of coffee (with milk and sugar) and dashed off to work. Digestion of milk proteins began immediately. The resulting amino acids were absorbed through the intestinal lining into Tom's blood. At the same time, enzymes split the sucrose (sugar) molecules, forming glucose and fructose, which also entered the blood. Caffeine molecules entered the blood immediately, without digestion.

Blood from Tom's intestine was carried via the [1]_____ to the liver. Liver cells absorbed some of the amino acids, sugars, and caffeine. The fructose was immediately converted into glucose. The blood carried some of the glucose through the liver and into general circulation, where it was distributed to nourish body cells. The rest of the glucose was converted to [2]_____ and stored in the liver. It would later be turned back into [3]_____, as Tom's blood sugar dropped between breakfast and lunch.

Some of the caffeine from the coffee made it through the liver and had the intended "wake-up" effect on Tom's brain as he was driving to work. Like other potentially harmful chemicals, the rest of the caffeine was [4]_____ by liver cells and the resulting less-toxic products were released back into the blood. All the caffeine would eventually be disposed of in this manner.

Some of the amino acids from the milk were carried by the blood to cells that could use them to make proteins. Other amino acids were broken down for energy or converted to other nutrients. Leftover amino groups were first changed into highly toxic [5]_____, but the liver cells quickly converted this into [6]_____, a much safer nitrogenous waste. This was also dumped into the blood exiting the liver.

The blood processed by the liver, with its cargo of glucose, amino acids, caffeine by-products, urea, and H^+ ions from the acid in the coffee, flowed through the hepatic vein to the [7]_____. This vessel carried the blood to the heart. The heart pumped the blood to the lungs and back, and then out through the aorta, whose branches distributed the blood to body tissues. About one-fifth of the blood pumped by each heartbeat flowed through the [8]_____ arteries to the kidneys.

In each of Tom's kidneys, branches of the renal artery delivered blood to millions of [9]_____, the kidney's tiny working units. As a drop of blood passed through a [10]_____, a porous knot of capillaries, blood pressure filtered some of the blood through the pores into a funnel-like structure called [11]_____. Blood cells and large protein molecules were left behind in the blood, while smaller molecules were filtered. The fluid that collected in Bowman's capsule, called [12]_____, contained some of the water, glucose, H^+, and amino acids from Tom's morning coffee, plus urea and caffeine by-products from the liver. But filtration was just the beginning of the work of the nephron. The processes of [13]_____ and [14]_____ had to occur before the blood could leave the kidney and waste products could be [15]_____ from the body in urine.

As Tom exited from the freeway, the walls of the proximal tubules were beginning to reabsorb the glucose and amino acids in the filtrate via the process of [16]_____. Uptake of these substances, and others like salt, caused [17]_____ to be reabsorbed by osmosis, especially in the long [18]_____ of Henle and the collecting duct. These materials, valuable to the body, would reenter the blood and leave the kidneys through the [19]_____.

Filtration is not usually enough to get rid of all the acid the body needs to excrete, so Tom's kidneys actually removed some H^+ from the blood and added it to the filtrate, a process known as [20]_____. Although some of the urea was reabsorbed, most of it, along with the caffeine by-products, remained in the refined filtrate, which is called [21]_____ when it leaves the kidney.

Because the caffeine in Tom's coffee is a stimulant that raises blood pressure, Tom's kidneys were filtering his blood at a slightly higher than normal rate as he drove into the company parking lot. Urine left the kidneys through the [22]_____ and accumulated in the [23]_____. Tom had just enough time to stop at the restroom on his way to his office.

Testing Your Knowledge

Multiple Choice

1. Which of the following is an endotherm?
   a. mouse
   b. iguana
   c. frog
   d. trout
   e. all of the above except a

2. Which of the following describes the route of urine out of the body after it leaves the kidney?
   a. renal vein–bladder–urethra–ureter
   b. urethra–bladder–ureter
   c. renal vein–ureter–bladder–urethra
   d. ureter–bladder–urethra
   e. ureter–urethra–bladder

3. Blood flows through the hepatic portal vessel
   a. from the aorta to the kidney.
   b. from the intestine to the liver.
   c. from the kidney to the inferior vena cava.
   d. from the liver to the inferior vena cava.
   e. from the liver to the intestine.

4. As your kidneys regulate your body fluid composition, which of the following is the largest?
   a. the volume of filtrate formed by the nephrons
   b. the volume of urine excreted
   c. the volume of blood flowing through the nephrons
   d. the volume of solutes added to the filtrate by secretion
   e. the volume of filtrate reabsorbed

5. A countercurrent heat exchanger enables an animal to
   a. produce more heat when needed.
   b. reduce loss of heat to the environment.
   c. slow metabolism when food is not available.
   d. increase heat loss by evaporation.
   e. absorb heat from the environment.

6. Uric acid is the nitrogenous waste excreted by birds, insects, and many reptiles. An advantage of excreting uric acid is that it _____ , but a disadvantage is that it _____ .
   a. saves water . . . costs energy
   b. saves energy . . . is highly toxic
   c. is not very toxic . . . wastes a lot of water
   d. is much more soluble in water than other wastes . . . costs energy
   e. saves water . . . is highly toxic

7. Which of the following is the most accurate and comprehensive description of the function of the kidneys?
   a. breaking down body wastes
   b. excreting wastes
   c. regulating body fluid composition
   d. filtering the blood
   e. producing urine

8. Which of the following happens first as a nephron processes blood?
   a. excretion
   b. osmosis
   c. secretion
   d. reabsorption
   e. filtration

9. On a cold day, blood vessels in the skin
   a. dilate, allowing blood to keep the skin warm.
   b. constrict, forcing blood to flow through vessels in the skin.
   c. constrict, reducing heat loss from blood at the surface.
   d. dilate, causing blood to pass through the cold skin more quickly.
   e. dilate, preventing blood flow to the surface.

10. The animals in which of these pairs have similar problems regulating water balance?
    a. freshwater fish–saltwater fish
    b. land animal–freshwater fish
    c. osmoconformer–freshwater fish
    d. salmon in fresh water–salmon in salt water
    e. saltwater fish–land animal

11. The filtrate formed by the nephrons in the kidney is not the same as urine. The filtrate is first refined and concentrated by the processes of _____ , forming the urine that leaves the body.
    a. filtration and secretion
    b. reabsorption and secretion
    c. reabsorption and excretion
    d. filtration and reabsorption
    e. secretion and excretion

12. Nitrogenous waste products are made from by-products of the breakdown of
    a. fats.
    b. starch.
    c. glucose.
    d. urea.
    e. proteins.

13. By definition, an ectotherm
    a. is cold-blooded.
    b. is warm-blooded.
    c. obtains most of its heat from its environment.
    d. derives most of its heat from its own metabolism.
    e. none of the above

14. Most aquatic animals excrete ammonia, while land animals excrete urea or uric acid. What is the most likely explanation for this difference?
    a. They have different diets.
    b. Land animals can get the energy needed to make urea or uric acid.
    c. Ammonia is very toxic, and it takes lots of water to dilute it.
    d. Land animals cannot afford the energy needed to make ammonia.
    e. Fish need to get rid of ammonia, but land animals need it to live.

15. A freshwater fish tends to _____ water by osmosis. As a consequence, its kidneys excrete _____ .
    a. gain . . . large amounts of dilute urine
    b. lose . . . small amounts of concentrated urine
    c. gain . . . large amounts of concentrated urine
    d. lose . . . large amounts of dilute urine
    e. gain . . . small amounts of concentrated urine

Essay

1. Describe how the skin can make adjustments that cool the body on a hot day and warm it on a cold day.

2. Explain how a goose can stand barefoot on ice without losing large amounts of body heat.

3. Compare the problems that freshwater and saltwater fish face in maintaining the water and solute balances of their body fluids. How does each kind of fish solve these problems?

4. Describe five behaviors by which animals control their body temperatures.

5. The kidneys regulate body fluid composition by means of filtration, reabsorption, secretion, and excretion. Where does each of these processes occur? How does each contribute to the formation of filtrate and urine?

Applying Your Knowledge

Multiple Choice

1. Which of the following primarily involves heat transfer by convection?
   a. You roll down the car window to let the cool breeze blow through.
   b. The water in the lake is so cold that your legs become numb.
   c. You sweat profusely as you mow the lawn on a hot summer day.
   d. After sunset, you can feel heat from the warm pavement.
   e. As you lie on the sand, you can feel the sun's warm rays on your skin.

2. Which needs to drink the smallest amount of water to maintain its water balance?
   a. a sparrow
   b. a saltwater fish
   c. a freshwater fish
   d. a dog
   e. both a and b drink very small amounts

3. Humid weather makes you feel warmer because humid air
   a. interferes with heat loss by conduction.
   b. holds warm water vapor.
   c. interferes with heat loss by evaporation.
   d. prevents countercurrent heat exchange from occurring.
   e. increases metabolic heat production.

4. Pound for pound, a kidney uses as much energy as the heart. What do you think the energy is used for?
   a. to produce pressure for filtration
   b. for water reabsorption
   c. for the breakdown and detoxification of harmful substances
   d. to pump urine to the bladder
   e. for reabsorption and secretion of solutes by active transport

5. Look at the diagram of kidney dialysis, Figure 25.12, in the text. For kidney dialysis to work properly, the dialyzing solution should contain
   a. a higher solute concentration than blood.
   b. a higher concentration of urea than blood.
   c. a lower glucose concentration than blood.
   d. a lower concentration of urea than blood.
   e. a much smaller volume of fluid than the blood passing through it.

6. Which would have the toughest time surviving over the long term in the environment given?
   a. an osmoconformer in seawater
   b. an endotherm in a warm environment
   c. an ectotherm in a cold environment
   d. an osmoregulator in seawater
   e. an ectotherm in a warm environment

7. The loops of Henle in the kidneys of a desert kangaroo rat are much longer than those in a white laboratory rat because
   a. the kangaroo rat lives in an environment where water is scarce.
   b. the white rat's diet is much less varied than the kangaroo rat's.
   c. the kangaroo rat cannot always find food.
   d. the kangaroo rat produces more body wastes.
   e. the kangaroo rat has less stress and lower blood pressure.

8. How does the filtrate produced by the filtration process of the glomerulus differ from urine? The filtrate
   a. contains very little water.
   b. contains a higher concentration of glucose.
   c. contains a much lower concentration of salt.
   d. contains a lower concentration of proteins.
   e. all of the above.

9. The kidney's filtration process is nonselective, so
   a. many valuable substances are lost in the urine.
   b. the proportions of substances in urine are the same as in blood.
   c. urine is much less concentrated than blood.
   d. it really has little control over body fluid composition.
   e. useful substances must be selectively reabsorbed.

10. Which of the following would be filtered from the blood but not normally found in urine?
    a. water
    b. red blood cell
    c. H⁺ ions
    d. amino acid
    e. urea

Essay

1. In terms of heat loss, why is the wind-chill factor given in a weather report always a lower figure than the air temperature?

2. An animal behaviorist found that a large part of a horned lizard's daily routine consisted of behavior related to thermoregulation. Very little of the lizard's time was occupied by searching for food. A pocket mouse in the same environment spent most of its time seeking food but very little time thermoregulating. Explain this difference in behavior.

3. Many kinds of poisons cause liver damage. Sniffing the solvents in glue can quickly injure the liver. Poisonous *Amanita* mushrooms also kill by destroying the liver. Over a longer period, alcohol causes liver damage. Why do you think the liver ends up being a "lightning rod" for all these different poisons?

4. You may have noticed that you never see pictures of antarctic crocodiles lying in ambush for unwary penguins or seals. Why are birds and mammals more successful in the Arctic and Antarctic than reptiles and amphibians?

5. Imagine that you are working with a team of physiologists sampling and comparing the blood entering and leaving the liver. How might the blood entering and leaving the liver differ in each of the following circumstances? After the subject eats a large candy bar. After the subject consumes two shots of whiskey. After a 4-hour period when no food was consumed.

Using Your Knowledge

1. For a patient with kidney failure, a kidney transplant is the key to an improved, more normal life. For an individual with liver failure, an immediate transplant is essential for survival. Why is a liver transplant so important? If this is the case, why are liver transplants performed much less often than kidney transplants?

2. Suppose a friend or relative needed a kidney transplant. Would you volunteer to donate a kidney if your tissue types were compatible? What might be some reasons for and against donating one of your kidneys?

3. Side effects of many medications include liver or kidney damage. Why? Have any of the prescription or over-the-counter medications you use been implicated in causing liver or kidney impairment? Try looking at some of the labels or package instructions. How else might you find out about this?

Chemical Regulation

As you read this, your blood is carrying dozens of different hormones from endocrine glands to target cells all over your body. These hormones are chemical signals that regulate the activities of every cell. Hormones coordinate the activities of body organs, maintaining homeostasis. Their effects on their targets range from tiny adjustments in chemistry to control of large-scale body processes, such as growth and reproduction. Hormones are produced in miniscule amounts; cells are capable of responding to very subtle signals. Hormones carry specific messages, and each cell is equipped to respond only to certain ones. Different target cells may even respond differently to the same hormone. This subtle and intricate hormone communication and control system is the subject of Chapter 26.

Organizing Your Knowledge

Exercise 1 (Module 26.1)

This module describes the roles of chemical signals in the function of the endocrine and nervous systems. Color and label the diagrams below, comparing the activities of endocrine cells, neurosecretory cells, and nerve cells. Label the following: **neurosecretory cell, hormone molecules, nerve signals, endocrine cell, neurotransmitter molecules, secretory vesicles, target cell, blood vessel, nerve cell.** Note that the diagrams are not in the same order as those in the text. Choose a color for each of the cells. Make hormone molecules from the endocrine cell green dots, hormone molecules from the neurosecretory cell blue dots, and neurotransmitter molecules red dots.

Exercise 2 (Module 26.2)

Just about everything you might want to know about the two mechanisms by which hormones affect their target cells is summarized in the two figures and the last paragraph in Module 26.2. To review them, color and label the diagrams below.

A. Choose your own colors for this diagram of a steroid hormone acting on a target cell, and label **steroid hormone, DNA, plasma membrane, new protein, receptor.**

B. Choose your own colors for this diagram of a nonsteroid hormone acting on a target cell, and label **nonsteroid hormone, second messenger, enzyme, receptor, plasma membrane, effect on cellular function.**

This module is an overview of the glands and hormones discussed in this chapter. You may want to refer back to the diagram and table as references as you read about glands and hormones in the modules that follow. Label the human endocrine glands on the diagram below, and color each one a different color. Choose from **thymus, testis, pineal gland, pituitary gland, adrenal glands, parathyroid glands, pancreas, thyroid gland, hypothalamus**, and **ovary**. Consulting Table 26.3B in the text, list under the name of each gland one of the hormones it produces.

Exercise 4   (Module 26.3)

Table 26.3B in the text is a useful summary of the glands, hormones, and hormone actions. Use it to complete the following chart by filling in the blanks. Note that the order of the glands is different from Table 26.3B. (You may want to skip this for now and come back to it as a review after you have studied the glands and hormones in the following modules in more detail.)

Gland	Hormone(s)	Action
1.	2. & 3.	Control metabolic rate
	Calcitonin	4.
Ovaries	5.	6.
	Progesterone	7.
8.	9. & 10.	Trigger "fight or flight" response
11.	Glucocorticoids	12.
	13.	Regulate mineral gain and loss by kidneys
Pancreas	Glucagon	14.
	15.	16.
17.	18.	Raises blood calcium level
19.	Melatonin	20.
21.	22.	Stimulates growth
	Thyroid-stimulating hormone	23.
	Prolactin	24.
	25.	Stimulates adrenal cortex
	Luteinizing hormone	26.
	27.	Stimulates egg and sperm production
Posterior pituitary	Oxytocin	28.
	29.	Promotes water retention by kidneys
Thymus	30.	31.
32.	Androgens	33.

Exercise 5   (Modules 26.4 – 26.5)

The hypothalamus and pituitary glands are closely related and tie the nervous and endocrine systems together. But the anterior and posterior lobes of the pituitary do different jobs and are controlled in different ways. State whether each of the following relates to the hypothalamus (H), anterior lobe of the pituitary (A), or posterior lobe of the pituitary (P).

____   1. Stores and secretes hormones actually made in the hypothalamus
____   2. Secretes releasing and inhibiting hormones that stimulate the anterior pituitary
____   3. Composed of nonnervous, glandular tissue
____   4. Part of the brain
____   5. Where the hormones oxytocin and ADH are released into the blood
____   6. Responds to releasing and inhibiting hormones from the hypothalamus
____   7. Secretes growth hormone, ACTH, and thyroid-stimulating hormone
____   8. It is the terminus of neurosecretory cells from the hypothalamus
____   9. Most hormones that it secretes influence other endocrine glands
____   10. Exerts master control over the endocrine system
____   11. Consists of nervous tissue and is actually an extension of the hypothalamus
____   12. Negative feedback usually acts on this part of the system
____   13. Blood vessels carry releasing hormones from here to the anterior pituitary
____   14. Along with the brain, it secretes the body's natural painkillers

Exercise 6   (Modules 26.4 – 26.6)

The thyroid gland is familiar to most people. Once you know that thyroid hormones stimulate metabolism, it is easy to remember the symptoms of over- and undersecretion and the effects of thyroid medication. Review the thyroid gland by matching each phrase on the right with a substance on the left.

A. Iodine

B. T_3

C. TSH

D. T_4

E. TRH

____   1. Secreted by hypothalamus; stimulates anterior pituitary
____   2. Secreted by anterior pituitary; stimulates thyroid
____   3. Hormone secreted by thyroid
____   4. Another hormone secreted by thyroid
____   5. Needed for the manufacture of thyroid hormones
____   6. T_3 and T_4 signal the anterior pituitary to stop making this
____   7. Goiter occurs if there is not enough of this in the diet
____   8. Also called thyroxine

Exercise 7   (Module 26.7)

Two opposing, or antagonistic, hormones control the balance of calcium in the blood. To review calcium homeostasis, study Figure 26.7 in the text. Use it to help you choose the correct italicized words to complete the following paragraph.

Calcium is important for nerve impulse transmission, muscle contraction, transport of molecules through cell membranes, and (1) *digestion, protein synthesis, blood clotting.* Blood calcium concentration is held in a narrow range by the thyroid and parathyroid glands. If blood calcium drops, the (2) *thyroid gland, parathyroid glands* increase(s) secretion of (3) *parathyroid hormone, calcitonin.* This causes (4) *increased, decreased* reabsorption of calcium as the kidneys form urine, (5) *increased, decreased* absorption of calcium from food, and (6) *release of calcium from, deposition of calcium in* bone. If blood calcium concentration climbs too high, (7) *parathyroid hormone, calcitonin* is secreted by the (8) *thyroid gland, parathyroid glands.* This causes (9) *increased, decreased* reabsorption of calcium in the kidneys, (10) *increased, decreased* absorption of calcium from food, and (11) *release of calcium from, deposition of calcium in* bone.

Exercise 8   (Modules 26.8 – 26.9)

To understand how the pancreas controls blood sugar, you need to remember only two things: (1) the pancreas makes two hormones, insulin and glucagon, and (2) diabetics have high blood sugar, so many of them take insulin. Given these two facts, you can figure everything else out: If diabetics have to take insulin, insulin must make blood sugar go down. It must make cells take sugar out of the blood and use it or store it. That means glucagon must make blood sugar go up, by causing cells to get it out of storage and put it into the blood. Finally, an increase in blood sugar must trigger insulin secretion (to make sugar go down), so a drop in blood sugar must trigger glucagon secretion (to make blood sugar go up).

To review control of blood sugar level, choose either the word *increase(s)* or *decrease(s)* to complete each of the following statements.

1. Eating a meal rich in carbohydrates causes blood glucose to _____.

2. When blood glucose _____, the pancreas secretes more insulin.

3. Insulin causes body cells to _____ their uptake and use of glucose.

4. Insulin also causes glycogen formation by the liver to _____.

5. Insulin therefore causes blood glucose to _____.

6. Between meals, blood glucose levels tend to _____.

7. When blood glucose _____, the pancreas secretes more glucagon.

8. Glucagon causes blood glucose to _____.

9. Glucagon _____ breakdown of glycogen in the liver and release of glucose to the blood.

10. Glucagon also triggers many cells to _____ their use of glucose and use fats instead.

11. As blood glucose rises toward the set point, secretion of glucagon _____.

12. In type I diabetes, blood sugar _____ because the body is unable to produce insulin.

13. In type II diabetes, there is a _____ in cells' ability to respond to insulin.

14. Hypoglycemia results from an excess of insulin, which causes a sudden _____ in blood glucose.

Exercise 9   (Modules 26.10 – 26.11)

The adrenal gland is similar to the pituitary, in that it is two glands in one. The parts have different jobs and send out different hormones. One of its parts (the medulla) is stimulated by nerve impulses to secrete its hormones, while the other part (the cortex) is stimulated by a hormonal signal. The hormones from the adrenal medulla and cortex help the body deal with stress. After reading the modules and studying Figure 26.10 in the text, try to match each of the phrases on the right with adrenal stress hormones, E, M, or G.

E. Epinephrine and   ____   1. Increase breathing rate
     norepinephrine   ____   2. Respond to short-term stress
M. Mineralocorticoids   ____   3. Secreted by adrenal cortex
G. Glucocorticoids   ____   4. Also secreted by the adrenal cortex
     ____   5. Increase blood volume and pressure in response to long-term stress
     ____   6. Triggered by nerve impulses from the hypothalamus
     ____   7. Secreted by the adrenal medulla
     ____   8. Cause proteins and fats to be broken down to make glucose
     ____   9. Triggered by ACTH from the pituitary
     ____   10. Also triggered by ACTH from the pituitary
     ____   11. Suppress the inflammatory response and immune system
     ____   12. Cause retention of sodium and water by kidneys
     ____   13. Dilate and/or constrict blood vessels, redirecting blood flow
     ____   14. Increase metabolic rate
     ____   15. Often prescribed to relieve pain from athletic injuries

Exercise 10   (Module 26.12)

Sex hormones are discussed in more detail in Chapter 27. After reading Module 26.12, try sketching on a separate sheet of paper a concept map for sex hormones. Refer to Chapter 3, Exercise 10, for a discussion of constructing a concept map. Include the following in your concept map: **testosterone, gonads, estrogens, hypothalamus, FSH and LH, progestins, testes, releasing factor, anterior pituitary, ovaries.**

Exercise 11   (Summary)

On a separate sheet of paper, briefly describe how the items in each of the following groups are related to one another. They may affect one another, oppose one another, or have something in common.

1. luteinizing hormone   estrogens   androgens   follicle-stimulating hormone
2. oxytocin   antidiuretic hormone
3. glucagon   glucocorticoids   epinephrine
4. glucocorticoids   ACTH
5. antidiuretic hormone   mineralocorticoids   parathyroid hormone   calcitonin
6. thyroxine   calcitonin T_3
7. glucagon   insulin
8. glucocorticoids   epinephrine   mineralocorticoids   norepinephrine
9. ACTH   LH   growth hormone   FSH   TSH   prolactin
10. calcitonin   parathyroid hormone
11. oxytocin   prolactin

Testing Your Knowledge

Multiple Choice

1. Another system that works closely with the endocrine system to control body processes is the
   a. circulatory system.
   b. immune system.
   c. digestive system.
   d. nervous system.
   e. muscular system.

2. Every time you eat a cookie or candy bar, your blood sugar increases. This triggers an increase in the hormone
   a. thyroxine.
   b. epinephrine.
   c. adrenocorticotropin (ACTH).
   d. glucagon.
   e. insulin.

3. Every hormone
   a. is a protein.
   b. is produced in response to stress.
   c. is under the control of the pituitary gland.
   d. enters a cell and interacts with DNA.
   e. acts as a signal between cells.

4. Researchers have found increased levels of hormones from the _____ in the blood of students preparing for final exams. These hormones are produced in response to stress.
   a. thyroid gland
   b. pineal gland
   c. posterior pituitary
   d. adrenal glands
   e. parathyroid glands

5. Which of the following hormones have antagonistic (opposing) effects?
   a. thyroxine and calcitonin
   b. insulin and glucagon
   c. growth hormone and epinephrine
   d. ACTH and glucocorticoids
   e. epinephrine and norepinephrine

6. What is the role of a "second messenger" in hormone action?
   a. It signals a cell to secrete a hormone.
   b. It informs a gland as to whether its hormones are having an effect.
   c. It relays a hormone's message inside a target cell.
   d. It stops hormone action when it is no longer needed.
   e. It carries a hormone while it is in the blood.

7. When a boy goes through puberty, the steroid hormone testosterone "puts hair on his chest" by
   a. interacting with DNA in the nuclei of cells.
   b. causing cells to change shape.
   c. altering the permeability of plasma membranes.
   d. triggering nerve impulses in cells.
   e. turning enzymes on.

8. A hormone from the parathyroid glands works in opposition to a hormone from the _____ to regulate _____ .
   a. posterior pituitary . . . metabolic rate
   b. thyroid gland . . . blood calcium
   c. pancreas . . . water reabsorption
   d. adrenal medulla . . . blood calcium
   e. thyroid gland . . . blood glucose

9. Some glands produce hormones that stimulate other endocrine glands. Which of the following hormones specifically acts to trigger secretion of hormones by another endocrine gland?
   a. thyroxine
   b. progesterone
   c. adrenocorticotropin (ACTH)
   d. antidiuretic hormone (ADH)
   e. melatonin

10. How is the level of thyroxine in the blood regulated?
    a. Thyroxine stimulates the pituitary to secrete thyroid-stimulating hormone (TSH).
    b. Thyroxine inhibits secretion of TSH from the pituitary.
    c. TSH-releasing hormone (TRH) inhibits secretion of thyroxine by the thyroid gland.
    d. Thyroxine stimulates the hypothalamus to secrete TRH.
    e. TSH inhibits secretion of TRH.

11. Steroid hormones are produced only by the
    a. adrenal medulla and pancreas.
    b. thyroid gland and pancreas.
    c. anterior and posterior pituitary.
    d. thyroid gland and sex organs.
    e. sex organs and adrenal cortex.

12. It takes much longer for sex hormones and other steroids to produce their effects than it takes nonsteroid hormones. Why?
    a. Steroids are bigger, slower molecules.
    b. Steroids usually must be carried longer distances by the blood.
    c. Steroids cause target cells to make new proteins, which takes time.
    d. Steroids must relay their message via a second messenger.
    e. It takes longer for endocrine cells to make and secrete steroids.

13. The pituitary is actually two glands. The anterior lobe of the pituitary secretes its hormones when stimulated by _____, and the posterior lobe of the pituitary secretes its hormones when stimulated by _____ .
    a. hormones from the adrenal cortex . . . hormones from the thyroid
    b. hormones from the hypothalamus . . . nerve impulses from the hypothalamus
    c. hormones from the hypothalamus . . . hormones from the thyroid
    d. nerve impulses from the hypothalamus . . . hormones from the hypothalamus
    e. hormones from the pineal gland . . . hormones from the pancreas

14. Injections of a hormone are sometimes given to strengthen contractions of the uterus during childbirth. What hormone might this be?
    a. adrenocorticotropin (ACTH)
    b. thyroxine
    c. oxytocin
    d. insulin
    e. follicle-stimulating hormone (FSH)

15. Which of the following hormones has the broadest range of targets?
    a. ADH
    b. prolactin
    c. TSH
    d. epinephrine
    e. calcitonin

Essay

1. The pituitary is often called the master gland. In what way is this true? In what way is it misleading?

2. Compare how the adrenal cortex and adrenal medulla deal with stress.

3. How are hormones and neurotransmitters alike? How are they different?

4. Briefly describe an example of a hormonal disease or abnormality in which
    a. too much hormone is secreted.
    b. not enough hormone is secreted.
    c. hormone is secreted in normal amount, but cells fail to respond.

5. Why are glucocorticoids useful in treating injuries? Why is prolonged use of glucocorticoids risky?

6. Describe two different situations in which a pair of hormones have opposite (antagonistic) effects on the body.

Applying Your Knowledge

Multiple Choice

1. Jet lag occurs when a person moves rapidly from one time zone to another, causing conflict between the body's biological rhythm and the new cycle of light and dark. Some scientists suspect that jet lag may result from disruption of a daily hormone cycle. Which of the following hormones do you think is the most likely suspect?
    a. epinephrine
    b. insulin
    c. melatonin
    d. estrogen
    e. prolactin

2. Which of the following hormones triggers secretion of the other two?
    a. thyroxine
    b. thyroid-stimulating hormone (TSH)
    c. TSH-releasing hormone (TRH)
    d. Any of the above can trigger secretion of the others.
    e. None of the above can trigger secretion of the others.

3. A tumor in an endocrine gland caused Jennifer to have weakened bones and unusually high levels of blood calcium. Which of the following was affected?
    a. anterior pituitary
    b. pancreas
    c. adrenal glands
    d. parathyroid glands
    e. thymus

4. Which of the following exerts control over all the others?
   a. adrenal cortex
   b. hypothalamus
   c. thyroid gland
   d. anterior pituitary
   e. testes

5. Because only the _____ gland uses iodine to make its hormones, radioactive iodine is often used as a treatment for tumors of this gland.
   a. pituitary
   b. pancreatic
   c. thyroid
   d. adrenal
   e. testicular

6. Diabetes insipidus is an inherited endocrine malfunction (unrelated to diabetes mellitus) in which the kidneys fail to reabsorb normal amounts of water. Victims of this disease produce gallons of urine each day, and their kidneys soon wear out. Treatment of this disease involves replacing a missing hormone. Which of the following do you think it is?
   a. glucagon
   b. epinephrine
   c. glucocorticoids
   d. antidiuretic hormone (ADH)
   e. thyroid-stimulating hormone (TSH)

7. In an experiment, researchers removed the ____ of young mice, and as a result, these mice were able to accept organ transplants without rejection.
   a. pineal glands
   b. thymus glands
   c. thyroid glands
   d. parathyroid glands
   e. adrenal glands

8. Tim once suffered a severe allergic reaction to a bee sting. The sting caused him to suffer a near-fatal drop in blood pressure called anaphylactic shock. Now he carries a kit containing a syringe of _____, which he can inject to speed up his heart if he reacts to a bee sting.
   a. insulin
   b. thyroxine
   c. testosterone
   d. calcitonin
   e. epinephrine

9. It has been found that certain salamanders fail to go through the normal transformation from tadpole to adult if there is a shortage of iodine in the pond water in which they live. This makes it impossible for the _____ to manufacture hormones necessary for normal development.
   a. thyroid gland
   b. posterior pituitary
   c. adrenal cortex
   d. pineal gland
   e. pancreas

10. As a young girl, Maria suffered a head injury that damaged her pituitary. An injury to the pituitary is particularly serious because of all the functions controlled by this gland. As Maria got older, she and her doctors found that all of the following except _____ were affected.
    a. metabolic rate
    b. growth
    c. her menstrual cycle
    d. milk production
    e. blood sugar level

Essay

1. Hypoglycemia is a condition in which blood sugar drops to abnormally low levels. What seems to be the cause of hypoglycemia? Why would it not be a good idea to try to correct this problem by eating more sugar?

2. One of the symptoms of severe diabetes mellitus is breath that has a sweetish, acetone smell. The smell comes from by-products of fat breakdown in the body. If another consequence of diabetes is excess sugar in the blood, why do the cells not just use the sugar instead of breaking down fat?

3. A tumor can cause enlargement of the thyroid gland. How could this result in abnormally high metabolic rate and body temperature? How could similar symptoms be produced by a tumor of the pituitary gland?

4. Some hormones, such as ADH, act on very specific targets (for ADH, the kidney). Other hormones, such as insulin, are able to affect every cell in the body. Based on what you have learned about how hormones exert their effects on target cells, speculate as to how some hormones can affect many targets, while other hormones affect only one.

5. A chemical called dioxin, or TCDD, is produced as a contaminant of some chemical manufacturing processes. Trace amounts of this substance were present in Agent Orange, a defoliant sprayed on vegetation during the Vietnam War. There has been a continuing controversy over its effects on soldiers exposed to it during the war. Animal tests have shown that dioxin can be lethal and can cause birth defects, cancer, liver and thymus damage, and immune system suppression. But its effects on humans are unclear, and even animal tests are uneven; a hamster is not affected by a dose that can kill a guinea pig. Researchers have discovered that dioxin exerts its effects like a steroid hormone. It enters a cell and binds to a receptor protein, which in turn attaches to the cell's DNA. How might this mechanism help explain the variety of effects of dioxin on different body systems and in different animals?

Using Your Knowledge

1. Compare communication in the body via nerve impulses and hormones with everyday communication by telephone and letter. What are the characteristics of each? In what situations is each the most useful?

2. A survey indicates that over a quarter million U.S. high school students have used anabolic steroids, synthetic derivatives of testosterone. This number is in addition to the thousands of college and professional athletes who have used these hormones. Why do athletes use anabolic steroids? What are some of the dangers of steroid use? Do you know anyone who has used steroids in this way?

3. Do you know someone who has an endocrine disorder, such as diabetes? How does that person deal with his or her condition? Have there been recent discoveries or developments that have improved treatment of this condition?

Reproduction and Embryonic Development **27**

How old are you? When is your birthday? You celebrate your birthday as the beginning of your life, but birth is not really the beginning. Your birth was the culmination of a process that started months, even years, before. A couple of months before you were conceived, some of your father's cells divided to form sperm. One of these sperm contributed half your genes. Much earlier, when your mother was a baby, eggs began to form in her ovaries. A couple of weeks before fertilization, one of those eggs was brought to maturity. Meanwhile, hormones prepared your mother's uterus for your new life. A sperm penetrated the egg. It divided and redivided, creating a ball of cells that implanted itself in the wall of the uterus. Cells moved and specialized, shaping tissues and organs. Within weeks, fingers and toes appeared, and a tiny heart started beating. After 9 months of growth, you entered the world. These processes—sexual reproduction and embryonic development—are the focus of this chapter.

Organizing Your Knowledge

Exercise 1   (Introduction and Module 27.1)

These modules introduce some terms and concepts basic to an understanding of reproduction. Write the correct term in the space next to each phrase below.

_____ 1. Male and female reproductive systems in the same body

_____ 2. Small gamete usually propelled by a flagellum

_____ 3. When one parent produces offspring

_____ 4. Diploid cell formed by union of egg and sperm

_____ 5. General name for egg or sperm

_____ 6. Regrowth of lost body parts

_____ 7. Shedding of gametes into water

_____ 8. Large gamete that is not self-propelled

_____ 9. Sexual intercourse

_____ 10. Splitting of new individuals from existing ones

_____ 11. When egg and sperm unite inside the female's body

_____ 12. Breakup of the parent body into several pieces

These modules introduce the human reproductive systems. The diagrams below show a front view of the reproductive system of the human female and a side view of the reproductive system of the human male. See how many of the following parts you can label without looking at the modules, and then consult the modules for the remainder: **seminal vesicle, ovary, scrotum, urethra, epididymis, vagina, oviduct, vas deferens, prepuce, uterus, follicle, erectile tissue, testis, endometrium, cervix, glans, prostate gland, bulbourethral gland, corpus luteum, wall of uterus.**

1. _____

2. _____

3. _____

4. _____

5. _____

6. _____

7. _____

8. _____

9. _____

10. _____

11. _____

12. _____

13. _____

14. _____

15. _____

16. _____

17. _____

18. _____

19. _____

20. _____

Exercise 3   (Module 27.4)

Both spermatogenesis and oogenesis produce haploid gametes, but there are several important differences in how they occur. Compare these processes in the chart below.

	Spermatogenesis	Oogenesis
Location of process	1.	2.
When primary cells form	3.	4.
Numbers and names of cells resulting from first meiotic division	5.	6.
Numbers and names of cells resulting from second meiotic division	7.	8.
Number of gametes produced from division of one primary cell	9.	10.
Total number of gametes produced per day or per month	11.	12.

Exercise 4   (Module 27.5)

This module explains the monthly ovarian and menstrual cycles that coordinate the activities of the ovaries and uterus. You can use Figure 27.5B in the text to get most of the words you need to complete the following story, which integrates the ovarian and menstrual cycles.

Jennifer wasn't consciously aware of it, but a slight change was occurring in her brain. Her menstrual period had barely started, but already her [1]_____ was starting to secrete tiny amounts of a releasing hormone, which in turn signaled the pituitary gland to boost its output of [2]_____ and [3]_____, restarting her ovarian and menstrual cycles. FSH stimulated a [4]_____, dormant in her [5]_____ since before she was born, to begin to grow and secrete its own hormones. As the follicle grew, it started to secrete small amounts of [6]_____. This hormone in turn signaled the [7]_____, the lining of the [8]_____, to begin to grow and develop in preparation for possible implantation of a fertilized [9]_____. The estrogen from the [10]_____ also exerted [11]_____ feedback on the hypothalamus, keeping levels of [12]_____ and [13]_____ low during this [14]_____ phase of the ovarian cycle.

As the follicle grew in Jennifer's ovary, it secreted more and more estrogen. The higher level of estrogen caused the hypothalamus to signal the [15]_____ to release bursts of [16]_____ and [17]_____. The surge of LH caused the mature follicle to rupture, releasing an ovum in a process called [18]_____. This occurred on the [19]_____ day of Jennifer's 28-day cycle. The ovum entered the [20]_____, where fertilization could occur, and began to travel toward the uterus. The [21]_____ phase of the ovarian cycle had begun.

One of the effects of luteinizing hormone was to transform the ruptured follicle into a yellow, glandular structure, called the [22]_____, which continued to secrete estrogen, plus another hormone, called [23]_____. Both these hormones stimulated further growth and thickening of the uterine lining, further preparing it for the fertilized egg. These hormones also exerted strong negative feedback on the hypothalamus and pituitary, suppressing secretion of FSH and LH and so preventing additional [24]_____ from growing in the ovary. Eventually, the drop in LH caused the corpus luteum to degenerate and reduce its output of estrogen and [25]_____. (Because Jennifer's egg was not [26]_____, there was no hormonal signal sent from a developing embryo to maintain the corpus luteum.) Estrogen and progesterone were maintaining the lining of the uterus. As levels of these hormones dropped, the [27]_____ began to slough off. The monthly cycles were coming to an end, and [28]_____ had begun.

Exercise 5   (Module 27.6)

Human sexual responses occur in four phases (A–D) listed below on the left. Number the phases in order, and match each with a descriptive phrase (P–S) from the list on the right.

Phase	Order	Description	Phrase
A. Orgasm	_____	_____	P. Return to relaxed state
B. Resolution	_____	_____	Q. Increased breathing and heart rate
C. Excitement	_____	_____	R. Rhythmic contractions of reproductive systems
D. Plateau	_____	_____	S. Erection of penis and clitoris

Exercise 6   (Module 27.7)

Review sexually transmitted diseases by completing this crossword puzzle.

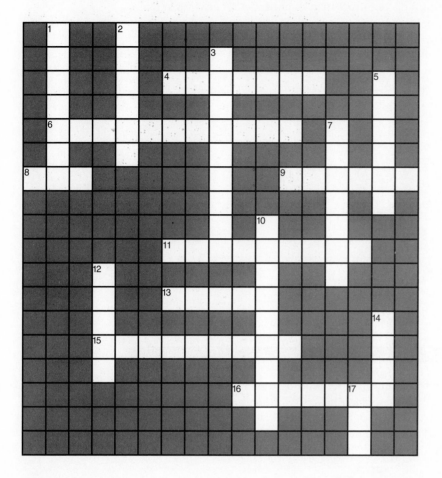

Across

4. Viral STDs are not ____, but bacterial STDs usually are.
6. Bacterial STDs are treated with ____.
8. ____ is short for "sexually transmitted disease."
9. A drug called acyclovir is used to treat symptoms of genital ____.
11. Chlamydial infections, syphilis, and ____ are caused by bacteria.
13. Chlamydial infections and gonorrhea may cause no symptoms in ____.
15. PID can cause pain, scarring, and ____.
16. Latex ____ can usually prevent the spread of STDs.

Down

1. STDs are spread by sexual ____.
2. Some STDs can lead to ____ inflammatory disease.
3. Trichomoniasis is caused by a ____.
5. Genital herpes has been linked to cervical ____.
7. AIDS and genital herpes are both caused by ____.
10. A genital discharge or painful ____ can be symptoms of STDs.
12. A fungus called Candida causes ____ infections.
14. A virus called HIV is responsible for ____.
17. ____ usually have no symptoms of candidiasis or trichomoniasis.

Exercise 7   (Module 27.8)

State whether each of the following methods of contraception prevents pregnancy by pre-venting release of gametes, preventing implantation, or preventing egg and sperm from meeting. Then rank the average effectiveness of the methods (as given in the table in Module 27.8), with 1 the most effective and 9 the least effective in preventing pregnancy—if used correctly.

Method	*How It Works*	*Effectiveness*
A. Birth control pill	_____	_____
B. Withdrawal	_____	_____
C. Spermicide alone	_____	_____
D. Vasectomy	_____	_____
E. IUD	_____	_____
F. Condom	_____	_____
G. Rhythm	_____	_____
H. Progestin implant	_____	_____
I. Diaphragm and spermicide	_____	_____

Exercise 8   (Module 27.9)

Embryonic development begins with fertilization. After reading the module and studying the illustrations, answer each of the following questions with one sentence.

1. How is the sperm able to get to the egg?

2. How does the sperm penetrate the egg?

3. In species whose eggs are fertilized externally, what keeps sperm of other species from fertilizing the egg?

4. Why is it important that only one sperm penetrate the egg?

5. What prevents more than one sperm from penetrating the egg?

6. In what ways does the egg change after fertilization?

Exercise 9 (Modules 27.10 – 27.12)

These three modules trace the development of an embryo from the zygote to development of rudimentary organs. The three stages in this process are cleavage, gastrulation, and organ formation. The illustrations below (combining development of a sea urchin and a frog) are in random order. Number them in the correct sequence (blanks A–K), and then label the following (blanks L–Z): **ectoderm, blastula, zygote, blastopore, digestive cavity (archenteron), blastocoel, endoderm, notochord, mesoderm, gastrula, neural tube, body cavity (coelom), somite.** Some terms are used more than once.

L. _____
M. _____
N. _____
O. _____
A. _____
B. _____
C. _____
P. _____
Q. _____
R. _____
S. _____
D. _____
E. _____
F. _____
T. _____
U. _____
V. _____
G. _____
H. _____
I. _____
W. _____
X. _____
Y. _____
Z. _____
J. _____
K. _____

Gastrulation produces a three-layered embryo. Each of the three tissue layers gives rise to particular structures. After reading the module and studying the table, name the embryonic tissue layer from which each of the following structures develops—**ectoderm, mesoderm,** or **endoderm.** (You can figure out most of them just by remembering that ectoderm is on the outside of the embryo, endoderm is on the inside, and mesoderm is in between.)

_____ 1. Lining of digestive tract
_____ 2. Circulatory system
_____ 3. Pancreas and liver
_____ 4. Nervous system
_____ 5. Skeleton
_____ 6. Muscles
_____ 7. Lining of respiratory system
_____ 8. Skin epidermis

These three modules describe some of the processes that shape an embryo. Each of the following statements describes an experiment that demonstrates one of the processes. Match the process with the experiment. Choose from **pattern formation, cell migration, cell aggregation, induction, differentiation, changes in cell shape.**

1. Ectoderm from the side of a frog gastrula normally develops into skin. If a piece of this ectoderm is removed from the side and placed on the back of the embryo in contact with the notochord, it turns into neural tissue. This experiment illustrates _____.

2. A certain chemical is painted on a limb bud of a mouse. An additional set of toes develops at the site, with the little toe where the chemical is most concentrated. This experiment illustrates

   _____.

3. The reproductive system develops from mesoderm, except for the germ cells that will actually divide to form eggs and sperm. Embryologists have dyed these cells and followed them as they "crawl" from a distant site to the developing gonads, an example of _____.

4. A sponge is placed in a blender and whipped into a "sponge shake," consisting of separated cells. Placed in a solution of the proper chemical composition, the cells attach to each other in the proper arrangement to re-form the sponge, showing _____.

5. If cells of a blastula are treated with chemicals that interfere with changes in the cells' internal cytoskeletons, the cells fail to fold inward properly to start the process of gastrulation. This experiment shows the importance of _____.

6. As a nerve cell matures and becomes specialized, it is no longer able to divide. Researchers are attempting to determine how genes are activated and deactivated in developing nerve cells. If turned-off genes could be turned back on, perhaps a nerve cell could be made to return to a dividing, embryonic state. This might be useful in repairing nervous system damage. These experimenters are attempting to reverse the process of _____.

Exercise 12   (Modules 27.16 – 27.18)

Many of the stages and processes seen in the development of other animals are also seen in human development and birth. Match the words and phrases below to review human development.

A. HCG
B. First trimester
C. Chorionic villi
D. Inner cell mass
E. Yolk sac
F. Fertilization
G. Prostaglandins
H. Blastocyst
I. Trophoblast
J. Fetus
K. Placenta
L. Labor
M. Gestation
N. Dilation
O. Oxytocin
P. Third trimester
Q. Allantois
R. Conception
S. Amnion
T. Chorion

_____ 1. Outer layer that allows embryo to implant
_____ 2. Carrying developing young in reproductive tract
_____ 3. Name for embryo after 9 weeks' gestation
_____ 4. Help oxytocin induce labor
_____ 5. Hormone secreted by chorion; prevents spontaneous abortion
_____ 6. Structure that taps the mother's blood supply
_____ 7. Period when embryo grows rapidly and prepares for birth
_____ 8. Membrane that becomes the embryo's part of the placenta
_____ 9. Another term for fertilization
_____ 10. Forms fluid-filled sac around embryo
_____ 11. Trimester in which the biggest changes occur
_____ 12. Equivalent of frog blastula
_____ 13. The first stage of labor
_____ 14. In birds and reptiles it is important in waste disposal
_____ 15. Portion of the blastocyst that actually forms the baby
_____ 16. A pituitary hormone that initiates labor
_____ 17. Produces the embryo's first blood cells
_____ 18. This process begins gestation
_____ 19. Outgrowths of the chorion that absorb nutrients and oxygen
_____ 20. Contractions of the uterus that bring about birth

Exercise 13   (Module 27.19)

In vitro fertilization, surrogate motherhood, and the possibility of sperm and embryo donation open a multitude of reproductive possibilities to infertile couples. A woman's eggs could be fertilized with her husband's sperm and implanted in her uterus. Alternatively, donated sperm or eggs could be used. Or the eggs could be implanted in another woman's uterus. How many combinations are possible? List as many as you can.

Testing Your Knowledge

Multiple Choice

1. Which of the following is a form of sexual reproduction?
   a. budding
   b. fission
   c. fragmentation
   d. regeneration
   e. hermaphroditism

2. Which of the following correctly traces the path of sperm from their site of production out of a man's body?
   a. seminiferous tubule–vas deferens–epididymis–urethra
   b. epididymis–urethra–seminiferous tubule–vas deferens
   c. seminiferous tubule–epididymis–vas deferens–urethra
   d. epididymis–seminiferous tubule–vas deferens–urethra
   e. vas deferens–epididymis–urethra–seminiferous tubule

3. A peak in _____ triggers ovulation on about the _____ day of the monthly cycle.
   a. progesterone . . . fourteenth
   b. FSH and LH . . . seventh
   c. FSH and LH . . . second
   d. FSH and LH . . . fourteenth
   e. estrogen . . . twentieth

4. External fertilization occurs mostly in
   a. land animals.
   b. insects.
   c. aquatic animals.
   d. animals that reproduce asexually.
   e. mammals.

5. Human eggs are usually fertilized in the
   a. vagina.
   b. oviduct.
   c. ovary.
   d. cervix.
   e. uterus.

6. On its way to fertilize a human egg, a sperm cell does *not* have to pass through which of the following?
   a. oviduct
   b. vagina
   c. ovary
   d. vas deferens
   e. cervix

7. Which of the following hormones is the first to increase significantly every 28 days or so and initiates the ovarian cycle?
   a. progesterone
   b. follicle-stimulating hormone
   c. estrogen
   d. luteinizing hormone
   e. human chorionic gonadotropin

8. The fertilization membrane around an egg
   a. prevents the egg from being fertilized by more than one sperm.
   b. attracts sperm to the egg from some distance away.
   c. enables the egg to obtain nourishment from its environment.
   d. ensures that the egg is fertilized by sperm of the proper species.
   e. enables the fertilized egg to implant in the uterine wall.

9. After ovulation occurs, the empty follicle
   a. can be recycled to produce more eggs.
   b. changes into the corpus luteum and makes hormones.
   c. quickly degenerates.
   d. immediately initiates menstruation.
   e. becomes the site of implantation of a fertilized egg.

10. The first stage of embryonic development is _____ . This process produces _____ .
    a. gastrulation . . . a three-layered embryo
    b. gestation . . . a gastrula
    c. ovulation . . . a zygote
    d. cleavage . . . a hollow ball of cells
    e. parturition . . . a fetus

Essay

1. Some animals, such as aphids, alternate between sexual and asexual reproduction. Describe environmental conditions under which each form of reproduction might be most advantageous to the animal.

2. Name and describe the functions of the extraembryonic membranes that surround a human embryo.

3. Compare sperm formation in the testes with egg formation in the ovaries. When in life does each process begin? What triggers each to occur or be completed? How many cells result from division of a primary oocyte? A primary spermatocyte?

4. Draw a series of simple sketches that show how a fertilized egg develops into a three-layered gastrula.

5. Tammy said, "I think I am going to stop using birth control pills. I have heard that the pill isn't safe over the long term." Her friend Melissa replied, "Actually, the pill is safer than pregnancy." Can you give reasons to back up both women's statements?

Applying Your Knowledge

Multiple Choice

1. In an early stage of development of the frog nervous system, the bits of tissue that will form the backs of the eyes bulge from the growing brain. If one of these bits of tissue is removed and transplanted under the ectoderm in the area that will become the frog's belly, a lens develops in front of it. This experiment illustrates the importance of _____ in shaping an embryo.
   a. pattern formation
   b. cell aggregation
   c. induction
   d. cell migration
   e. cleavage

2. Birth control pills contain synthetic estrogen and progesterone. How might these hormones prevent pregnancy?
   a. They trigger premature ovulation, before an egg is mature.
   b. They cause the lining of the uterus to be sloughed off.
   c. They cause the corpus luteum to degenerate.
   d. They keep the pituitary from secreting FSH and LH, so ovulation does not occur.
   e. They prevent monthly development of the uterine lining.

3. Suppose you wanted to get started in the animal breeding business. You could start out with only one of most of the following animals, but you would have to start with at least two
   a. sea anemones.
   b. frogs.
   c. rotifers.
   d. whiptail lizards.
   e. sea stars.

4. Pregnancy tests detect a hormone in a woman's urine that is present only when an embryo is developing in her uterus. This hormone is secreted by
   a. the ovary.
   b. the chorion.
   c. a follicle.
   d. the fetus.
   e. the endometrium.

5. One difference between the blastula and gastrula stages of development is
   a. blastula cells are more differentiated than gastrula cells.
   b. there are many more cells in a blastula.
   c. the blastula consists of more cell layers.
   d. the blastula is a solid ball of cells, but the gastrula is hollow.
   e. there is an opening from the cavity inside the gastrula to the outside.

6. Which of the following systems begin to take shape first in a frog?
   a. digestive and nervous systems
   b. nervous and muscular systems
   c. circulatory and digestive systems
   d. nervous and skeletal systems
   e. skeletal and circulatory systems

7. Doctors hoping to increase the chances of implantation of embryos fertilized in vitro might treat the recipient with
   a. progesterone.
   b. follicle-stimulating hormone.
   c. oxytocin.
   d. prolactin.
   e. all of the above.

8. Human embryonic development can be disrupted by diseases (such as rubella), drugs, alcohol, and radiation. These factors have the most severe effects on the baby if they are present during the first 2 months of gestation because at this time
   a. substances can most easily enter the embryo through the placenta.
   b. the embryo has not yet implanted in the wall of the uterus.
   c. the mother's hormone levels are highest.
   d. the most drastic and rapid changes are occurring in the embryo.
   e. the cells of the embryo are not yet fully activated.

9. If you wanted to examine the extraembryonic membranes surrounding a lizard or mouse embryo, you would have to cut through which of the following to see all the others?
   a. yolk sac
   b. amnion
   c. endoderm
   d. allantois
   e. chorion

10. The drug RU 486 was developed in France and is widely used in Europe as a method of birth control. Its introduction to the United States has been controversial, however. The drug is taken after sexual intercourse. It blocks the implantation of the developing blastocyst. Opponents of the drug say it is an "abortion pill." Its defenders point out that it works the same way as
   a. the familiar birth control "pill" used for the last 30 years.
   b. spermicidal foam.
   c. an intrauterine device (IUD).
   d. tubal ligation.
   e. a cervical cap.

Essay

1. Some populations of tropical fish called mollies consist of males and females that reproduce sexually, the way most fish do. Other populations consist only of females, whose eggs develop without fertilization. Which of these populations do you think would adapt most successfully to a change in their environment? Why?

2. A man who wore tight blue jeans every day was found to be suffering from temporary sterility. It seems that his sperm production had been slowed down considerably. Explain why this might have occurred. (Note: This phenomenon cannot be relied on as a method of birth control.)

3. In general, land animals engage in more elaborate courtship behaviors before mating than do aquatic animals. Why might this be more important for land animals?

Using Your Knowledge

1. Most mammals have an estrous cycle. Estrus is the period when ovulation occurs; the female is sexually receptive and mating occurs during this time. For some mammals, like mice, estrus occurs every few days. For dogs and cats, it might occur several times a year. Some wild animals, such as seals, go through estrus only once per year, during the mating season. Humans do not have estrus, and sexual intercourse can occur at any time during the ovarian and menstrual cycles. How would human life be different if women did have estrus? What do you think are possible reasons that estrus has been "lost" during human evolution?

2. How would our lives be different if we reproduced asexually, like the lizards described at the beginning of this chapter? If we were hermaphroditic, like earthworms?

Nervous Systems

28

As your eyes perceive these words, the light and dark patterns printed on the page stimulate light-sensitive cells in your eyes. Changes occur in the membranes of nerve cells in the optic nerve, causing ions to pass in and out of the cells. The changes are transmitted like waves along the nerve cells all the way to visual areas in the back of the brain, in the cerebral cortex. In association areas of the cortex, these words are interpreted, compared with images and memories, and perhaps stored in memory themselves. At the end of each line of text, the brain's motor cortex sends nerve impulses to tiny muscles to move the eyes back to the beginning of the next line. At the same time, other areas of the brain monitor body temperature and blood pressure. They send impulses out by way of the spinal cord to your heart, blood vessels, and sweat glands to maintain homeostasis. In this way, your nervous system senses the environment, interprets it, and directs responses to your muscles and glands. Your nervous system will enable you to read and understand Chapter 28, which is all about nervous systems.

Organizing Your Knowledge

Exercise 1 (Module 28.2)

Review the structure of a neuron by labeling and coloring this diagram. Label the **cell body, axon, myelin sheath, dendrites, synaptic knobs,** and a **node of Ranvier.** Color the dendrites green, the cell body blue, the axon red, and the myelin sheath yellow.

These modules describe the structures and functions of nervous systems and neurons. Review them by completing this crossword puzzle.

Across

2. A ____ is a cell specialized for carrying signals.

5. Motor ____ is conduction of signals from a processing center to effectors.

6. ____ is the interpretation of signals within processing centers.

9. The ____ nervous system (PNS) consists of nerves outside the CNS.

10. ____ neurons carry signals from sensory receptors to the CNS.

14. ____ carry signals to the neuron cell body.

15. The CNS consists of the ____ and spinal cord.

16. Nervous systems consist of neurons and ____ cells.

18. A ____ sheath insulates a neuron.

20. The ____ carries signals away from the neuron cell body.

Down

1. ____ in the central nervous system interpret data.

3. The ____ nervous system is abbreviated "CNS."

4. ____ neurons carry signals from the CNS to effector cells.

7. A ____ is a bundle of axons and dendrites wrapped in connective tissue.

8. Motor commands control muscles and ____.

10. The myelin sheath is destroyed in multiple ____.

11. Signals go faster when they jump along a neuron, between ____ of Ranvier.

12. ____ are clusters of cell bodies belonging to neurons making up a nerve.

13. A synaptic ____ at the end of an axon signals another cell.

17. Sensory ____ is conduction of signals from sensory receptors to the brain.

19. The nucleus of a neuron is in the cell ____.

Exercise 3   (Modules 28.4 – 28.6)

How do neurons carry signals? What exactly is a nerve signal? These three modules explain how nerve cells transmit nerve signals, which are called action potentials. Number in sequence each of the following steps (A–E) in the generation and transmission of a nerve signal. Then match each of the steps with an explanation (P–T).

Sequence   Explanation

_____   _____   A. Action potential occurs
_____   _____   B. Resting potential
_____   _____   C. Stimulus affects neuron
_____   _____   D. Action potential moves
_____   _____   E. Cell repolarizes

P. Na^+/K^+ pump moves Na^+ out of cell, K^+ into cell; K leaks out but Na^+ can't get in. Membrane is + outside and – inside.

Q. Na^+ gates open, Na^+ leaks into cell, and cell depolarizes a bit; membrane becomes somewhat less + outside and less – inside.

R. Na^+ spreads out and causes adjacent area of membrane to depolarize.

S. Na^+ gates close, K^+ gates open, and K^+ leaves cell. Membrane becomes + outside and – inside.

T. Threshold reached and more Na^+ gates open, allowing Na^+ to rush into cell. Membrane becomes + inside and – outside.

Exercise 4   (Module 28.7)

Synapses are relay points where nerve impulses are transmitted from one neuron to another. The following paragraph describes how this occurs. Complete the description by filling in the blanks.

Synapses are junctions between [1]_____. Some are [2]_____ synapses, where action potentials pass from one neuron to the next. Most are chemical synapses, where chemical signals called [3]_____ pass between neurons. A transmitting neuron and a receiving neuron are separated by a narrow gap, called the synaptic [4]_____. Neurotransmitter is contained in small vesicles in the synaptic knobs at the end of the [5]_____ of the transmitting cell. When an action potential arrives at the end of the transmitting cell's axon, chemical changes occur that cause the vesicles to fuse with the [6]_____. The neurotransmitter molecules are released into the synaptic cleft, rapidly diffuse across the cleft, and bind to [7]_____ molecules in the plasma membrane of the [8]_____. This opens ion [9]_____ in the receiving cell's membrane, allowing ions to diffuse through the membrane and initiating new [10]_____ in the receiving cell. The neurotransmitter is quickly broken down by an [11]_____, and the ion channels close. This ensures that the transmission of a signal from cell to cell will be brief and precise.

Exercise 5   (Module 28.8)

Some neurotransmitters cause receiving cells to transmit signals, and other neurotransmitters inhibit the receiving cell's transmission of signals. Neuron X, diagrammed below, adds up all the excitatory and inhibitory impulses it receives. This particular neuron transmits its own impulses only when at least *two* neurons send it excitatory impulses at the same time. Each neuron sending it inhibitory impulses can cancel out the effect of one neuron sending excitatory signals. Given the inputs outlined below, state whether neuron X would or would not transmit nerve impulses in each case.

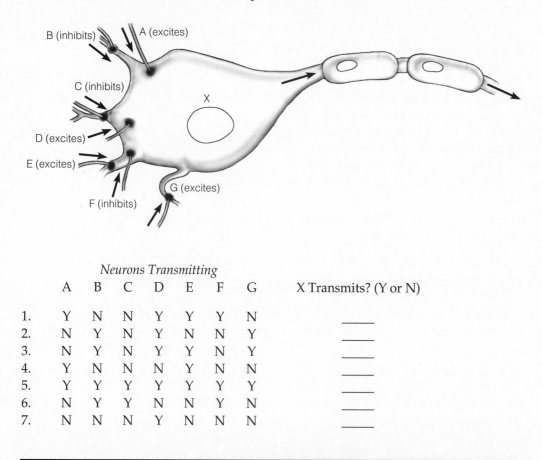

	Neurons Transmitting							X Transmits? (Y or N)
	A	B	C	D	E	F	G	
1.	Y	N	N	Y	Y	Y	N	_____
2.	N	Y	N	Y	N	N	Y	_____
3.	N	Y	N	Y	Y	N	Y	_____
4.	Y	N	N	N	Y	N	N	_____
5.	Y	Y	Y	Y	Y	Y	Y	_____
6.	N	Y	Y	N	N	Y	N	_____
7.	N	N	N	Y	N	N	N	_____

Exercise 6   (Modules 28.9 – 28.10)

These modules describe the roles of neurotransmitters on mood, and the effects of stimulants, depressants, and antidepressants on the nervous system. Imagine a neuron in your brain that transmits impulses to other neurons that elevate mood. This "feel good" neuron transmits when it is stimulated by serotonin from nearby neurons. It is inhibited by GABA from its neighbors. With this in mind, explain how each of the following might affect your mood.

1. Hereditary shortage of serotonin

2. A drug, such as alcohol, that enhances the effects of GABA

3. A drug that stimulates serotonin receptors

4. A chemical that blocks the enzymes that normally break down GABA after it has docked with receptors in the receiving neuron

5. A drug that blocks receptors so that they cannot respond to GABA

6. A drug that blocks the enzymes that normally break down serotonin after it has docked with receptors in the receiving neuron

Exercise 7   (Modules 28.11 – 28.12)

Different animals have different kinds of nervous systems. Write the name of the part or division of the vertebrate nervous system that goes with each of the phrases below.

_____ 1. Master control center

_____ 2. Division of the peripheral nervous system (PNS) that transmits information to the central nervous system (CNS)

_____ 3. Subdivision of the motor division that regulates involuntary activities

_____ 4. Fluid-filled spaces in brain

_____ 5. Division of the PNS that carries impulses from the CNS to effectors

_____ 6. Portion of the motor division that carries motor impulses to skeletal muscles

_____ 7. Myelinated axons and dendrites in the CNS

_____ 8. Fluid that circulates around CNS

_____ 9. All nerves and ganglia outside brain and spinal cord

_____ 10. Nerve tissue of the brain and spinal cord consisting mainly of cell bodies

_____ 11. Brain and spinal cord

_____ 12. Part of the CNS inside the vertebral column

_____ 13. Nerves that carry signals to or from the spinal cord

Exercise 8   (Module 28.13)

What is the overall effect of activity of the parasympathetic nervous system on the body? The sympathetic nervous system? One way to keep their actions straight is to remember that the sympathetic system triggers the "fight or flight" response to stress, gearing up the body for action. Effects of the parasympathetic system are best summarized in the phrase "rest and digest." It slows the body down, for the most part, but stimulates the digestive organs. With these effects in mind, see if you can state the effect of each system on the structures in the chart below.

Body Structure	Parasympathetic Effect	Sympathetic Effect
Stomach	1.	2.
Bronchi (lungs)	3.	4.
Genitals (male)	5.	6.
Heart	7.	8.
Salivary gland	9.	10.
Pupil	11.	12.

Exercise 9   (Module 28.14)

Sketch the paths of nerve signals responsible for the knee jerk reflex and pain reflex on the diagrams below. For each reflex, draw and label a **sensory neuron, sensory receptor, motor neuron, effector organ,** and **interneuron,** and show the routes of nerve signals with arrows. Use different colors for sensory neurons, motor neurons, and interneurons.

Spinal cord

Exercise 10 (Modules 28.15 – 28.16)

Name each of the structures identified by numbers on this diagram of the human brain. Then match each structure or area with its function (A–F).

A. Coordination, balance, position
B. Breathing and heart rate, swallowing
C. Eye reflexes, sleep, arousal
D. Body temperature, eating, drinking, mating
E. "Thinking"
F. Relays information to cerebral cortex

Name   *Function*

1. _____   _____
2. _____   _____
3. _____   _____
4. _____   _____
5. _____   _____
6. _____   _____

Exercise 11 (Modules 28.17 – 28.21)

These modules outline activities of various parts of the cerebrum. Review these activities and other aspects of the nervous system by filling in the blanks in the following story.

Liz appeared to be sleeping quietly, but she was dreaming, and her mind was active. Beneath her closed eyelids, her eyes darted back and forth rapidly, showing that she was in a state called [1]_____ sleep. Outside her window, a car squealed around a corner. An area in Liz's [2]_____ triggered arousal, and Liz woke up. She opened her eyes and looked at the clock—3:00 A.M. Then she remembered her dream. She had been riding her bike between two rows of trees as fluffy clouds passed overhead. "That's it! The idea for my art project," she thought. She fumbled for the light switch, grabbed a pencil and sheet of paper from the bookcase, and scribbled *clouds*. As she wrote, the movement of her hand was controlled by the [3]_____ cortex in the [4]_____ lobe of her cerebrum, which sent nerve impulses to the muscles of her arm via [5]_____ neurons. The [6]_____ helped the cerebrum by assessing her position, then making the movements smooth and coordinated. Liz clicked off the light and soon went back to sleep.

The next morning, when she saw her scrawled *clouds* note, Liz felt happier and more energized than she had in days. She hadn't realized how the looming deadline for her term art project had been getting her down. (She also didn't realize that the [7]_____ system, a group of centers in the forebrain, was actively involved in shaping these emotions.) Now she knew what she was going to paint for the project—a design based on clouds! She could picture the painting perfectly in her mind (actually, mostly in the [8]_____ cerebral hemisphere, as studies of [9]_____-brain patients has shown). She started to make a mental list of the supplies she would need to do the painting.

In the art department of the campus bookstore, Liz gathered the materials she needed. She tested the texture of several watercolor papers with her fingertips. The nerve impulses from her fingers traveled along [10]_____ neurons to her [11]_____, which transmitted them up to her brain. In the brain, the [12]_____ sorted the sensory impulses and relayed them to Liz's cerebral cortex. The [13]_____ cortex in the [14]_____ lobe of Liz's cerebrum interpreted the nerve impulses from her fingers, inferring from them the textures of the papers. The sensory information from her right hand was sent to the [15]_____ side of the brain, but Liz could check it against her mental image because the nerve impulses were able to cross from one side of the brain to the other via the [16]_____, a thick band of nerve fibers connecting the two cerebral hemispheres.

One of the shelves was empty, and Liz thought the missing paper might be just the one she needed. She read the stock number on the shelf, repeated it to herself a couple of times, and kept the number in her [17]_____-term memory just long enough to ask a clerk whether the store had more in stock. No luck. Liz recalled the paper she had used for a couple of assignments the previous term. This information was stored in [18]_____-term memory, in the form of structural or chemical changes in the [19]_____ of her cerebral cortex. Because she had used the paper several times, her [20]_____ and [21]_____, two parts of the limbic system, had selected the name of the paper for storage in memory.

Liz found the paper, paints, and a brush that she needed and headed for the checkout line, adding up the prices in her head (actually on the [22]_____ side of her head, because [23]_____ areas in the [24]_____ cerebral hemisphere seem to be primarily responsible for mathematical calculations). The total was a bit more than she had anticipated, and a moment of anxiety about her budget caused Liz's heart and breathing to speed up a bit, under orders from the [25]_____ and [26]_____, two areas of the hindbrain.

With her purchases tucked under her arm, Liz set off for home on her ten-speed. She crossed the park and turned onto a bike path flanked by tall maple trees. Fluffy clouds sailed overhead. Liz thought the scene seemed very familiar.

Testing Your Knowledge

Multiple Choice

1. Which of the following best describes a nerve signal?
   a. flow of electricity along a neuron
   b. passage of ions through the membrane of a neuron
   c. flow of neurotransmitter chemical along a neuron
   d. movement of tiny filaments of protein inside a neuron
   e. change in a neuron so that the inside becomes more negatively charged

2. The part of a neuron that carries nerve impulses toward the cell body is called
   a. a nerve.
   b. white matter.
   c. a neurotransmitter.
   d. a dendrite.
   e. an axon.

3. Which of the following maintains resting potential, the difference in electrical charge inside and outside a neuron membrane that enables the cell to transmit a signal?
   a. charges that pull sodium and potassium through the membrane
   b. opening of sodium and potassium gates in the membrane
   c. the myelin sheath, which prevents ions from entering or leaving
   d. transport and leakage of sodium and potassium into and out of the cell
   e. the mutual repulsion of sodium and potassium ions

4. The _____ contains association areas for speech, language, and calculation.
   a. right cerebral hemisphere
   b. medulla oblongata
   c. midbrain
   d. left cerebral hemisphere
   e. cerebellum

5. A stimulus triggers an action potential by
   a. causing sodium ions to leak into the neuron.
   b. triggering the release of neurotransmitter.
   c. causing potassium ions to leak out of the neuron.
   d. activating the sodium-potassium pump.
   e. causing sodium ions to leak out of the neuron.

6. Squids are useful animals in nervous system research because
   a. their nervous systems are so complex.
   b. they easily learn to respond to various stimuli.
   c. their nerve fibers are so large.
   d. they have a brain and spinal cord very similar to those of humans.
   e. they are easy to breed and raise in the laboratory.

7. The autonomic nervous system
   a. is the part of the nervous system outside the brain and spinal cord.
   b. controls and coordinates voluntary movements.
   c. regulates the internal environment.
   d. integrates all sensory information from the environment.
   e. consists of the brain and spinal cord.

8. You touch a hot frying pan and your hand jerks back. Which of the following describes the path of nerve impulses responsible for this reflex?
   a. motor neuron–sensory neuron–interneuron
   b. sensory neuron–motor neuron–interneuron
   c. interneuron–sensory neuron–motor neuron
   d. sensory neuron–interneuron–motor neuron
   e. motor neuron–interneuron–sensory neuron

9. The limbic system is involved in
   a. emotion and memory.
   b. speech and hearing.
   c. vision.
   d. sleep and wakefulness.
   e. control of heartbeat and respiration.

10. Which of the following is part of the central nervous system?
    a. cranial nerve
    b. spinal nerve
    c. spinal cord
    d. sympathetic nerve
    e. ganglion

Essay

1. Sketch a chemical synapse and use your sketch to explain how a nerve impulse crosses a chemical synapse from the transmitting neuron to the receiving neuron.

2. Compare the functions of the parasympathetic and sympathetic divisions of the autonomic nervous system.

3. Briefly describe three major trends in the evolution of the vertebrate brain.

4. What are the three major functions of nervous systems? How are they illustrated as you answer this question?

5. A neuron can receive inputs from thousands of transmitting neurons, but it can respond in only one of two ways—it can either transmit action potentials or not transmit action potentials. How do transmitting cells signal the receiving cell, and what determines whether the receiving cell will or will not transmit action potentials itself?

Applying Your Knowledge

Multiple Choice

1. The axons of which of the following end in the central nervous system?
   a. motor neurons
   b. sensory neurons
   c. interneurons
   d. a and c
   e. b and c

2. A man was admitted to the hospital suffering from abnormally low body temperature, loss of appetite, and extreme thirst. A brain scan showed a tumor located in the
   a. hypothalamus.
   b. cerebellum.
   c. pons.
   d. right cerebral hemisphere.
   e. corpus callosum.

3. A drug that causes potassium to leak out of a neuron, increasing the positive charge on the outside, would
   a. make it easier to trigger action potentials in the neuron.
   b. cause the cell to release its neurotransmitter.
   c. speed up nerve signals traveling the length of the cell.
   d. act as a stimulant.
   e. inhibit transmission of nerve signals by the neuron.

4. What is the difference between a neuron and a nerve?
   a. One is sensory in function, the other motor.
   b. Nerves are found only in the central nervous system.
   c. They consist of different numbers of cells.
   d. Neurons are made of white matter, nerves of gray matter.
   e. Neurons are found only in vertebrates.

5. Which of the following animals is *least* cephalized?
   a. clam
   b. ant
   c. flatworm
   d. human being
   e. fish

6. Which of the following includes all the others?
   a. autonomic nervous system
   b. motor division
   c. somatic nervous system
   d. peripheral nervous system
   e. sensory division

7. Alex became so dehydrated while playing tennis that his blood pressure started to drop. His _____ detected the pressure drop and sent signals via _____ to speed up the heart, compensating for the drop in pressure.
   a. hypothalamus . . . parasympathetic neurons
   b. medulla oblongata . . . sympathetic neurons
   c. cerebellum . . . sympathetic neurons
   d. medulla oblongata . . . parasympathetic neurons
   e. cerebellum . . . parasympathetic neurons

8. The gray matter of the cerebral cortex, where most higher-level "thinking" occurs, is composed mostly of
   a. interneuron cell bodies.
   b. myelinated axons of neurons.
   c. sensory neuron cell bodies.
   d. Schwann cells.
   e. motor neuron cell bodies.

9. Which of the following neurons would both receive nerve signals from other neurons and transmit signals to other neurons?
   a. a sensory neuron from the fingertip to the spinal cord
   b. a motor neuron from the spinal cord to the leg
   c. an interneuron in the brain
   d. all of the above
   e. both the sensory and motor neurons

10. Sensory receptors in the ear can detect a whisper or a shout and can transmit nerve signals about these sounds to the brain. How would sensory neurons relay information to the brain about the *loudness* of loud and soft sounds?
    a. The action potentials would travel at different speeds.
    b. The size of action potentials would vary.
    c. The number of action potentials would vary.
    d. The frequency of action potentials would vary.
    e. Action potentials would be routed to different parts of the brain.

Essay

1. Many nervous system drugs and poisons act at synapses. Explain how each of the following would alter the transmission of nerve impulses.
    a. Curare, a substance used on poison arrows by South American Indians, competes with acetylcholine for receptor sites on receiving neurons.
    b. Botulism toxin inhibits the release of acetylcholine.
    c. Diisopropyl fluorophosphate (DF) is used in warfare as a nerve gas. It blocks the enzyme that breaks down acetylcholine after it has crossed the synapse to the receiving neuron.

2. Why is transmission of a nerve signal along a neuron sometimes compared with the toppling of a row of dominoes? How does it differ from the toppling of dominoes?

3. Andrew's back was broken in an auto accident, and he was paralyzed below the shoulders. Explain why he is able to move his arms but not his legs. Like many victims of such injuries, Andrew cannot feel his legs, but he does respond when the doctor tests his knee-jerk reflex. Explain how.

4. A victim of a severe head injury may live for years in a "persistent vegetative state"—unconscious but still alive and breathing. Explain how a person can continue to live even though the cerebrum, the largest part of the brain, ceases to function.

Using Your Knowledge

1. Most adults require about 8 hours of sleep per night, teenagers more than 9 hours. How much do you get? Researchers are beginning to think that sleep deprivation is a major health problem in the United States. It may impair thinking, alertness, attention span, creativity, and judgment. Signs of sleep deprivation include needing an alarm clock to wake up, falling asleep within 5 minutes of going to bed (a well-rested person goes to sleep in 10 to 15 minutes), napping at will, and nodding off at work or in classes. Do you think you might be short of sleep? If you don't think you are getting enough sleep, consider adjusting or lightening your class or work schedule, taking naps, and giving sleep higher priority on your "things to do" list.

2. Have you taken a college psychology course? How do the facts and concepts in this chapter relate to what you learned in psychology? How was the coverage of the nervous system similar in biology and psychology? How did the coverage in the two courses differ? What are some reasons for possible differences?

The Senses

We take it for granted that humans experience the world through the five senses—sight, hearing, touch, taste, and smell. Sometimes we say that an acutely intuitive individual has a "sixth sense," but in reality we all have a sixth sense, and a seventh, and an eighth. The sense of touch is actually a complex of senses—touch, pressure, and temperature. We even have separate sensory receptors for heat and cold. Our ears respond not only to sound but also to changes in body position. Additional senses monitor the interior of the body, collecting information on blood pressure and the positions of muscles and joints. Other animals possess sensory ranges that we can only imagine. A dog is aware of sounds and smells beyond our sensory range. Bees see ultraviolet light; rattlesnakes see infrared. Some creatures can even feel the Earth's magnetic field. This chapter is about the senses and how they reveal the world to us.

Organizing Your Knowledge

Exercise 1   (Modules 29.1 – 29.2)

As you read these words, several different processes occur in your eyes and brain before you understand what the words mean. Match each of these processes (A–F) with its description (L–Q), and number the processes in the order in which they might occur.

Order   Description

_____   _____   A. Transduction

_____   _____   B. Sensation

_____   _____   C. Transmission

_____   _____   D. Perception

_____   _____   E. Reception

_____   _____   F. Adaptation

L. Conscious understanding of sensory data

M. Conversion of a stimulus into electrical signals

N. Sending action potentials to the brain

O. Drop in sensitivity of receptors when stimulated repeatedly

P. Awareness of sensory stimuli by the brain

Q. Detection of stimuli by sensory cells

Exercise 2 (Module 29.3)

There are five general kinds of sensory receptors—pain receptors, thermoreceptors, mechanoreceptors, chemoreceptors, and electromagnetic receptors. To review them, complete this chart. Give the general category of receptors to which each example belongs, the kind of stimulus it responds to, and where you might find one of the receptors.

Example	General Receptor Type	Stimulus It Responds To	Locations
Infrared receptor	1.	Infrared radiation	2.
Touch receptor	3.	Contact and movement	4.
5.	6.	Electric currents in water	7.
8.	9.	10.	Eyes
Pain receptor	11.	Danger, injury, disease	Many body locations
12.	Mechanoreceptor	13.	Human ear
Heat receptor	14.	15.	16.
17.	18.	Body position	19.
Smell receptor	20.	21.	Nose

Eyes can be simple or complex. The vertebrate eye is a complicated and sensitive sensory organ. To familiarize yourself with the major parts of the eye, label and color the diagram below. Include the **sclera, choroid, retina, pupil, iris, lens, cornea, vitreous humor, aqueous humor, optic nerve,** and the **muscle** attached to the lens.

1. _____
2. _____
3. _____
4. _____
5. _____
6. _____
7. _____
8. _____
9. _____
10. _____
11. _____

Now match each of the following functions with a part of the eye by placing its letter next to the correct label on the diagram.

A. Carries nerve signals to the brain
B. Covers the front of the eye but lets light in
C. Supplies nutrients and oxygen to the lens, iris, and cornea
D. Photoreceptor cells here transmit nerve signals to the brain
E. Regulates the size of the pupil to let more or less light into the eye
F. Bends light rays and focuses them on the retina
G. Changes the shape of the lens

Exercise 4   (Module 29.6)

The lenses in the eyes of a fish move back and forth to bring images into focus, but the lenses in human eyes focus by changing shape. After studying Figure 29.6, sketch the path of light rays and the shape of the lens when the eye focuses on a nearby object and a distant object.

Nearby object   Distant object

Exercise 5   (Module 29.7)

There are a number of possible reasons for less-than-perfect vision. Some people are near-sighted, others farsighted. Some suffer from astigmatism. After reading about these vision problems, sort them out by completing the following chart.

Diagram			
Problem name	1.	2.	3.
Nearby or distant objects clear	4.	5.	6.
Nearby or distant objects unclear	7.	8.	9.
Cause(s)	10.	11.	12.
Method of correction	13.	14.	15.

Exercise 6   (Module 29.8)

This module discusses rods and cones, the sensory receptors of the retina. Indicate below the statements that relate to rods and those that relate to cones by placing check marks in the appropriate column. (Some statements relate to both.)

	Rods	Cones
1. Come in three types		
2. Stimulated by bright light		
3. Function in night vision		
4. Most numerous in the fovea		
5. Located in the back layer of the retina		
6. Contain a visual pigment called rhodopsin		
7. Can see different colors		
8. Stimulus transducers		
9. Absent from the fovea		
10. Most numerous at the outer edges of the retina		
11. Contain pigments called photopsins		
12. See only shades of gray		
13. Transmit signals to the visual cortex		
14. Defective in color blindness		
15. Synapse with retinal neurons		
16. Most numerous receptors		
17. Most sensitive receptors		

Exercise 7   (Module 29.9)

The ear is a mechanical marvel, catching and amplifying sound waves and transforming them into nerve impulses. As you review the structure and function of the ear, label the following parts on the diagram at the top of the next page: **auditory nerve, eardrum, cochlea, oval window, anvil, semicircular canals, inner ear, auditory canal, stirrup, outer ear, pinna, hammer, eustachian tube,** and **middle ear**. Color all structures of the outer ear red, the middle ear yellow, and the inner ear blue.

Now match each of the following functions with a part of the ear by placing its letter next to the correct label on the diagram.

A. Its hair cells transduce motion into nerve signals
B. Equalizes pressure outside and inside eardrum
C. Other animals, such as dogs and horses, can turn it toward sounds
D. Transmits action potentials from ear to brain
E. Transmits vibrations to middle-ear bones
F. Collects and channels sound waves to eardrum
G. Transmits vibrations from eardrum to cochlea
H. Function in balance

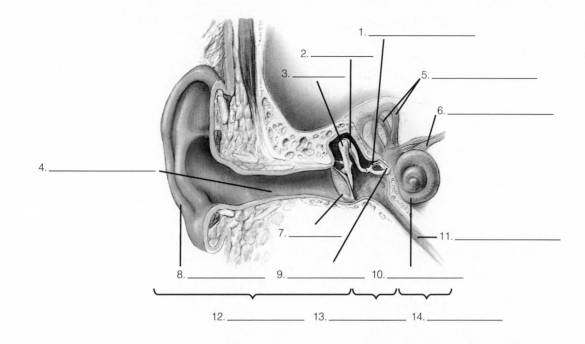

1. _____
2. _____
3. _____
4. _____
5. _____
6. _____
7. _____
8. _____
9. _____
10. _____
11. _____
12. _____
13. _____
14. _____

Exercise 8   (Modules 29.10 – 29.13 and Summary)

Modules 29.10 – 29.12 discuss the senses of balance, smell, and taste, and Module 29.13 reviews the role of the senses in the overall function of the nervous system. Review the senses (including vision and hearing) by matching each of the phrases on the right with a sensory structure from the list on the left. Some answers are used more than once.

A. Retina
B. Semicircular canal
C. Utricle and saccule
D. Hair cell
E. Odor receptor
F. Taste bud
G. Organ of Corti
H. Cone

1. ____ Receptors here detect changes in movement
2. ____ Receptors here sense sour, salty, sweet, bitter
3. ____ Chemicals dissolve in mucus and bind to its cilia
4. ____ Site of receptors for hearing
5. ____ Light receptor of the retina
6. ____ Receptors here sense position of head relative to gravity
7. ____ Receptor of hearing
8. ____ Site of photoreceptors
9. ____ Balance receptor
10. ____ Chemoreceptor of the nasal cavity

Testing Your Knowledge

Multiple Choice

1. Insect eyes are much better at seeing _____ than are our eyes.
   a. colors
   b. in dim light
   c. movement
   d. fine detail
   e. distant objects

2. A thermoreceptor in the skin converts heat energy into nerve impulses. This conversion process is called
   a. sensation.
   b. transduction.
   c. reception.
   d. integration.
   e. perception.

3. Which of the following correctly traces the energy of sound waves into the ear?
   a. auditory canal–eardrum–ear bones–cochlea
   b. eardrum–auditory canal–cochlea–ear bones
   c. auditory canal–ear bones–eardrum–cochlea
   d. eardrum–auditory canal–ear bones–cochlea
   e. eardrum–ear bones–auditory canal–cochlea

4. Which of the following are *not* correctly paired?
   a. mechanoreceptor–stretch receptor
   b. electromagnetic receptor–photoreceptor
   c. chemoreceptor–taste bud
   d. mechanoreceptor–touch receptor
   e. electromagnetic receptor–hair cell

5. When you focus your eyes on a nearby object, the lenses
   a. decrease in diameter.
   b. move forward.
   c. become more rounded.
   d. become flatter.
   e. move backward.

6. As light passes into the eye, it goes through which of the following first?
   a. lens
   b. pupil
   c. aqueous humor
   d. cornea
   e. vitreous humor

7. The brain determines the loudness of a sound from
   a. the part of the organ of Corti stimulated by the sound.
   b. the rate of nerve impulses received.
   c. the size of air pressure changes in the middle ear.
   d. the part of the brain receiving nerve impulses from the ear.
   e. the size of the nerve impulses received.

8. Josh is color blind, so he has a lot of trouble picking out clothes. Which sensory structures below are defective in a color-blind person?
   a. organs of Corti
   b. cones
   c. hair cells
   d. rods
   e. utricle and saccule

9. Which of the following is *not* a mechanoreceptor?
   a. taste bud
   b. hair cell
   c. touch receptor
   d. pressure receptor
   e. stretch receptor

10. Eating carrots really is good for your eyes. Carrots contain vitamin A, which is used to make a substance called rhodopsin, which
    a. is a visual pigment that absorbs light.
    b. provides energy for the function of rods and cones.
    c. colors the iris of the eye.
    d. stimulates the neurons in the retina to form branches and connections.
    e. keeps the lens clear and transparent.

Essay

1. Bats can navigate in total darkness by listening to the echoes of their high-pitched clicks. In a few sentences, describe three other examples of animal sensory capabilities beyond the range of human senses.

2. Explain the difference between sensation and perception.

3. What is farsightedness? What eye defect causes it? How is farsightedness corrected?

4. Explain how a stimulus (light, chemical, temperature change, movement) triggers a nerve impulse in a sensory receptor.

5. How are the sense of smell and sense of taste similar? What are the differences between the senses of taste and smell?

Applying Your Knowledge

Multiple Choice

1. You look all over for your glasses and then find them pushed back onto your forehead. This error could be attributed to
   a. perception.
   b. transduction.
   c. sensation.
   d. sensory adaptation.
   e. accommodation.

2. Michael and Laura are both looking at an illustration in her psychology textbook. Michael says, "It is an old woman looking down toward the left." Laura says, "I see a young woman looking away from us." How can they

look at the same picture and see different things?

a. Perhaps one of them has astigmatism.

b. Their sensations differ.

c. They sense the same thing, but their perceptions differ.

d. It may take some time before their eyes accommodate to the picture.

e. They perceive the same thing, but their sensations differ.

3. Whales and dolphins are known to send out clicking sounds and listen to the echoes. This suggests that they might find their prey in the same way

a. a rattlesnake finds a mouse in total darkness.

b. a salmon locates its home stream.

c. an owl locates a mouse in a dark barn.

d. an electric eel finds its prey in a muddy river.

e. bats navigate in dark caves.

4. As you read this sentence, how do the photoreceptors of your retina tell the brain whether an area is light (the paper) or dark (the print)?

a. Either rods or cones send signals to the brain.

b. Nerve signals are sent to different areas of the brain.

c. Signals are sent to the brain at a faster (light) or slower (dark) rate.

d. Accommodation adjusts for light and dark areas.

e. Larger (white) or smaller (dark) action potentials are sent to the brain.

5. A fish detects vibrations in the water around it by means of its "lateral lines," rows of sensory receptors along each side of the body. Based on what you know about sensory receptors, the lateral line receptors are probably which of the following?

a. hair cells

b. rods

c. chemoreceptors

d. infrared receptors

e. taste buds

6. Damage to the nerve from the saccule and utricle to the brain could result in

a. loss of sense of taste.

b. blindness.

c. dizziness.

d. loss of sense of smell.

e. deafness.

7. There may be as many as 50 different kinds of smell receptors, each activated by different molecules. We can differentiate among thousands of different smells because each substance stimulates a different combination of these receptors. In much the same way

a. our eyes can distinguish many different colors.

b. our ears can differentiate between sounds of many different pitches.

c. pain receptors can distinguish between mild and severe pain.

d. our eyes can adjust to light of differing intensities.

e. our ears can determine how loud a sound is.

8. It is said that if you are seasick, it is better to look out at the water than at the boat. Why?

a. It fools your brain into thinking that you are not really moving.

b. This stimulates the saccule and utricle, organs of equilibrium.

c. Seeing that you are moving reduces conflict between vision and equilibrium.

d. Keeping your head level reduces activity of the semicircular canals.

e. Actually, looking at the water just makes seasickness worse.

9. The eyes of a nocturnal animal, such as an owl, could be expected to have a larger proportion of _____ than do our eyes.

a. chemoreceptors

b. cones

c. rods

d. hair cells

e. mechanoreceptors

10. The carotid body is a structure in the wall of the main artery leading to the head. In it are special receptors that monitor blood pressure. These receptors would belong to which of the following groups?

a. chemoreceptors

b. thermoreceptors

c. photoreceptors

d. mechanoreceptors

e. electromagnetic receptors

Essay

1. Theresa is suffering from a mysterious disorder called Meniere's disease. Her symptoms are quite debilitating: loud roaring sounds and dizziness. Why do these symptoms occur together? What part of the body must be affected by Meniere's disease?

2. The eyes contain three kinds of cones—green, blue, and yellow. Which are stimulated when we see blue? Yellow? White? Black?

3. Some researchers believe that human beings might have an ability to sense direction that is distinct from the traditional "five senses." It has been suggested that perhaps we have some ability to sense the Earth's magnetic field. Describe an experiment that could test this suggestion.

4. Most sensory receptors are sensitive to a sudden change in a stimulus, but soon adapt—become less sensitive—to a continuing stimulus. For example, thermoreceptors quickly adapt, and a very hot bath soon feels comfortable. How might this be useful? Pain receptors are an exception; they do not readily adapt. Why do you think this is the case?

5. There are many different kinds and causes of deafness. Explain how each of the following would result in deafness: damage to hair cells from repeated loud sounds, brain injury, arthritis of the middle-ear bones, a torn eardrum, buildup of earwax in the auditory canal.

Using Your Knowledge

1. Anosmia is loss of the sense of taste or smell. It sometimes results from a brain injury or as a side effect of medication. Do you think anosmia could be inconvenient or dangerous in any way? Do you think it would be worse to lose your sense of smell or your sense of taste? Why?

2. Have you ever felt your ears ringing after listening to loud music from a stereo or at a concert? What do you think about the possibility that listening to loud music could permanently damage your hearing? Are your friends aware of this danger to their hearing? What, if anything, should be done to warn people about this danger or protect them from it?

How Animals Move

30

Look at the palm of your hand. Primates are the only mammals that can turn their arms palm-up. Our lower arm bones are able to pivot at the elbow and cross, so we are able to rotate our hands through almost a full circle. The arrangement of bones and muscles that allows us to do this is a heritage passed down from our tree-living ancestors. Although other mammals cannot do the palm-up trick, their skeletons and muscles show equally remarkable adaptations to other methods of movement. A horse runs on very long middle toes. A sea lion's legs have become flattened paddles, and its muscles can store extra oxygen for extended dives. Every hop of a kangaroo stores energy in elastic leg tendons; rebound of the tendons helps power the next hop. Running, swimming, and hopping are all movements produced by muscles pulling on bones. Skeletons, muscles, and movement are the subjects of this chapter.

Organizing Your Knowledge

Exercise 1   (Module 30.1)

This module introduces various forms of animal locomotion. Test your knowledge of locomotion by filling in the blanks.

Mammals range in size from tiny shrews to great whales. They have become adapted to a variety of life-styles on land, in the water, and in the air.

Most mammals, from mice to musk oxen, walk or run on land. The surrounding air offers little resistance to movement, so [1]_____ is not much of a problem, even for the fastest runners. But a mammal that walks or runs must expend considerable effort supporting itself against the force of [2]_____. To maintain balance and stability when it walks, an animal employs the principle of the [3]_____, keeping three feet on the ground most of the time. The fastest land animal, the cheetah, is of course a mammal. When it runs, all of its feet may leave the ground at once, but its [4]_____ stabilizes its position.

Some mammals crawl or burrow—moles and gophers, for example. Unlike the earthworm, which moves by waves of muscular contractions called [5]_____, burrowing mammals use their legs for digging. The mole's coat is velvety smooth, reducing [6]_____ as it tunnels underground.

Millions of years ago, a group of shore-dwelling mammals took up life in the sea. Their descendants are the whales and dolphins. Because of their buoyancy, [7]_____ is no problem for these aquatic animals, and some have grown to enormous size. Water is much more dense than air, but a whale's tapered, [8]_____ shape enables it to slip through the water with little effort. The whale's tail forms a pair of broad, flat flukes, and the whale bends its body [9]_____ to push it through the water.

Bats are the only flying mammals. A bat's wings are modified arms, with skin stretched between elongated fingers. The bat's wings, like those of airplanes and birds, are [10]_____. As a bat beats its wings, air travels farther over the top surface than the bottom surface, making the pressure underneath the wings [11]_____ than the pressure on top. This generates [12]_____ and allows the bat to overcome the pull of [13]_____.

Exercise 2   (Module 30.2)

There are three main kinds of skeletons. Summarize your knowledge of skeletons by completing this chart.

	Hydrostatic Skeleton	*Exoskeleton*	*Endoskeleton*
Example animals	1.	2.	3.
Description of skeleton	4.	5.	6.
Materials skeleton is made of	7.	8.	9.
Functions of skeleton	10.	11.	12.
Drawbacks of skeleton	13.	14.	15.

Exercise 3   (Module 30.3)

The human skeleton is unique in several ways, but it has the same overall pattern as the skeleton of a mouse or a duck. You may find it useful to know the names of some of the major bones of the skeleton. Label these bones on the diagram below: **humerus, sternum, femur, tarsals, tibia, carpals, radius, phalanges, ulna, vertebrae, skull, clavicle, metacarpals, ribs, scapula, fibula, pelvic girdle, metatarsals, shoulder girdle,** and **patella**. (Note: The letters relate to Exercise 4 below.) Before you leave this exercise, color the bones of the axial skeleton green and the appendicular skeleton yellow.

Exercise 4   (Module 30.3)

Which of the letters in Exercise 3 identifies each of the following types of joints? (Three are from the module; you will have to figure out the other three yourself.)

1. Ball-and-socket joints: _____

2. Hinge joints: _____

3. Pivot joints: _____

Exercise 5   (Modules 30.4 – 30.6)

Review bone structure and growth by matching each phrase on the right with a word or phrase from the left. Some answers are used more than once.

A. Spongy bone
B. Cartilage
C. Estrogen
D. Red bone marrow
E. Blood vessel
F. Fibrous connective tissue
G. Yellow bone marrow
H. Compact bone
 I. Matrix
 J. Tendon

_____ 1. Fat stored in central cavity
_____ 2. Protein fibers and hard calcium salts
_____ 3. Outer hard layer of bone
_____ 4. Cushions ends of bone at joints
_____ 5. Decrease after menopause may cause osteoporosis
_____ 6. Covers outside of bone
_____ 7. Shafts of long bones made of this at first
_____ 8. Delivers food and hormones to bone
_____ 9. Produces blood cells
_____ 10. Site of red marrow
_____ 11. Bone grows until plates of this material are replaced by bone
_____ 12. Makes up the shaft of a long bone like the tibia
_____ 13. Connects a muscle to a bone

Exercise 6   (Modules 30.8 – 30.9)

A muscle is composed of many muscle fibers. Each fiber contains smaller parts, which are responsible for muscle contraction. Review the contractile machinery of muscle by labeling these parts in the diagram below: **muscle, thin filament, muscle fiber, myofibril, thick filament, sarcomere, Z line.**

Exercise 7   (Modules 30.8 – 30.10)

Try to visualize the inner workings of a muscle by filling in the blanks below.

Your new mission is to view contracting muscle cells close up. You step into the Microtron, and a moment later, you are working your way against the current in a small arm vein. Ahead are smaller, thin-walled [1]_____ that exit from the muscle and join to form the vein. You stay close to the wall and enter the smallest vessel.

Through the translucent wall of the capillary, you can make out a nearby surface that seems to be made of shiny white ropes. This is the [2]_____ that connects the end of the biceps to the lower arm bones. Also visible is a smoother, shinier cylindrical structure—one of the [3]_____ that controls contraction of the muscle. Ahead, you see the wrinkled blob of a white blood cell slipping through a gap in the vessel wall, and you follow it into the intercellular fluid. Close up, the nerve you saw a moment ago appears frayed—its neurons branch and rebranch to control every muscle cell. You follow one of the neurons; it branches to about 50 cells. The neuron and the cells it controls make up a [4]_____—a group of muscle fibers that work together.

Following one of the axons to its end, you finally arrive at a [5]_____, where the axon synapses with a muscle cell—called a muscle [6]_____. You are deep inside the muscle now, so you radio for more light and ask for the subject to tense her arm.

Suddenly a cloud of particles sprays from the end of the axon, the membrane of the muscle fiber seems to shimmer, and you are tossed by a surge of pressure into a nearby tangle of connective tissue fibers. She's moving the arm just a little bit, but to you it feels like an earthquake! You contact your colleagues: "Hey, take it easy! Try it again, but gently!" This time you hang on tightly. Again there is a spray of particles—the release of [7]_____ from the end of the axon, signaling the muscle fiber to contract. When the neurotransmitter molecules contact the muscle fiber membrane, the shimmer you saw before is repeated—an [8]_____ spreading across the cell. Another earthquake, but this time you are ready.

You grip the membrane of the muscle fiber. It is pocked with numerous openings, where infoldings of the membrane form [9]_____ that carry action potentials deep into the cell. You swim inward, following the action potentials, as the cell continues to receive impulses and contract. You can press your powerful spotlight against the tubule wall and illuminate the inside of the fiber. Inside are numerous [10]_____, large bundles of parallel protein filaments. Transverse [11]_____ lines, made of protein, separate repeating [12]_____, the basic contractile units of the muscle fiber. The filaments themselves are of two types: The [13]_____ filaments look like twisted strings of beads, the beads themselves being globular [14]_____ molecules. The [15]_____ filaments are made of elongated [16]_____ molecules, each with a head that can reach out and pull against the thin filaments.

You watch the thick and thin filaments [17]_____ along each other with each volley of nerve impulses. As each action potential travels into the cell via the tubules, you see a cloud of [18]_____ ions released from storage in the endoplasmic reticulum. These ions quickly attach to the thin filaments, freeing up binding sites for the myosin [19]_____ on the adjacent thick filaments. Meanwhile, the myosin molecules of the thick filaments are writhing like a bundle of worms. Their heads use energy from [20]_____ molecules to detach from the thin filaments, straighten, attach, and bend, several times per second. The hundreds of myosin heads in one thick filament pull on all the surrounding thin filaments. This creates the pull that causes the thick and thin filaments to slide together, [21]_____ the sarcomere, and on a larger scale, the muscle fiber.

The nerve impulses cease. You can see [22]_____ ions being pumped out of the cytoplasm and back into the [23]_____. As the binding sites on the thin filaments close up, the myosin heads let go, and the muscle stops contracting.

Exercise 8   (Modules 30.11 – 30.12)

Combine the topics of the last two modules: robots and the connection between structure and function. Imagine that you are a writer for a news magazine. You have been assigned to do an article about a new generation of robots designed to perform specific jobs. Write three article titles that state how the function of these robots emerges from their structure.

Testing Your Knowledge

Multiple Choice

1. _____ connects a muscle to a bone.
   a. Cartilage
   b. A neuromuscular junction
   c. A tendon
   d. A myofibril
   e. A motor unit

2. Inside a muscle fiber, _____ triggers contraction and _____ provides the energy.
   a. myosin . . . actin
   b. calcium ions . . . ATP
   c. actin . . . myosin
   d. calcium ions . . . myosin
   e. ATP . . . calcium ions

3. Rheumatoid arthritis
   a. is a disease in which joints are attacked by the immune system.
   b. results from the wear and tear of old age.
   c. is deterioration of joints that results from a hormone imbalance.
   d. occurs most frequently in women after menopause.
   e. can be prevented by adequate calcium in the diet.

4. Muscles are arranged in pairs,
   a. so if one is injured, the other can take over.
   b. doubling their strength.
   c. because one pulls while the other pushes.
   d. enabling them to perform opposing movements.
   e. so they can take turns contracting and resting.

5. Which of the following is the last part of a long bone to harden?
   a. the central shaft
   b. the heads
   c. cartilage between the heads and the shaft
   d. the central cavity
   e. cartilage at the very ends

6. Which of the following correctly pairs an example of a joint with its general type?
   a. ball and socket–elbow
   b. pivot–shoulder
   c. hinge–hip
   d. hinge–elbow
   e. ball and socket–knee

7. Which of the following skeletons works poorly on land?
   a. endoskeleton
   b. hydrostatic skeleton
   c. exoskeleton
   d. a and b
   e. b and c

8. A stronger muscle contraction occurs when the brain
   a. activates muscle cells more quickly.
   b. sends stronger nerve impulses to the muscle.
   c. activates the motor units of the muscle one at a time.
   d. signals a larger number of motor units to contract.
   e. sends nerve impulses to the muscle one at a time instead of in bursts.

9. Which of the following animals has an endoskeleton?
   a. clam
   b. sea star
   c. sponge
   d. a and b
   e. b and c

10. Which of the following is part of the human appendicular skeleton?
    a. vertebral column
    b. scapula
    c. rib
    d. sternum
    e. skull

Essay

1. How do the shape of its wings and their movement through the air lift a bird off the ground?

2. What are the three major functions of a skeleton? Describe each, using parts of your own skeleton as illustrations.

3. What kinds of animals have jointed exoskeletons? What are the major advantages and disadvantages of exoskeletons, compared with endoskeletons?

4. What two forces must be overcome by a moving animal? Which is more of a problem on land? In the water?

5. Without looking it up, make a sketch showing the arrangement of thick and thin filaments in a muscle cell when the cell is relaxed and when the cell is contracting.

Applying Your Knowledge

Multiple Choice

1. Which of the following shortens when a muscle fiber contracts?
   a. thin filament
   b. myosin molecule
   c. sarcomere
   d. actin molecule
   e. thick filament

2. Which of the following ranks the parts in order, from largest to smallest?
   a. muscle–myofibril–filament–fiber
   b. fiber–muscle–myofibril–filament
   c. muscle–fiber–filament–myofibril
   d. myofibril–muscle–fiber–filament
   e. muscle–fiber–myofibril–filament

3. His doctor suspects that Jon might be suffering from leukemia. The doctor orders a sternal puncture, a test that samples red marrow. This test is performed to determine
   a. whether the marrow holds sufficient fat reserves.
   b. how fast Jon is growing.
   c. the rate of blood flow through the bone tissues.
   d. whether blood cell production is occurring normally.
   e. how fast cartilage is forming in the bones.

4. For every bone in the arm and hand, there is a corresponding bone in the leg and foot. Which of the following matches corresponding bones?
   a. humerus–femur
   b. metacarpals–tarsals
   c. radius–femur
   d. carpal–patella
   e. humerus–tibia

5. An animal that crawls or burrows faces some of the same problems faced by an animal that
   a. swims.
   b. runs.
   c. flies.
   d. walks.
   e. hops.

6. A scallop escapes from danger by clapping its shells together, which shoots out a stream of water and causes the scallop to hop backward. This is most like the movement of
   a. an earthworm.
   b. a water beetle.
   c. a squid.
   d. a bird (but under water instead of in the air).
   e. a whale.

7. Smooth muscle is a type of muscle found in the internal organs—the walls of blood vessels and the intestine, for example. It is called "smooth" because the cells of smooth muscle lack the alternating light and dark bands seen in the muscle fibers that move the skeleton. Which of the following do you think might best explain this difference? Smooth muscle
   a. does not require nervous stimulation to contract.
   b. does not contain the regular, repeating filaments of skeletal muscle.
   c. cells are much larger than the cells of skeletal muscle.
   d. motor units are much smaller than those of skeletal muscle.
   e. requires much less blood flow than skeletal muscle.

8. Which of the following drugs would cause muscle spasms or cramps (uncontrolled contractions)? A drug that
   a. blocks the release of calcium ions from endoplasmic reticulum.
   b. prevents the release of acetylcholine from motor neurons.
   c. blocks neurotransmitter receptors on muscle fiber membranes.
   d. blocks the enzyme that breaks down acetylcholine after contraction.
   e. prevents attachment of myosin heads to thin filaments.

9. If you were to cut through a muscle fiber and look at the cut end under a powerful microscope, what would the contractile parts of the cell look like?
   a. many crisscrossing lines
   b. thousands of tiny dots
   c. thousands of overlapping circles
   d. numerous irregular splotches
   e. many overlapping, parallel lines

10. A wildlife biologist came across the carcass of a deer. He suspected that the deer had died of starvation. To check this guess, he could
    a. examine the cartilage at the ends of the bones for signs of wear.
    b. crack open a bone and check the yellow bone marrow.
    c. have the deer's tissues tested for signs of rheumatoid arthritis.
    d. look for signs of spongy bone in the ends of long bones.
    e. evaluate the amount of fat stored in the red bone marrow.

Essay

1. Look at the skeleton of the frog in Module 30.2. In what ways is the frog's skeleton modified for its way of life?

2. Both roundworms and earthworms possess hydrostatic skeletons. The body of a roundworm is a single elongated sac, filled with fluid. The roundworm is able to move only by thrashing from side to side. Compare this with the structure and movement of an earthworm, and explain the difference.

3. The wings of birds come in several different shapes. Many small songbirds have short, stout wings. The wings of an eagle are long and broad, while those of a gull are long and narrow. How are these different wing shapes (structures) well suited to the activities and environments (functions) of these different birds?

4. A map of the human cerebral cortex shows that the area of the brain that controls muscle activity of the hands is equal in size to the area that controls all the muscles below the neck. The total size of the muscles moving the hands is much smaller than all the muscles of the trunk, legs, and so on. Explain in terms of the motor units that make up the muscles why so much nerve tissue is required to control the hands.

5. Early reconstructions of dinosaur skeletons showed them with their legs splayed out to the sides, like present-day alligators and lizards. Most dinosaur experts now think that the legs of dinosaurs were directly under the torso, like those of a dog or elephant. Why do you think they consider this important?

Using Your Knowledge

1. There are actually two main kinds of muscle tissue: slow (red) muscle and fast (white) muscle. Slow muscle contracts slowly in response to nerve impulses. It is able to store oxygen and keep working with more endurance for extended periods. White muscle contracts quickly but has much less endurance. Most muscles contain a mixture of both types, in different proportions in different muscles. With this information in mind, speculate about the following: Which muscle type would you expect to predominate in the muscles that are moving your eyes as you read this? In your thigh muscles? Why is the breast muscle of a duck dark (red) meat but a chicken breast white meat? Which type do you think has been found in higher-than-average proportions in the leg muscles of world-class marathoners?

2. Millions of Americans regularly engage in body-building exercise at health clubs, weight rooms, and gyms. Millions of others engage in aerobic activities like swimming, running, cycling, and dance aerobics. How does body building differ from aerobic exercise? Do you participate in either of these activities? What is their value, in your opinion? What effects does each have on muscle tissue, respiration, and the cardiovascular system? The experts say that 20 minutes of aerobic exercise at least three times a week is important for overall cardiovascular fitness, weight control, and stress reduction. How much exercise do you get each week? Is it enough? If you would like to get more exercise, how could you make it a higher priority?

Plant Structure, Reproduction, and Development

31

Tulips are natives of Turkey, not Holland. Tulip bulbs were first brought to the Netherlands in 1590. The Dutch were enchanted by the beauty of the flowers and by stories of their immense value to Turkish sultans. Demand quickly exceeded supply, and growing and selling bulbs became a big business. Greed and speculation drove prices up. Family heirlooms and land were sold to buy bulbs. Fortunes were made overnight. The frenzy was known as "tulipmania." At its peak, one bulb was traded for "two loads of wheat, four loads of rye, four fat oxen, eight fat pigs, 12 sheep, two barrels of butter, 1000 pounds of cheese, two hogsheads of wine, four barrels of beer, a silver beaker, a suit of clothes, and a bed." In 1637, the market collapsed, thousands of businesses were ruined, and much of Europe was thrown into a depression. Nowadays, flowers do not attract this kind of interest from investors. But flowering plants are important to us in many ways. Just how are plants important to us? What kind of flower is a tulip? What is the function of a flower? These and many other questions are answered in this chapter on plant structure, reproduction, and development.

Organizing Your Knowledge

Exercise 1   (Introduction and Module 31.1)

Look around you and reflect a moment on all the products and "services" supplied by plants. In the space below, list six specific examples from your surroundings illustrating how we use plants. Name the plants if you can.

Exercise 2   (Module 31.2)

The figures below show differences between the two major groups of flowering plants—monocots and dicots—but the figures are scrambled. List the figures characteristic of monocots and dicots, and briefly describe the characteristic shown in each.

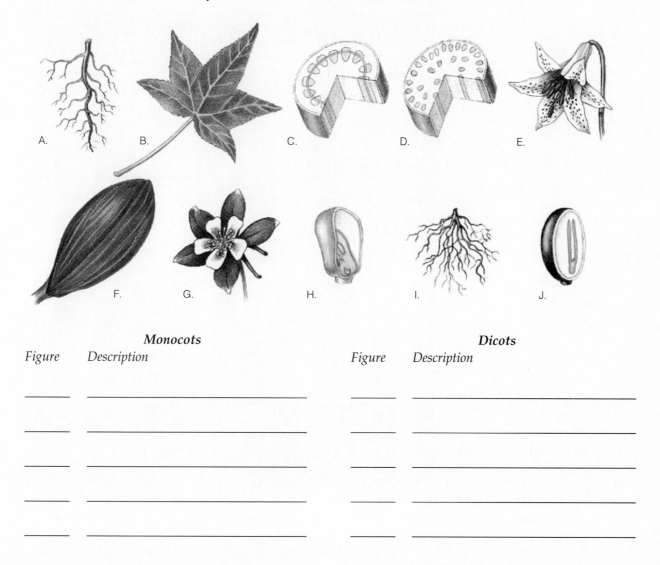

A.   B.   C.   D.   E.

F.   G.   H.   I.   J.

	Monocots		*Dicots*	
Figure	*Description*		*Figure*	*Description*
_____	_____		_____	_____
_____	_____		_____	_____
_____	_____		_____	_____
_____	_____		_____	_____
_____	_____		_____	_____

Exercise 3   (Modules 31.3 – 31.6)

These modules introduce plant structure—cells, tissues, organs, and systems. Review the vocabulary of plant structure by completing this crossword puzzle. All answers are in **bold** type in Modules 31.3 – 31.6.

Across

3. The ____ system consists of stems, leaves, and flowers.

7. The ____ is ground tissue that makes up most of a root.

10. ____ dominance is inhibition of axillary buds by the terminal bud.

11. A ____ is the point where a leaf is attached to a stem.

12. ____ is the parenchyma cells in a leaf specialized for photosynthesis.

15. ____ tissues consist of a single cell type.

17. ____ cells are the most abundant cells in most plants.

20. A ____ is a long, slender sclerenchyma cell.

22. A ____ is a vascular bundle in a leaf.

24. A ____ is an irregular cell that makes nutshells or seeds hard.

25. A waxy coating called the ____ helps the plant retain water.

26. ____ fills the central portion of a stem and may function in food storage.

27. Root ____ increase the absorptive surface of a root.

28. ____ cells regulate air flow through stomata.

Down

1. A ____ is an enlarged rhizome that stores food.

2. A ____ is a horizontal stem.

3. A ____ is a tiny pore in a leaf.

4. The ____ tissue system makes up much of a young plant and has many functions.

5. The ____ bud is the growth point at the tip of a stem.

6. Any tissue consisting of more than one cell type is called a ____ tissue.

8. ____ carries water up from the roots.

9. ____ buds form where leaves join a stem and are usually dormant.

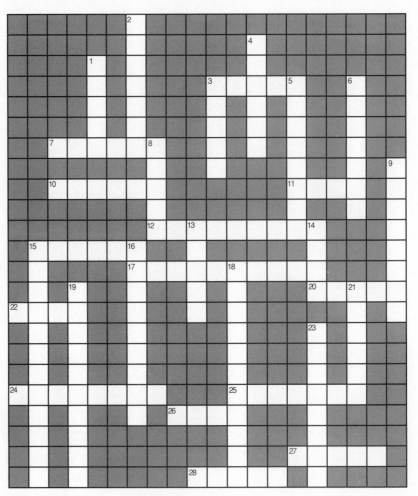

13. A ____ plate is the perforated end of a food-conducting cell.

14. The ____ is the main site of photosynthesis in most plants.

15. ____ cells are the dead, hardened supporting cells that form wood.

16. The ____ covers and protects a plant.

18. ____ cells support the growing parts of a plant.

19. The ____ forms a barrier that regulates flow into root vascular tissue.

21. A vascular ____ is a strand of vascular tissue in the stem.

23. The ____ tissue system is composed of xylem and phloem.

Exercise 4   (Module 31.6)

The cross-sectional diagrams below will help you to visualize the internal structures of roots, stems, and leaves. Label the following: **dicot stem, monocot stem, root, leaf, cortex, cuticle, endodermis, stoma, pith, epidermis, guard cell, mesophyll, vascular bundle, phloem, xylem, vein, root hair.** Color the epidermis and endodermis yellow, mesophyll green, pith and cortex orange, xylem red, and phloem blue.

a. _____

b. _____

c. _____

d. _____

Exercise 5   (Module 31.7)

These modules discuss plant growth. There are different kinds of growth and different kinds of plant life cycles. After reading the modules, match each term on the right with a phrase on the left.

A. Lengthwise growth
B. Makes up the wood of a tree or shrub
C. Growth point at the tip of a shoot or root
D. Thick, waxy cells that protect woody stem
E. A plant that completes its life cycle in one year
F. Tissue just outside the vascular cambium of a tree trunk
G. Ceasing to grow after reaching a certain size
H. Meristematic tissue that forms cork
 I. A plant that takes 2 years to complete its life cycle
J. Growth in diameter
K. Older layers of secondary xylem in a tree trunk
L. Pushed outward by development of secondary phloem
M. A plant that continues to live for many years
N. Young secondary xylem that still conducts water
O. Rapidly dividing cells between xylem and phloem
P. Growing throughout life
Q. First water-conducting vascular tissue of a plant

_____ 1. Primary xylem
_____ 2. Biennial
_____ 3. Secondary growth
_____ 4. Indeterminate growth
_____ 5. Perennial
_____ 6. Primary phloem
_____ 7. Sapwood
_____ 8. Primary growth
_____ 9. Cork
_____ 10. Secondary phloem
_____ 11. Vascular cambium
_____ 12. Annual
_____ 13. Determinate growth
_____ 14. Heartwood
_____ 15. Secondary xylem
_____ 16. Cork cambium
_____ 17. Apical meristem

Exercise 6   (Module 31.7)

Label the following on this diagram of a root tip: **root cap, root hair, vascular cylinder, cortex, epidermis, apical meristem.** Color the root cap red, the area of greatest cell division orange, the zone of cell elongation yellow, and the area of cell differentiation blue.

Exercise 7 (Module 31.7)

Complete this flow chart showing how different tissues develop at the tip of a growing root. (A similar process of differentiation occurs in a shoot tip.) Some tissues are named. Choose the remaining ones from this list: **primary phloem, outer cylinder, apical meristem, inner cylinder, cortex.**

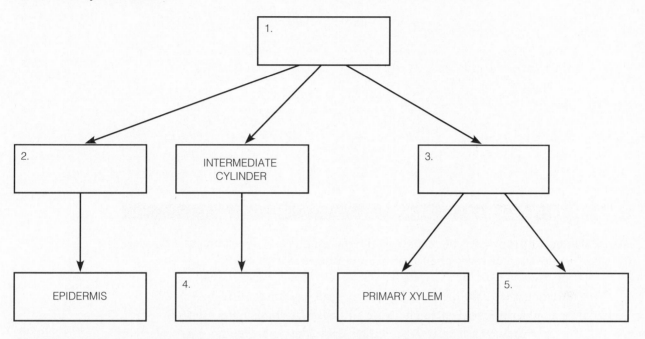

Exercise 8 (Module 31.8)

The diagram below will help you visualize the secondary growth of a tree trunk. Label each of the following on the diagram: **vascular cambium, secondary phloem, sapwood, secondary xylem, heartwood, cork with cambium,** and **bark.** Color the vascular cambium green. Color the oldest xylem red, the newest xylem pink, the oldest phloem dark blue, and the newest phloem lighter blue.

Exercise 9   (Module 31.9)

In the space below, sketch a flower and label the following parts: **sepal, stigma, stamen, carpel, anther, style, petal, ovary, ovule, pollen grain,** and **filament.** Color the female parts of the flower pink and the male parts blue.

Exercise 10   (Modules 31.9 – 31.13)

The following story will help you review plant reproduction and development. Complete the story by choosing the correct italicized word or phrase in each set.

     The spring rains have begun. A blackberry seed has lain undisturbed on the ground near a rail fence since September. Now (1) *sunlight, warmth, water* causes the seed to swell. The process of (2) *pollination, germination, fertilization* is beginning. The protective (3) *endosperm, carpel, seed coat* splits open, a tiny root emerges, and the leaves of the tiny (4) *ovule, pollen grain, embryo* unfurl toward the sun. At first the plant's leaves are pale; its seed leaves, or (5) *cotyledons, anthers, sepals,* have absorbed the (6) *pollen, endosperm, sperm* stored within the seed to obtain food for the plant's first few days of growth.

     As the days grow longer and warmer, apical meristems add new cells to the tips of the blackberry's roots and stems. By midsummer, the plant has produced several branches and dozens of leaves and has begun to climb up on the fence. Clusters of white flowers, resembling small wild roses, sprout at lateral buds. The blackberry plant represents the diploid (7) *gametophyte, sporophyte* phase of the blackberry's life cycle. The flowers are the reproductive structures where (8) *mitosis, germination, meiosis* produces haploid (9) *seeds, spores, fruits,* which undergo mitosis to become haploid (10) *gametophytes, sporophytes.*

     Each flower begins as a bud, enclosed by green, protective (11) *petals, carpels, sepals.* As a flower opens, white (12) *petals, carpels, sepals* attract bees. The bees swarm over the flowers, collecting nutrient-rich pollen grains from the (13) *carpels, sepals, anthers* at the tips of the (14) *stamens, carpels, stigmas,* the male reproductive structures of the flowers. As they buzz from flower to flower, the bees deposit some of the pollen grains on the sticky (15) *stigmas, anthers, stamens* at the tips of the female reproductive structures, called (16) *anthers, carpels, sepals.* Each blackberry flower has numerous stamens and carpels, so the bees are busy.

     Each pollen grain is actually a tiny (17) *male, female* gametophyte, containing a tube nucleus and two (18) *egg, seed, sperm* nuclei. (19) *Fertilization, Pollination, Meiosis* occurs as a bee rubs a pollen grain onto a stigma. The pollen grain germinates, and a pollen tube grows down through the carpel into the swollen (20) *anther, ovary, sepal* at its base. The two (21) *egg, sperm, tube* nuclei now travel down the pollen tube to (22) *a sepal, a carpel, an ovule,* which contains the (23) *stigma, stamen, embryo sac,* the tiny haploid female (24) *gametophyte, sporophyte.* The sperm nuclei enter the ovule, and one of them joins with the (25) *egg, anther, pollen grain* within, forming a diploid (26) *ovule, embryo sac, zygote.* This is fertilization. The other sperm nucleus combines with a large diploid cell, forming a triploid cell. This cell will divide to form the (27) *pollen, endosperm, ovule,* which will become enclosed in the seed and nourish the developing embryo. The joining of two sperm nuclei with an egg cell and another diploid cell is called double (28) *fertilization, germination, pollination* and is unique to flowering plants.

In the ovary at the base of each carpel, each ovule with its zygote now begins to develop into a (29) *seed, stamen, fruit.* The zygote divides, and one of the cells formed becomes the (30) *flower, endosperm, embryo.* Because the blackberry is a dicot, the embryo sprouts two (31) *sepals, cotyledons, seed coats,* which will later help the embryo absorb nutrients stored in the seed. The outer coating of the ovule dries and thickens and becomes the seed coat.

Each blackberry flower contains multiple carpels. The flower petals shrivel and fall away, and the (32) *sepal, anther, ovary* at the base of each carpel grows and ripens. Eventually, the seeds end up in a cluster of ripened ovaries, which form (33) *a simple fruit, an aggregate fruit, a multiple fruit,* called a blackberry.

The blackberries are hard, green, and sour at first, but they become soft, dark purple, and sweet as (34) *starch, organic acid, sugar* accumulates in them. The fruits protect the seeds and aid in their dispersal. A blackbird perches on a swaying cane, twists off a fat berry, and gulps it down. The soft flesh of the blackberry is easily digested, but the seeds pass through the bird's digestive system unharmed, protected by their tough (35) *endosperm, seed coats, sepals.* A few hours after eating the berry, the bird perches on a fence and deposits the seeds in its droppings. The fence is an ideal trellis for the next generation of blackberries.

Exercise 11 (Modules 31.14 – 31.15)

Match each phrase with a type or method of vegetative reproduction or propagation.

A. Cuttings
B. Root sprouts
C. Plant cloning
D. Fragmentation
E. Runners

_____ 1. Grass slowly spreads into a garden bed.
_____ 2. Pieces of a house plant grow roots in water.
_____ 3. A clump of tulips expands each year.
_____ 4. Many small trees grow from a stump.
_____ 5. House plants grow from cells in a test tube.

Testing Your Knowledge

Multiple Choice

1. Which of the following is *not* a characteristic of dicots?
   a. two seed leaves
   b. parts of flowers in fours or fives
   c. taproot
   d. vascular bundles arranged in a ring
   e. veins in leaves usually parallel

2. Most of the photosynthesis in a plant is carried out by _____ in the leaves.
   a. collenchyma cells
   b. water-conducting cells
   c. parenchyma cells
   d. sclerenchyma cells
   e. food-conducting cells

3. How do cells in a meristem differ from other cells in a plant?
   a. They continue to divide.
   b. They photosynthesize at a faster rate.
   c. They are growing.
   d. They are differentiating.
   e. They store food.

4. "Angiosperm" is another name for
   a. plant.
   b. pollen.
   c. stored food in a seed.
   d. flowering plant.
   e. fruit.

5. In the process of pollination, pollen grains are transferred from the _____ to the _____ .
   a. ovary . . . anther
   b. stigma . . . ovary
   c. anther . . . sepal
   d. carpel . . . stigma
   e. anther . . . stigma

6. Which of the following is *not* a method of vegetative reproduction?
   a. runners
   b. germination
   c. stump sprouts
   d. cuttings
   e. fragmentation

7. What is endosperm?
   a. male reproductive cells in plants
   b. stored food in a seed
   c. cells that make up the bulk of a pollen grain
   d. the fleshy part of a fruit such as an apple or strawberry
   e. plant chromosomes

8. After fertilization, the _____ develops into a seed, and the _____ develops into a fruit.
   a. ovule . . . ovary
   b. pollen grain . . . ovule
   c. ovary . . . ovule
   d. egg . . . ovule
   e. egg . . . ovary

9. Lengthwise growth of a root into the soil results mainly from
   a. cell division in the apical meristem.
   b. elongation of cells.
   c. cell division in the vascular cambium.
   d. differentiation (specialization) of root cells.
   e. pulling by root hairs.

10. Which of the following is closest to the center of a daisy stem?
    a. cortex
    b. phloem
    c. pith
    d. epidermis
    e. xylem

11. Cell division would be slowest in which of these tissues?
    a. apical meristem of root
    b. cork cambium
    c. epidermis
    d. vascular cambium
    e. apical meristem of terminal bud

Essay

1. Briefly describe four functions of a plant's roots. Which of these is primarily a job of the epidermis? The vascular cylinder? The cortex?

2. Sketch a cross section of a young dicot stem, showing pith, xylem, phloem, cortex, and epidermis. What do these layers have to do with the three tissue systems in the stem? How does the arrangement of tissues in a dicot root differ from their arrangement in a stem?

3. What is the sporophyte stage of a flowering plant? What reproductive structures does it produce? What is the gametophyte stage? What reproductive cells do gametophytes produce? Which of the preceding are haploid and which are diploid?

4. What are meristems and where would you find them in an apple tree? Which are responsible for primary growth? Secondary growth?

5. Briefly explain how the shape and structure of the following cells are well suited to their functions: sclerenchyma cells in a stem, water-conducting cells in a root, epidermal cells in a leaf.

6. Explain how the vascular cambium enables a tree to grow in diameter.

7. Starting with pollination, briefly explain how fertilization occurs in a flowering plant. What is the major difference between this process and fertilization in animals?

8. Briefly describe six different ways in which roots, stems, or leaves are modified for storage, reproduction, or other specialized functions.

Applying Your Knowledge

Multiple Choice

1. The vascular cambium in the trunk of a large, woody rhododendron shrub lies between
   a. secondary phloem and secondary xylem.
   b. secondary xylem and pith.
   c. primary xylem and primary phloem.
   d. secondary phloem and cortex.
   e. primary phloem and pith.

2. Which of the following is correctly matched with its tissue system?
   a. xylem–ground tissue system
   b. phloem–epidermis
   c. cortex–ground tissue system
   d. pith–vascular tissue system
   e. all of the above are correctly matched

3. The shoot system of a beavertail cactus consists of broad paddlelike structures covered with spines. The spines are modified _____ , so the flat green paddles must be modified _____ .
   a. buds . . . leaves
   b. buds . . . stems
   c. leaves . . . stems
   d. stems . . . roots
   e. stems . . . leaves

4. Cell division in the vascular cambium adds to the girth of a tree by adding new _____ on the inside of the cambium layer and _____ on the outside.
   a. phloem . . . xylem
   b. xylem and phloem . . . bark
   c. pith . . . xylem and phloem
   d. xylem . . . phloem
   e. xylem . . . cortex

5. A vandal killed a historic oak tree on the village green by "girdling" it with a chain saw. He cut through the bark and into the sapwood all the way around the tree. Why did the tree die?
   a. The leaves could not get water.
   b. Oxygen could not get to the roots.
   c. The roots could not get food.
   d. The leaves could not get food.
   e. The roots could not absorb water.

6. Artichoke hearts are tender and tasty. The leaves are tasty too, but most of an artichoke leaf is fibrous and impossible to chew. The leaves must contain lots of
   a. parenchyma cells.
   b. phloem.
   c. meristematic tissue.
   d. sclerenchyma cells.
   e. epidermal cells.

7. Based on your own experience with these plants, which do you think is correctly paired with the word describing its life cycle?
   a. apple tree–perennial
   b. rose bush–annual
   c. marigold–perennial
   d. oak tree–biennial
   e. tulip–annual

8. Plants growing in harsh environments such as deserts, sand dunes, and arctic tundra often reproduce vegetatively. This is because
   a. there are few animals available to pollinate them.
   b. they are members of plant families that only reproduce asexually.
   c. fruits would freeze or dry out in these environments.
   d. vegetative reproduction is not as risky or costly as making seeds.
   e. seeds would be eaten by hungry animals in these environments.

9. Which of the following would be least useful in figuring out whether a plant is a monocot or a dicot?
   a. size when mature
   b. numbers of flower parts
   c. pattern of veins in leaves
   d. number of seed leaves
   e. arrangement of vascular tissue in stem

10. A cross section of part of a plant exposes epidermis, a thick cortex, and a central cylinder of xylem and phloem. This part is a
    a. fruit.
    b. seed.
    c. stem.
    d. root.
    e. bud.

Essay

1. Amy told Paul that a giant sequoia is the largest living thing, bigger than ten blue whales. Paul replied, "Yes, the sequoia is bigger, but it is an unfair comparison because *most of a tree is not alive.*" Do the facts back up Paul's assertion? Explain.

2. Imagine pounding a large spike into a tree. Name (in order) all the tissue layers it would pass through as it penetrates into the heartwood.

3. What part of a plant are you eating when you consume each of the following: garlic, walnut, cabbage, carrot, cauliflower, artichoke, asparagus, cucumber, broccoli, rice, potato?

4. When he was a boy, Grandpa nailed a horseshoe to a maple tree, about 5 feet from the ground. The tree has since grown much taller, and the diameter of its trunk has quadrupled, but the horseshoe is still about 5 feet up. Explain this in terms of how trees grow.

5. Palms are monocots. Do you think that you could kill a palm tree by girdling it (as in multiple choice question 5 in the "Applying Your Knowledge" section)? Why or why not?

6. Examine the diagram of a growing root, Figure 31.7B in the text. Why does it make sense that lateral roots must start growing from the inside

of the main root rather than simply projecting from the outside?

7. Most food for humans is derived from the seeds of plants. Why does it make sense that seeds are a good source of nutrition?

Using Your Knowledge

1. Do you like to grow house plants? Are house plants products of controlled breeding, or are they free-living somewhere in the wild? Have you ever wondered where they originally came from? What do you think their natural environment might be like? What are the functions of their varied shapes, colors, and textures? If you are interested in the answers to these questions, there are many gardening and horticulture books that discuss the origins of common house plants.

2. Many individuals are concerned that wild plants be conserved and protected, especially in the tropics, where plant diversity is greatest and wild plants appear to be most endangered by human activities. They argue that humans use only a tiny number of plants for food and that there are undoubtedly unused and undiscovered species that could be better sources of food than any plants now in use. Other people reply that people are set in their ways. The fact that we use so few plants for food shows that people are unwilling to try new ones. While it might be important to preserve wild plants, it is unlikely that a new plant will become an important food source. What do you think? Are any new food plants becoming available? Have a significant number of people anywhere in the world taken up a new staple crop in the recent past? Give an example if you know of one.

3. Keep a list of the plants and plant products you use in a single average day. Which species are represented? How many species? What categories of functions do the plants serve in your life? Do you think that the number of plants and plant products used in everyday life has increased or decreased in the last century? Is the number likely to increase or decrease in the future? Why or why not?

Plant Nutrition and Transport 32

Is there a house plant where you live that you have had for a long time? Maybe a spider plant or a Boston fern or a Swedish ivy? Have you bought other plants that soon died? Why did one plant thrive and the other wither? Was it the temperature in the room or the amount of sunlight or the amount of water you provided? Did you under-water or over-water? Why can too much water kill a plant? Was the soil the right texture? What is the proper texture? Were the proper amounts of the right nutrients available in the soil? What nutrients do plants need? Which nutrients do they obtain from soil, from water, and from the air around them? Once a plant absorbs these nutrients, how are they transported to where they are needed? These are questions relevant to any plant, and they are the subject of this chapter.

Organizing Your Knowledge

Exercise 1   (Module 32.1)

Review the nutrients plants require by filling in the blanks in the paragraph below.

Plants need four main kinds of nutrients to survive and grow:
1_____, 2_____, 3_____, and
4_____. Most of a plant's mass is made of organic molecules built from
5_____, which is obtained from the 6_____. Hydrogen for
organic molecules comes from 7_____, which is obtained from the
8_____ and split during the process of photosynthesis. Many substances
manufactured by the plant require elements other than carbon, hydrogen, and
oxygen. Protein, for instance, contains 9_____. Chlorophyll contains
10_____ and 11_____, and ATP and nucleic acids contain
12_____ and 13_____. These 14_____ are
absorbed from the soil by the plant's roots. In addition, roots also absorb
15_____, which is necessary for aerobic respiration.

Exercise 2 (Module 32.2)

On the diagram below, trace the uptake of water and minerals by a root. Show water moving from cell to cell (the intracellular route) with a blue arrow. Trace water moving along cell walls, between cells (the extracellular route), with a red arrow. The Casparian strip forces all water and solutes to pass through endodermal cells, allowing the endodermis to control movement of water and solutes into the vascular tissues. Color the endodermis and Casparian strip yellow. Label the following: **endodermis, root hair, epidermis, cortex, xylem vessels, Casparian strip, intracellular route, extracellular route, plasmodesma.**

Exercise 3 (Modules 32.3 – 32.4)

Review the movement of water and ions up a plant stem by filling in the missing words in the phrases below, then numbering each of the phrases in order. Start with what happens in the roots and continue with what pulls water up from the top of the plant.

_____ A. Water molecules _____ from cells bordering the air spaces within the leaf.

_____ B. Inorganic ions are pumped into the _____ of the roots; this causes water to enter these cells by _____.

_____ C. Water molecules leaving leaf cells are stuck to nearby water molecules, a phenomenon known as _____. They also stick to the cellulose molecules in the walls of xylem cells, a phenomenon known as _____. As water molecules diffuse from leaf cells, they tug on adjacent water molecules.

_____ D. Water accumulating in the xylem of roots pushes xylem sap upward a few meters, a phenomenon known as _____.

_____ E. As water molecules in the leaf pull upward on adjacent water molecules, they in turn tug on water molecules in the xylem, all the way down to the roots. This occurs because water molecules stick to each other by _____ bonds.

_____ F. _____ cells in leaf epidermis take up K^+ ions. Water enters by _____, and the cells swell. This causes the _____ to open, and water evaporates from the leaf, a process called _____.

_____ G. Water molecules are pulled up from the roots, bit by bit, as water molecules evaporate from the leaves. This explanation for the ascent of xylem sap is called the _____ mechanism.

Exercise 4   (Module 32.5)

Phloem transports sugar from where it is produced to where it is used via a pressure-flow mechanism. Review this mechanism, paying particular attention to Figure 32.5B in the text. Then test your understanding by stating whether each of the following refers to a sugar source or a sugar sink.

_____ 1. Where phloem sap has its highest sugar concentration

_____ 2. A beet root or potato tuber during the summer

_____ 3. Any photosynthesizing part of a plant

_____ 4. Phloem sap flows away from this area

_____ 5. Where sugar is pumped into the phloem by active transport

_____ 6. Where water enters phloem sap by osmosis

_____ 7. A bulb in early spring

_____ 8. Where water (hydrostatic) pressure is highest

_____ 9. Where sugar leaves the phloem by active transport

_____ 10. A leaf on a summer day

_____ 11. Where water (hydrostatic) pressure is lowest

_____ 12. Where phloem sap has its lowest sugar concentration

_____ 13. Phloem sap flows toward this area

_____ 14. A growing bud in early spring

Exercise 5   (Modules 32.6 – 32.7)

Sixteen essential nutrients are required by plants: iron (Fe), manganese (Mn), nitrogen (N), hydrogen (H), calcium (Ca), phosphorus (P), zinc (Zn), potassium (K), molybdenum (Mo), copper (Cu), carbon (C), oxygen (O), magnesium (Mg), boron (B), chlorine (Cl), and sulfur (S). Which of these nutrients are the answers to each question below?

_____ 1. Which nine nutrients are macronutrients, needed in large quantities?

_____ 2. Which six macronutrients are the major ingredients of organic compounds?

_____ 3. Which seven nutrients are micronutrients, needed in smaller amounts as cofactors that help enzymes?

_____ 4. Shortage of which nutrient is the most common deficiency of plants?

_____ 5. Which macronutrient combines with proteins to "glue" cells together?

_____ 6. Which macronutrient is the main solute for osmotic regulation, causing opening and closing of stomata by guard cells, for example?

_____ 7. Which micronutrient is a metallic component of the cytochromes involved in electron transport?

_____ 8. Deficiency of which nutrient is the second-most common plant deficiency?

_____ 9. Which macronutrient is part of chlorophyll and a cofactor of several enzymes?

Exercise 6   (Module 32.8)

Match each of the phrases on the right with a component of soil from the list on the left.

A. + ions, like Ca²⁺
B. B horizon
C. C horizon
D. Clay particles
E. Air
F. Water
G. Bacteria and fungi
H. H⁺ ions
I. Topsoil
J. Humus

_____ 1. Fine clays and nutrients dissolved in water accumulate in this soil layer
_____ 2. Roots exchange these for nutrient ions bound to clay particles
_____ 3. Break down organic matter
_____ 4. Partially decomposed organic matter
_____ 5. Negative charges keep positively charged ions from draining away
_____ 6. Where most decomposing organic material and organisms are found
_____ 7. Mainly composed of slightly weathered rock
_____ 8. Adhere to clay particles
_____ 9. Humus and organisms form spaces for this
_____ 10. Held in tiny spaces between soil particles

Exercise 7   (Modules 32.9 – 32.10)

State how each of the following would be harmful or helpful to the soil or the environment.

1. Applying an excess of commercially produced inorganic fertilizer to a field

2. Using herbicides to kill weeds, rather than tilling the soil each year

3. Plowing and planting in rows up and down hills

4. Flooding fields with irrigation water and blocking flow to prevent runoff

5. Using fertilizers that contain biological material

6. Using perforated pipes to drip water onto soil

Review the adaptations of plants for obtaining nitrogen by filling in the blanks in the following story.

[1]_____ is often the hardest to obtain of all plant nutrients but is essential to plants for making [2]_____, [3]_____, and other organic molecules. Ironically, the [4]_____ is 80% N_2 gas, but plants cannot use nitrogen in that form. Most plants must obtain nitrogen from the soil in the form of [5]_____ (NH_4^+) or [6]_____ (NO_3^-) ions. Certain soil bacteria, called [7]_____ bacteria, are able to convert N_2 to ammonium. Other bacteria, called [8]_____ bacteria, decompose organic matter in the soil and produce additional ammonium.

A small amount of ammonium is used by plants directly, but it is difficult to absorb because it is positively charged and tends to stick to [9]_____ particles in the soil. A third kind of soil bacteria, called [10]_____ bacteria, convert soil ammonium to [11]_____, which is easier for plants to absorb.

In some plants, roots may form mutually beneficial associations with fungi, called [12]_____, which assist in absorption of nutrients. This is a [13]_____ arrangement—both plant and fungus benefit. The fungus is good at absorbing water and inorganic ions, some of which are transferred to the plant. The fungus may secrete [14]_____, which increases the solubility of some elements, and it may also help protect the plant from certain diseases. In return, some of the plant's [15]_____ products nourish the fungus.

Other specializations for obtaining nutrients have evolved in various groups of plants. Some plants, such as [16]_____ and [17]_____, are parasitic. They simply tap into the vascular systems of other plants and steal food and nutrients. [18]_____ plants, such as the sundew and [19]_____, often grow in bogs where the soil is poor in [20]_____. These plants obtain the nitrogen they need by trapping and digesting [21]_____.

The roots of plants in the [22]_____ family, such as peas, beans, and alfalfa, have swellings called [23]_____ that house [24]_____ bacteria. Again, the arrangement is mutually beneficial. The legume plant provides the bacteria with [25]_____, and the bacteria convert atmospheric [26]_____ into ammonium. Excess ammonium may actually enter the soil, and for this reason, farmers often [27]_____ crops, alternating between legumes and nonlegumes such as [28]_____.

The most common form of malnutrition in humans is [29]_____ deficiency. Because most people have a primarily [30]_____ diet, plant scientists want to improve the quality and quantity of protein in plants. New varieties of corn, rice, and wheat are enriched in protein, but they require more [31]_____, which is expensive. Plant scientists hope to improve the output of nitrogen-fixing [32]_____ and perhaps use the techniques of [33]_____ to create nitrogen fixers that can infect nonlegumes or perhaps insert [34]_____ for nitrogen fixation directly into crop plants, increasing their protein production without need for fertilizer.

Testing Your Knowledge

Multiple Choice

1. Nitrogen fixation is
   a. using nitrogen to build molecules such as proteins and nucleic acids.
   b. converting nitrogen in the air into a form usable by plants.
   c. recycling nitrogen from organic matter in the soil.
   d. absorbing N_2 from the soil.
   e. an unhealthy interest in nitrogen.

2. Guard cells
   a. control the rate of transpiration.
   b. push water upward in a plant stem.
   c. protect the plant's roots from infection.
   d. control water and solute intake by roots.
   e. protect nitrogen-fixing bacteria in root nodules.

3. Which of the following is a macronutrient?
   a. hydrogen
   b. calcium
   c. phosphorus
   d. nitrogen
   e. all of the above

4. Soil can easily become deficient in _____ , because these ions are negatively charged and do not stick to negatively charged clay particles.
   a. potassium
   b. calcium
   c. magnesium
   d. nitrate
   e. ammonium

5. Which of the following would trigger opening of stomata?
   a. extreme heat
   b. loss of potassium by guard cells
   c. nightfall
   d. swelling of guard cells due to osmosis
   e. all of the above

6. The sundew plant has to digest insects because
   a. it obtains nitrogen from their flesh that it cannot get from the soil.
   b. it has lost the ability to perform photosynthesis.
   c. it lives in a dry environment and needs the moisture in their bodies.
   d. it needs to get rid of insects that accidentally get stuck in its hairs.
   e. its flowers are fertilized by pollen in their digestive tracts.

7. Ammonifying bacteria in the soil
   a. convert ammonium to nitrate.
   b. fix nitrogen.
   c. convert nitrogen in organic molecules into ammonium.
   d. change nitrate into ammonium.
   e. use nitrate to make amino acids that plants can use.

8. Mycorrhizae are
   a. nutrients required by plants in relatively small amounts.
   b. plants such as mistletoe that parasitize other plants.
   c. medium-sized soil particles.
   d. cells that control evaporation of water from leaves.
   e. associations of roots with beneficial fungi.

9. Which of the following is a sugar source?
   a. a green leaf
   b. a developing fruit
   c. a growing root
   d. a growing shoot
   e. a tree trunk

10. What is the main source of energy that moves water upward in the trunk of a tree?
    a. musclelike contraction of xylem cells
    b. evaporation of water by the sun
    c. pressure exerted by root cells
    d. breakdown and release of energy of sugar molecules
    e. osmotic changes caused by alterations in salt content

11. All water and solute molecules must pass through _____ before they can enter the vascular system and move upward to the leaves.
    a. a stoma
    b. a root hair cell
    c. an endodermal cell
    d. an epidermal cell
    e. a cortex cell

12. A plant does *not* obtain which of the following substances from soil?
    a. magnesium
    b. nitrogen
    c. carbon
    d. oxygen
    e. phosphorus

Essay

1. Explain what causes water to enter the xylem of a tree root and move upward in the xylem of the trunk to the leaves.

2. Describe an experiment to find out whether the element molybdenum is an essential plant nutrient.

3. In a growing potato plant, what causes sugar to flow from a leaf to the potato? Early in the spring, what causes sugar to flow from the potato to the growing shoot?

4. Explain how root hairs obtain ions that cling to the surfaces of clay particles in the soil.

5. Organic fertilizers such as manure and compost and inorganic commercial fertilizers contain the same nutrient ions needed by plants. In addition to this, what does the humus in organic fertilizer contribute to the soil?

6. How might agricultural scientists be able to alter crop plants to improve their production of protein?

Applying Your Knowledge

Multiple Choice

1. The roots of many aquatic plants have special structures that project above the surface of the water. For example, cypress trees (which grow in swamps) have "knees" that extend upward above water level. Which of the following is the most logical function of these structures?
   a. obtaining carbon dioxide for photosynthesis
   b. nitrogen fixation
   c. obtaining oxygen for the roots
   d. transpiration
   e. absorbing trace minerals

2. Helen had a terrarium on her windowsill containing various house plants. She wondered why the glass was often fogged with water droplets. Her friend Sara, who has had a biology class, tried to impress Helen by explaining, "The water evaporates from the leaves—it's a process called _____."
   a. root pressure
   b. adhesion
   c. photosynthesis
   d. pressure flow
   e. transpiration

3. In an apple tree, sugar might flow from _____ to _____ .
   a. a developing apple . . . a leaf
   b. the trunk . . . a leaf
   c. a growing root . . . a growing shoot tip
   d. a leaf . . . a developing apple
   e. a growing shoot tip . . . the trunk

4. Soil could be deficient in any of the following nutrients. If you had to supply one of them, which would be needed in the smallest amount?
   a. sulfur
   b. phosphorus
   c. nitrogen
   d. potassium
   e. iron

5. Which of the following is a difference between transport by xylem and transport by phloem?
   a. Active transport moves xylem sap but not phloem sap.
   b. Transpiration moves phloem sap but not xylem sap.
   c. Phloem carries water and minerals; xylem carries organic molecules.
   d. Xylem sap moves up; phloem sap moves up or down.
   e. Xylem sap moves from sugar source to sink, but phloem sap does not.

6. What keeps the force of gravity from pulling water molecules out of leaves?
   a. upward pressure from the roots
   b. high water pressure in the leaves
   c. the Casparian strip blocks them from moving out
   d. movement of water toward a sugar sink
   e. cohesion and adhesion of water molecules

7. Jon is performing a chemical analysis of xylem sap. He should not expect to find much of which of the following?
   a. nitrogen
   b. sugar
   c. phosphorus
   d. water
   e. potassium

8. A botanist discovered a mutant plant that was unable to produce the material that forms the Casparian strip. This plant would be
   a. unable to fix nitrogen.
   b. unable to transport water or solutes to the leaves.

c. able to exert greater root pressure than normal plants.

d. unable to control the amounts of water and solutes it absorbs.

e. unable to use its roots as a sugar sink.

9. Normally when an aphid feeds by puncturing plant tissues, it does not have to suck the sap out. An inexperienced aphid, however, accidentally inserted its feeding tube in the wrong place and found the fluid in its gut being sucked out through the feeding tube. It had punctured

a. the Casparian strip.

b. a root nodule.

c. a xylem cell.

d. a phloem tube.

e. a stoma.

10. Professor Timothy Schmidlap claims to have discovered a new macronutrient required for plant growth. Most of his colleagues are skeptical of his claim. Why might they consider it unlikely?

a. All the nutrients required for plant growth have already been found.

b. It is very difficult to prove that a plant needs a certain nutrient.

c. Plants need thousands of nutrients; a new one is not significant.

d. Any nutrient needed in large amounts has probably been found already.

e. His colleagues are jealous and want to claim credit for the discovery.

Essay

1. Farmers in Lower Quasiland found swellings full of bacteria on the roots of their bean plants, so they sprayed the plants with antibiotics. The swellings disappeared, but the plants no longer grew well. Examining the plants more closely, the farmers concluded that the cause of the difficulty was fungi associated with the roots. They treated the plants with a fungicide, but then the plants did even worse. Explain what is happening.

2. Sometimes on very hot days, water evaporates so quickly from the leaves of a plant that empty "bubbles" occur in xylem tubes. This can be harmful to the plant. Explain why.

3. The jade plant, a familiar house plant, has a peculiar adaptation to the hot, dry environment where it grows naturally. Its stomata open at night, carbon dioxide enters the leaves, and the carbon dioxide is stored in a chemical form. The stomata remain closed during the day, when the stored carbon dioxide is used in photosynthesis. How does this differ from most other plants? How might it be advantageous in the jade plant's environment?

4. A plant does not use potassium for a building material or energy source, but it stops doing photosynthesis and dies if it runs out of potassium. Why?

5. Denise and Roger dreamed of starting an organic homestead in southeast Alaska. They bought a plot of forested land, cut down the fir and spruce trees, built a cabin, and then had the soil tested. An accumulation of acid from decaying conifer needles, in combination with the heavy rainfall, had caused the soil to become quite deficient in a particular macronutrient. What nutrient was most likely to be in short supply? Why was it washed from the soil and not other nutrients? How might they enrich the soil with this nutrient without using commercial fertilizers?

Using Your Knowledge

1. More and more supermarkets are stocking organic produce. What does "organic" mean? Is there a legal definition of "organic" in your state? (It varies.) Do you buy organic fruits and vegetables when you have the chance? Why or why not? What are some reasons people buy organic produce? What are some reasons they don't buy more?

2. Do you have a "green thumb"? Planting a garden or growing house plants can be an interesting and relaxing hobby. Some people save money by growing their own vegetables. Others like to surround themselves with fresh flowers. Some people take a more scientific interest in growing ferns or cacti. Wherever you live there is somewhere to grow plants—a sunny window, a balcony, pots along the sidewalk. You may even want to find out whether community garden plots are available in your city.

Control Systems in Plants

33

While most of us worry about a shortage of the ozone that protects us from ultraviolet light, agricultural researchers are concerned that an excess of ozone in smog is killing crops. The U.S. government figures that ozone is responsible for 90% of crop losses from air pollution, at an estimated cost of 1 to 5 billion dollars per year. Biologists in England have discovered that plants exposed to a burst of ozone produce twice as much ethylene—a plant hormone—than normal plants. Interestingly, it is the ethylene, not the ozone, that seems to be responsible for most of the damage to the plants. This discovery may point the way to a method for preventing crop losses caused by air pollution. Ethylene is just one of many plant hormones. This chapter is about how these hormones regulate plant growth and development.

Organizing Your Knowledge

Exercise 1   (Module 33.1)

A series of experiments demonstrated that phototropism is due to the effect on shoot elongation of a chemical (named auxin) formed in the shoot tip. The following figures (on the next page) illustrate some of those experiments, and others. Predict whether each of the numbered shoots will grow to the left, right, straight up, or show no growth in response to the experimental conditions shown. Assume that each seedling has germinated in the dark and has just been placed under the experimental conditions.

Light

Control

1. _____

Tip separated by gelatin block

2. _____

Tip covered by opaque cap

3. _____

After exposure to light, tip removed, rotated 180°, and replaced

4. _____

Light

Tip covered by transparent cap

5. _____

Tip removed

6. _____

Base covered by opaque shield

7. _____

Tip separated by mica

8. _____

Excised tip placed on agar block

Agar block with chemical from shoot tip

Tip removed (control)

9. _____

Block without chemical (control)

10. _____

Control

11. _____

Offset block with chemical

12. _____

Block with chemical

13. _____

Offset block with chemical

14. _____

Offset block without chemical

15. _____

Exercise 2   (Modules 33.2 – 33.4)

Module 33.2 introduces the classes of plant hormones, while the next two modules focus on auxins and cytokinins, two important hormone classes. Review the effects of auxins and cytokinins by filling in the blanks in the following statements.

A. A number of different [1]_____, which promote elongation of shoots, have been identified in plants. Others have been synthesized artificially.

B. [2]_____ promote cell division. Several are natural; some are produced in the lab.

C. The best-known natural auxin is [3]_____.

D. Most auxin synthesis occurs in the [4]_____ at the tip of a shoot. It moves down the stem and stimulates cells to elongate.

E. Cytokinins are produced in growing tissues: [5]_____, [6]_____, and [7]_____. Those that originate in roots affect stems by traveling upward in [8]_____.

F. [9]_____ promotes stem elongation, perhaps by weakening [10]_____ and allowing cells to stretch.

G. The same concentration of IAA that causes shoots to elongate [11]_____ elongation of roots. A lower concentration [12]_____ root elongation.

H. Auxin transported down the stem from the terminal bud [13]_____ growth of axillary buds, causing the plant to grow straight up without branching.

I. If the terminal bud is pinched off, [14]_____ buds become active, forming lateral branches, and the plant becomes [15]_____. [16]_____ moving upward from the roots are thought to activate the axillary buds.

J. The ratio of [17]_____ and [18]_____ probably shapes the complex growth pattern of most plants. Auxins from the terminal buds [19]_____ branching, while cytokinins from the roots [20]_____ it. The [21]_____ axillary buds usually begin to grow before those closer to terminal buds, reflecting the higher ratio of [22]_____ to auxins in the lower parts of the plant.

K. The ratio of auxins and cytokinins also helps to coordinate root and shoot growth. A growing [23]_____ system produces cytokinins, which cause the [24]_____ system to branch.

Exercise 3   (Modules 33.5 – 33.6)

Gibberellins and abscisic acid each have a number of functions. State whether each of the following refers to gibberellin (G) or abscisic acid (ABA).

_____ 1. Inhibits both primary and secondary growth

_____ 2. Stimulates growth of stems and leaves

_____ 3. Signals buds to form protective scales

_____ 4. Responsible for maintaining seed dormancy

_____ 5. Produced by fungi, it causes "foolish seedling disease"

_____ 6. Causes dwarf plants to grow tall

_____ 7. Opposes the effects of gibberellins in seeds

_____ 8. Acts as "stress hormone," to help a plant cope with harsh conditions

_____ 9. When released from an embryo, it signals a seed to germinate

_____ 10. If it is washed from seeds, they will germinate

_____ 11. Acts antagonistically to ABA in seeds

_____ 12. Causes fruit to develop

_____ 13. Higher level causes dormancy of buds during winter

Review the maturation and ripening of fruit by filling in the blanks in the following paragraphs.

Although auxins and gibberellins cause fruits to develop and mature, a gas called [1]_____ triggers ripening. It is also involved in other processes related to [2]_____. Very high auxin levels apparently trigger the release of ethylene in fruit. It diffuses through the fruit in the air spaces between cells and brings on changes in the fruit such as [3]_____ or changes in color. As cells age, they produce more ethylene, speeding up the [4]_____ process. Ethylene can also diffuse through the air to a fruit. [5]_____ are often picked green and ripened by exposure to ethylene. Ripening of apples is slowed by flooding storage areas with [6]_____, which prevents buildup of [7]_____.

Ethylene probably triggers autumn color changes, drying, and fall of leaves from [8]_____ trees such as maples and oaks. Dropping its leaves allows a tree to save [9]_____. Fall colors are [10]_____ made in the fall and other substances revealed by the breakdown of green [11]_____.

The signals for leaf drop are the [12]_____, [13]_____ days of autumn. As a leaf ages, it produces less [14]_____. Cells in the [15]_____ layer, at the base of the leaf stalk, begin to produce ethylene. The ethylene causes enzymes to digest cell walls in the abscission layer. The wind and the weight of the leaf cause the stalk to break away, and the leaf falls.

Review use of plant hormones in agriculture by writing the name of the hormone next to the summary of its uses.

_____ 1. Can make fruits develop without fertilization of seeds; useful for producing seedless fruits. Induce biennials to seed during the first year of growth.

_____ 2. Triggers ripening of stored fruit. May be used to thin fruit or loosen it for picking.

_____ 3. In low doses, prevent trees from dropping fruit. In larger amounts, used to cause fruit drop for thinning. A synthetic variant, 2,4-D, used as herbicide.

_____ 4. Hormone spray used to keep cut flowers fresh.

Exercise 6   (Modules 33.2 – 33.8)

Review the actions of the five classes of plant hormones by matching the hormones with their overall functions.

A. Stimulate cell division and growth, affect root growth; oppose auxin by stimulating branching

B. Stimulate stem elongation, root growth, fruit development, tropisms

C. Inhibits growth, maintains dormancy, closes stomata during stress

D. Promote seed and bud germination, stem elongation, leaf growth, flowers and fruits

E. Promotes fruit ripening, leaf drop, aging; opposes some auxin effects

_____ 1. Abscisic acid
_____ 2. Cytokinins
_____ 3. Gibberellins
_____ 4. Ethylene
_____ 5. Auxins

Exercise 7   (Module 33.9)

Tropisms enable plants to adapt to environmental circumstances. Complete this chart comparing tropisms.

Name	Description	Mechanism
Phototropism	1.	2.
3.	Growth movement in response to gravity	4.
5.	6.	Contact triggers greater growth on opposite side, causing tendril to bend toward support

Exercise 8   (Modules 33.10 – 33.13)

Short-day plants flower when the night exceeds a certain minimum length. They are inhibited from flowering when red light (or white light, which includes red) shortens a long night. (Far red light following red or white light reverses this effect.) Long-day plants flower when night length is shorter than a certain maximum. They flower when red or white light shortens a long night. The graphs below are similar to those in the text. Test your understanding of flowering and day (night) length: Under each graph, write "yes" if the plant will flower under the conditions described. Write "no" if the plant will not flower.

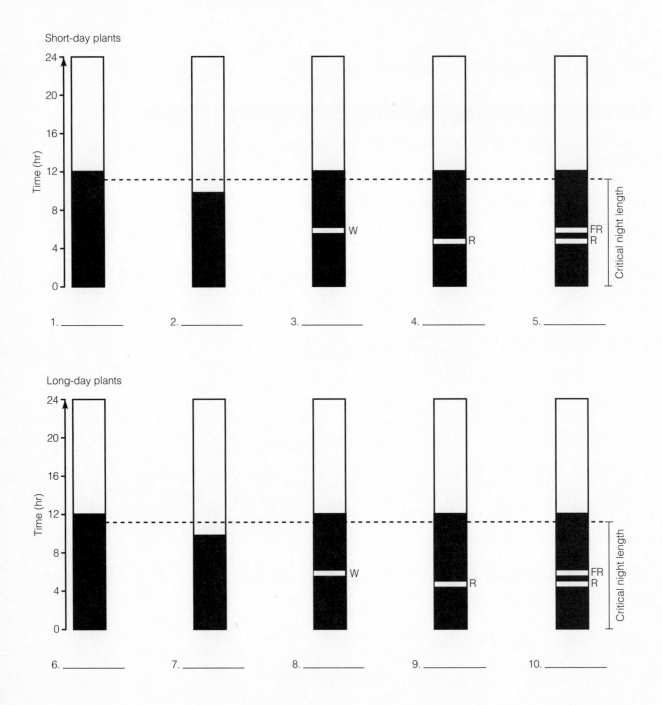

Short-day plants

1. _____   2. _____   3. _____   4. _____   5. _____

Long-day plants

6. _____   7. _____   8. _____   9. _____   10. _____

Testing Your Knowledge

Multiple Choice

1. Most plants flower when
   a. the soil reaches a certain temperature.
   b. they deplete soil nutrients.
   c. the days are the right length.
   d. a certain number of days have passed since they last flowered.
   e. the nights are the right length.

2. Hormones called _____ appear to be responsible for tropisms.
   a. auxins
   b. tropogens
   c. gibberellins
   d. cytokinins
   e. phytochromes

3. The abscission layer
   a. causes a shoot to bend toward light.
   b. secretes cytokinin.
   c. is the location of the biological clock in a plant.
   d. detects light and measures photoperiod.
   e. is where a leaf separates from a stem.

4. _____ stimulate fruit development, so they are sprayed on crops to produce seedless fruits or vegetables.
   a. Ethylene and auxins
   b. Auxins and gibberellins
   c. Gibberellins and abscisic acid
   d. Ethylene and cytokinins
   e. Abscisic acid and auxins

5. Branching is inhibited by _____ from the tip of a growing shoot, but this effect is countered by _____ from the roots.
   a. cytokinins . . . auxins
   b. gibberellins . . . ethylene
   c. auxins . . . cytokinins
   d. gibberellins . . . abscisic acid
   e. auxins . . . abscisic acid

6. A synthetic auxin called 2,4-D is used
   a. to make fruit ripen.
   b. as a weed killer.
   c. to produce seedless fruits.
   d. to make plants produce larger numbers of seeds.
   e. to keep fruit from ripening.

7. A chemical change in a substance called phytochrome
   a. causes a plant to bend toward light.
   b. triggers fruit drop.
   c. enables a plant to measure day length.
   d. is responsible for gravitropism.
   e. allows a plant to deal with stresses like shortage of water.

8. A biological cycle with a period of about 24 hours is called
   a. thigmotropism.
   b. a circadian rhythm.
   c. photoperiod.
   d. abscission.
   e. a biological clock.

9. Seeds of many desert plants will not germinate until a heavy rain washes away their
   a. phytochrome.
   b. abscisic acid.
   c. gibberellins.
   d. auxins.
   e. ethylene.

10. A rapid loss of water in specialized cells in the sensitive plant *Mimosa pudica* causes
    a. germination of seeds.
    b. the plant to bend toward light.
    c. stomatal opening so photosynthesis can begin.
    d. stress that results in delayed flowering.
    e. leaves to droop.

Essay

1. When a plant is tilted, it bends back toward a vertical position, even in the dark. How does a plant "know" which way is up, and what do hormones have to do with this?

2. How do we know that circadian rhythms are internal and not responses to environmental cycles? In what way are environmental cues important in circadian rhythms?

3. A farmer's almanac says, "To keep your stored apples crisp all winter, don't put bruised or wormy apples in the same barrels with the good ones." State the scientific reason why this is good advice.

4. Describe the changing balance of hormones that causes the leaves of deciduous trees to fall in the autumn. What does this change do to the structure of the leaf?

5. Explain the difference between long-day and short-day plants. How do plants "know" how long the day is?

Applying Your Knowledge

Multiple Choice

1. As leaf lettuce matures, the basal edible leaves suddenly send up a tall flowering shoot. After the plant "bolts" like this, it no longer produces broad, tasty leaves. Suppose you wanted to prevent bolting so that you could harvest lettuce longer. You might look for some way to interfere with the effects of
   a. abscisic acid.
   b. gibberellins.
   c. cytokinins.
   d. ethylene.
   e. gravitropism.

2. If the seeds in a cantaloupe are poorly fertilized and fail to develop, the cantaloupe itself will develop poorly and will probably drop off the vine. How does lack of seeds produce this effect?
   a. There is not enough ethylene in the cantaloupe for ripening to occur.
   b. The auxin concentration of the cantaloupe is too high for fruit growth.
   c. Abscisic acid causes the cantaloupe to shrivel and drop off.
   d. There is not enough auxin to stimulate development and stop fruit drop.
   e. Increased cytokinins cause the vine to break at the abscission layer.

3. According to the graph in Figure 33.3B in the text, an auxin concentration of 10^{-6}g/L would
   a. inhibit stem elongation and stimulate root elongation.
   b. stimulate stem elongation and inhibit root elongation.
   c. inhibit both stem and root elongation.
   d. stimulate both stem and root elongation.
   e. It is impossible to predict this from the graph.

4. If you wanted your plants to branch more, you could spray them with
   a. cytokinins.
   b. auxins.
   c. gibberellins.
   d. ethylene.
   e. abscisic acid.

5. When a plant structure such as a leaf is injured, it produces _____, which may cause the part to age and drop off.
   a. cytokinins
   b. phytochrome
   c. auxins
   d. abscisic acid
   e. ethylene

6. An Alaskan trapper was worried about being attacked by grizzly bears, so he left the lights in his cabin on all the time. Plants near the cabin flowered a month early. Which of the following best explains this?
   a. It was due to phototropism.
   b. They must have been long-night plants.
   c. The lights must have emitted far-red light.
   d. They must have been long-day plants.
   e. They must have been short-day plants.

7. Once a flower is pollinated, changes occur that make it less attractive to insects. Its petals, for example, shrivel and fall off. Pollination must
   a. increase the output of cytokinins in the flower.
   b. block the flow of auxins from the roots.
   c. trigger the release of ethylene in the flower.
   d. increase the formation of phytochrome, which sets the biological clock.
   e. trigger the formation of gibberellins in the flower.

8. Which of the following seedlings will probably bend toward the light?
   a. tip covered with a cap made of black plastic
   b. tip separated from base by a gelatin block
   c. tip cut off
   d. tip separated from base by aluminum foil
   e. tip cut off; agar with auxin placed on cut surface on side toward light

9. A biologist growing plant cells in a laboratory dish wanted to cause them to _____ , so he treated them with cytokinins.
   a. enlarge
   b. become dormant
   c. grow roots
   d. produce auxins
   e. divide

10. A plant flowers only if days are shorter than 10 hours. Which of the following would cause it to flower?
    a. 8 hours light, 8 hours dark, flash of white light, 8 hours dark
    b. 12 hours light, 6 hours dark, flash of white, 6 hours dark
    c. 8 hours light, brief dark period, 8 hours light, 8 hours dark

d. 8 hours light, 8 hours dark, flash of white, flash of far red, 8 hours dark

e. 6 hours light, 6 hours dark, 6 hours light, 6 hours dark

Essay

1. During the 1800s, when natural gas was used for lighting, it was found that leaking gas from gas mains sometimes caused shade trees to drop their leaves. Based on what we now know about plant hormones, explain why this probably occurred.

2. A plant scientist hoped to make asparagus shoots grow faster by spraying them with auxin. She found in the lab that a 0.0001 M concentration of auxin increased elongation slightly. To increase its effect, she increased the auxin concentration to 0.01 M (100 times stronger) for spraying in the field. The results were not at all what she expected. The growth rate was less than normal! Explain why.

3. Predict the effects on a shoot of grafting it onto a much larger root system. What would be the role of hormones in these effects?

4. If you pinch off a leaf but leave the leaf stalk, the stalk soon falls from the stem. Pinching off the leaf must upset the balance between which two hormones? Which hormone, now in higher concentration in the stalk, causes the stalk to drop? Which hormone could you smear on the stalk to test whether this hypothesis is correct? What do you predict would happen?

5. In an experiment, soybean plants were planted at weekly intervals from the beginning of May through June. All the plants flowered at the same time—early September. Why?

6. Sometimes when grains are exposed to moisture they begin to germinate and are no longer usable. If you wanted to develop a hormone treatment to prevent this, what would you start with? Why?

7. Shoots exhibit positive phototropism and negative gravitropism. How could you find out which effect is stronger?

Using Your Knowledge

1. Plans for a space colony include growing plants for food and oxygen. In low Earth orbit, there will be weightless conditions and a sunrise and sunset every 90 minutes. In what ways might these conditions create difficulties in raising plants? Can you suggest ways in which plant hormones might be used to overcome any of these problems?

2. How do nurseries and greenhouses use plant hormones to propagate and grow plants? How do they get plants to flower in the proper season or out of season? If there is a nursery in your area, you may want to arrange a visit to find out how the control systems of plants are manipulated to get the plants to do what a grower wants them to do.

The Biosphere: An Introduction to Earth's Diverse Environments

34

Much of biology, and much of your textbook, deals with what goes on inside individual organisms—systems, cells, genes, and biological molecules. But there are biological levels of organization larger and more inclusive than the individual organism—populations, communities, and ecosystems. These higher levels are the realm of ecology—the study of interactions between organisms and their environment. Some ecological questions are simple: How much rainfall does a maple tree need to survive? Others are complex: What processes affect the concentrations of oxygen and carbon dioxide in the atmosphere? This chapter introduces the ecological view of life, which is also the subject of Chapters 35–38.

Organizing Your Knowledge

Exercise 1   (Modules 34.1 – 34.3)

These modules introduce the subject of ecology. Review the relationships between the various levels of study in ecology by completing the concept map on the next page. Place the following terms on the map: **abiotic factors, biosphere, ecology, groups belonging to same species, populations, interactions, habitats, ecosystems, organisms, biotic factors, nutrients.**

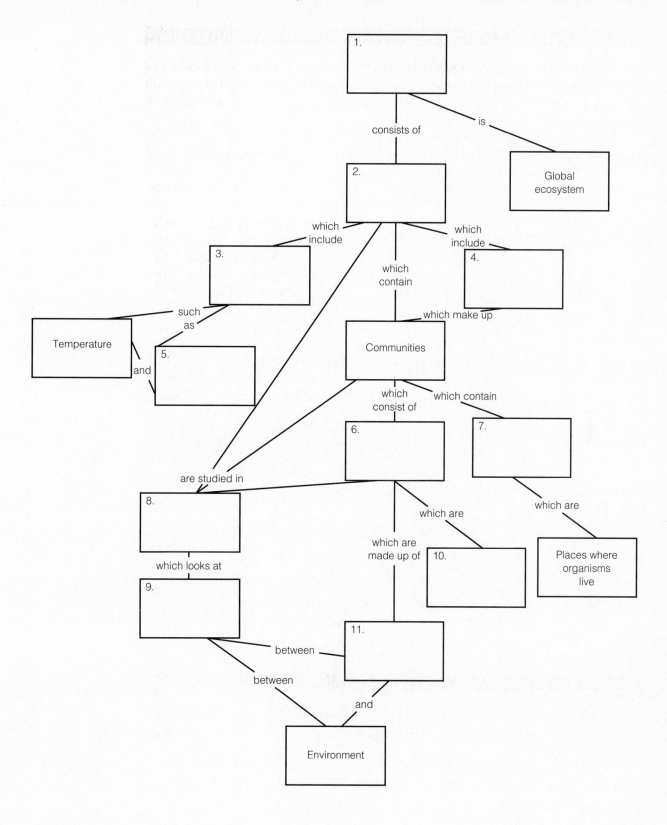

1.

consists of

is

Global
ecosystem

2.

which
include

which
include

which
contain

3.

4.

such
as

which make up

Communities

Temperature

5.

and

which
consist of

which contain

6.

7.

are studied in

which are

which are

8.

which are
made up of

10.

Places where
organisms
live

which looks at

9.

11.

between

between

and

Environment

Exercise 2   (Modules 34.4 – 34.5)

Complete the following chart listing abiotic factors that shape an ecosystem. In each case, list an organism, the abiotic factor that affects it, and how the organism is adapted to that aspect of its environment. Some examples are taken from the text modules; others are new examples.

Organism	Abiotic Factor	Adaptation
Cactus	1.	Thick cuticle
2.	Antarctic cold	3.
4.	5.	Hollow hairs
Pine tree	6.	Cones release seeds only after being heated
Desert rat	7.	Comes out of burrow only at night
Arctic poppy	8.	Grows low to ground, next to rocks
Lungfish	9.	Lungs enable it to breathe air, if necessary

Exercise 3   (Module 34.6)

Practice using concepts and terminology of climate by filling in the blanks in the following paragraphs.

The continent of North America stretches almost from the North Pole to the equator. Its life forms—from polar bears to iguanas—reflect this diversity in climate. The climate is warmest nearest to the equator, where the sun's rays strike the Earth most 1_____. Near the 2_____, the sun's rays strike at more of an angle, so their energy is spread out over a larger area. The 3_____ of the Earth in relation to the sun causes the seasons. In 4_____, the Northern Hemisphere is tilted toward the sun. In 5_____, North America tilts away from the sun.

In addition to temperature differences, wind and rain are important aspects of climate. The sun's heat [6]_____ air and evaporates [7]_____ near the equator. The warm air is less dense than cold air, so it [8]_____. The air cools as it ascends, and its moisture condenses and falls as [9]_____, watering the tropical forests of Guatemala, Costa Rica, and Panama. After losing their moisture, the high-altitude air masses spread away from the equator. Cool air is denser than warm air, so the air [10]_____ around [11]_____ degrees north and south of the equator. This dry air is responsible for the [12]_____ landscape of Mexico and the Southwest United States. Some of the dry air spreads back toward the equator, and some picks up [13]_____ and spreads toward temperate latitudes. This air will cool and drop its moisture as it spreads farther from the equator, into areas such as the eastern United States.

The rise and fall of air masses and their resulting spread from place to place is deflected by the [14]_____ of the Earth to create the [15]_____ winds. Wind rushes back to replace rising air at the equator, creating the [16]_____, which dominate the tropics. The Earth's rotation causes the trade winds to blow from the [17]_____ in the Northern Hemisphere.

Ocean [18]_____ are created by the combined effects of [19]_____, the Earth's [20]_____, and the locations of the [21]_____. Currents profoundly affect local climate. The cold Japanese current creates the foggy conditions that water the redwood forests of northern California. The [22]_____ warms the United States southeast, as well as Western Europe.

Landforms shape local climate. Altitude influences temperature and [23]_____. Moisture may be captured as air masses move over mountain ranges, creating dry [24]_____ downwind. The sagebrush landscape of Nevada lies behind the Sierra Nevadas, for example.

Exercise 4   (Modules 34.7 – 34.8)

All aquatic communities are shaped by similar environmental factors. Review words and concepts related to marine and freshwater environments by completing this crossword puzzle.

Across

2. A _____ is an ecosystem intermediate between aquatic and terrestrial ones.
4. The bottom of the sea, a lake, or a river is called the _____ zone.
7. The _____ zone is flooded by tides every 12 hours.
8. Light does not reach the _____ zone.
12. Motile animals like fishes and whales live in the _____ zone.
13. _____ is only possible in the photic zone.
14. Nitrogen and _____ may be limiting nutrients in lakes and ponds.
17. Near the _____ of a river, the water is colder and lower in nutrients.
18. The major difference between fresh and salt water is _____.
19. _____ are drifting animals of the pelagic zone.

Down

1. Drifting algae and cyanobacteria make up the _____.
3. The availability of _____ has a major effect on organisms in the sea.
5. An _____ is a coastal area where fresh and salt water meet.
6. Seasonal _____ in a lake brings nutrients to the surface and oxygen to the bottom.
9. The most productive parts of oceans are on _____ shelves.
10. _____ vents are densely populated communities of the deep sea.
11. Near a river's mouth, the water may be murky with _____.
12. The _____ zone is the area near the surface where light can penetrate.
15. Decomposition on a lake bottom may deplete the water of _____.
16. Near a river's _____, the water is warmer and moves more slowly.

Exercise 5   (Modules 34.9 – 34.18)

As you read about the biomes, a good way to learn their characteristics is to make a chart that compares them. Using a large sheet of paper, list the *name* of each biome in the left column, and lay out columns for each of their characteristics across the top: *location, description, temperature, rainfall, special environmental conditions* (fire, etc.), *organisms*, and *human impact*. Once you have covered all the biomes, do the next few exercises, which compare them in various ways.

Exercise 6   (Modules 34.9 – 34.18)

Temperature and precipitation are the most important factors shaping biomes. The figure below is a "climograph" outlining the temperature and rainfall limits within which each of the biomes occurs. Label the biomes on the graph. You do not need to know absolute temperatures and rainfall amounts; just ask yourself which biomes are hotter or colder, wetter or drier. Choose from **desert, tundra, temperate grassland, taiga, temperate forest**, and **tropical forest**. Then color the areas on the graph to match the colors on the map in Module 34.9.

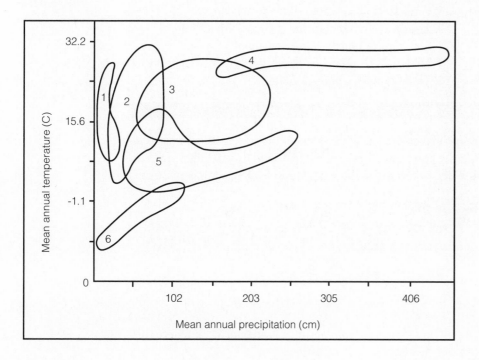

Exercise 7   (Modules 34.9 – 34.18)

Match each of the phrases on the right with one of the biomes from the list on the left.

A. Tundra
B. Desert
C. Tropical forest
D. Temperate decidu-
    ous forest
E. Chaparral
F. Taiga
G. Temperate grassland
H. Savanna

_____ 1. Grows around the Mediterranean
_____ 2. Greatest diversity of species
_____ 3. Prairies, pampas, veldts, and steppes
_____ 4. Coniferous forest of Canada, Siberia
_____ 5. May be hot or cold but always is dry
_____ 6. Permafrost occurs here
_____ 7. May occur in rain shadow
_____ 8. Characterized by broad-leafed deciduous trees
_____ 9. Richest farmland in the United States
_____ 10. Antelope, zebras, lions, and cheetahs live here
_____ 11. Closest to the North Pole
_____ 12. Unlike other biomes, this one is growing in size
_____ 13. Characterized by mild, rainy winters and hot, dry summers with fires
_____ 14. Some characteristics of savannas but in colder areas than savannas
_____ 15. Straddles the equator
_____ 16. Grassland with scattered trees
_____ 17. The biome of most of the eastern United States
_____ 18. Scrubland of dense, spiny shrubs
_____ 19. Moose, elk, snowshoe hares, beavers, bears, and wolves live here
_____ 20. The warmest, rainiest biome

Exercise 8   (Modules 34.9 – 34.18)

Color and label this map to show the distribution of these major biomes: **tropical forest, savanna, desert, chaparral, temperate grassland, temperate deciduous forest, taiga**, and **tundra**. (If you want to go into more detail, you can include polar and high-mountain ice areas.)

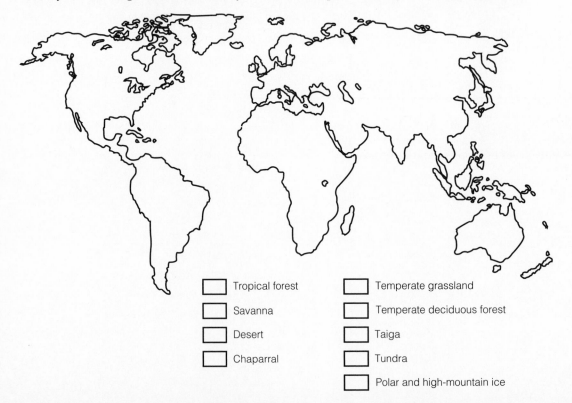

☐ Tropical forest   ☐ Temperate grassland

☐ Savanna   ☐ Temperate deciduous forest

☐ Desert   ☐ Taiga

☐ Chaparral   ☐ Tundra

                      ☐ Polar and high-mountain ice

Testing Your Knowledge

Multiple Choice

1. Which of the following is the most complex?
   a. community
   b. individual
   c. species
   d. ecosystem
   e. population

2. Change of seasons is caused by
   a. change in the distance of the Earth from the sun.
   b. the tilt of the Earth.
   c. variations in output of solar energy.
   d. changes in wind patterns.
   e. gravitational pull of the sun and moon.

3. The kind of tropical forest in a given area—thorn forest, deciduous forest, or rain forest—depends mostly on
   a. rainfall.
   b. average temperature.
   c. temperature difference between day and night.
   d. distance from the equator.
   e. prevalence of fire.

4. The tundra biome gets little precipitation, but during the short summer season it is very wet, with many marshy areas, ponds, and bogs. Why?
   a. Water cannot soak into the frozen ground.
   b. It does not rain much, but the tundra gets a lot of snow.
   c. The ground is so flat that the water cannot run downhill.
   d. It is flooded by melting snow from the surrounding mountains.
   e. The tundra does not look wet; it is a cold desert.

5. If you were to compare the source (beginning) of a river to its mouth (end), you would probably find that
   a. the water is colder near the mouth.
   b. there are more nutrients in the water near the source.
   c. there are more phytoplankton in the water near the mouth.
   d. the current is swifter near the mouth.
   e. silt accumulates near the source.

6. Which of the following occurs in areas a bit warmer than taiga and a bit too dry for temperate forest?
   a. desert
   b. tropical thorn forest
   c. temperate grassland
   d. chaparral
   e. tundra

7. Which of the following is dominated by plants that drop their leaves yearly to conserve water?
   a. desert
   b. taiga
   c. tropical rain forest
   d. temperate deciduous forest
   e. savanna

8. Winds tend to blow toward the equator because
   a. this is a very rainy area.
   b. equatorial air is heated by the sun and rises.
   c. air flows "downhill" from the North Pole.
   d. the Earth's rotational speed is fastest at the equator.
   e. the oceans are broadest at the equator.

9. Which of the following describes the climatic conditions characteristic of the taiga?
   a. mild with occasional drought and fires
   b. cold and snowy
   c. hot or cold but always rainy
   d. cold and dry
   e. mild, rainy winters and hot, dry summers

10. The primary ecological factor determining the distribution of deserts is
    a. windiness.
    b. elevation.
    c. rainfall.
    d. temperature.
    e. fertility of soil.

11. Which of the following is *not* an abiotic factor that shapes ecosystems?
    a. soil minerals
    b. predators
    c. fire
    d. rainfall
    e. volcanic eruptions

12. Which of the following would you expect to find in a rain shadow?
    a. a desert
    b. an ocean
    c. a rain forest
    d. a river system
    e. tundra

Essay

1. Ecology is the study of the interaction between organisms and their environment. We tend to think in terms of the environment affecting organisms. But interaction works both ways. Briefly describe two situations in which organisms (other than humans) change their environment.

2. In the post-World War II period (the 1940s and 1950s) there was a widespread belief in the United States that technology could provide a simple solution to many of our human problems. How and why has that feeling changed?

3. Explain why rain forests occur along the equator, while deserts are found 30 degrees north and south of the equator.

4. Which biomes in North America have been most affected by human activities? How have they been changed, and what kinds of activities have changed them?

5. Using two specific biomes as examples, explain the conditions that determine why a particular biome occurs in a specific area.

Applying Your Knowledge

Multiple Choice

1. Which of the following shows how biotic environmental factors can affect an organism?
   a. Mice have the highest reproductive rate of any common mammal.
   b. Maple trees will not grow in waterlogged soil.
   c. Some shrubs grow only where forest fires scorch their seeds.
   d. Trout will not live in shallow, warm water.
   e. Monarch butterflies live only where there are milkweed plants for food.

2. An ecologist studying how cattle grazing affects population dynamics of native animals like pronghorns and prairie dogs is focusing on
   a. populations.
   b. the biome.
   c. the ecosystem.
   d. the community.
   e. individuals.

3. Which of these biomes has been most altered by human activities in North America?
   a. temperate deciduous forest
   b. taiga
   c. desert
   d. tundra
   e. None of the above has been altered much in North America.

4. Which of the following might eat phytoplankton?
   a. great white shark
   b. beaver
   c. seal
   d. shrimp
   e. sea gull

5. Which of the following lists only abiotic environmental factors?
   a. food, temperature, fire, wind
   b. soil minerals, oxygen level, light, predators
   c. food, parasites, predators, competitors
   d. wind, temperature, soil minerals, light
   e. light, rainfall, food, temperature

6. In the foothills of California, between the flat valley floors and mountain forests, there is a zone of grassland with scattered oak trees. Although it is too narrow to show on our biome map, this "oak woodland" would probably be classed as an example of
   a. taiga
   b. savanna
   c. temperate deciduous forest
   d. chaparral
   e. tropical deciduous forest

7. Which of the following countries is *not* paired with a biome that occurs there?
   a. Ecuador—tundra
   b. Australia—desert
   c. Argentina—temperate grassland
   d. Panama—taiga
   e. France—temperate deciduous forest

8. Which of these biomes generally lies between tundra and temperate deciduous forest?
   a. savanna
   b. taiga
   c. desert
   d. tropical forest
   e. chaparral

9. Most _____ must live in both the benthic and photic zones.
    a. sharks
    b. seaweeds
    c. clams
    d. whales
    e. phytoplankton

10. Bob was reading the newspaper. "More people burned out of their homes in Southern California," he remarked. His wife, Barb, looked up from her reading. "That's a shame," she said. "But I guess you have to expect a fire once in a while if you live in the _____ ."
    a. temperate grassland
    b. tundra
    c. chaparral
    d. taiga
    e. savanna

Essay

1. Name five abiotic factors that might be important to the life of a tree in a mountain forest. Name one biotic factor that might affect the tree. How does the tree adapt to changes in these factors on a day-to-day basis? How has the tree species become adapted to these factors over many generations?

2. In general, temperature is an important abiotic factor shaping biomes on land, but it is less important in the ocean. In the ocean, light is an important abiotic factor, but it is less important on land. Explain why these factors differ in importance on land and in the sea.

3. Sailing ships going from Europe to North America generally sailed south, then west, and on the way back to Europe first sailed north, then east. From the information given in Module 34.6, can you explain why?

4. Climatologists are becoming more and more certain that human activities that produce carbon dioxide and other "greenhouse gases" are creating a "global warming" trend. Predicting the effects of global warming on the climate and organisms in a particular place stretches even the most complex computer models. That said, in very general terms, what effect will global warming probably have on the distribution of biomes? Which might shrink? Which might grow? In which directions will they probably expand or shrink?

5. A marine biologist was asked to speak on the topic of "Feeding the Growing Human Population by Harvesting the Sea." She said, "We can't do it. The sea is about as productive as a desert. The photic zone and the nutrients in the benthic zone are too far apart in most places." Explain what she meant.

Using Your Knowledge

1. Have you ever flown in an airplane and looked out the window at the landscape passing below? Large-scale ecological patterns become much more obvious from the air. Next time you fly, look for interactions between organisms and their environment. Transitions from one biome to another are readily apparent. Also look for effects of large bodies of water, elevation, rain shadows, rock and soil type, and human activities.

2. When was the last time you visited the zoo? Zoos and aquariums are getting away from the old "one-of-each-kind-of-animal-in-a-cage" format to a much more ecological approach, with large groups of animals and plants in natural settings. Exhibits are very educational, with lots of information on the roles and interactions of the animals in their ecosystems. The zoo may be the only place you will ever get to see the creatures of the tundra, or a South American rain forest, or the African savanna. At most zoos and aquariums, you can work as a volunteer guide or "explainer," or even take part in animal behavior research projects.

Population Dynamics

A marine biologist observes humpback whales in the Gulf of Alaska, zoologists band migrating shorebirds in Brazil, and a primatologist tracks orangutans in the forests of Borneo. Foresters map the spread of gypsy moths in a New England forest, a Peace Corps volunteer monitors crop pests in Kenya, wildlife biologists watch seal turtles lay their eggs on a Costa Rican beach, and United Nations public health workers compile statistics on the growth of the world's human population. The theme that unites all these activities is the study of population dynamics. We need to know where living things live; how numerous they are; the way they are distributed on land or in water; their birth rates, death rates, and life spans; the environmental conditions that affect them; and how they are adapted to these conditions. The many aspects of population dynamics are the subject of this chapter.

Organizing Your Knowledge

Exercise 1 (Modules 35.1 – 35.2)

This exercise will allow you to work with the concepts of population density, dispersion pattern, and sampling. The map represents a meadow on the edge of the city of Mapleton. It is surrounded by developed and farmed land but has remained relatively undisturbed. Developers plan to build a subdivision that would cover the meadow. The Mapleton Open Space Alliance would like the meadow to remain as public open land. They note that the dwarf hawthorn, an uncommon shrub, is found in the meadow. It is considered a "sensitive species" by the state conservation department. The city council has asked for a construction delay until the status of the shrub is determined. You have been sent to determine the density of the hawthorn population in the meadow, as well as that of a deer mouse that may also be present. Use the map of hawthorn distribution on the next page for your survey, and answer the following questions.

The area of the meadow is 16.8 hectares. (A hectare is a metric unit of area equal to about 2.2 acres, so the meadow totals about 40 acres.) This is too big an area to count every shrub, so you will have to look at sample plots. On the ground, this would be done with ropes and measuring tapes. You can choose random samples by merely dropping a penny on the map, drawing a circle around it, and counting the "shrubs" inside. On the scale of the map, the area covered by a U.S. penny equals 0.2 hectare.

1. Take ten samples. How many hectares does this total? _____

2. What is the total number of shrubs in the ten samples? _____

3. What is the density of hawthorns in shrubs per hectare? _____

4. What is the total number of hawthorns in the meadow? _____

5. How could you make your count more accurate? Why not do this?

6. Look at the map again. What is the pattern of dispersion of the shrubs? What might cause this pattern of dispersion?

You would also like to know the number of deer mice in the meadow. For this, it will probably work best to use the mark-recapture method.

7. Why does this work better than the method used to count the plants?

8. One night, you trap 40 mice, mark them, and let them go. Two nights later, you again trap 40 mice, and ten of them are marked. What is the total number of mice in the meadow? _____

9. What is the population density of mice in the meadow, in animals per hectare? _____

10. What do you have to assume about the mice and your method for your results to be valid? Could you be wrong? Why or why not?

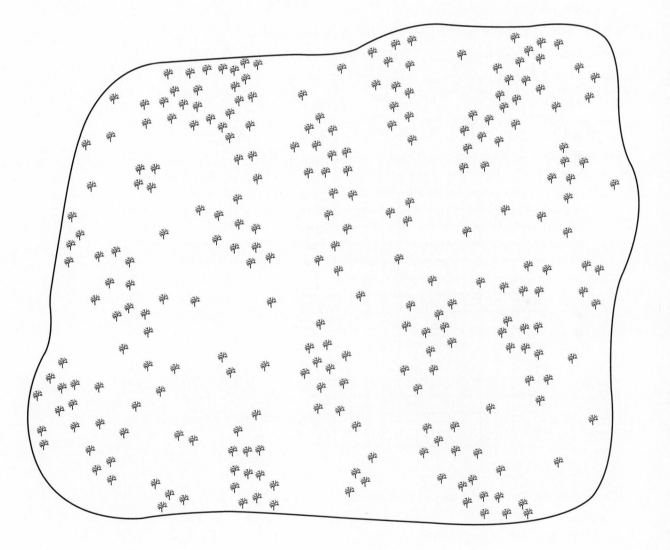

Exercise 2   (Module 35.3)

Models devised by ecologists describe two kinds of population growth. Exponential growth is described by this equation: $G = rN$. The rate of growth, G, depends on N, the size of the population, multiplied by r, the population's intrinsic rate of increase. r is calculated by subtracting the death rate from the birth rate. Exponential growth is unregulated. The bigger the population, the faster it grows. This cannot be sustained for long in real populations, but it is interesting as a theoretical possibility. Populations of fast reproducers like bacteria and insects can grow at near-exponential rates for short periods.

Let's calculate and graph the exponential growth of a population of aphids for which $r = 40\%$ per week. Remember that $G = rN$. If there are 10 aphids to start with, the number of aphids added by the end of the first week (G) is equal to rN, or 0.4 x 10, which equals 4. So the total population (N) after one week is $10 + 4 = 14$.

1. Starting with the new total (N) of 14, how many aphids will be added (G) in the second week? _____ (Round off fractions.)

2. What will the total population (N) be at the end of the second week? _____

3. Aphids added in third week? _____ Total after third week? _____

4. Aphids added in fourth week? _____ Total? _____

5. Total at end of fifth week? _____ sixth week? _____ seventh week? _____ eighth week? _____ ninth week? _____ tenth week? _____

6. Graph the size of the aphid population (N) versus time (in weeks) below. Population size was 10 at time = 0. Label the axes of the graph.

7. How would you describe the shape of this graph?

8. Could this kind of growth continue indefinitely? Why or why not?

Real environments will not support exponential growth. Populations are limited by space, food supply, or other factors that slow growth. The population levels off at a density the environment can maintain—carrying capacity. This so-called logistic growth is described by the equation $G = rN(K − N)/K$, where K represents carrying capacity.

The following data chart the growth of a population of deer on a small protected island off the coast of Alaska, recorded over a 50-year period:

1940	92	1970	814
1945	151	1975	765
1950	295	1980	688
1955	603	1985	740
1960	861	1990	729
1965	920		

9. Graph the growth of the deer population below. Label the axes. How would you describe the overall shape of the graph?

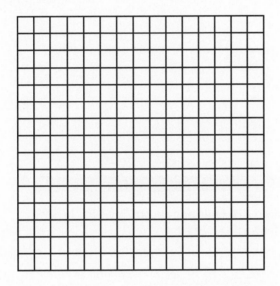

10. What happened to the population during the 1960s and early 1970s?

11. What may have caused the population density to level off?

12. What is your estimate of the carrying capacity of the island for deer?

13. What is the biological term for this kind of population growth?

Exercise 3   (Modules 35.4 – 35.5)

There are two major kinds of factors that limit population growth. Density-dependent factors, predicted by the logistic model, have a greater effect as the population grows. The effects of density-independent factors are the same no matter how big or small the population is. State whether each of the following is a characteristic or example of a density-dependent (DD) or density-independent (DI) factor.

_____ 1. Freezing weather
_____ 2. Has a greater effect on a larger population
_____ 3. A volcanic eruption
_____ 4. Predation
_____ 5. Floods
_____ 6. Food supply
_____ 7. May limit the population before it even gets close to carrying capacity
_____ 8. Disease or parasites
_____ 9. Decreases when the population is below carrying capacity, increases when the population exceeds carrying capacity
_____10. Fire
_____11. Intensity of effect no greater for larger population, no less for smaller population
_____12. Clear-cut examples are not often seen in nature
_____13. May cause cyclical changes in lynx and hare populations
_____14. Storms
_____15. Tends to be a biotic factor (Look at the examples above.)
_____16. Tends to be an abiotic factor

Exercise 4   (Module 35.6)

Opportunistic and equilibrial life histories are the two ends of a spectrum of life history patterns. Compare these two patterns by completing this chart.

Characteristic	*Opportunistic*	*Equilibrial*
Life history emphasis	1.	Stability near carrying capacity
Relative body size	2.	3.
Number of offspring per reproduction	4.	5.
Age at first reproduction	Younger	6.
Controlled mainly by ____ factors	7.	8.
Emphasis on ____ of offspring	9.	Quality and care
Examples	10.	11.

Exercise 5   (Module 35.7)

Check your understanding of life tables and survivorship by matching each phrase on the right with a term on the left. Answers may be used more than once.

A. Life table
B. Survivorship curve
C. Type I survivorship
D. Type II survivorship
E. Type III survivorship

_____ 1. Graph of percent alive at the end of each age interval
_____ 2. Tabulation of deaths and chance of surviving
_____ 3. Most young die, but a few live to old age
_____ 4. Originally used to set life insurance rates
_____ 5. Characteristic of clams and insects
_____ 6. Death rate constant over life span
_____ 7. Characteristic of *Hydra* and gray squirrels
_____ 8. Most offspring live a long life and die of old age
_____ 9. Characteristic of humans and elephants

Exercise 6   (Modules 35.8 – 35.10)

Review your knowledge of human population growth by filling in the blanks in the following essay.

The world's human population now stands at over ^{1}_____ billion. We are the most numerous large animal on earth, and our huge population is the result of thousands of years of accelerating growth, a phenomenon unprecedented in the history of life.

Whether a population grows or shrinks depends on the balance between ^{2}_____ rate and ^{3}_____ rate. When ^{4}_____ replaced a lifestyle of hunting and gathering about 10,000 years ago, the human ^{5}_____ rate jumped, and our numbers started increasing. Growth was slow and steady until the seventeenth century, except for a slight dip when the ^{6}_____ wiped out about a fourth of the population of Europe (in the fourteenth century). In 1650 the global human population probably stood at around ^{7}_____ million. Improved nutrition, health care, and living conditions, brought about during the ^{8}_____ Revolution, caused a drop in ^{9}_____ rates, and population growth accelerated. The population doubled to 1 billion by 1850, again to 2 billion by 1930, and again to 4 billion by 1975. World population is growing at a rate of ^{10}_____% per year and is projected to reach 6.25 billion by the year 2000.

Of course, rates of population growth differ in different parts of the world. In the industrialized countries, population growth is relatively ^{11}_____—0.3% per year in Europe, 0.7% in the United States. Graphs of population age structure for these countries show an even distribution of prereproductive, ^{12}_____, and ^{13}_____ individuals, with the exception of a bulge representing the post–World War II ^{14}_____. This even age distribution occurs because in each generation, the average couple has two children—just enough to replace themselves. The "baby boomers" in the United States are actually having ^{15}_____ children than their parents, but because there are so many of them having children, the U.S. birth rate continues to exceed the death rate.

In the ^{16}_____ countries of Asia, Africa, and Latin America, populations are growing much faster. The rate of increase is 2.2% per year for India, 2.1% for

Latin America, and 2.9% for Africa. In these countries, the graph of [17]_____ is broad-based; each generation of children outnumbers their parents. The [18]_____ age group is the largest. At an annual growth rate of 2.9%, the population of Africa, now around 650 million, could double to 1.3 billion in only 25 years!

What does the future hold? Could the world sustain twice as many Africans? What is *K*, the Earth's [19]_____? Could we have already exceeded the Earth's ability to sustain us? Some people, like ecologist Paul Ehrlich, are pessimistic about the future. They doubt that our increasingly damaged biosphere will be able to support many more people. Others, such as economist Julian Simon, count on continuing advances in technology to provide for us.

Population growth will continue to challenge us well into the next century. One way or another, the overall human [20]_____ rate and [21]_____ rate will come into balance. The countries with high growth rates will struggle to provide their growing populations with [22]_____ and [23]_____. The industrialized countries will face other challenges. The U.S. population is growing much more slowly than the populations of less-developed nations, but an American baby has a much greater impact on the global environment than an African or Latin American child will ever have. The United States, with only one-twelfth the population, consumes more energy than Africa, Latin America, India, and China combined! In addition, the industrialized nations will need to confront the problem of a small number of young people providing services such as [24]_____ and [25]_____ to an increasing proportion of "senior citizens."

There is no question that the human birth rate will eventually come into line with the death rate. The questions that remain are as follows: How large will the population get before this happens? Will the balance come about by choice, or will it be forced on us? Will it occur through a decrease in the [26]_____ rate or an increase in the [27]_____ rate? Finally, what kind of life will there be in the future for all the people who share the Earth?

Testing Your Knowledge

Multiple Choice

1. In wild populations, individuals most often show a _____ pattern of dispersion.
   a. random
   b. density-dependent
   c. equilibrial
   d. uniform
   e. clumped

2. A population would grow exponentially
   a. if it were limited only by density-dependent factors.
   b. until it reaches carrying capacity.
   c. if there were no limiting factors.
   d. if it were a population with an equilibrial life history.
   e. if it showed logistic growth.

3. Which of the following could be a density-dependent factor that might limit population growth?
   a. fire
   b. storms
   c. drought
   d. food supply
   e. cold

4. The effects of which of the following environmental factors would probably *not* change as a population grows?
   a. disease
   b. limited food supply
   c. competition for nesting sites
   d. weather
   e. predation

5. Human population growth was slow and gradual for a long period, but it turned sharply upward
   a. after the development of agriculture, which decreased the death rate.
   b. during the last half-century, when the birth rate increased.
   c. as a result of the Black Death, which left more food for the survivors.
   d. during the Industrial Revolution, when the death rate dropped.
   e. as a result of the post-World War II "baby boom."

6. A broad-based, pyramid-shaped age structure is characteristic of a population that is
   a. growing rapidly.
   b. at carrying capacity.
   c. stable.
   d. limited by density-dependent factors.
   e. shrinking.

7. A population grows rapidly at first and then levels off at carrying capacity if it is
   a. limited by density-dependent factors.
   b. limited by density-independent factors.
   c. an opportunistic species.
   d. limited by density-dependent or density-independent factors.
   e. relatively unaffected by limiting factors.

8. Which of the following is the most accurate comment on the Earth's carrying capacity (*K*) for people?
   a. *K* is smaller now than it was a thousand years ago.
   b. The human population is still a long way from *K*.
   c. Our technology has allowed us to keep increasing *K*.
   d. When it comes to humans, the concept of *K* is irrelevant.
   e. The human population has already exceeded *K*.

9. Which of the following would be true of a species with an opportunistic life history?
   a. Members of the species take a relatively long time to reach reproductive age.
   b. They are regulated mostly by density-dependent factors.
   c. They produce large numbers of offspring.
   d. The population usually stabilizes near carrying capacity.
   e. They give their young lots of care.

10. In the models that describe population growth, *r* stands for
    a. population density.
    b. a time interval.
    c. total number of individuals in the population.
    d. growth rate.
    e. carrying capacity.

11. A biologist wants to use the mark-recapture method to census a population of desert mice. This will not work if
    a. the mice are territorial.
    b. the mice become wary of the traps.
    c. the biologist waits too long before resampling the population.
    d. the mice deliberately return because the traps contain food.
    e. any of the above.

12. When needed resources are unevenly distributed, organisms often show a(n)_____ dispersion pattern.
    a. density-dependent
    b. clumped
    c. exponential
    d. random
    e. uniform

Essay

1. Describe the exponential and logistic population growth curves, explain their shapes, and explain why they are different.

2. Describe some factors that might affect carrying capacity.

3. Describe three density-dependent and three density-independent factors that might restrict population growth. Explain how the effects of these two kinds of factors on populations differ.

4. What has enabled the Earth's human population to continue to grow for a long period at near-exponential rates? How does this differ from other species?

5. What are population cycles? What kinds of hypotheses have been suggested to explain them?

6. Compare opportunistic and equilibrial life histories, and give examples of each.

Applying Your Knowledge

Multiple Choice

1. A particular species of tropical fish has only a few offspring and takes care of them for an extended period. We might also expect the fish population
   a. to be controlled mostly by density-independent factors.
   b. to show exponential growth.
   c. to live in a harsh environment.
   d. to start reproducing very young.
   e. to be relatively stable, near carrying capacity.

2. Chimpanzees have a relatively low birth rate. They take good care of their young, and most chimps live a long life. The chimp survivorship curve would look like
   a. a line that slopes gradually upward.
   b. a relatively flat line that drops steeply at the end.
   c. a line that drops steeply at first, then flattens out.
   d. a line that slopes gradually downward.
   e. a horizontal line.

3. Locust populations go through periods of sudden explosive growth, followed by sudden decline in numbers. Their numbers are probably regulated by
   a. predation.
   b. density-dependent factors.
   c. logistic growth.
   d. random dispersion.
   e. density-independent factors.

4. Sea gulls fiercely defend the areas around their nests in their clifftop breeding colonies. Within the colony they would show a _____ dispersion pattern.
   a. uniform
   b. random
   c. dense
   d. density-independent
   e. clumped

5. A wildlife biologist is trying to predict what will happen to a bear population if bear hunting is banned. He had the equations all worked out but then realized that he had grossly underestimated the amount of food available to the bears. To make his predictions more accurate, he will have to go back to his equations and
   a. decrease N.
   b. increase N.
   c. decrease K.
   d. increase K.
   e. decrease r.

6. An ecologist would suspect a population is growing rapidly if it
   a. contains many more prereproductive than reproductive individuals.
   b. is near its carrying capacity.
   c. is limited only by density-dependent factors.
   d. shows a clumped pattern of dispersion.
   e. is far below its carrying capacity.

7. An oak tree produces thousands of acorns, but very few grow into mature oak trees. The oak tree exhibits a _____ survivorship curve.
   a. Type I
   b. Type II
   c. Type III
   d. Type I or II
   e. Type I or III

8. When birth rate equals death rate
   a. a population grows rapidly.
   b. the size of a population remains constant.
   c. density-dependent limiting factors do not affect the population.
   d. a population is in danger of extinction.
   e. a population goes through up-and-down cycles.

9. If you wanted to see what percentage of the population of Thailand is under 10 years old, you could look at
   a. a logistic curve for the population.
   b. the population's age structure.
   c. a life table for the population.
   d. a plot of population density.
   e. the population's survivorship curve.

10. To determine the density of a rabbit population, you would need to know the number of rabbits and
    a. the factors that limit population growth.
    b. their birth rate.
    c. the area in which they live.
    d. their population growth rate.
    e. their pattern of dispersion.

Essay

1. A fisheries biologist is interested in determining the population density of smallmouth bass in a lake. Using a net, she captures and tags 100 bass. A week later, she again catches 100 bass, and out of these fish, 5 are tagged. How many smallmouth bass are there in the lake?

2. Tim said, "It says here that Japan has a much greater population density than the United States." Carl replied, "That can't be right. There are a lot of people in Japan, but surely the population of the United States is much larger." What would you say to clear this up for them?

3. Biologists figure that the carrying capacity of a particular river is 100,000 salmon. They want to manage the catch of salmon so that the salmon population is at a size where replacement of the fish that are caught will happen at the fastest rate. Would you recommend that they catch fish until the salmon population stands at 80,000, or 50,000, or 10,000? Explain why. (Hint: They are assuming logistic growth of the salmon population.)

4. The temperature seldom dips below freezing along the Pacific coast, so robins can stay there all winter long. But during rare spells of subfreezing weather, the birds have a hard time finding food and shelter. Birds that find a warm place sheltered by evergreen branches have the best chance of making it through a cold snap, but there are only a limited number of such places. After a week of cold, only a fraction of the robin population may remain. Explain how both density-dependent and density-independent factors affect the robin population in this case.

5. Population explosions are often seen when animals are introduced into a new area. Examples include the spread of starlings across North America, a plague of rabbits in Australia, and the spread of cane toads in Florida. Why are these kinds of population explosions more likely in areas where the animals are not native?

6. The most intensive and rigorous studies of populations have been undertaken on islands and in lakes. Why are populations easier to study on islands or in lakes?

Using Your Knowledge

1. If you bend a wire coat hanger into a rough square, the area enclosed by the square is roughly 0.05 square meter. You can use this "quadrat" to survey population densities of plants or slow-moving animals. You might toss it onto a lawn, for example, and count the plants inside to compare densities of weeds in the shade and in the sun, or before and after treatment with weed killer. How can you make it more likely that your sample quadrats are truly random samples?

2. Many ideas related to population dynamics are idealized, theoretical constructs that are difficult to see clearly in real populations. Some examples are logistic and exponential growth, and the difference between opportunistic and equilibrial populations. Are these ideas useful, even if they are hard to find? Or do they blind us to what is really going on in nature?

Communities and Ecosystems

36

In his classic book *The Closing Circle*, ecologist Barry Commoner explained what he saw as "the four laws of ecology": First, "Everything is connected to everything else." In a biological community, organisms are connected in numerous ways. They compete, they prey on and parasitize one another, and sometimes they even help each other through beneficial symbiotic relationships. Second, "Everything must go somewhere." Matter and energy cannot simply disappear. In an ecosystem, chemicals are recycled and energy from the sun escapes as heat. Third, "Nature knows best." So far, our efforts to alter or create ecosystems (through agriculture, logging, spraying pesticides) have produced unexpected results, and the systems have often functioned less well after our interference. Finally, "There is no such thing as a free lunch." Because all parts of an ecosystem are connected, everything must come from someplace; every gain is balanced by a loss. It will be helpful to keep these simple laws in mind as you study this chapter about the connections in communities and ecosystems.

Organizing Your Knowledge

Exercise 1   (Module 36.1)

This module introduces four major characteristics of biological communities. Each of the italicized phrases in the essay below relates to one of these four characteristics. State whether each phrase relates to diversity (D), prevalent form of vegetation (V), stability (S), or trophic structure (T), by writing the corresponding letter in the parentheses.

The tropical rain forest may hold a *greater variety of species* (^{1}_____), but no place on Earth can rival the great *conifer forests* (^{2}_____) of the Pacific Northwest for the mass of living things they contain. The major tree species of these forests are *Douglas fir, Sitka spruce, western hemlock, western red cedar, and coastal redwood* (^{3}_____). The trees may be 2 or 3 meters in diameter and more than 60 meters tall. There are groves of redwoods whose tops are 100 meters above the forest floor. A ten-story building would fit comfortably beneath their lowest branches. Along streams and where the tall trees have fallen, a *tangle of bushes and smaller trees* (^{4}_____) reach upward for sunlight, but in the shade under the dense canopy the ground is covered mostly by a thick blanket of needles. Sunlight slants through the canopy and illuminates scattered shrubs and ferns.

The forest may seem quite silent on a warm summer day because much of its activity takes place high in the canopy or in the thick layer of litter on the ground. In fact, many animals spend their entire lives on the trunks and branches of tall Douglas firs. *As many as 1500 species of invertebrates* (^{5}_____)—insects, spiders, mites, and their relatives—have been counted on a single tree. The red tree vole never leaves the canopy, where *it feeds on the needles of Douglas fir* (^{6}_____). The flying squirrel nests in hollow trees but glides to the forest floor *to eat lichens and fungi* (^{7}_____). Some of the fungi combine with tree roots and help trees absorb water and nutrients; the flying squirrel and other small mammals help to spread the spores of these fungi to new areas during their nocturnal feeding forays.

Many birds inhabit the forest. One of the most common is the Steller's jay, a black-and-dark-blue cousin of the eastern blue jay. Brown creepers and red-breasted nuthatches *pry insects from crevices in bark* (^{8}_____), while the pileated woodpecker, the largest commonly seen woodpecker, chisels ants from tree trunks. In all, this ancient forest is *home to more bird species than any other habitat north of Mexico* (^{9}_____).

Small mammals like tree voles and flying squirrels are *preyed on by carnivores such as the spotted owl* (^{10}_____). In fact, flying squirrels make up the bulk of a spotted owl's diet. Some other predators of this ancient forest are the northern goshawk and the pine marten, a relative of the weasel.

Periodic fires scar the trunks of the big trees. Their thick bark and the large amount of water stored in their trunks usually protect the trees, but perhaps *once every few centuries a stand is destroyed by fire* (^{11}_____). Landslides and floods also can wipe the slate clean. Plants from the surrounding forest gradually invade the clearing, and over a period of hundreds of years a stand of big trees may become reestablished. Natural disasters have given the natural forest a patchy appearance, with groves of ancient trees adjoining meadows and stands of immature trees.

Now the biggest cause of change in the ancient forest is not fire or flood but the *disturbance caused by logging* (^{12}_____). The knot-free, fine-grained wood of the big trees is valuable, and the trees are clear-cut for lumber and plywood. Only about 4% of the ancient forests of the Pacific Northwest remain undisturbed. The planted forests that replace them may not have *the variety of plant and animal life seen in the natural old stands* (^{13}_____). *Species such as the spotted owl are endangered by the disappearance of the old trees* (^{14}_____) that they depend on for food and nesting sites. Many people are working to preserve undisturbed stands of ancient forest so that the organisms that live there can survive and so that we can continue to marvel at the most massive biological community on Earth.

Exercise 2   (Modules 36.2 – 36.5)

The structure of a community is shaped by interactions among the populations making up the community. The most important kinds of interactions are predator/prey relationships, competition, and three kinds of symbiosis—parasitism, commensalism, and mutualism. State which of these five interactions is described in each of the examples below.

_____   1. Small fish called remoras accompany sharks and dine on scraps left over when sharks feed.

_____   2. Sheep liver flukes feed on bile and can weaken or kill their hosts. They are passed on to other sheep in the animals' droppings.

_____   3. Grazing by introduced mountain goats has reduced the numbers of alpine wildflowers in Olympic National Park.

_____   4. Pest-control specialists have brought in a destructive moth to eat tansy ragwort, a poisonous weed.

_____   5. Mistletoe obtains nutrients from a tree host.

_____   6. A small shrimp takes shelter inside a sponge, which is apparently unaffected by its tenant.

_____ 7. Mycorrhizal fungi associated with roots obtain carbohydrates from a tree, while enabling the tree to absorb water and minerals more efficiently.

_____ 8. In many parts of North America, the starling has displaced the bluebird from its nest sites.

_____ 9. A bee pollinates a tropical orchid by being tricked into "mating" with the flower; the bee uses a perfume from the flower to attract a mate.

_____ 10. The influenza virus attacks the lining of the respiratory tract and is passed from person to person by contact or airborne droplets.

_____ 11. Red-winged blackbirds arrive earlier on the breeding grounds but are forced to the edges of a marsh by larger, later-arriving yellow-headed blackbirds.

_____ 12. Lions hunt large herbivorous mammals such as zebras and wilde-beest.

Exercise 3   (Modules 36.2 – 36.5)

You can think of an organism's niche as its "role" or "job" in the community. The niche includes the sum of the organism's functions, abilities, and tolerances. It is possible to describe the niche as a sort of "job description" for a species, as you might see in a classified ad: "Applicant will be required to travel in herd, drink through nose, and knock down trees for food" could only describe the job of an elephant!

Identify the organism whose niche is outlined in each of the following job descriptions. Some are from the text; others are not but will probably (like the elephant) be familiar to you.

_____ 1. "Ability to build trap to catch flying insects. May or may not need to devour mate."

_____ 2. "Ability to cling tightly to rocks, withstand wave action, and out-last competitors at low tide."

_____ 3. "Will be traveling and working outdoors in cold weather. Must have the patience to wait long periods to catch and eat seals through hole in ice. Some swimming ability and camouflage helpful."

_____ 4. "Will be stabilizing tide pool community through predation. Must like mussels."

_____ 5. "Must be able to withstand coastal storms and forest fires while maintaining species' reputation as world's tallest tree."

_____ 6. "Must live in South American rain forest, eat insects, and have poison glands and bright coloration."

_____ 7. "Will work closely with legume. Will be required to fix nitrogen in exchange for daily carbohydrate allowance."

Exercise 4   (Module 36.6)

Ecosystems develop and recover from disturbance through a process called ecological succession, described in this module. State whether each of the following represents a relatively early stage in succession or a relatively late stage. (Hint: Ask yourself if the community were left untouched, whether it would look the same or different in a hundred years. If it would look different, it is an early stage—there are later stages to come.)

_____   1. Lichen-covered rocks near a melting glacier in Alaska
_____   2. The Pacific Northwest forest described in Exercise 1
_____   3. A vacant lot near your college
_____   4. An oak-maple-beech forest in Ohio
_____   5. A lava flow on the island of Hawaii
_____   6. A lawn in a suburb in New Jersey
_____   7. A cornfield in Virginia

Exercise 5   (Modules 36.7 – 36.9)

The trophic structure of an ecosystem is the pattern of feeding relationships by which energy and chemicals flow through the system, from tropic level to trophic level. Name the trophic level of each of the organisms in the following description of a freshwater marsh food web: primary producer (PP), primary consumer (1C), secondary consumer (2C), tertiary consumer (3C), quaternary consumer (4C), or detritivore (D). (Note that a consumer can function on more than one level, depending on what it eats.)

Marshes and other wetlands are among the most endangered of habitats. They are productive "nurseries" for many wildlife species, but many of our wetlands have been drained for agriculture or filled for development.

The freshwater marsh food web starts with plants like cattails, arrowleaf, and various floating or submerged "water weeds" (^{1}_____). They provide food for muskrats (^{2}_____) and mallard ducks (^{3}_____), both of which may in turn be eaten by hawks (^{4}_____) or mink (^{5}_____). Microscopic algae (^{6}_____) make much of the food in the marsh. Small shrimplike crustaceans (^{7}_____) and insect larvae (^{8}_____) graze on the algae. The insects are eaten by ducks (^{9}_____), frogs (^{10}_____), and sunfish (^{11}_____). A frog or sunfish might be eaten by a larger yellow perch (^{12}_____), a great blue heron (^{13}_____), a water snake (^{14}_____), or a mink (^{15}_____). The heron (^{16}_____) also eats perch and snakes, and the hawk (^{17}_____) will also occasionally devour a snake.

This is a highly simplified description of a marsh food web. There might be hundreds of species of large plants and animals making up the community of a marshy roadside pond, not to mention a swarm of microscopic creatures. In addition to this grazing food web, many inconspicuous worms, insect larvae, and snails (^{18}_____) get their food from the dead material produced by the other plants and animals of the marsh.

On a separate sheet of paper sketch the marsh food web. It is *not* the same as the example in Module 36.9. Write the names of the organisms at their appropriate trophic levels, and connect the names with arrows. (Where in your diagram should the producers be? The top consumers? Which way do the arrows point? Why?)

Exercise 6   (Modules 36.10 – 36.11)

This flow chart illustrates the movement of energy through an ecosystem. The boxes represent the total mass of organisms at each trophic level. The arrows show the amount of energy passing through each trophic level. Energy enters the primary producer level as sunlight. Some of this energy is stored in molecules produced in photosynthesis. Energy enters each of the consumer trophic levels when the consumers feed on the level below. Much of the energy in the food entering any level is used to power life processes; the food is used as fuel in cellular respiration, and its energy ends up as heat. Some energy is wasted; it is lost to the detritus food web in the form of dead leaves or droppings. A small portion of the energy is stored up in tissue when organisms grow or reproduce; this production—about 10% of energy intake at any trophic level—is the only energy passed on to the next level. Label and color the trophic levels on the diagram: **primary producers** (green), **primary consumers** (blue), **secondary consumers** (pink), **tertiary consumers** (orange). Label and color the pattern of energy flow: **sunlight** (yellow), **production energy** (red), **energy used in cellular respiration** (red), and **energy in wastes** (brown).

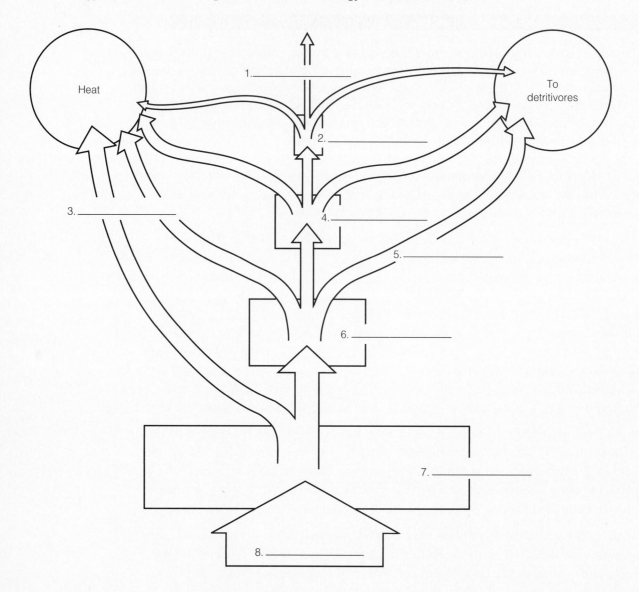

Exercise 7   (Modules 36.12 – 36.17)

The biosphere receives a constant supply of energy from the sun, uses this energy for a while, then loses it to space as heat. Unlike energy, the chemicals necessary for life are present on Earth in fixed amounts, and these chemicals are used over and over. These chemicals, such as water, carbon, and nitrogen, occur in various forms and are changed from one form to another by various physical and chemical processes. The story below traces a nitrogen atom as it moves through the various reservoirs and processes of the nitrogen cycle. Fill in the blanks as you follow its journey.

The nitrogen atom, N, had been in the atmosphere for more than 2 years. It was paired with another identical nitrogen atom, forming a molecule of [1]_____ gas, which makes up about 80% of the air. During its time in the atmosphere, N had circled the Earth several times, from the skies over the Philippines, to Africa, to the Antarctic, to South America, and now over a sand dune in North Carolina. There it was captured by [2]_____ in a nodule on the root of a legume called a beach pea. There N was split away from its partner and combined with hydrogen atoms, eventually ending up in an amino group in a protein molecule built by the plant. The protein was stored in one of the peas, which ripened, dried and fell on the ground, and was eaten by a mouse. The amino acid from the pea was incorporated into a [3]_____ molecule in a leg muscle of the mouse. On a moonlit night a month or so later, a great horned owl caught the mouse, and N became part of a protein molecule in one of the owl's feathers. And there it remained until the following spring.

In the spring, as every spring, the owl molted some of its feathers. The small feather containing N ended up in the litter under a pine tree. [4]_____ in the soil broke down the feather over a period of several months, and N was eventually released into the soil, in the form of [5]_____, NH_3. Other bacteria, called [6]_____ bacteria, then attached N to three oxygen atoms, forming [7]_____, which was taken up by a huckleberry bush and used to build another protein. The huckleberry was eaten by a cardinal, which broke down the protein and eventually excreted N in the uric acid in its droppings. Decomposers in the forest soil changed this waste product to [8]_____, which nitrifying bacteria quickly converted into [9]_____.

When the patch of forest was cut for lumber, most of the nitrate ions in the soil were washed downhill. A flood deposited the nitrate ion containing N in the soil of a flat river valley. N was soon absorbed by the roots of a buttercup plant—a process called [10]_____. The buttercup used N to make an [11]_____, which was used to build a protein molecule, which ended up in a pollen grain. A beetle collected some of this pollen, and soon N was part of the insect's body. The beetle flew into a bog and landed on the leaf of a Venus flytrap, an insect-eating plant. The leaf snapped shut, and the beetle was slowly digested; the plant used N to make its own proteins.

About a year later, the rotting trunk of a dead tree toppled into the swamp and buried the Venus flytrap. As the plant decomposed, bacteria again incorporated N into ammonia and then nitrate. But this time, in the low-oxygen conditions of the mud, [12]_____ bacteria were able to break down the nitrate ions, releasing N to the air as part of an N_2 molecule. The wind swept the molecule out over the ocean, where . . .

You might find it fun and educational to continue the story from here. You might also find it helpful to make up similar stories for a carbon atom, a phosphorus atom, and a water molecule. This is a good review to do in a study group.

These modules contrast processes that occur in natural, "healthy" ecosystems with processes that take place in "unhealthy" systems that have been damaged by human activities. Each of the statements below describes the situation in a healthy ecosystem. Briefly describe the corresponding situation in an unhealthy system affected by human disturbance.

1. The amount of CO_2 released into the atmosphere by respiration equals the CO_2 used in photosynthesis.

2. Forest organisms and soil contain a large store of mineral nutrients such as calcium and phosphorus, which are continuously recycled. A small amount of nutrients are lost each year, but this is balanced by the flow of nutrients into the ecosystem.

3. Over time, nutrients from the surrounding land gradually accumulate in a lake and the lake becomes more productive, a process called eutrophication.

4. National parks and wilderness areas are surrounded by buffer zones where sustainable agriculture and other human uses are allowed on a controlled and managed basis.

5. Much of the water that falls on tropical forests is returned to the atmosphere by transpiration.

Testing Your Knowledge

Multiple Choice

1. All of the organisms in a particular area make up
   a. a food chain.
   b. a population.
   c. a community.
   d. a niche.
   e. an ecosystem.

2. When you eat an apple, you are a
   a. primary consumer.
   b. secondary producer.
   c. primary producer.
   d. secondary consumer.
   e. tertiary consumer.

3. An organism's "trophic level" refers to
   a. the rate at which it uses energy.
   b. where it lives.
   c. what it eats.
   d. whether it is early or late in ecological succession.
   e. the intensity of its competition with other species.

4. The relationship between species A and species B is described as commensalism. This means that
   a. both species suffer.
   b. one species benefits and the other species suffers.
   c. both species benefit.
   d. one species benefits and the other species is unaffected.
   e. any of the above is possible in commensalism.

5. Most plants get nitrogen from
   a. nitrates in the soil.
   b. N_2 gas in the air.
   c. proteins
   d. ammonia in the soil.
   e. rainfall

6. The energy for nearly every organism in nearly every ecosystem ultimately comes from
   a. minerals in the soil.
   b. the sun.
   c. heat from the earth.
   d. respiration.
   e. decomposition.

7. Why is a diagram of energy flow from trophic level to trophic level shaped like a pyramid?
   a. Organisms at each level store most of the energy and pass little on.
   b. There are more producers than primary consumers, and so on.
   c. Organisms eventually die as they get older.
   d. Most energy at each level is lost, leaving little for the next.
   e. Secondary consumers are larger than primary consumers, and so on.

8. The main decomposers in an ecosystem are
   a. plants and animals.
   b. bacteria and viruses.
   c. fungi and bacteria.
   d. bacteria and plants.
   e. plants and fungi.

9. Bacteria are especially important in
   a. the water cycle.
   b. the nitrogen cycle.
   c. ecological succession.
   d. the phosphorus cycle.
   e. recycling of energy.

10. The biggest difference between the flow of energy and the flow of chemical nutrients in an ecosystem is
    a. the amount of energy is much greater than the amount of nutrients.
    b. energy is recycled, but nutrients are not.
    c. organisms always need nutrients, but they don't always need energy.
    d. nutrients are recycled, but energy is not.
    e. organisms always need energy, but they don't always need nutrients.

11. Which of the following is *not* a property of a community?
    a. prevalent vegetation
    b. diversity
    c. stability
    d. trophic structure
    e. density

12. The niche of an animal is
    a. the number of individuals of the species the environment will support.
    b. the same as its habitat.
    c. the way the animal fits into its environment.
    d. the specific place in the habitat where the animal lives.
    e. its position in the food chain.

Essay

1. Explain in ecological terms why a given area of farmland can support more people if they eat plants rather than meat.

2. What is decomposition? What are the two major kinds of decomposers in most ecosystems, and how is their role important to the ecosystem?

3. Describe an example of coevolution between predator and prey. Are there ecological relationships other than predator/prey interactions that are shaped by coevolution between two species? Give an example.

4. State whether each of the following is a primary producer, primary consumer, secondary (or higher) consumer, or detritivore: squirrel, oak tree, mosquito, great white shark, moose, cheetah, mushroom, spider, phytoplankton, grass, vulture.

5. There is probably a river or stream near where you live. Describe the place in the global water cycle of that river or stream. Where did its water come from? Where is it going?

6. Trace a carbon atom through the carbon cycle. In what chemical form is carbon in the air? How does a carbon atom enter the food chain? In what chemical form might the carbon atom be obtained by a consumer? What chemical process would put the carbon atom back into the atmosphere?

Applying Your Knowledge

Multiple Choice

1. A lichen is actually composed of two organisms—a fungus and an alga. They depend on each other for survival. The most specific term that describes their relationship is
   a. parasitism.
   b. predation.
   c. commensalism.
   d. symbiosis.
   e. mutualism.

2. Which of the following describes mimicry?
   a. An insect's bright colors warn a predator that it tastes bad.
   b. The mottled pattern on a fish looks like dead leaves on the bottom of a pond.
   c. Two species of mice live in the same area and eat the same kinds of seeds.
   d. A harmless frog resembles a poisonous frog.
   e. Both kangaroo rats and jackrabbits hop erratically when escaping from predators.

3. When goats were introduced to an island off the California coast, the goats lived in the same areas and ate the same plants as the native deer. The deer population dwindled, and the deer finally disappeared. This is an example of
   a. commensalism.
   b. succession.
   c. a food chain.
   d. coevolution.
   e. competitive exclusion.

4. Suppose you wanted to establish a self-sustaining ecosystem by sealing some sterilized soil, water, and air in a glass container with a few organisms. You would be most likely to succeed with which of the following?
   a. aphids, bacteria, and spiders
   b. bacteria and ants
   c. clover, bacteria, and grasshoppers
   d. beetles, fungi, and bacteria
   e. spiders, grasshoppers, and grass

5. After clear-cutting, timber companies cannot afford to wait for the long process of _____ to occur naturally; they plant trees right away.
   a. mutualism
   b. succession
   c. coevolution
   d. decomposition
   e. competitive exclusion

6. In an ecosystem the _____ is always greater than the _____ .
   a. number of primary producers . . . number of primary consumers
   b. biomass of secondary consumers . . . biomass of primary producers
   c. energy used by primary consumers . . . energy used by secondary consumers
   d. biomass of primary producers . . . biomass of primary consumers
   e. energy used by primary consumers . . . energy used by primary producers

7. Under which of the following circumstances would interspecific competition be most obvious?
   a. when resources are most abundant
   b. in a stable, long-established ecosystem
   c. when organisms have quite different ecological niches
   d. among species whose trophic levels are different
   e. when a foreign organism is introduced to a community

8. Milkweed plants produce bad-tasting and poisonous compounds that deter most plant-eaters. But the caterpillars of monarch butterflies are able to eat milkweed leaves without being harmed. In fact, the chemicals obtained from milkweed actually protect the monarch from insect-eating birds. This example illustrates
   a. coevolution.
   b. competitive exclusion.
   c. mutualism.
   d. the effect of a keystone predator.
   e. succession.

9. Two species of cuckoo doves live in a group of islands off the coast of New Guinea. Out of 33 islands, 14 have one species, 6 have the other, 13 have neither, and none has both. What might best explain this? The two species of birds could
   a. be on different trophic levels.
   b. have similar niches.
   c. be detritivores.
   d. have different niches.
   e. be keystone predators.

10. If you wanted to determine whether succession is occurring in a certain area, it might be useful to
    a. determine how many trophic levels are represented.
    b. diagram its food web.
    c. look at its species diversity.
    d. look at old pictures of the area.
    e. measure its primary productivity.

11. All of the following are found in the soil. Which category includes all the others?
    a. fungi
    b. decomposers
    c. scavengers
    d. detritivores
    e. bacteria

12. An ecosystem is unlikely to be limited by the supply of _____ because it is obtained from the air.
    a. water
    b. carbon
    c. phosphorus
    d. calcium
    e. nitrogen

Essay

1. A researcher noted that in many small ponds, fish species A preyed on several smaller species, B, C, D, E, and F. He suspects that A may be a keystone predator in the pond communities. What kind of experiment could you suggest to test whether this hypothesis is valid? If A is a keystone predator, how would you expect the experiment to turn out?

2. A forested area of a new subdivision has been set aside for a park. The developer cuts down most of the trees and plants a lawn. How would you expect the park to compare with the original forest, in terms of prevalent form of vegetation, diversity, trophic structure, and stability?

3. When Gause cultured *Paramecium aurelia* and *Paramecium caudatum* together, *P. caudatum* was invariably driven to extinction (Module 36.2). When *P. aurelia* and *P. bursaria* were cultured together in another experiment, however, both were able to coexist for months. What kinds of similarities and differences must there be between the three species? What do you think would happen if *P. caudatum* and *P. bursaria* were cultured together? Why?

4. Elmer Q. Retchfickle, a civil-defense expert, has developed a design for a fallout shelter to be used during a nuclear war. Elmer proposes to have the inhabitants of the shelter eat mushrooms, which would be grown in the dark on the body wastes of the residents of the shelter. He thinks that this system would allow shelter residents to remain underground indefinitely. Aside from the questionable nutritional balance of an all-mushroom diet, do you see any problems with this plan from an energy standpoint? Explain.

5. An experimental fish farm consists of algae, crustaceans that eat algae, and fish that eat crustaceans. If the algae grow at a rate of 20 kilograms of algae per square meter of pond per year, how big will the pond have to be to harvest 1000 kilograms of fish per year? (Assume that energy content is proportional to weight.)

6. In the mountains of California, Stanford ecologist Craig Heller found that in most areas the least chipmunk lived in sagebrush areas and the yellow pine chipmunk lived in higher areas of mixed sagebrush and pinon pines. If the yellow pine chipmunk was absent from an area, the least chipmunk lived in both the sagebrush and sagebrush-pinon pine zones. If the least chipmunk was absent, the yellow pine chipmunk distribution was unchanged. Explain this difference in terms of the concepts discussed in this chapter.

7. The push to develop and settle the Amazon Basin has resulted in the first extensive contact between many groups of Indians and the outside world. Many of these native people have died from diseases that are seldom fatal in the developed countries, such as measles, chickenpox, and mumps. What kind of coevolution must have occurred between the viruses causing these diseases and their hosts in the developed countries? What will probably occur in the Amazon?

Using Your Knowledge

1. Imagine that you have been chosen as the biologist for the design team for Biosphere III, a self-contained habitat to be assembled in orbit. It will be stocked with organisms you choose, to form a self-sustaining ecosystem that will support you and three other people for 2 years. What are the main functions that you would expect to be performed by the organisms that you choose? What species would you select, and why? How many of each would you start with, and why?

2. Are you a vegetarian? Have you ever followed a vegetarian diet or considered following such a diet? Do you know anyone who is a vegetarian? What are some advantages and disadvantages of a vegetarian diet? What are some reasons for deciding to be a vegetarian? Are there ecological reasons? What are they?

Behavioral Adaptations to the Environment 37

Animal behavior is one of the most accessible areas of biology. To see it, you don't need to dive to a coral reef or travel to the Serengeti. You don't even need a microscope. There are examples around you every day: A spider spins its intricate web. How does it know what to do? A robin sings its song from a fence post. Why does it sing, and why only from certain perches? A moth circles a light bulb, but a cockroach scurries away. Why do they react in different ways? A cat pounces on a ball, and a dog perks up its ears at the sound of its master's voice. What kinds of behaviors are they born with? What kinds of behaviors are learned? These everyday examples are reminders of the interconnectedness of animals and their adaptations to their environments. These are the kinds of topics discussed in this chapter. By studying the chapter, you will gain some tools you can use to understand the animal behavior you see every day.

Organizing Your Knowledge

Exercise 1 (Modules 37.1 – 37.2)

Every animal behavior has a proximate cause, related to the animal's interaction with its environment, and an ultimate cause, related to natural selection and the behavior's survival value. Some aspects of behavior are genetically programmed, and some are acquired through experience. For each of the behaviors described below, briefly give what appear to be both the proximate and ultimate causes of the behavior, and tell whether the behavior seems to be primarily genetic or related to experience.

1. A frog will automatically attempt to catch any small, moving, insect-sized object. It will starve if given live insects that cannot move.

2. Gulls that nest close together in large colonies are able to remember the intricate patterns of spots on their own eggs and return only their own eggs to the nest if they roll out or are removed.

3. On its first flight each morning, a bee circles the hive. After foraging, it can find its way back. If the hive is moved even a short distance after the "orientation flight," the bee will not be able to locate the hive when it returns.

4. Male red-winged blackbirds defend territories (with food and nest sites) and respond aggressively—with threats and attacks—to the red wing patches of other males. If a male's wing patches are painted over, it can enter territories of other males and is ignored by them.

5. A monkey can soon figure out how to unbolt a door and escape from a cage.

6. After it has been stung by a bee, a toad will not try to catch other bees.

7. Monarch butterflies breed in the United States, but they are basically a tropical species, unaccustomed to much cold weather. Monarchs that hatch from eggs in the United States unerringly fly to a small wintering ground in the mountains of Mexico.

8. Newly hatched chicks will run for cover if a cardboard cutout of a hawk "flies" over the barnyard

Exercise 2   (Module 37.3)

An innate (genetically programmed) behavior usually takes the form of an automatic, stereotyped response—a fixed action pattern—to a specific stimulus—called a sign stimulus. The following chart lists some examples of innate behaviors from Module 37.3 and from the previous exercise. Fill in the blanks for species, sign stimulus, and fixed-action pattern.

Species	Sign Stimulus	Fixed Action Pattern
Cuckoo chick	1.	2.
3.	Flying shape	4.
5.	Open mouth of chick	6.
7.	Red wing patches	8.
9.	10.	Catches insect with tongue
11.	12.	Smile
13.	Egg near nest	14.

Exercise 3   (Modules 37.4 – 37.7)

There are several kinds of learning, from simple to complex. State whether each of the following describes an instance of habituation, imprinting, association, imitation, or innovation. Some examples are from the modules; others are new.

_____   1. A bird pecks at a caterpillar and gets a beakful of stinging hairs. After a couple of such experiences, the bird leaves this kind of caterpillar alone.

_____   2. A salmon is found stuck in a pipe, trying to get into a hatchery pond.

_____   3. Gulls learn to identify their own young and keep them separate from the chicks in nearby nests.

_____   4. A hermit crab will at first withdraw into its shell if you tap on the glass of its aquarium. After a few taps it no longer responds.

_____   5. A chimp learns to reach bananas by either hitting them with a stick or stacking up boxes. When the bananas are placed too high for either the stick or the stacked boxes, the chimp uses both the boxes and the stick.

_____   6. If a bird does not hear its species' song shortly after hatching, it will not be able to sing it later.

_____   7. A dolphin learns the commands "Throw ball" and "Retrieve ring." The first time the trainer says, "Retrieve ball," the dolphin does it correctly.

_____   8. A cat learns to come running into the kitchen when it hears the electric can opener.

_____   9. A monkey washes dirt from his food. Other monkeys observe this and start washing their food, too.

_____   10. At first, young ground squirrels dash into their burrows at any unusual sight or sound. Later they learn to run only from real danger.

_____   11. When their mother was killed, ducklings on a farm learned to follow a tractor.

_____   12. Young monkeys see and hear their parents give a warning call when they see a snake. Next time the young monkeys give the warning call, too.

Exercise 4   (Modules 37.3 – 37.8)

To help you better understand innate and learned behaviors, sketch a concept map on a separate sheet of paper. Constructing a concept map is discussed in Chapter 3, Exercise 10.

Exercise 5   (Modules 37.9 – 37.11)

How do biologists figure out whether a worm displays negative phototaxis or a mouse has a bioclock? These modules describe how animals keep track of time and how they orient themselves and find their way. They also describe some experiments and observations that have given us evidence about their rhythmic and navigational abilities. The following describe some observations and hypotheses about animal behavior. Briefly describe the kind of observations or experiments that could be done to test each hypothesis. Some examples are from the chapter, and others are new.

1. Turtles seen off the coast of Brazil are thought to lay their eggs on Ascension Island, in the middle of the Atlantic Ocean.

2. A kind of flatworm lives under rocks in streams and ponds and comes out at night to feed. It is probably negatively phototactic.

3. A crab is most active at high tide and becomes inactive at low tide, roughly 6 hours later. It is thought that this is not due to an internal bioclock but rather that the crab can somehow sense the tidal pull of the moon.

4. Homing pigeons travel even on cloudy nights. It is thought that they do not need to use vision to find their way to the vicinity of their home.

5. Birds become restless in the fall. It is thought that an environmental change other than a drop in temperature makes them ready to migrate.

6. The kangaroo rat is a nocturnal desert rodent. It may possess an internal bioclock that signals when to emerge from its burrow.

Review concepts of feeding and social behavior by filling in the blanks in the following story.

The gray wolf is the top predator of forests, grasslands, and tundra of North America. Wolves are social animals, cooperating in packs to hunt large mammals such as deer, moose, and caribou. Thus they are neither food [1]_____ nor "generalists" but somewhere in between. A typical wolf pack consists of 5 to 20 animals, led by a [2]_____ "alpha" male and his mate. Unlike lone hunters such as the cougar, wolves do not ambush their prey. The pack tracks an animal over long distances, finally exhausting it, attacking from all directions, and eventually dragging their victim down for the kill. The wolf pack is successful in about one chase in ten. Deer often outrun the pack, and moose often fight wolves off.

In the summer, when larger prey are harder to corner, wolves often change their [3]_____ and seek out smaller animals, such as beaver and hares. These smaller prey are easier to catch, and when wolves are alone, they must pursue smaller prey. Why then do wolves work in a pack to go after (usually unsuccessfully) prey sometimes much larger than they are, only to share their kill with other members of the pack? Apparently it is more [4]_____ to cooperate in a pack and tackle large prey than to catch smaller prey working alone. This is an example of [5]_____— feeding behavior that provides maximum energy gain for minimal energy expense.

The wolf pack constantly roams and defends a [6]_____ of from 100 to 1000 square kilometers. Like cheetahs and many other large territorial mammals, wolves use [7]_____ markers to stake their territorial claim. When on the move, they stop to urinate every few minutes. The territory assures a [8]_____ supply for the wolf pack and a safe place to raise their [9]_____.

The dominant feature of social behavior within the pack is a [10]_____ hierarchy. The alpha male leads the pack in chasing prey and in territorial defense. A dominance hierarchy may help to [11]_____ the wolf population. By allocating resources unequally, the hierarchy ensures that at least some of the wolves will receive an adequate food supply. It also minimizes the amount of energy spent in [12]_____ within the pack, energy that could more efficiently be used in hunting, defense, and reproduction.

An essential feature of social behavior is [13]_____, the exchange of information between individuals. Wolves use a varied repertoire of barks, yips, whines, and howls, similar to those of dogs. Body postures signal status in the pack hierarchy. The ears and tail of the alpha male are held erect. An ears-back, tail-down posture signifies submissiveness and lower rank. Thus wolves use three major channels of communication: [14]_____ signals to mark territory, and [15]_____ displays and [16]_____ within the pack.

The mating season, from late fall to late winter, is a time of turmoil for the wolf pack. In most cases, the dominant male and female are the only individuals in the pack who will mate. There is an increase in [17]_____ behavior, both threats and actual fighting, as wolves jockey for position in the hierarchy. Much of the behavior is in the form of [18]_____—mostly growls and posturing. Even if the dominant male and female are not challenged, lower-ranking individuals may test each other, seeking an opportunity to move up in rank.

Unlike many species that engage in a lengthy [19]_____ ritual, mating between the dominant male and female in the pack is performed with the minimum of fuss required to neutralize [20]_____ behavior. The pair may mate several times a day over a 2-week period. In the spring, the female gives birth to four to seven pups. They are given much food and attention by all members of the pack, not just their parents. This helping or self-sacrificing behavior is an example of [21]_____. Many of the individuals in the pack are aunts, uncles, even siblings of the pups. Their caring behavior could be explained in terms of [22]_____ selection. Even though most of the adults in the pack do not get to mate, their sharing and helping improve the chances that copies of their [23]_____ will be passed on. The pups nuzzle and lick the muzzles of adults, a signal that causes the adults to regurgitate food. (This behavior is also seen between adults, even those who are not related. This might be [24]_____ altruism, where the favor of food sharing may be repaid later by the beneficiary.) In 3 to 5 months, the cubs are able to travel with the pack. Later some will leave the pack, but most stay. Thus, social ties in the wolf pack are also family ties.

Exercise 7　(Module 37.19)

There are three main channels of animal communication: chemicals (scents), movements (visual displays or touch), and sounds (vocalizations, etc.). Communication in a number of species is described in several of the modules in this chapter. A list of species is given below. Most are from the chapter; you are probably familiar with most of the others. Place the name of each species on the chart to show how you think that species communicates. For example, grasshoppers communicate almost totally by means of sound; "grasshopper" goes in the "sounds" corner. House cats use sounds, movements, and scents; they are placed near the center of the triangle, roughly equally distant from the names of all three communication channels. Add the following to the chart: **honeybee, ring-tailed lemur, wolf, cricket, human, domestic dog, common loon, frog,** and **white-crowned sparrow**. It is not important in this exercise for you to be sure of the "right" answer. The important thing is to think about the variety of forms of animal communication.

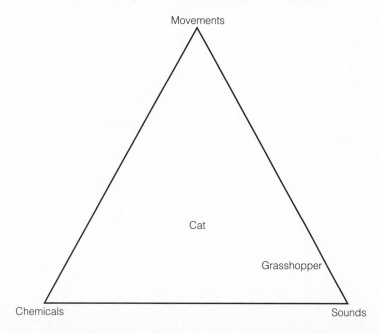

Review some of the vocabulary of animal behavior by matching each of the phrases on the right with a term from the list on the left. Each answer is used once.

A. Reciprocal altruism

B. Optimal foraging

C. Sociobiology

D. Kin selection

E. Courtship ritual

F. Social behavior

G. Culture

H. Territory

I. Agonistic behavior

J. Dominance hierarchy

K. Altruism

L. Search image

M. Communication

____ 1. Study of the biological basis of social behavior

____ 2. Signaling or exchange of information

____ 3. Threats or combat

____ 4. May reinforce inherited social behavior in humans

____ 5. When self-sacrifice of relatives leads to altruism

____ 6. Helps an animal look for food

____ 7. Most efficient feeding behavior

____ 8. Help or sacrifice repaid later

____ 9. Minimizes agonistic behavior before mating

____ 10. Interaction between individuals, usually of same species

____ 11. Area defended against members of same species

____ 12. Self-sacrifice

____ 13. Ranking by social interactions

Testing Your Knowledge

Multiple Choice

1. Which of the following is considered to be the most sophisticated kind of learning?
   a. innovation
   b. kinesis
   c. habituation
   d. imprinting
   e. trial and error

2. Scientists think that some birds may find their way during migration by
   a. imprinting on faint odors.
   b. listening to high-pitched sounds.
   c. sensing the magnetism of the Earth.
   d. seeing wavelengths of light that we cannot see.
   e. sensing the gravitational pull of the sun and moon.

3. Kineses and taxes are simple, automatic movements in response to stimuli. Movement of a moth toward a light is positive phototaxis. What is the difference between kineses and taxes?
   a. Taxes are directed toward a stimulus, kineses away.
   b. Kineses are innate, and taxes are learned.
   c. Kineses are more random, and taxes are more directed.
   d. Kineses are performed much faster than taxes.
   e. Taxes are innate, and kineses are learned.

4. Which of the following is a fixed-action pattern?
   a. A goose rolls an egg back into its nest.
   b. A hamster becomes active at the same time each evening.
   c. A wolf tracks its prey.
   d. A robin eats a distasteful bug, spits it out, and never eats one again.
   e. A sea gull returns to the same island breeding grounds each year.

5. Ducklings learn to follow their mothers. This is an example of
   a. habituation.
   b. insight.
   c. altruism.
   d. imprinting.
   e. agonistic behavior.

6. Which of the following appears to be the primary method by which worker bees communicate the location of nectar to other members of the colony?
   a. They secrete a particular combination of chemical signals.
   b. They produce complex, high-pitched sounds.
   c. They fly to the nectar, and other bees follow.
   d. They perform a sequence of movements in the hive.
   e. They lay down a chemical trail to the source of food.

7. Birds sing, squirrels chatter, jaguars deposit their feces in prominent places, and wolves urinate to
   a. warn other members of the group of danger.
   b. mark their territories.
   c. attract mates.
   d. assert dominance.
   e. defend themselves from predators.

8. In ____ , an animal learns to associate one of its behaviors with reward or punishment.
   a. anthropomorphism
   b. trial-and-error learning
   c. fixed-action pattern learning
   d. habituation
   e. imprinting

9. Which of the following have researchers found particularly difficult to explain?
   a. anthropomorphism
   b. altruism
   c. habituation
   d. dominance
   e. imprinting

10. When animals engage in ____ , they often perform displays that make them look as large and dangerous as possible.
    a. courtship rituals
    b. altruism
    c. kin selection
    d. kineses
    e. agonistic behavior

11. The key idea of sociobiology is that
    a. human social behavior is the same as animal social behavior.
    b. social behavior evolves.
    c. little behavior is inherited; most is learned in social situations.
    d. almost all human and animal behavior is preprogrammed.
    e. animal behavior and human behavior are so different they cannot be compared.

12. Anthropomorphism is
    a. the study of animal behavior under controlled laboratory conditions.
    b. a productive and helpful point of view for studying animal behavior.
    c. the study of human beings' closest relatives—monkeys and apes.
    d. attributing human characteristics to animals.
    e. communication by means of visual signals.

Essay

1. Describe the functions of imprinting in the lives of birds. How does it increase fitness?

2. What are the advantages of a dominance hierarchy to a high-ranking individual? What are the advantages of the hierarchy to the group?

3. Describe four functions of courtship rituals.

4. Explain why communication is essential to social behavior, and give five examples of animal communication.

5. Explain three methods birds use to find their way during their long-distance migrations.

6. What are the advantages of fixed-action patterns over learned behavior? In what situations do fixed-action patterns seem to be especially important?

7. Name three species of animals that cooperate in social groups, and give three general benefits or advantages of social cooperation in these animals.

Applying Your Knowledge

Multiple Choice

1. A frog may at first be startled by tree branches swaying in the wind, but it soon gets used to these kinds of unimportant changes in its environment. This is an example of
   a. a fixed-action pattern.
   b. imprinting.
   c. altruism.
   d. habituation.
   e. trial-and-error learning.

2. Bees can see colors we cannot see, and they can detect minute amounts of chemicals we cannot sense. Unlike many insects, bees cannot hear very well. An ethologist would probably give which of the following as the ultimate explanation of their poor hearing?
   a. Bees are too small to have functional ears.
   b. Good hearing must not contribute much to a bee's reproductive success.
   c. If a bee could hear, its tiny brain would be swamped with information.
   d. This is an example of altruism.
   e. If bees could hear, the noise of the hive would drive them crazy.

3. Chimpanzees usually maintain their rank in the dominance hierarchy with charging displays. Sometimes actual combat occurs briefly, but usually aggression is ritualized. Why don't chimps fight more?
   a. Chimps are not strong enough to inflict much injury.
   b. Dominant females keep order and prevent fights from escalating.
   c. Natural selection favors avoiding injury and saving energy.
   d. Chimps spend most of their time mating; they don't have time to fight.
   e. Chimps are too hard-pressed finding enough to eat to fight much.

4. Western fence lizards are sometimes called bluebellies because the undersides of the males are bright blue. If one male invades another's territory, the territory owner protests vigorously, doing "pushups" to display his belly and, if necessary, pushing the invader out. If an intruder's blue belly is first painted brown, he is completely ignored by the territory owner. The blue belly of the intruder apparently acts as a _____ , triggering _____ .
   a. search image . . . optimal foraging
   b. sign stimulus . . . altruism
   c. sign stimulus . . . agonistic behavior
   d. search image . . . altruism
   e. fixed-action pattern . . . imprinting

5. An aquaculture facility hatched salmon eggs and released young fish into a river leading to the ocean. The fish fed and grew in the ocean, and in a few years they returned to the facility. Because the number of returning fish was low, a scientist suggested that the facility add a chemical to the water draining from the fish ponds into the river. She hoped the chemical would increase the number of returning fish in a few years by
   a. decreasing their agonistic behavior.
   b. enabling them to imprint on the facility.
   c. stimulating habituation in the fish.
   d. improving the salmons' search image.
   e. promoting optimal foraging.

6. Every morning Brian turns on the light in the room and then feeds the fish in his aquarium. After a couple of weeks of this routine, he noticed that the fish came to the surface to feed as soon as he turned the light on, whether or not any food was present. This illustrates
   a. habituation.
   b. positive phototaxis.
   c. imprinting.
   d. association.
   e. innovation.

7. An animal would be most likely to show a circadian rhythm with regard to which of the following?
   a. sleeping
   b. fighting
   c. building a nest
   d. fleeing from a predator
   e. migrating

8. Which of the following best illustrates optimal foraging?
   a. A robin will repeatedly attack any red object near its territory.
   b. Musk oxen will form a circle to fend off a wolf attack.
   c. Bats emerge to feed at about the same time each night.
   d. A blackbird will warn others in the flock if it senses danger.
   e. A sunbird will more fiercely defend flowers that produce more food.

9. Which of the following suggests that some animals have an internal bioclock?
   a. Some animals can sense changes in the Earth's magnetic field.
   b. A crab that is moved away from the shore can still sense the tides.
   c. An animal kept in constant light stops ovulating.
   d. Many animals become active at dawn and settle down at sunset.
   e. Rats kept in constant darkness show a daily rhythm of activity.

10. Which of the following sayings best summarizes the idea of reciprocal altruism?
    a. "A rolling stone gathers no moss."
    b. "You scratch my back and I'll scratch yours."
    c. "A penny saved is a penny earned."
    d. "A bird in the hand is worth two in the bush."
    e. "Beauty is in the eye of the beholder."

Essay

1. Some animals, such as wolves and lions, hunt in groups, while others, like foxes and house cats, are loners. Every method of getting food involves trade-offs; there is no one best way to

do it. What do you think are the advantages of hunting in a group? The disadvantages?

2. If cuckoos are raised by foster parents of a different species (Module 37.3), how do you think they are able to later find mates and engage in appropriate courtship and mating behaviors? How could you test this hypothesis?

3. Bats are the major predators of a species of moth. When one of these moths hears the high-pitched sound of a bat's sonar, it folds its wings and dives to the ground. How would an ethologist describe the proximate cause of the moth's behavior? The ultimate cause?

4. European birds called wagtails eat insects. During the winter, when food is scarce, a wagtail may defend a territory, on which it captures an average of 20 insects per hour. Or the wagtail may join a flock that ranges widely over the countryside. A bird in a flock averages 25 insects per hour. If this is the case, why don't the birds always travel in flocks? Discuss in terms of energy gain and expenditure.

5. Cancer patients often suffer nausea and loss of appetite as a side effect of the powerful drugs used in chemotherapy. They have a particularly tough time eating foods that they have in the past eaten just before drug treatment, even if the foods are subsequently given to them days or weeks after the effects of the drugs have worn off. Explain this "food aversion" in terms of your knowledge of learning in animals. How might this kind of learning be helpful to an animal?

6. Researchers have found that workers who must rotate from one shift to another have a much easier time staying up to adjust to a later shift than getting up earlier to move to an earlier shift. What do we know about animal and human circadian rhythms that might help explain this?

7. Many bird species use special alarm calls to signal the approach of danger. Biologists ask, "What possible advantage to a bird is there in warning other members of the flock? Isn't the bird that gives the warning call putting itself in danger? If so, how could this behavior possibly continue to exist? How could it improve a bird's fitness? Wouldn't the genes for warning other members of the flock be eliminated by natural selection?" The fact is, alarm calls and other forms of altruism—self-sacrifice—continue to exist. How might altruism improve the alarm-giver's fitness? List the two general kinds of explanations that biologists give for altruistic behavior.

8. Most people think babies are "cute." They like to hold them, play with them, and protect them. How might a sociobiologist explain this?

Using Your Knowledge

1. Sometimes people ask, "Why are scientists wasting time and money studying animal behavior? Why aren't we putting our resources into research more relevant and useful to humans?" What do you think? Is animal behavior research relevant or useful? Can you cite specific examples?

2. Chimpanzees, ants, lions, and wolves sometimes attack other groups and kill individuals of their own species, but human beings are the only animals that carry out "warfare" in an organized way on a mass scale. Does our knowledge of animal behavior offer any insight into human aggression and violence? Are humans more aggressive than other animals? Do other animals control their aggression in some way that we have left behind? What role does our technology play in the scale and ferocity of our wars? Or is knowledge of animal behavior irrelevant when it comes to the subject of war?

3. As mentioned in the introduction to this study guide chapter, animal behavior is one of the aspects of biology most easily observed in everyday life. Just for fun, keep a list of all the examples of animal behavior that you observe in a day. What kinds of behavior did you see? How many species were included?

Human Evolution and Its Ecological Impact 38

Their footprints, preserved in 3.7-million-year-old volcanic ash, tell a story: Two upright-walking early humans walked across the dusty African plain. Their tracks cross, and are crossed by, the tracks of other animals. Some may have been their prey, and some perhaps preyed on them. At one point the walkers stopped and perhaps looked back. Did they hear something? What were they thinking? More than any of the animals around them, these small creatures lived by their wits. Did they have a concept of who they were? Did they wonder where they came from? These early humans were our ancestors, but they certainly had little idea where their path had started or where it might lead. Perhaps they did not wonder about their origins, but we do. We also speculate about our effect on the Earth and how it will shape our future. The origin of human beings and our influence on the ecology of our planet are the subjects of this chapter.

Organizing Your Knowledge

Exercise 1 (Module 38.1)

Summarize this module in a sentence of *exactly* 25 words.

Exercise 2 (Module 38.2)

Review the characteristics of primates. Then explain which of these characteristics enables you to do each of the following.

1. Reach in and find a quarter among a pocketful of coins.

2. Catch a ball that is thrown to you.

3. Throw the ball back.

4. Pick up a postage stamp from your desk top.

5. Do a handspring.

6. Thread a needle.

7. Shoot a basketball.

Exercise 3 (Modules 38.2 – 38.4)

Complete this diagram to show the relationships among the various kinds of primates. Fill in the blanks, using the following names: **apes, Old World monkeys, primates,** **tamarins, bonobo, prosimians, hominids, langurs, orangutan, New World monkeys, aus-tralopithecines, lemurs, baboon, modern human, chimpanzee, loris, anthropoids, mon-keys, gorilla,** and **gibbons.**

Exercise 4 (Modules 38.4 – 38.7)

Review human evolution by matching each of the phrases on the right with a name from the list on the left. Answers may be used more than once.

A. *Homo sapiens*

B. *Homo habilis*

C. Neanderthal

D. *Australopithecus africanus*

E. *Homo erectus*

F. *Australopithecus afarensis*

G. *Australopithecus robustus*

_____ 1. *Homo sapiens* type that disappeared

_____ 2. First hominid known to walk upright

_____ 3. First hominid associated with stone tools

_____ 4. "Lucy"

_____ 5. First to use fire

_____ 6. Thought to be the first hominid to leave Africa

_____ 7. Left footprints 3.7 million years ago

_____ 8. Probably between *Australopithecus* and *Homo erectus*

_____ 9. Some believe they were intermediate between *Homo erectus* and modern humans

_____ 10. Modern humans

_____ 11. An early relative, but not thought to be our direct ancestor

_____ 12. Another early relative not thought to be our direct ancestor

_____ 13. Lived 1.6 million to 300,000 years ago

These two evolutionary trees show alternative interpretations of the place of *Australopithecus afarensis* and Neanderthals in our ancestry. Fill in the blanks in the two trees, using the following names: *Homo sapiens, Homo habilis,* Neanderthal, *Australopithecus africanus, Homo erectus, Australopithecus afarensis, Australopithecus robustus.*

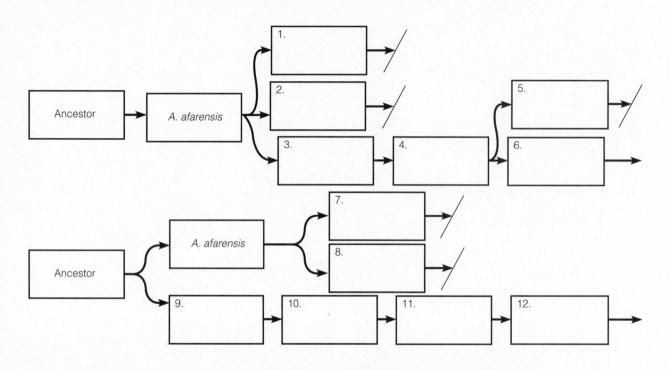

Exercise 6   (Modules 38.8 – 38.11)

Our culture has enabled humans to alter nature. Which of the three stages in cultural evolution is reflected in each of the phrases below?

_____ 1. Permanent settlements first arose

_____ 2. The current stage in cultural evolution

_____ 3. Still practiced by Australian aborigines and African Bushmen

_____ 4. Probably caused the extinction of saber-toothed cats and other large animals

_____ 5. Began 10 thousand to 15 thousand years ago

_____ 6. Led to significant acceleration in population growth

_____ 7. Turned areas of the Middle East and North Africa into desert

_____ 8. People migrated to cities in search of factory jobs

_____ 9. The way of life of *Homo habilis* and *Homo erectus*

_____10. First allowed people to specialize in different occupations

Exercise 7   (Modules 38.12 – 38.15)

Today life on Earth faces an environmental crisis. Summarize some current environmental problems by filling in the remaining blanks in the chart below.

Problem/Cause	Consequences	Solutions
Sulfur and nitrogen emissions from burning coal	1.	2.
3.	Global warming, climate change, flooding	4.
Oil spills	5.	6.
7.	Death of top predators—ospreys, eagles, etc.	8.
Destruction of tropical forests	9.	10.
11.	12.	Reduce use of chlorofluorocarbons
13.	Decline of whale populations	14.

Exercise 8   (Modules 38.12 – 38.16)

You have been asked to speak to a high school biology class on the topic of "The Global Environmental Crisis." At the end of your talk, the students have the following questions. Give a short, succinct answer to each question.

1. Why do the industrialized countries have to worry about population growth? Aren't populations growing faster in developing countries?

2. Is destroying tropical rain forests a more serious problem than cutting down forests in the United States? Why?

3. Why should I care what happens in the rain forest? How can it affect me?

4. I heard that the greenhouse effect is a good thing. Is that so? If so, why are people so worried about it?

5. A friend told me that burning fossil fuels is destroying the ozone layer. Is this true?

Testing Your Knowledge

Multiple Choice

1. The lemurs of Madagascar are examples of
   a. Old World monkeys.
   b. anthropoids.
   c. prosimians.
   d. New World monkeys.
   e. apes.

2. Which of the following spends the most time in the trees?
   a. gibbon
   b. baboon
   c. gorilla
   d. orangutan
   e. chimpanzee

3. The human population of the Earth is now more than
   a. 1 billion.
   b. 3 billion.
   c. 5 billion.
   d. 8 billion.
   e. 10 billion.

4. Which of the following is least likely to be a direct ancestor of modern humans?
   a. *Homo habilis*
   b. *Australopithecus afarensis*
   c. *Homo erectus*
   d. *Australopithecus africanus*
   e. Neither a nor d is thought to be our direct ancestor.

5. The earliest human fossils so far discovered are _____ years old.
   a. 5–10 million
   b. 100–500 thousand
   c. 3–4 million
   d. 1–2 million
   e. 60–65 million

6. Fossil evidence suggests that humans evolved in
   a. Europe.
   b. South America.
   c. Asia.
   d. the Middle East.
   e. Africa.

7. Which of the following steps in human evolution appears to have occurred first?
   a. development of language
   b. large brain
   c. erect stance
   d. development of culture
   e. use of tools

8. About _____ of all species of plants and animals are thought to live in tropical rain forests.
   a. 10%
   b. 25%
   c. 50%
   d. 80%
   e. 97%

9. Most endangered species are victims of
   a. greenhouse warming.
   b. habitat destruction.
   c. overhunting.
   d. competition with introduced species.
   e. pesticide pollution.

10. Which were the first hominids to use fire?
    a. *Homo erectus*
    b. *Homo habilis*
    c. *Australopithecus africanus*
    d. the Neanderthals
    e. *Australopithecus afarensis*

Essay

1. List the specialized characteristics of primates.

2. Which primate is the closest living relative of humans? What is the evidence for this close relationship?

3. Describe the characteristics that separate each of the paired groups:
   a. anthropoids—prosimians
   b. Old World monkeys—New World monkeys
   c. monkeys—apes
   d. apes—humans

4. What were the two most important physical changes that occurred during human evolution?

5. The industrial revolution of the last three centuries has had major impacts on the biosphere. Did the earlier scavenger-gatherer-hunter and agricultural stages of cultural evolution cause any widespread, lasting environmental changes?

6. Compare the rates of human biological and cultural evolution over the last 100,000 years. How has the evolution of culture contributed to our success? How has it caused problems?

Applying Your Knowledge

Multiple Choice

1. Which of the following is least closely related to the others?
   a. human
   b. lemur
   c. gorilla
   d. squirrel monkey
   e. baboon

2. At the zoo, Tom saw a species of primate he had never even heard of before. He said, "It was called a white-faced saki. It had long, dark, spiky fur, a long fluffy tail, forward-facing brown eyes, a white forehead, and yellow cheeks." Based on this information, the white-faced saki could *not* be
   a. an ape.
   b. a prosimian.
   c. a New World monkey.
   d. an Old World monkey.
   e. an anthropoid.

3. Why does the birth of a German baby have a greater impact on the environment than the birth of an Indonesian baby?
   a. The population of Germany is smaller, so one baby is a bigger increase.
   b. Europe is closer to environmental disaster than Indonesia.
   c. A German baby will use more resources and produce more pollution.
   d. An Indonesian baby is more likely to die young.
   e. A German is more likely to travel widely.

4. Trina saw an old Tarzan movie on television. The movie supposedly took place in Africa, but Trina easily spotted that it was not really filmed there. Which of the following could have tipped her off?
   a. Chimps like Cheetah do not live in Africa.
   b. The monkeys in the jungle all had prehensile tails.
   c. There were no prosimians shown in the forest.
   d. Tarzan wrestled with a gorilla.
   e. Only Old World monkeys were shown.

5. Which of the following categories includes all of the others?
   a. apes
   b. Old World monkeys
   c. anthropoids
   d. hominids
   e. New World monkeys

6. Which of the following probably coexisted for a time with early modern humans?
   a. *Australopithecus robustus*
   b. *Homo erectus*
   c. *Australopithecus afarensis*
   d. *Homo habilis*
   e. *Australopithecus africanus*

7. Environmentalists suggest that driving more efficient cars, which produce less CO_2, might reduce global warming. It is uncertain what effect this would have because
   a. it takes hundreds of years for CO_2 to affect temperature.
   b. burning fossil fuels actually adds very little CO_2 to the atmosphere.
   c. cutting of tropical forests might compensate for added CO_2 anyway.
   d. scientists aren't sure that CO_2 and temperature rise are really related.
   e. more efficient cars actually produce more ozone.

8. Much more is known about the early agriculture of the Middle East than other areas where agriculture originated because
   a. farmers there kept better records of their activities.
   b. Middle Eastern farmers stayed in one place for longer periods.
   c. traces of early farms were soon obliterated by cities in other areas.
   d. agriculture started there much more recently.
   e. farmers there grew all of the food crops we grow today.

9. Which of these countries would you suppose consumes the most aluminum per person?
   a. France
   b. Mexico
   c. Indonesia
   d. India
   e. Tanzania

10. Paleontologists have found fossils of several kinds of australopithecines. What is their place in human evolution?
    a. They are all thought to be ancestors of modern humans.
    b. They are all extinct side branches of the human family tree.
    c. Some evolved into humans, others into apes.
    d. They are not thought to be related to modern humans.
    e. Some may have been our ancestors, others offshoots of our family tree.

Essay

1. The top predators in an ecosystem are often regarded as "indicator species"—indicators of the health of the entire system. For this reason, researchers monitor jaguars in the rain forest of Central America and spotted owls in the coniferous forest of the Pacific Northwest. Why do top predators reflect the integrity of their ecosystems?

2. Many factors—both natural and caused by humans—affect the greenhouse effect. State whether you think each of the following would tend to increase or decrease the possibility of global warming, and why.
    a. Conversion from coal-fired to solar and hydroelectric power plants
    b. Large-scale tree-planting programs
    c. Deforestation
    d. Increasing the average efficiency (gas mileage) of automobiles
    e. Slowing world population growth
    f. Warming of climate, speeding up decomposition of organic matter

3. Finding which of the following would require anthropologists to change their current ideas about human evolution the most? Why?
    a. A large-brained, knuckle-walking hominid 2.5 million years old
    b. A small-brained, bipedal hominid 4.0 million years old

4. Marine algae need iron in trace amounts, and their growth seems to be limited, especially in the Antarctic Ocean, by the lack of iron in seawater. Biologists have suggested that fertilizing the sea with iron compounds might help reduce global warming. How would this work? Do you think we ought to try it? Why or why not?

5. What are some of the debated and unanswered questions about human evolution?

Using Your Knowledge

1. What aspects of your lifestyle might contribute to global warming? Would you be willing to change the way you live to lessen your contribution to the problem? What kinds of changes could you make that would lessen global warming?

2. Some experts suggest that we are unable to deal effectively with environmental problems because our Stone Age minds are unable to see and cope with the dangers created by our accelerating technology. Neurobiologist Robert Ornstein and ecologist Paul Ehrlich present this idea in their book *New World, New Mind.* They think that evolution has shaped our brains to be better at identifying obvious, immediate dangers like saber-toothed tigers than subtle changes in the environment like pollution. Does anthropological evidence support this view? Can you think of other ways in which our behavior lags behind our technology? How might we get around this limitation?

3. What do you think human life will be like if the Earth's population reaches 8 billion in the next quarter century? In what ways might your life change? How might life change for a citizen of Bangladesh or Bolivia?

Answers

Chapter 1

Organizing Your Knowledge

Exercise 1: 1. A 2. CDE 3. B 4. E 5. D 6. D 7. E 8. ABC 9. AB 10. C 11. A

Exercise 2: 1. related 2. fossils 3. Charles Darwin 4. ancestor 5. adaptation 6. population 7. varied 8. heritable 9. predators 10. reproduce 11. offspring 12. natural selection 13. variations 14. environmental 15. reproductive 16. species 17. evolution 18. time 19. theme

Exercise 3: 1. observation 2. question 3. hypothesis 4. test 5. experimental group 6. control group 7. variable 8. controlled experiment 9. prediction 10. repetition 11. theory

Exercise 4: 1. All of the organisms of a particular area, plus the nonliving environment that affects them 2. Community 3. A group of interbreeding organisms of one species 4. Organism 5. Organ system 6. A body structure consisting of several tissues, performing particular functions 7. Tissue 8. Cell 9. Chemical

Exercise 5: A college student is a living thing. She has *precise structural organization*—everything from the arrangement of her limbs and facial features, to the cellular organization of the brain, to the structure of individual cells, cell membranes, and molecules. These features were shaped during fetal development, childhood, and adolescence, and they reflect the student's *capacity for growth and development*. She *takes in energy* in her food *and uses it* to perform work—walking, talking, and studying. In performing these actions, she *responds to stimuli* from the environment. Like other emergent properties, her *ability to reproduce* results from the precise arrangement of her molecules, cells, and organ systems. All of these human characteristics have been shaped by natural selection at the population level, reflecting our *ability to evolve*.

Exercise 6:

Crossword answers (across and down): LIFE, EMERGENT, CHEMICAL, FUNCTION, ECOSYSTEM, SPECIALIZED, DNA, GENE, NUCLEUS, BACTERIA, SELECTION, REGULATION, REPRODUCTION, DECOMPOSE, ORGANISM, SYSTEMS, CELLS, PLANTS, SHIELD

Exercise 7: See Module 1.14 for ideas. Recall that science is seeking to explain nature through observation and experiment. Technology is the application of scientific knowledge.

Testing Your Knowledge

Multiple Choice: 1. c 2. b 3. d 4. d 5. c 6. d 7. e 8. e 9. a 10. b 11. a 12. c

Essay: 1. Bacteria are in Kingdom Monera. They differ from other forms of life in that their cells lack a nucleus. Protists—algae and protozoans—have more complex cells (like plants, animals, and fungi), but most of them are single-celled. Plants, animals, and fungi are all multicellular. Plants, such as pine trees and rosebushes (Kingdom Plantae), differ from the others in that they are photosynthetic and have cells with rigid walls made of cellulose. The fungi of Kingdom Fungi, such as mushrooms and molds, decompose the remains of dead organisms and absorb the nutrients. Animals (Kingdom Animalia), from worms to human beings, have cells without rigid walls, eat other organisms, and most move actively in search of food.

2. A hypothesis is an idea, an educated guess proposed as a tentative explanation for a spe-

cific phenomenon, which may be tested in an experiment. A theory is a broad explanatory idea that is supported by a large body of evidence. Thus a hypothesis is more narrow and much less certain than a theory.

3. Science deals with questions about natural phenomena that have natural causes. Some examples of questions that potentially could be answered by science are, What particles make up an atom? What makes a cell divide? How do birds find their way when they migrate? Science does not deal with questions that are not concerned with the natural world and its laws: What is the purpose of human existence? Why is there a universe? These are valid questions for philosophy and religion.

4. Chemical structure—the arrangement of atoms and molecules—is the fundamental level that underlies all the properties of life. The specific properties of a cell, organ, or organism arise from the combinations and arrangements of the molecules of which the organism is composed. The structure of DNA—most notably the particular sequence of nucleotides that make up the DNA molecule—provides a blueprint for growth and development and information that cells need to make the molecules involved in their structure, energy use, and response to stimuli. Replication, or copying, of DNA underlies the organism's ability to reproduce.

Applying Your Knowledge

Multiple Choice: 1. e 2. c 3. d 4. b 5. c 6. c

Essay: 1. The squirrel depends on the oak tree for shelter, an escape route, and food. It may live in a cavity hollowed out by a woodpecker. It buries nuts, adding organic matter to the soil that might be consumed by insects and worms. Sometimes the buried acorns sprout and grow into trees. Hawks and owls prey on squirrels, and various parasites depend on them. Chipmunks sometimes compete with squirrels for food.

2. The ancestors of beavers may have been rat-like animals that occasionally ventured into the water to escape from predators or obtain food. There may have been heritable variation among the rodents in the population with regard to the shapes of their tails and feet. Those rodents with more flattened tails and more webbed feet were able to swim more efficiently, thus escaping predation and gathering food more effectively, making them more likely to leave offspring than more ratlike individuals. Their

offspring inherited their flattened tails and webbed feet, and the proportion of individuals with flattened tails and webbed feet increased over time, until eventually all the individuals in the population looked like beavers. Natural selection occurred as heritable variations were exposed to environmental factors that favored the reproductive success of some individuals over others.

3. Jason's test of MegaGro was not scientifically valid because it was not a controlled experiment. There are any number of factors that might have caused this year's crop of tomatoes to exceed last year's—rainfall, seed type, sunny weather, location of plants, and so on. Jason needs to plant two groups of tomato plants side by side at the same time. Only one factor—the amount of MegaGro—should be allowed to differ between the experimental (MegaGro) and control (no MegaGro) plants. If one group of plants then produces more tomatoes than the other, Jason can safely conclude that the fertilizer made the difference.

4. Set up an obstacle course for oilbirds to fly through. Divide the birds to be tested into two groups. Plug the ears of the birds in one group so that they cannot hear. Then compare how these experimental birds navigate the obstacle course with the performance of control birds that can hear.

5. The long, narrow, flexible fingers and sensitive nerve endings in the fingertips enable us to grasp and manipulate objects. The flat shape of a leaf and its orientation perpendicular to the rays of the sun allow it to capture sunlight efficiently for the process of photosynthesis. The hawk's strong, curved, sharp beak is used for tearing meat. A frog's long, muscular legs, with their webbed feet, are used in jumping and swimming.

6. You are an organism composed of organ systems such as the circulatory and skeletal systems. Each system is made up of organs, such as the heart or an artery. An organ is made of tissues, and a tissue is made of cells, the smallest living units. Each cell is made up of molecules, which are made of atoms. A cell is alive; it uses energy, responds to stimuli, grows, and reproduces. The molecules that make up a cell are not alive, and they do not have all these properties.

7. Living things possess precise structural organization. They take in energy and use it to perform work, and they are able to respond to

stimuli. They also grow, develop, reproduce, and evolve. Some nonliving things possess some of these properties, but only living things are characterized by all of them.

8. Human, chimpanzee, and goldfish DNA are all made of the same chemical subunits; they all have the same fundamental double-helix chemical structure; and they all use the same alphabet to encode hereditary information. The four nucleotides in the DNA of chimps and humans are arranged in similar sequences and spell out similar genetic messages. The nucleotides of human DNA and goldfish DNA are less similar in sequence and spell out rather different messages.

Chapter 2

Organizing Your Knowledge

Exercise 1: 1. holistic 2. reductionist 3. reductionist 4. holistic

Exercise 2: 1. cell (2) 2. system (5) 3. tissue (3) 4. molecule (1) 5. organism (6) 6. organ (4)

Exercise 3: 1. Mg T 2. O L 3. Zn T 4. H L 5. Cu T 6. I T 7. C L 8. Ca M 9. P M 10. N L 11. Na M 12. Fe T

Exercise 4: 2. N 7 14 7 7 7 3. Cl 17 35 17 18 17 4. O 8 16 8 8 8 5. O 8 17 8 9 8

Exercise 5: 1. The calcium atom gives away 2 electrons, one to each of 2 Cl atoms, producing a Ca^{++} ion and 2 Cl^- ions. These ions form $CaCl_2$.

2. The nitrogen atom shares its electrons with 3 hydrogen atoms, forming ammonia, NH_3.

Exercise 6: 1. chemical 2. cohesion 3. surface tension 4. oxygen 5. hydrogen 6. share 7. covalent 8. nonpolar 9. electronegative 10. negative 11. positive 12. polar 13. neutral 14. negatively 15. positively 16. hydrogen 17. four 18. solid 19. liquid 20. gas 21. solvent 22. aqueous 23. solutes 24. heat 25. temperature 26. cup 27. bathtub 28. hydrogen bonds 29. less 30. releases 31. absorbs 32. releases 33. fast 34. evaporate 35. boiling point 36. evaporative 37. less 38. floats

Exercise 7: 1. 4 2. 7 3. 4 4. 7 5. 6 6. 11 7. 2 8. 3 9. 13 10. 8 11. 6 12. 2

Exercise 8: $2H_2O$ (reactant) ——> $2H_2O + O_2$ (products)

Exercise 9:

Testing Your Knowledge

Multiple Choice: 1. b 2. c 3. e 4. d 5. b 6. b 7. e 8. a 9. c 10. c 11. e 12. b

Essay: 1. See Table 2.2A in the text.

2. Iron is an element, and there are atoms of iron. Water is a compound, containing two elements. The smallest particle of water is a molecule, composed of two hydrogen atoms and one oxygen atom. There are no atoms of water; if a water molecule is broken up into hydrogen and oxygen atoms, it is no longer water.

3. Temperature is the average speed of molecules in a body of matter. Heat is the total amount of energy due to movement of molecules. Molecules in a teakettle move faster, but there are many more in a swimming pool, so even though the molecules in a swimming pool are moving slower, their total amount of energy is greater.

4. Acid precipitation forms when air pollutants from burning fossil fuels—sulfur and nitrogen oxides—combine with water vapor in the air to form sulfuric and nitric acids. Acids can break down chemicals in animals, and low pH can interfere with chemical processes, especially in the eggs and young of aquatic organisms. Acid in the soil can lead to mineral imbalances that affect plants.

5. The shared electrons in a water molecule are pulled more strongly toward the oxygen atom than the hydrogen atoms. This makes the oxygen atom partially negative and the hydrogen atoms partially positive, causing the water mole-

cule to be polar. The + and − charged regions on each water molecule are attracted to oppositely charged regions on adjacent molecules; these attractions are called hydrogen bonds. When water is heated, the heat energy first disrupts the H bonds and then makes the molecules move faster. Because heat is absorbed as bonds break, water can absorb and store much heat while warming only a small amount.

6. Water evaporates when the fastest molecules move so rapidly that they escape from the liquid and form water vapor. This leaves the molecules still in the liquid with a lower average speed, and as a result the remaining liquid has a lower temperature.

Applying Your Knowledge

Multiple Choice: 1. a 2. e 3. a 4. c 5. e 6. d 7. b 8. a 9. c 10. a

Essay: 1. It will form 2 covalent bonds, because it needs to share 2 electrons with other atoms to have a complete outer shell of 8 electrons. H_2S.

2. Give a plant water containing a radioactive isotope of oxygen H_2O, and look to see whether the sugar made contains the radioactive oxygen $C_6H_{12}O_6$. Carry out a similar test with CO_2.

3. It takes a large amount of heat to break the hydrogen bonds in water before the water warms up much, so water can absorb much heat with a small rise in temperature. As the rock loses heat it cools off a lot, but this same amount of heat, absorbed by the water, does not raise the temperature of the water much.

4. Water molecules can form hydrogen bonds with the + and − charged atoms in a sugar molecule, so the water molecules can surround individual sugar molecules. Water molecules are not attracted to oil molecules; the water molecules stick to each other and do not surround the oil molecules.

5. Each atom is characterized by a certain number of electrons, arranged in energy shells. For most atoms, the outermost shell can hold up to 8 electrons. Carbon has only 4 electrons in its outer shell. It tends to share electrons—form bonds—with four other atoms, filling its outer shell. Thus it tends to form four bonds, and methane is CH_4, not CH_6.

6. Seattle is much closer to the ocean, which stores heat and stabilizes climate. During the summer, the water absorbs heat, cooling Seattle, and during the winter the water releases stored heat to the air, warming the city.

7. Water molecules are more polar than those of most other substances. They are attracted to one another, bonded by hydrogen bonds. These bonds are hard to break, so water must absorb more heat to vaporize than most other substances. The easier-to-vaporize alcohol molecules tend to evaporate first.

Chapter 3

Organizing Your Knowledge
Exercise 1:

1.

Carbon has 4 electrons in an outer shell that holds 8. It completes its outer shell by sharing with 4 other atoms.

2. C_4H_{10}

3.

or

C_5H_{10}

4.

or

C_5H_{12}. They are isomers.

5.

Without double bond: 10 Hs. With double bond: 8 Hs

Exercise 2: 4 hydroxyls, 3 carbonyls, 3 carboxyls, 5 aminos

Exercise 3: 1. D 2. D 3. H 4. H 5. D 6. D 7. H 8. D 9. H 10. H

Exercise 4: 1. monosaccharides 2. polysaccharides 3. proteins 4. fats 5. starch 6. glucose 7. dehydration 8. water 9. hydrolysis 10. disaccharide 11. cellulose 12. linked 13. digesting 14. fiber 15. disaccharide 16. glucose 17. fructose 18. monosaccharide 19. isomers 20. $C_6H_{12}O_6$ 21. dehydration 22. glycogen 23. starch 24. glucose

Exercise 5:

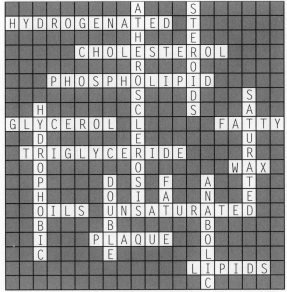

Exercise 6: 1. transport 2. contractile 3. defensive 4. structural 5. signal 6. storage 7. enzyme

Exercise 7: See Figures 3.12 and 3.13 in the text.

Exercise 8: 1. quaternary DF 2. tertiary AIJ 3. secondary EH 4. primary BCG

Exercise 9: 1. H 2. G 3. J 4. D 5. F 6. B 7. C 8. D 9. J 10. I 11. A 12. E

Exercise 10: 1. fats 2. steroids 3. energy 4. glycerol 5. unsaturated 6. plant oils 7. heart disease 8. butter 9. corn oil 10. cell membranes

Testing Your Knowledge

Multiple Choice: 1. c 2. a 3. c 4. e 5. d 6. c 7. e 8. c 9. d 10. d 11. b 12. d

Essay: 1. Structural proteins form body parts such as fibers in tendons and ligaments. Contractile proteins in muscles are responsible for movement. Storage proteins such as albumin store amino acids. Defensive proteins called antibodies fight infection. Transport proteins such as hemoglobin carry substances. Some hormones are signal proteins, which convey messages from cell to cell. An enzyme is a protein that catalyzes chemical changes.

2. Animal fats are saturated; their fatty acids contain no double bonds and can pack together tightly and solidify at room temperature. Plant oils are unsaturated; their fatty acids contain double bonds, which cause kinks that prevent them from packing together and solidifying.

3. The secondary and tertiary levels of protein structure are maintained mostly by relatively weak hydrogen bonds and ionic bonds. These weak bonds are easily disrupted by changes in the protein's environment.

4. Glucose: one circle. Maltose: two joined circles. Starch: a chain of linked circles. See Figures 3.4, 3.5, and 3.7 in the text. Maltose is a disaccharide. Starch is a polysaccharide.

5. See Figures 3.15–3.18 in the text.

Applying Your Knowledge

Multiple Choice: 1. e 2. d 3. d 4. d 5. d 6. b 7. e 8. a 9. a 10. e 11. a 12. a

Essay: 1. Every starch molecule is a polymer of glucose monomers, which are all the same, so starch molecules differ only in length. A protein molecule can be composed of 20 kinds of amino acid monomers. These can be arranged in chains of an almost infinite number of different sequences and lengths, producing many different kinds of proteins.

2. A protein is folded in a specific three-dimensional shape, and a protein's function usually depends on its ability to bind to some other molecule because of its specific shape. The shape of these enzyme molecules allows them to fit a starch molecule. The glucose monomers of cellulose are linked together in a different orientation, forming a rod instead of a coil. Apparently the enzymes do not fit this structure.

3. The hydrogen and ionic bonds that maintain the three-dimensional shape of a protein are fairly weak. A slight change in the environment

can alter these bonds, change the shape of the protein, and impair its function, even if the covalent bonds that hold the amino acids together in the protein are unaffected. The main function of starch is to store glucose; the 3-D shape of the molecule is relatively unimportant. A much more drastic change is needed to break the covalent bonds in the starch molecule, separate the glucose subunits, and alter its function.

4. See Figure 2.1 in the text. Make sure every C forms 4 bonds and every H forms 1 bond. Isomers have the same molecular formulas, but their structures differ.

5. The body stores excess glucose in the form of glycogen, a polysaccharide. We can draw on this store of sugar between meals. Since Fred cannot produce glycogen, he cannot store as much sugar and must eat more often to provide his cells with glucose "fuel."

6. The number of possible tripeptides is $20 \times 20 \times 20 = 8000$. The number of possible polypeptides 100 amino acids long is 20^{100}—a very large number indeed!

Chapter 4

Organizing Your Knowledge

Exercise 1: 1. visible light 2. visible light 3. electrons 4. eye lens 5. glass lenses 6. electromagnets 7. none 8. $1000\times$ 9. $100,000\times$ + 10. dust particle 11. smallest bacterium 12. large atom 13. N/A 14. 0.2 micrometer 15. 0.2 nanometer 16. no magnification 17. limited magnification 18. cannot be used on living specimens

Exercise 2: *Cell 1:* Surface area = 96 μm^2. Volume = 64 μm^3. Surface/volume = 1.5. *Cell 2:* Surface area = 384 μm^2. Volume = 512 μm^3. Surface/volume = 0.75. 1. Cell 2. 2. Cell 2. 3. Cell 1. 4. Cell 1.

Exercise 3: See Figure 4.4 in the text.

Exercise 4: *Bacterial cell:* prokaryotic, small, plasma membrane, cell wall, cytoplasm, ribosomes, bacterial flagellum. *Plant:* eukaryotic, large, internal membranes, plasma membrane, cell wall, cytoplasm, ribosomes, nucleus, rough ER, smooth ER, Golgi, lysosome, microbody, mitochondrion, chloroplast, central vacuole, cytoskeleton. *Animal:* eukaryotic, large, internal membranes, plasma membrane, cytoplasm, ribosomes, nucleus, rough ER, smooth ER, Golgi, lysosome, microbody, mitochondrion, cytoskeleton, flagellum, centriole.

Exercise 5: 1. D 2. I 3. A 4. E 5. B 6. D

7. G 8. A 9. G 10. F 11. C 12. B 13. E 14. G 15. H 16. I 17. F 18. E 19. A 20. B

Exercise 6: See Figure 4.14 in the text. 1. rough ER—vesicle—Golgi—vesicle—outside 2. rough ER—vesicle—Golgi—lysosome 3. rough ER—vesicle—Golgi—vesicle—plasma membrane 4. smooth ER—vesicle—Golgi—vesicle—outside

Exercise 7: *Chloroplast:* Found in plants and some protists; carries out photosynthesis; converts energy of sunlight into chemical energy in sugar. *Mitochondrion:* Found in all eukaryotes; carries out cellular respiration; converts chemical energy of foods into chemical energy in ATP.

Exercise 8: *Microfilaments:* solid rods, actin, help change shape, muscle contraction. *Microtubules:* hollow tubes, tubulin, help change shape, move chromosomes, act as tracks, give cell shape, in cilia, in flagella, in centrioles.

Exercise 9: 1. E 2. D 3. A 4. B 5. C

Exercise 10: See Figures 4.5 and 4.20 in the text.

Testing Your Knowledge

Multiple Choice: 1. e 2. b 3. b 4. e 5. a 6. b 7. e 8. c 9. b 10. b 11. a

Essay: 1. An electron microscope has greater resolving power and is capable of greater magnification than a light microscope, enabling it to reveal finer details than a light microscope. An electron microscope cannot be used to view living specimens, because they must be held in a vacuum and all air and liquids must be removed. A light microscope is preferable for living specimens.

2. Prokaryotic cells are smaller and much simpler than eukaryotic cells. Prokaryotes lack a nucleus and other membranous organelles that compartmentalize eukaryotic cells.

3. There are three major structures present in plant cells but lacking in animal cells. Rigid cell walls containing cellulose surround and support plant cells. Chloroplasts perform photosynthesis. Plant cells usually contain central vacuoles that may help carry out metabolic processes or store water, food, or other chemicals.

4. Membranes separate metabolic processes and allow processes that require different conditions to be carried out simultaneously. They also increase the total surface area of membranes where many metabolic processes occur.

5. Chloroplasts carry out photosynthesis, capturing the energy of sunlight and storing it in sugar molecules. Mitochondria carry out cellular

respiration, converting the chemical energy in sugar and other foods into the chemical energy of ATP, which serves as a cellular fuel.

Applying Your Knowledge

Multiple Choice: 1. d 2. d 3. c 4. a 5. a 6. b 7. b 8. c 9. a 10. e

Essay: 1. The TEM would be best for examining fine details within cell organelles, because it is capable of greater magnification and resolving power and uses a beam of electrons that passes through thinly sliced sections. The SEM would be best for studying bumps on the cell surface, because its electron beam scans the surface and forms an image of the outside of the cell. Electron microscopes are not suitable for observing living specimens because they must be placed in a vacuum and the air and liquids must be removed, so a light microscope would be best for observing changes in the nucleus and cell shape.

2. Surface area = 150 mm^2. Volume = 125 mm^3. Surface/volume = 1.2. Surface area = 600 mm^2. Volume = 1000 mm^3. Surface/volume = 0.6. The smaller cell has the greater surface-to-volume ratio, which would enable it to exchange materials more efficiently with its environment via its membrane.

3. The amylase molecules are produced by ribosomes attached to the rough ER and deposited inside the ER. Transport vesicles filled with the protein bud from the ER and fuse with the Golgi apparatus, where the proteins may be modified. Transport vesicles then bud from the Golgi apparatus and fuse with the plasma membrane, dumping amylase outside the cell.

4. The mitochondrion is a sort of cellular "power plant," converting the energy in food molecules into the energy of ATP, which in turn provides energy for many cellular activities. Interfering with the mitochondria could reduce the supply of ATP fuel and thus impair many cellular processes.

5. As a muscle cell grows, ribosomes, ER, and Golgi would work together to produce more protein and cell membrane. Microfilaments, made of the protein actin, are responsible for the contractile activity of muscle and would increase in number in a growing muscle cell. Smooth ER stores calcium that is involved in the contraction process, and it too would grow. Mitochondria provide energy for muscle contraction and other cellular activities and would perhaps increase in number.

Chapter 5

Organizing Your Knowledge

Exercise 1: 1. work 2. kinetic 3. thermodynamics 4. first 5. transformed 6. potential 7. heat 8. second 9. entropy 10. more 11. endergonic 12. exergonic 13. releases 14. ATP 15. ADP 16. phosphate 17. potential 18. hydrolysis 19. phosphate 20. ADP 21. work 22. phosphorylation 23. heat 24. endergonic 25. coupling 26. activation 27. barrier 28. protein 29. catalyst 30. increases 31. changed 32. lower 33. metabolism

Exercise 2: 1. Kinetic energy is energy that is doing work, and potential energy is stored energy. 2. Exergonic reactions release energy, and endergonic reactions consume energy. 3. In a chemical reaction, reactants are changed into products. 4. ADP is phosphorylated (a phosphate group is added), producing ATP. When ATP is hydrolyzed, a phosphate group is removed from ATP, and ADP is produced. 5. Enzymes lower the energy of activation for a reaction, so the reaction with an enzyme requires less energy to get it started than the reaction without the enzyme. 6. In photosynthesis, the energy of sunlight is used to make glucose from carbon dioxide and water. In cellular respiration, oxygen and glucose combine, and the energy in glucose is used to make ATP. 7. First law: Energy can be transferred or transformed but not created or destroyed. Second law: Energy changes are accompanied by an increase in disorder, or entropy.

Exercise 3: 1. See Figure 5.6 in the text. Substrate should fit enzyme active site. 2. Shape of enzyme altered by heat or pH, deforming active site so substrate will not fit. 3. Competitive inhibitor is shaped like substrate and blocks active site. 4. Noncompetitive inhibitor can be any shape; attaches to enzyme and alters shape of active site so substrate will not fit.

Exercise 4: 1. phospholipid 2. glycoprotein 3. carbohydrate I.D. tag 4. glycolipid 5. extracellular fluid 6. transport protein 7. cholesterol 8. receptor protein 9. enzyme 10. cytoplasm

Exercise 5: 1. E 2. B 3. A 4. F 5. DGH 6. G 7. C 8. B 9. H 10. E 11. CF 12. D 13. I 14. D

Exercise 6: 1. Out 2. Out 3. In 4. None 5. In 6. Out 7. In 8. In 9. None 10. Out

Exercise 7: 1. phospholipid 2. nonpolar 3. phospholipid 4. protein 5. fluid 6. glycolipids 7. glycoproteins 8. identity 9. receptor

10. diffusion 11. passive 12. concentration
13. more 14. less 15. isotonic 16. hypertonic
17. out of 18. selectively 19. osmosis
20. transport 21. energy 22. facilitated 23. active 24. ATP 25. endocytosis 26. phagocytosis
27. vesicle 28. lysosome 29. exocytosis

Testing Your Knowledge

Multiple Choice: 1. a 2. c 3. b 4. b 5. d
6. c 7. b 8. a 9. e 10. b 11. d 12. b

Essay: 1. Large (and polar) molecules cannot easily diffuse through cell membranes. Transport proteins provide pores by which these substances can pass through membranes.

2. The enzyme is shaped in such a way that its active site forms a "pocket" that holds the substrate.

3. Molecules slow down at lower temperatures, slowing the rate of contact between reactant molecules and the enzyme's active site. Higher temperatures denature the enzyme, altering its shape and impairing its specific function.

4. Larger molecules contain more covalent bonds, so they have more potential energy stored in their complex structure. It takes energy to form these bonds, so the reactions by which amino acids join to form proteins are endergonic. The cell obtains energy for endergonic reactions by coupling them to exergonic reactions, most important the hydrolysis of ATP.

5. A cell will gain water by osmosis if the solution outside the cell is hypotonic, that is, if the outside solution has a lower total concentration of solutes than the cytoplasm. It will lose water if the solution outside is hypertonic, that is, if it has a higher total concentration of solutes than the cytoplasm. Water gain by a plant cell is least serious; the cell wall prevents bursting.

Applying Your Knowledge

Multiple Choice: 1. b 2. e 3. b 4. b 5. a
6. d 7. e 8. b 9. d 10. d 11. e 12. e

Essay: 1. Paper does not burn spontaneously (at room temperature) because energy is required to get this reaction started—energy of activation. This creates an energy barrier that prevents the reaction from happening spontaneously. A flame supplies energy of activation.

2. The hypertonic solution would cause water to move out of the stomach cells by osmosis, causing dehydration.

3. You can't win: The first law of thermodynamics says that energy cannot be created or destroyed, so you can't come out ahead in the energy game. You can't break even: The second law of thermodynamics says that energy changes are accompanied by an increase in disorder, or entropy. Heat, a form of disorder, is created during energy conversions; some energy is always lost to the surroundings as heat, so you can't break even.

4. The insecticide molecule is the same shape as the substrate molecule. It fits into the active site of the enzyme, preventing the enzyme from acting on its normal substrate, and thus interfering with nervous system function.

5. The lecithin molecules coat the droplet, with their nonpolar tails embedded in the fat and the polar heads projecting into the surrounding water.

Chapter 6

Organizing Your Knowledge

Exercise 1: 1. respiration 2. oxygen 3. carbon dioxide 4. cellular respiration 5. glucose
6. oxygen 7. carbon dioxide 8. water 9. ATP
10. 2200 11. aerobic 12. moderate 13. anaerobic 14. lactic acid 15. ache 16. fatigue
17. yeast 18. aerobic 19. glucose 20. oxygen
21. carbon dioxide 22. anaerobic 23. glucose
24. ethanol 25. carbon dioxide

Exercise 2:

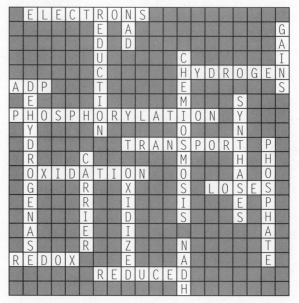

Exercise 3: 1. high-energy electrons carried by NADH 2. NADH and FADH$_2$ 3. glycolysis
4. glucose 5. pyruvic acid 6. Krebs cycle

7. electron transport chain and chemiosmosis
8. cytoplasmic fluid 9. CO_2 10. CO_2 11. mitochondrion 12. ATP 13. ATP 14. ATP

Exercise 4: 1. I 2. H 3. E 4. C 5. C 6. D
7. B 8. G 9. F 10. A

Exercise 5: 1. NAD^+ 2. NADH 3. pyruvic acid
4. acetyl CoA 5. CO_2 6. coenzyme A 7. CO_2
8. NADH 9. ATP 10. $FADH_2$

Exercise 6: 1. final 2. ATP 3. chemiosmosis
4. electron 5. inner 6. NADH 7. Krebs cycle
8. O_2 9. H_2O 10. H^+ ions 11. active transport
12. potential 13. ATP 14. inner compartment of
the mitochondrion 15. ATP synthases 16. ADP
17. ATP 18. some

Exercise 7: See Figure 6.13 in the text.

Exercise 8: 1. E 2. G 3. G 4. G 5. E 6. K
7. E 8. E 9. K 10. E 11. E

Exercise 9: 1. facultative 2. yeast 3. pyruvic
4. muscle 5. oxygen 6. NADH 7. ethanol
8. lactic 9. ATP 10. glycolysis 11. CO_2
12. anaerobes

Exercise 10: 1. polysaccharides 2. fats 3. proteins 4. sugars 5. glycerol 6. fatty acids
7. amino acids 8. glucose

Exercise 11: See Figure 6.17 in the text.

Testing Your Knowledge

Multiple Choice: 1. e 2. b 3. c 4. b 5. d
6. c 7. c 8. a 9. b 10. e 11. e 12. a

Essay: 1. Breathing allows our lungs to supply cells with the oxygen used in cellular respiration and expel the carbon dioxide produced by cellular respiration.

2. Cellular respiration is more efficient (it produces more ATP per molecule of glucose consumed), but it must take place in an oxygen-containing (aerobic) environment. Fermentation requires no oxygen, but it is much less efficient (it produces much less ATP per glucose).

3. Chemiosmotic phosphorylation occurs in the third (electron transport–chemiosmosis) stage of cellular respiration. Substrate-level phosphorylation occurs in glycolysis and the Krebs cycle. In chemiosmosis, ATP synthase harnesses the flow of H^+ ions down their concentration gradient to make ATP. In substrate-level phosphorylation, an enzyme transfers a phosphate group from an organic molecule to ADP to make ATP. Chemiosmosis makes the most ATP under aerobic conditions. Substrate-level phosphorylation makes the most ATP under anaerobic conditions.

4. Rotenone prevents electrons from passing

from an electron carrier molecule to the next. Cyanide and carbon dioxide block the passage of electrons to oxygen. In either case, the energy of moving electrons cannot be captured. Oligomycin blocks the passage of H^+ ions through ATP synthase, so ATP cannot be made. DNP makes the membrane of the mitochondrion leaky to H^+ ions, thus abolishing the H^+ concentration gradient whose energy is harnessed to make ATP.

5. These processes provide small organic molecules that the cell can use in the synthesis of organic molecules not obtained in food.

Applying Your Knowledge

Multiple Choice: 1. b 2. a 3. c 4. c 5. d
6. d 7. c 8. d 9. b 10. d

Essay: 1. Organisms that rely on aerobic cellular respiration are limited to environments where oxygen is available. Fermenters are able to live in anaerobic environments such as mud, manure piles, and sealed wine vats, free from competition with aerobic organisms.

2. Oxygen picks up electrons at the end of the electron transport chain. When oxygen runs out, there is no place for electrons to go. They "pile up," and electron transport stops. When electron transport stops, NADH and $FADH_2$ pile up, and NAD^+ and FAD are not regenerated. Without NAD^+ and FAD to pick up electrons and hydrogens, the Krebs cycle must stop.

3. Because FAD and NAD^+ are constantly recycled, the cell requires only a small supply of them. Without NAD^+ and FAD to pick up electrons and hydrogens from glycolysis and the Krebs cycle and deliver them to the electron transport chain, cellular respiration would stop.

4. If the victim had suffocated, the lack of oxygen would have caused electrons to pile up, and all the electron carriers would have held electrons. They would have been in the reduced state. Because the carriers were in the oxidized state—lacking electrons—Mickleberry knew that oxygen had been available in the cells to pick them up, so the victim did not suffocate.

5. At first the yeasts in the flask were able to produce sufficient ATP via aerobic cellular respiration. As the yeast population grew, the oxygen in the flask was used up, and they turned to fermentation to produce ATP. Because fermentation is less efficient than aerobic respiration, the yeasts had to consume glucose at a faster rate to maintain their supply of ATP. Ethanol is a waste product of fermentation.

Chapter 7

Organizing Your Knowledge

Exercise 1: pine tree, green bacterium, rosebush, seaweed, moss, alga, grass

Exercise 2:

```
                             P
    P P           W A T E R              H
C H L O R O P L A S T     V              O
    H T     I                 F          T
    C T T G       T H Y L A K O I D S    O
    E O S H           E           X      S
    N N Y T       G R A N A       A      Y
    T   Y N C         G           T      S
    E L E C T R O M A G N E T I C        T
    R   H     R       H           O      E
    E L E C T R O N     M           N    M
    S         T         E
    C A L V I N         T     S T R O M A
    U   S               E             O
    T             C H L O R O P H Y L L
    O                   I       H
S T O M A T A         D         Y
    R                 S         L
    O                           L
    P
    H
```

Exercise 3: $6\ CO_2 + 12\ H_2O \longrightarrow C_6H_{12}O_6 + 6\ H_2O + 6\ O_2$ The C and O in $C_6H_{12}O_6$ come from CO_2. The H in $C_6H_{12}O_6$ comes from H_2O. The O in O_2 comes from H_2O. The H in H_2O comes from H_2O and the O comes from CO_2. H_2O is oxidized and CO_2 is reduced.

Exercise 4: 1. light 2. H_2O 3. outer membrane of chloroplast 4. CO_2 5. light reactions 6. $NADP^+$ 7. Calvin cycle 8. ADP + P 9. stroma 10. granum 11. thylakoids 12. electrons 13. ATP 14. NADPH 15. O_2 16. sugar

Exercise 5: 1. B 2. A 3. H 4. C 5. B 6. A 7. D 8. H 9. G 10. F 11. C 12. E 13. I 14. D 15. A 16. A 17. H 18. G

Exercise 6: a. 5 b. 9 c. 2 d. 6 e. 3 f. 7 g. 8 h. 4 i. 1 Gamma waves have the most energy. Blue and red light are used in photosynthesis.

Exercise 7: 1. photosynthesis 2. chloroplasts 3. mesophyll 4. light 5. chlorophyll 6. thylakoids 7. Calvin 8. stroma 9. carbon 10. photons 11. blue 12. red 13. green 14. carotenoids 15. reaction center 16. noncyclic 17. electrons 18. primary electron acceptor 19. electron transport 20. NADPH 21. cyclic 22. photons 23. electrons 24. photosystem I 25. H^+ 26. stroma 27. thylakoids 28. gradient 29. ATP synthase 30. ATP 31. photophosphorylation 32. water 33. water 34. oxygen 35. oxygen 36. Calvin 37. photosynthesis 38. stroma 39. carbon dioxide 40. ATP 41. NADPH 42. G3P 43. glucose 44. respiration 45. ATP

Exercise 8: 1. C_3 2. C_4 3. C_4 4. CAM 5. C_3 6. C_3 7. C_4 or CAM 8. C_4 or CAM 9. CAM 10. C_3

Exercise 9: 1. warm 2. cool 3. cool 4. cool 5. warm 6. cool 7. cool 8. warm

Testing Your Knowledge

Multiple Choice: 1. b 2. a 3. a 4. c 5. e 6. b 7. b 8. d 9. e 10. b 11. e 12. b

Essay: 1. If a plant is given carbon dioxide containing radioactive ^{18}O, no labeled oxygen atoms appeared in the O_2 produced by the plants, showing that the oxygen in O_2 does not come from CO_2. If the plants are given water containing ^{18}O, the O_2 gas they produce does contain labeled O, showing that the oxygen in O_2 comes from H_2O.

2. See Figure 7.5 in the text.

3. Photosynthesis allows producers to manufacture organic food molecules from inorganic reactants. This process produces virtually all the organic matter required as food by plants and all other living things on Earth.

4. Under hot, dry conditions, a C_3 plant closes its stomata. This saves water, but it causes the plant's CO_2 supply to drop and O_2 to build up in the leaf. Under these conditions, the first enzyme in the Calvin cycle fixes O_2 instead of CO_2, making a compound that cannot be used to make sugar, slowing the growth of the plant. In a C_4 plant, a different enzyme fixes CO_2 into a 4-carbon compound—which subsequently delivers the carbon to the Calvin cycle—instead of the 3-carbon molecule made by a C_3 plant. This enzyme cannot fix O_2, so the C_4 plant can fix CO_2 even when stomata close and the CO_2 concentration drops. A CAM plant opens its stomata and admits CO_2 at night. This CO_2 is fixed into a 4-carbon compound that is stored and used to make sugar during the day, while the stomata remain closed.

5. Increased burning of fossil fuels (such as oil and gas) releases CO_2 into the atmosphere. Deforestation reduces global photosynthesis, reducing removal of CO_2 from the atmosphere by plants. Both of these activities contribute to an increased greenhouse effect, because more CO_2 in the atmosphere could trap more heat. We could reduce our contribution to the greenhouse effect by using fossil fuels more efficiently and replanting trees on deforested land.

Applying Your Knowledge

Multiple Choice: 1. c 2. d 3. e 4. e 5. a 6. c 7. c 8. a 9. c 10. c

Essay: 1. Plants would grow most slowly under green lights, because green light is not absorbed well by photosynthetic pigments. Green is reflected, which is why plants are green.

2. Carotenoids absorb all colors except yellow and orange. Because carotenoids absorb green light, they are able to capture some light energy not absorbed by chlorophyll and channel this energy to chlorophyll for use in photosynthesis.

3. Energy from the movement of excited electrons along a chain of electron carriers is used to pump H^+ through the membrane into the thylakoid space. This buildup of H^+ lowers the pH of the solution in the thylakoid space.

4. In photosynthesis, high-energy electrons come from chlorophyll whose electrons are excited by light. These electrons are replaced by splitting water. In cellular respiration, glucose is the source of high-energy electrons. In both processes, movement of the electrons through electron carriers is harnessed to move H^+ through a membrane; this energy is ultimately used to make ATP via chemiosmosis. In photosynthesis, the high-energy electrons are also used to make glucose. In cellular respiration, the electrons combine with O_2 and H^+ to form water. Photophosphorylation takes place in the thylakoid membranes of the chloroplast; oxidative phosphorylation takes place along the inner membrane of the mitochondrion. In both processes, ATP synthase harnesses flow of H^+ through the membrane to make ATP.

5. In the light reactions of photosynthesis, sunlight excites chlorophyll in a leaf, causing it to give up excited electrons. High-energy electrons move through electron carriers, which generate an H^+ gradient, which in turn is harnessed to make ATP via chemiosmosis. The electrons and hydrogen are picked up by $NADP^+$, forming NADPH. In the Calvin cycle, energy from the ATP and hydrogen and electrons from NADPH are used to reduce CO_2, forming sugar. When you consume the plant, the sugar is "burned" in cellular respiration. The glucose is oxidized, and its hydrogen and electrons are transferred to NAD^+ and FAD, forming NADH and $FADH_2$. These molecules deliver the high-energy electrons to the electron transport system in the mitochondrion. The electrons move through electron carriers, which generate an H^+ gradient, which in turn is harnessed to make ATP via chemiosmosis. The ATP is then hydrolyzed to ADP and P to power muscle movement.

Chapter 8
Organizing Your Knowledge

Exercise 1: 1. like 2. asexual 3. DNA 4. DNA 5. chromosomes 6. identical 7. mother 8. binary fission 9. circular 10. smaller 11. simpler 12. membrane 13. wall 14. egg 15. sperm 16. fertilized 17. variation

Exercise 2:

Exercise 3: 1. interphase, activity between divisions 2. G_1, cell growth 3. S, DNA synthesis 4. G_2, activity between DNA synthesis and division 5. mitotic phase, mitosis plus cytokinesis 6. M 7. mitosis, division of nucleus and chromosomes 8. cytokinesis, division of cytoplasm

Exercise 4: *Nucleus:* I—membrane-bound, P—envelope breaks down, M—none, A—none, T—daughter nuclei and envelopes form. *Spindle:* I—none, P—spindle forms from microtubules, M—fully formed, A—fibers pull chromosomes, T—breaks down. *Chromosomes:* I—duplicated but dispersed as chromatin, P—chromatin coils to form shorter, thicker pairs of chromatids joined at centromeres, M—line up on metaphase plate, A—sister chromatids separate, T—chromosomes uncoil to form chromatin. *Cell size and shape:* I—rounded, doubles in volume, P—rounded, M—rounded, A—spindle elongates cell, T—elongation continues, cytokinesis splits cell into two smaller rounded cells.

Sketch: See Module 8.7 in the text.

Exercise 5: 1. interphase 2. telophase 3. metaphase 4. prophase 5. anaphase

Exercise 6: 1. E 2. A 3. C 4. B 5. D

Exercise 7: For example: In animals, microfilaments produce a cleavage furrow that pinches the cell apart, while in plants, vesicles align in a cell plate, where a cell wall grows to split the cell.

Exercise 8: 1. asexual 2. growth 3. muscle 4. nerve 5. liver 6. skin 7. digestive tract 8. replaced 9. numbers 10. types 11. mitosis 12. cultures 13. surface 14. density-dependent inhibition 15. growth factors 16. G_1 17. restriction point 18. cancer 19. 20 20. 50 21. radiation 22. chemotherapy 23. cell division 24. microtubules

Exercise 9: 1. Homologous pairs of chromosomes called autosomes are found in the somatic cells of both males and females. Sex chromosomes are an additional pair of chromosomes—homologous in females (who have 2 *X* chromosomes) but not homologous in males (who have *X* and *Y* chromosomes). 2. The two chromosomes of a homologous pair carry genes for the same inherited traits at the same place, or locus. One is inherited from the father and one from the mother. 3. The two chromatids of a single chromosome are exactly identical, having been produced via the process of DNA replication during S phase of interphase. 4. A diploid cell contains two homologous sets of chromosomes. A haploid cell contains one chromosome set. Diploid cells divide to form haploid cells in meiosis. 5. Gametes are sperm and eggs. Somatic cells are body cells other than gametes. In humans, somatic cells are diploid. 6. An egg is haploid. A zygote, formed by fusion of a sperm and egg, is diploid. 7. In meiosis, a diploid cell divides to form haploid cells. In fertilization, haploid cells (egg and sperm) fuse to form a diploid cell (zygote). 8. Mitosis produces two cells that are genetically identical to the mother cell. Meiosis is the division of a diploid cell to produce four haploid cells that are not genetically like the mother cell or each other.

Exercise 10: See Module 8.15 in the text for appearance of chromosomes, homologous chromosomes, sister chromatids, and crossing over. 1. interphase I 2. prophase I 3. metaphase I 4. anaphase I 5. telophase I 6. prophase II 7. metaphase II 8. anaphase II 9. telophase II 10. meiosis I 11. meiosis II

Exercise 11: 1. produces daughter cells identical to parent cell 2. involves two cell divisions 3. produces four daughter cells 4. sister chromatids of each chromosome separate 5. homologous pairs (tetrods) line up at metaphase plate 6. crossing over occurs between homologous chromosomes 7. produces identical daughter cells for asexual reproduction, growth, replacement, repair

Exercise 12:

1.

2.

3.

8 combinations are possible

Testing Your Knowledge

Multiple Choice: 1. b 2. a 3. d 4. b 5. a
6. a 7. d 8. b 9. a 10. c 11. c

Essay: 1. In asexual reproduction, the cells or buds that give rise to offspring are produced by mitosis. Because in this process genetic information is copied and passed on to identical daughter cells, the offspring produced by asexual reproduction are exactly like the single parent and each other. In sexual reproduction, each offspring inherits a unique combination of genes from both parents. Because of this genetic recombination, the offspring produced by sexual reproduction are not exactly like either parent or each other.

2. In mitosis, the nucleus and duplicated chromosomes divide and are equally distributed to daughter cells. In cytokinesis, the cytoplasm divides in two and is distributed to the daughter cells.

3. Cancer cells escape from the control mechanisms that regulate division in normal cells. Cancer cells do not need to be anchored to a solid surface to grow. They seem unaffected by density-dependent inhibition and will pile up on one another in cell culture. They either don't require growth factors or manufacture their own growth factors. Unlike normal cells, cancer cells can divide indefinitely. If they stop dividing, they stop at random points in the cell cycle, not at the restriction point.

4. Mitosis produces two cells that are genetically identical to the mother cell. Meiosis is the division of a diploid cell to produce four haploid cells that are not genetically like the mother cell or each other. In humans, mitosis is responsible for producing diploid somatic cells for growth, replacement of worn-out cells, or body repair. Meiosis produces haploid gametes (eggs and sperm) for sexual reproduction. Many (but not all) somatic cells undergo mitosis; special cells in ovaries and testes undergo meiosis.

5. Because sister chromatids are produced when DNA is duplicated, the genes of sister chromatids are identical. Homologous chromatids carry genes for the same traits (eye color, for example) at corresponding places, or loci, but they might carry different versions of the gene (blue or brown) at these loci.

6. The two chromosomes—maternal and paternal—of a homologous pair carry different genetic information. Because the two chromosomes of each pair line up independently of other pairs at metaphase I of meiosis, many different combinations of chromosomes are possible in the gametes produced by meiosis. Also in meiosis, homologous chromosomes exchange corresponding segments in the process of crossing over, creating chromosomes with a variety of gene combinations. Finally, many combinations of eggs and sperm are possible at fertilization, further increasing genetic variation.

Applying Your Knowledge

Multiple Choice: 1. d 2. b 3. c 4. c 5. a
6. b 7. a 8. d 9. d 10. d 11. d 12. b

Essay: 1. Cancer is uncontrolled cell division. Chemotherapy (and other cancer therapies like radiation) slows tumor growth by interfering with cell division. Cell division is most rapid in the skin and digestive tract lining, where it replaces lost cells, so the side effects of chemotherapy are greatest on skin and intestinal lining.

2. Cells with between 100 and 200 units were in S phase, in the process of duplicating their DNA.

3. These data suggest that prophase is the longest stage of mitosis, making up about half the process of mitosis. Telophase is also fairly long, lasting about a third of mitosis. Metaphase and anaphase take a relatively short time to occur.

4. There would be 39 different kinds of chromosome—39 homologous pairs. (In female mammals, the two sex chromosomes, as well as each pair of autosomes, look alike.)

5. Possible alignments: *ABC* and *abc*, *ABc* and *abC*, *Abc* and *aBC*, and *AbC* and *aBc*. These same eight combinations are possible in eggs or sperm. There are $8 \times 8 = 64$ combinations possible in the zygotes.

6. Parental type gametes: *QR* and *qr*.
Recombinant gametes: *Qr* and *qR*.

Chapter 9

Organizing Your Knowledge

Exercise 1: 1. S 2. G 3. C 4. Q 5. U 6. K
7. W 8. L 9. A 10. T 11. H 12. X 13. D
14. E 15. V 16. B 17. P 18. M 19. N 20. F
21. I 22. O 23. J 24. R

Exercise 2: 1. a. green, yellow b. *GG, gg* c. all
Gg 2. a. 1 *GG*:2 *Gg*:1 *gg* b. 3 green:1 yellow
3. a. both *Bb* b. white *bb*

Exercise 3: 1. *BBSS* 2. *bbss* 3. *BS* 4. *bs*
5. *BbSs* 6. *BS* 7. *Bs* 8. *bS* 9. *bs* 10. *BS* 11. *Bs*
12. *bS* 13. *bs* 14. *BBSS* 15. *BBSs* 16. *BBSs*
17. *BbSS* 18. *BBss* 19. *BbSS* 20. *BbSs* 21. *BbSs*
22. *BbSs* 23. *BbSs* 24. *Bbss* 25. *bbSS* 26. *Bbss*
27. *bbSs* 28. *bbSs* 29. *bbss* 30. ⁹⁄₁₆ 31. brown,
short-haired 32. ³⁄₁₆ 33. brown, long-haired
34. ³⁄₁₆ 35. white, short-haired 36. ¹⁄₁₆ 37. white,
long-haired

Exercise 4: 1. a. white b. brown 2. a. *bb* b. *Bb*
or *BB* 3. white 4. *Bb* 5. all brown, 0 white
6. heterozygous 7. homozygous

Exercise 5: 1. 0 2. 1 3. 1 4. ⅛ 5. ⅝ 6. independent 7. product 8. ⅛ 9. ⅛ 10. ¹⁄₃₆ 11. multiplication 12. ½ 13. ½ 14. homozygous
15. multiplication 16. ½ 17. ½ 18. ¼ 19. addition 20. sum 21. ¹⁄₃₆ 22. ¹⁄₃₆ 23. ²⁄₃₆ (¹⁄₁₈) 24. ½
25. ½ 26. ½ 27. ½ 28. ¼ 29. ¼ 30. ¼ 31. ¼
32. ½

Exercise 6: 1. none 2. *Ww* and *Ww* 3. ½ 4. ¾

Exercise 7: 1. B 2. A 3. D 4. B 5. C 6. C
7. E 8. D 9. E 10. A

Exercise 8: 1. *B* 2. *b* 3. *S* 4. *s* 5. *B* 6. *b*
7. *S* 8. *s* 9. *B* 10. *S* 11. *b* 12. *s* 13. *S* 14. *B*
15. *S* 16. *B* 17. *s* 18. *b* 19. *s* 20. *b* 21. *b*
22. *B* 23. *S* 24. *s* 25. *b* 26. *B* 27. *S* 28. *B*
29. *b* 30. *S* 31. *B* 32. *s* 33. *S* 34. *b* 35. *S*
36. *b* 37. *s* 38. *B* 39. *s* 40. *B*

Exercise 9: 1. B 2. C 3. E 4. A 5. D

Exercise 10: 1. E 2. C 3. F 4. D 5. B 6. A

Exercise 11: 1. *X* 2. dominant 3. $X^C Y$ 4. $X^c Y$
5. daughters 6. sons 7. *X* 8. *Y* 9. carriers
10. $X^C X^c$ 11. half 12. half 13. half 14. all
15. father 16. carriers 17. mother 18. *X* 19. *c*

Testing Your Knowledge

Multiple Choice: 1. c 2. e 3. d 4. b 5. e
6. b 7. b 8. a 9. d 10. e 11. c 12. a 13. b

Essay: 1. Mendel studied peas, which are easy
to grow and come in many readily distinguish-
able varieties. He studied seven traits that each
occur in two distinct forms. He first made sure
that he had true-breeding varieties. Mendel carefully controlled which plant mated with which
and kept scrupulous records of his matings and
their results. Finally, he analyzed and interpreted his results mathematically, devising hypotheses that he tested and retested experimen-
tally.

2. Each flip is an independent event, with a
probability of 1/2. The probability of getting
two heads is the product of the individual probabilities, $1/2 \times 1/2 = 1/4$. The probability of
coin 1 coming up heads and coin 2 coming up
tails is also the product of the individual probabilities, $1/2 \times 1/2 = 1/4$. But there is another
way that you can get one head and one tail—
coin 1 could come up tails and coin 2 could
come up heads. The combined probability of
this occurring is also $1/2 \times 1/2 = 1/4$. There
are two different ways to get one head and one
tail. The probability of an event that can occur
two different ways is the sum of the separate
probabilities of the different ways, in this case
$1/4 + 1/4 = 1/2$.

3. Peas and fruit flies are easy to raise in a
small space at little expense. They have many
easy-to-spot traits, and it is easy to control matings between different varieties. They both produce many offspring in a relatively short time.

4. Humans who inherit *X* chromosomes from
both mother and father *(XX)* are female.
Humans who inherit an *X* chromosome from
the mother and a *Y* chromosome from the father *(XY)* are male. A gene on the *Y* chromosome appears to trigger testis development. In
the absence of this gene, an individual develops
ovaries and is female. In grasshoppers and
crickets, an individual with one sex chromosome
(XO) is male, and one with two sex chromosomes *(XX)* is female. Sperm cells with or without sex chromosomes determine the sex of the
offspring. In certain fishes, butterflies, and birds,
a *ZZ* individual is male, a *ZW* individual is female, and eggs determine the sex of offspring.
In ants and bees, diploid individuals are female
and haploid individuals are male.

Applying Your Knowledge

Multiple Choice: 1. c 2. d 3. c 4. b 5. a
6. d 7. c 8. c 9. c

Essay: 1. The parents are both heterozygous, *Aa*.
The daughter is homozygous recessive, *aa*. The
probability is 1/4.

2. ⁹⁄₁₆ purple green, ³⁄₁₆ purple yellow, ³⁄₁₆ white
green, ¹⁄₁₆ white yellow

3. The man is heterozygous, *Ff*. His wife is homozygous recessive, *ff*. The children with freckles are *Ff*; those without freckles are *ff*.

4. Pink × pink yields ¼ red, ½ pink, ¼ white. Pink × white yields ½ pink, ¼ white.

5. The man is *I^A^i*, the woman *I^B^i*. The probability of a type B child is ¼.

6. All four individuals have about the same skin color. Couple 2 could have children with the widest range of skin colors, ranging from *AABBCC* to *aabbcc*. (Couple 1 could have children ranging only from *aaBBCC* to *aabbCC*.)

7. Prediction: ¼ red eyes and straight wings, *RrSs*; ¼ red eyes and curled wings, *Rrss*; ¼ pink eyes and straight wings, *rrSs*; and ¼ pink eyes and curled wings, *rrss*. The actual results of the cross indicate that the genes for eye color and wing shape are linked. In the heterozygous parent, *R* and *S* are on one chromosome, and *r* and *s* on the homologous chromosome. Most gametes were *RS* and *rs*, but crossing over resulted in a small percentage of *Rs* and *rS* gametes being produced.

8. Genes *s* and *b* are farther apart than *h* and *b*. The order of the genes on the chromosome is *shb* (or *bhs*).

Chapter 10

Organizing Your Knowledge

Exercise 1: 1. N 2. D 3. DR 4. SPB 5. B 6. I 7. D 8. SP 9. ACGT 10. A 11. K 12. L 13. S 14. D 15. ACGU 16. C 17. E 18. B (or N) 19. L 20. F 21. J 22. O 23. H 24. Q (or D or R) 25. U

Exercise 2: 1. sugar-phosphate backbone 2. phosphate group 3. sugar (deoxyribose) 4. double helix 5. complementary base pair 6. adenine (A) 7. guanine (G) 8. thymine (T) 9. hydrogen bond 10. cytosine (C) 11. nucleotide 12. polynucleotide 13. pyrimidine bases 14. purine bases

Exercise 3: See Figure 10.4A in the text.

Exercise 4:

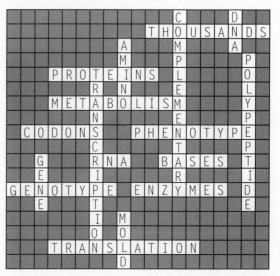

Exercise 5: 1. nucleus 2. RNA polymerase 3. promoter 4. nucleotides 5. terminator 6. messenger RNA 7. cytoplasm 8. amino acid 9. transfer RNA 10. amino acid 11. codon 12. enzyme 13. anticodon 14. ribosomes 15. ribosomal RNA 16. polypeptide 17. initiator 18. start codon 19. polypeptide 20. amino acid 21. amino acids 22. stop codon 23. ribosome 24. ribosome 25. protein

Exercise 6: 1. transcription 2. translation 3. DNA 4. mRNA 5. RNA polymerase 6. amino acid 7. tRNA 8. anticodon 9. large ribosomal subunit 10. initiator tRNA 11. initiation 12. small ribosomal subunit 13. mRNA 14. start codon 15. polypeptide 16. peptide bond 17. elongation 18. codons 19. termination 20. polypeptide 21. stop codon

Exercise 7: 1. Met-Pro-Asp-Asn-Ile-Lys 2. Met-Pro-Asp-His-Ile-Lys 3. one base changed 4. substitution (10th base changed from A to C) 5. one amino acid changed (Asn to His) 6. Met-Pro-Asp-Glu-Tyr 7. one base changed 8. Insertion (10th base, G, inserted between C and A) 9. two amino acids changed and chain shortened by one amino acid, due to change to stop codon 10. effect of mutation in mRNA 3 greatest, due to reading frame shift and alteration of all codons following insertion

Exercise 8: 1. E 2. G 3. B 4. O 5. R 6. S 7. U 8. A 9. J 10. D 11. V 12. U 13. T 14. I 15. F 16. L 17. H 18. M 19. U 20. P 21. N 22. L 23. C 24. Q 25. K

Testing Your Knowledge

Multiple Choice: 1. a 2. c 3. c 4. b 5. c 6. e 7. a 8. b 9. a 10. d 11. c 12. c 13. b 14. e

Essay: 1. See Figure 10.3D in the text.

2. A single-ringed pyrimidine, such as T, must pair with a double-ringed purine, such as A. Two pyrimidines—A and C—would not be large enough to reach across the double helix. T specifically pairs with A and not G because T and A have complementary chemical side groups that form hydrogen bonds.

3. Twenty amino acids are used in building proteins. There are only 4 nucleotides in DNA, so a one-base code could specify only 4 amino acids. A two-base code could specify only 4^2, or 16 amino acids. In a triplet code, 4^3, or 64, combinations of bases are possible, more than enough to code for 20 amino acids.

4. A mutation is a change in the nucleotide sequence of DNA. Some mutations are spontaneous, but many are caused by X rays, ultraviolet light, and chemicals. Most mutations are harmful because they alter protein amino acids sequences and impair the function of proteins. Some mutations do not alter amino acid sequence; others lead to an improved protein or one with new capabilities that enhance an organism's success. Mutations create genetic diversity that makes evolution by natural selection possible.

5. A base substitution can result in no change to a protein or, most often, a change in one amino acid. Effects on the organism may be minimal. A base deletion alters the sequence of triplet groupings in a gene, causing a drastic change in the amino acid sequence downstream from the deletion. This can have a profound effect on the protein and the organism.

6. Messenger RNA (mRNA) is transcribed from a gene and carries the instructions for making a particular polypeptide. A ribosome is the site of translation, the process of making a polypeptide according to the mRNA message. Transfer RNAs act as translators, each kind matching a particular amino acid with a particular codon in the mRNA. The ribosome "reads" the mRNA one codon at a time, and tRNAs deliver their amino acids, which are added to the polypeptide chain one at a time.

Applying Your Knowledge

Multiple Choice: 1. a 2. a 3. d 4. e 5. c 6. d 7. b 8. a 9. a 10. c

Essay: 1. Extract DNA from dead type A *E. coli* and mix with live type V *E. coli*. If the type V *E. coli* absorb the DNA and genes are made of DNA, the type V bacteria will be able to grow on the simple medium without the vitamin. For comparison, extract protein from dead type A *E. coli* and mix with live type V *E. coli*. If the type V *E. coli* absorb the protein and genes are made of protein, the type V bacteria will be able to grow on the simple medium without the vitamin. Prediction: DNA will transform the type V bacteria, but protein will not.

2. In the DNA double helix, T and A bases pair up; wherever there is a T nucleotide in one strand, there is a complementary A nucleotide in the other. Thus the amount of T is equal to the amount of A. The specific nucleotide sequences of goldfish and human DNA are different, so the number of A-T pairs is different in the two species.

3. They are both right. Eric is talking about phenotype and Renee about genotype. The nucleotide sequences of your genes—genotype—determine the amino acid sequences of proteins that cause your hair to be curly or straight—your phenotype.

4. Met-Glu-Leu-Ser-Ile-Asp (Remember that translation always starts at the AUG start codon.)

5. Because a 3-base codon specifies each amino acid, the message coding for the polypeptide would consist of 233×3, or 699 nucleotides (actually 702, if the stop codon is included). The rest of the nucleotides are extras. Translation of protein does not have to start and stop at the ends of the mRNA. It starts at a start codon (AUG) and stops at a stop codon (UAA, UAG, or UGA). The nucleotides from the start codon to the stop codon are the only ones that spell out the amino acid sequence of the polypeptide.

6. The virus may be able to insert its genes into a nerve cell's DNA and remain latent within the cell as a provirus. From time to time the provirus may begin reproducing complete viruses, causing disease symptoms.

Chapter 11

Organizing Your Knowledge

Exercise 1: *lac* operon: 1. E 2. G 3. A 4. B 5. F 6. C 7. H 8. D *trp* operon: 1. E 2. A 3. G 4. B 5. F 6. C 7. H 8. D

Exercise 2: *Stomach gland cell:* Krebs cycle enzyme gene, digestive enzyme gene. *Hair follicle cell:* keratin gene, Krebs cycle enzyme gene. *Stem cell:* hemoglobin gene, Krebs cycle enzyme gene.

Exercise 3:

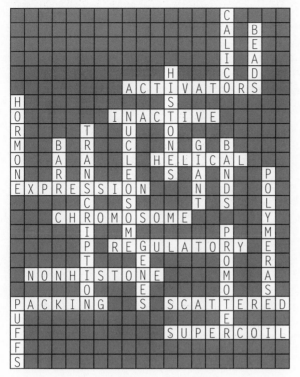

Exercise 4: 1. E 2. D 3. G 4. B 5. C 6. A 7. F

Exercise 5: 1. DNA packing 2. transcription 3. splicing/editing 4. addition of cap and tail 5. transport 6. mRNA degradation 7. translation 8. cleavage/modification/activation 9. protein degradation

Exercise 6: For example: Homeotic genes are master control genes containing nucleotide sequences called homeoboxes, which code for polypeptides that bind to DNA and activate or repress transcription of groups of genes.

Exercise 7: 1. growth 2. division 3. tumors 4. surgery 5. radiation 6. chemicals 7. carcinomas 8. carcinogens 9. radiation 10. oncogenes 11. proto-oncogenes 12. growth 13. division 14. mutation 15. tumor-suppressor

Exercise 8: 1. C 2. F 3. D 4. B 5. C 6. D 7. C 8. G 9. A 10. E

Exercise 9: 1. W 2. G 3. W 4. G 5. B 6. G 7. B 8. W 9. W

Testing Your Knowledge

Multiple Choice: 1. d 2. a 3. c 4. a 5. c 6. e 7. d 8. a 9. b 10. e 11. b 12. e

Essay: 1. Differentiated cells can be made to dedifferentiate and regrow the entire organism. Differentiation apparently does not irreversibly change the DNA.

2. Most genes in plants and animals consist of regions that code for polypeptides, called exons, interrupted by long noncoding segments, called introns. Both introns and exons are transcribed from DNA into RNA, then the introns are removed and the remaining exons linked together—a process called RNA splicing. Introns may contain nucleotide sequences that regulate gene activity. The splicing process may help control flow of mRNA from nucleus to cytoplasm. In some cases splicing can occur in more than one way, producing different mRNA molecules from the same transcript. Introns make DNA longer, increasing the number of points where crossing over can occur and thus increasing recombination and genetic diversity.

3. A homeotic gene is a master control gene that functions during development. It produces a polypeptide that acts as an activator or repressor for a battery of genes, switching them on or off together and thus shaping large-scale aspects of body plan.

4. Flies, frogs, and humans have evolved separately for many millions of years. Since they share homeoboxes, this suggests that homeoboxes arose early in the history of life and have remained remarkably unchanged for a long time.

5. Cancer cells escape from normal controls that regulate division and growth and multiply excessively. A cell becomes cancerous when a mutation occurs in a proto-oncogene, a gene that normally makes a protein growth factor that stimulates cell division. The mutation changes the gene into an oncogene that codes for a hyperactive protein—one that stimulates cell division more than normal—or excess production of protein. A second mutation usually occurs in a tumor suppressor gene, which keeps a protein that normally blocks cell division from being made. The combined effect of these mutations is the uncontrolled cell division characteristic of cancer.

6. To avoid cancer a person should avoid carcinogens—UV radiation, unnecessary X rays, tobacco, alcohol, and so on. A high-fiber, low-fat diet, with foods high in beta-carotene, vitamin C, and vitamin E, also reduces the risk of cancer.

7. Aging may be due to the cumulative effects of mutations and other damage to cells that cause the cells to lose function gradually. It is also possible that aging is part of the development process, an innate property of cells programmed by changes in gene expression.

Applying Your Knowledge

Multiple Choice: 1. c 2. b 3. c 4. a 5. b 6. e 7. c 8. a 9. c 10. c

Essay: 1. Normally, when lactose is absent, the repressor binds to the operator site, blocking gene transcription, and no enzymes for using lactose are made. When lactose is present, it binds to the repressor and changes the shape of the repressor in such a way that it can no longer bind to the operator. Genes are then transcribed and enzymes for using lactose are made. If the mutation altered the repressor in such a way that it could no longer bind to the operator, the genes would be transcribed and enzymes for using lactose would be made all the time, whether or not lactose is present.

2. Many answers are possible, but the following is an example: Liver cells are small and metabolically active, making and breaking down many substances. Muscle cells are long and thin and have the ability to contract or shorten. Salivary gland cells form saclike clusters and secrete saliva. Liver cells make fibrinogen, a blood-clotting protein. Genes for making fibrinogen would be active in liver cells but not in the other cells. The genes for building contractile proteins would be active in muscle cells, and the gene for making amylase, a digestive enzyme, would be active in salivary gland cells. All the cells perform glycolysis, the first process in breaking down sugar, so genes that code for glycolysis enzymes would be active in all the cells. The gene that codes for the blood protein hemoglobin would not be active in any of the cells.

3. The minimum number of nucleotides needed to code for the protein is $258 \times 3 = 774$ (or 777, if you wish to include the 3 nucleotides of the stop codon). There are numerous nucleotides, including a "cap" and "tail," upstream from the AUG start codon and downstream from the stop codon. Only the 777 nucleotides from start to stop actually code for the protein. The mRNA is much shorter than the gene because noncoding introns in the gene are cut out and the remaining exons are joined together to form the final mRNA, in the RNA splicing process that occurs in the nucleus.

4. A gene might code for a repressor or activator protein that binds to the DNA of other chromosomes and turns off or enhances transcription of genes at those sites. A gene may code for production of a hormone that travels through the bloodstream to a distant site, enters a cell, and similarly affects gene transcription.

5. Different carcinogens can cause the same kind of cancer because they all might cause mutations in bone-marrow cells that change proto-oncogenes into oncogenes. The oncogenes, no matter how they are activated, have the same effect—to stimulate the uncontrolled cell division characteristic of this kind of cancer.

Chapter 12

Organizing Your Knowledge

Exercise 1: 1. B 2. D 3. B 4. D 5. C 6. A 7. B 8. A 9. A 10. D 11. C

Exercise 2: 1. J 2. H 3. G 4. A 5. D 6. C 7. E 8. I 9. C 10. E 11. B 12. I 13. F 14. B

Exercise 3: 1. vectors 2. gene 3. restriction 4. sequences 5. recognition 6. sticky 7. plasmids 8. complementary 9. ligase 10. replication 11. recombinant 12. transformation 13. cloning 14. library 15. shotgun 16. viruses 17. introns 18. reverse 19. exons 20. retrovirus 21. bacteria 22. genomic 23. expressed 24. protein 25. probe 26. UUCACAUC 27. radioactive 28. nucleotides

Exercise 4: 1. Isolation of plasmid from *E. coli*. 2. Obtaining copies of gene and protein from cloned bacteria. 3. Cloning recombinant DNA. 4. Using a molecular probe to find a gene. 5. Cutting DNA with restriction enzyme. 6. Extracting DNA from a eukaryotic cell. 7. Joining plasmid and DNA fragment using DNA ligase. 8. Inserting a plasmid into a bacterium via transformation. 9. Using reverse transcriptase to make an artificial gene. 10. Mixing plasmids and DNA fragments with sticky ends.

Exercise 5:

```
   F I B R O S I S       M U T A N T S
           M                 I
       T H E R A P Y   T   P L A N T
         E   L       R   R     R
         P   L     L A B O R A T O R Y
         A   P   A   N   T       G E
 G   A N T I B O D I E S   E     E A
 E   I     X   D   G I   M   N   S
 N   T     S   E N   U   N     T
 E C O L I     P     N     L
 S         S   L     I     T   B
         P H A R M A C E U T I C A L
   C     S           C   C
 V R     M     G E N O M E   T I
 A I N S U L I N         L   E
 C P   D   E T H I C A L     R
 C P   S           U       I
 I L   H E M O P H I L I A
 N E E D L E   G       A
 E D     H     S E C R E T E
```

Testing Your Knowledge

Multiple Choice: 1. d 2. a 3. c 4. d 5. d
6. a 7. d 8. e 9. c 10. a 11. c

Essay: 1. Transposons are segments of DNA that are able to move from one site in an organism's genome to another. They often insert in the middle of other genes, disrupting their normal functions. This could alter genes in such a way as to increase genetic diversity, contributing to evolution.

2. A restriction enzyme is used to cut DNA at a particular nucleotide sequence, called a recognition sequence. It cuts the two strands of DNA unevenly, producing DNA fragments with single-stranded ends, called "sticky ends." The sticky ends produced by a restriction enzyme are complementary, and complementary fragments produced by the enzyme can hydrogen bond to one another by their single-stranded ends. An enzyme called DNA ligase can join these loosely linked pieces of DNA by catalyzing the formation of covalent bonds between adjacent nucleotides. A fragment of DNA can be spliced into a plasmid, a small, circular bacterial DNA molecule. A bacterium can take up the plasmid from solution by transformation. When the bacterium replicates its larger chromosome, it also replicates the plasmid. It divides repeatedly, producing a clone of cells containing the recombinant plasmid.

3. Bacteria and yeasts have plasmids that can be used as gene vectors, and they can rapidly and cheaply be grown in large quantities. They will secrete protein products into their growth medium, where they can be collected and then purified.

4. Recombinant DNA technology can be used to produce transgenic crop plants able to fix nitrogen, resist disease, or grow on poor soil. Growth hormone genes might cause farm animals to grow faster. Recombinant bacteria might produce useful drugs, hormones, and vaccines.

5. Several human hormones, such as insulin and growth hormone, are produced by recombinant bacteria and yeasts and used to treat diseases. Genetically engineered viruses and bacteria are also used in the production of vaccines against diseases. Monoclonal antibodies are used in clinical tests and cancer treatment. Transgenic mice, engineered to be susceptible to HIV, are used in AIDS research. Recombinant DNA technology is beginning to be used in gene therapy—alteration of genes in human cells to remedy genetic diseases such as cystic fibrosis.

6. Genetic engineering could create hazardous pathogens that could infect humans, animals, or plants. New organisms could also transfer genes to other species, harming them or making them harmful. Recombinant organisms could also compete with and displace wild species, with unpredictable consequences. Strict laboratory procedures prevent accidental infection and escape of genetically engineered organisms. The organisms themselves are genetically crippled, so they cannot survive outside the laboratory. Some dangerous experiments are banned.

7. The potential dangers of recombinant DNA technology present some ethical questions: Are possible benefits worth the risks? How do we evaluate the risk:benefit ratio? How do we feel about altering species? About adding new species to an already precarious environment? Do we have a right to forgo recombinant DNA technology if it can solve environmental and medical problems? Should we alter our own genes? What uses will be made of genetic information?

Applying Your Knowledge

Multiple Choice: 1. b 2. d 3. b 4. c 5. b 6. b
7. a 8. c

Essay: 1. Some bacteria are combining their DNA with the DNA from the other strain, producing bacteria with the characteristics of both strains. It could be that some bacteria are dying, and their DNA is being taken up from the medium by living bacteria—a process called transformation. Bacterial genes could be transferred from bacterium to bacterium by a bacteriophage—transduction. Or the bacteria could be undergoing conjugation—bacterial mating, in which DNA is transferred from one bacterium to another.

2. Knowing the amino acid sequence of EGF, you could use the genetic code chart to figure out the corresponding DNA nucleotide sequence, and then program the machine to synthesize a gene coding for the protein. Use a restriction enzyme to cut the gene and plasmid, producing complementary sticky ends. Combine the gene and plasmid and use ligase to bond them. Introduce the plasmid into *E. coli* via transformation. The bacteria will replicate their DNA and divide in the growth medium, producing a clone of bacteria with the EGF gene. When you have grown a sufficient quantity of the bacteria, extract the EGF.

3. Shotgun cloning might produce clones of bacteria with altered genes and new combinations

of genes, giving them unpredictable and potentially dangerous properties. For example, bacteria could be produced that are capable of infecting human cells, or producing a dangerous toxin, or containing cancer genes.

4. She could use a molecular probe—an RNA or single-stranded DNA with the complementary base sequence. Such a probe might be an RNA fragment with the sequence UACCGAUAG, for example. This probe contains nucleotides labeled with radioactive isotope. She could mix it with each of the clones. It bonds to the desired complementary base sequence, and the radioactivity tags the clone with the sequence she is trying to find.

5. This bacterium could grow only under laboratory conditions. If it accidentally escaped from the lab, it would not be able to survive and thus would not cause harm in the outside environment.

Chapter 13

Organizing Your Knowledge

Exercise 1: 1. 46 2. 23 3. 3 billion 4. 3 5. 300,000 6. 2 7. 100,000 8. 2000 9. 1000 10. 80 11. 10

Exercise 2: 1. E 2. G 3. B 4. A 5. F 6. H 7. C 8. D

Exercise 3:

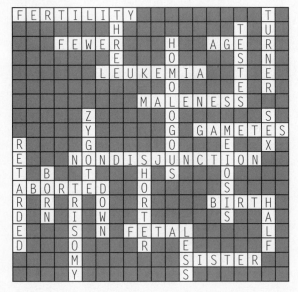

Exercise 4: 1. C, X 2. D, W 3. B, Z 4. A, Y

Exercise 5: 1. ss 2. ? 3. ? 4. ? 5. Ss 6. Ss 7. Ss 8. ss 9. ? 10. ss 11. Ss 12. ? 13. Ss 14. SS 15. ? 16. ? 17. half colored 18. half colored 19. ? 20. half colored 21. fully colored 22. ? 23. ? 24. half colored 25. half colored 26. blank 27. fully colored 28. fully colored

Exercise 6: 1. T 2. F (one-fourth) 3. F (over a thousand) 4. F (normal parents) 5. T 6. F (hemophilia) 7. F (not evenly distributed) 8. F (X chromosome) 9. T 10. F (not necessarily more common) 11. T 12. T 13. F (more common in males) 14. F (There is debate among geneticists.) 15. T 16. F (none of their sons but all of their daughters) 17. T 18. T

Exercise 7: 1. dominant 2. 50% 3. counselor 4. pedigree 5. 25% 6. 4 7. recombinant 8. genetic marker 9. cure 10. Tay-Sachs 11. recessive 12. 25% 13. phenylketonuria 14. amino acid 15. diet 16. amniocentesis 17. amniotic 18. karyotype 19. Down 20. chorionic villi 21. proliferating 22. ultrasound 23. fetoscopy 24. 35

Exercise 8: 1. D 2. L 3. F 4. S 5. H 6. J 7. Q 8. C 9. E 10. K 11. M 12. A 13. P 14. G 15. N 16. O 17. B 18. I 19. R

Exercise 9: 1. correcting 2. bone marrow 3. blood 4. protein 5. gene 6. virus 7. infections 8. cystic fibrosis 9. turn on 10. multiply 11. nerve 12. testes 13. ovaries 14. eugenics 15. nucleotide 16. genome 17. map 18. order 19. distances 20. restriction fragment 21. markers 22. physical 23. restriction 24. sequence 25. 3 26. *E. coli* 27. embryonic 28. ethical

Exercise 10: There is a wide range of possible answers for this exercise, such as the following: "Insurance Companies to Require Genetic Testing" and "Murder Suspect Sues—Claims Genetic Test Is Invasion of Privacy."

Testing Your Knowledge

Multiple Choice: 1. d 2. a 3. e 4. e 5. b 6. d 7. b 8. a 9. b 10. a 11. c 12. e

Essay: 1. Sometimes a pair of homologous chromosomes fails to separate during meiosis I. One of the cells resulting from this division ends up with 24 chromosomes, the other with 22 chromosomes. All of the gametes produced by meiosis II have abnormal numbers of chromosomes. Some have 24 chromosomes—one of each kind plus an extra of one kind. Others have 22 chromosomes—one of each kind but one missing. Sometimes meiosis I is normal, but a pair of sister chromatids stays together during meiosis II. In this case, some gametes have the normal number of chromosomes, but some have one extra and some are one short.

2. An extra copy of chromosome 21 is usually

not fatal, but it does produce Down syndrome. In most other situations in which an abnormal number of chromosomes occurs, the offspring is spontaneously aborted, so these abnormalities must be so serious as to be fatal.

3. John's father is undoubtedly heterozygous, and his mother is almost certainly normal (homozygous recessive), so there is a 50% probability that John has inherited the HD gene from his father. Many individuals at risk for Huntington's choose not to be tested because there is no treatment or cure for the disease.

4. Restriction enzymes cut DNA into fragments at specific sequences called restriction sites. The cut-up DNA is placed on a gel and pulled through the gel by the attraction of the negatively charged phosphate groups on the DNA to a positive electrode. This method is called electrophoresis. Smaller fragments move through the gel faster, and this separates fragments of different sizes. Fragments of different sizes accumulate in bands in the gel, and the resulting DNA fingerprint looks a bit like a supermarket bar code. Each individual has a unique DNA fingerprint, because the nucleotide sequence of each person's DNA is unique. This means that restriction sites differ among different individuals, and restriction enzymes cut the DNA into different-sized fragments, which produce a different RFLP band pattern.

5. The most difficult ethical question concerning human gene therapy is whether we should correct genes in human germ cells—the cells in ovaries and testes that give rise to eggs and sperm. This could eliminate genetic defects in future generations, but it could alter the genetic makeup of the human gene pool and alter our evolution in unexpected ways.

6. The human genome project will (1) produce a linkage map of the human genome, showing the order of linked genes on chromosomes and the distances between them, (2) produce a physical map of restriction fragments, (3) determine the nucleotide sequence of the human genome, and (4) compare the human genome with the genomes of other species. The data will give us insight into human development and evolution. It will also aid in understanding, diagnosis, treatment, and prevention of genetic disorders, as well as understanding of the relationship between genetic makeup and cancer, heart disease, and other health problems.

7. DNA data might be used to deny health insurance or specific jobs or positions to individuals or certain ethnic groups. It could be used to coerce people into medical tests or treatment. It could also lead to ill-conceived eugenics schemes—efforts to "improve" the human species.

Applying Your Knowledge

Multiple Choice: 1. d 2. d 3. b 4. d 5. c 6. b 7. a 8. d 9. c 10. b 11. e 12. d

Essay: 1. *XXY, XYYY,* and *XXYY* would produce male phenotypes; *XXXX* and *X* would produce female phenotypes. A single *Y* chromosome produces "maleness"; the absence of a *Y* chromosome produces "femaleness."

2. Most human genetic disorders are recessive, and the genes that cause them are unknowingly carried by heterozygotes. One in 25 Caucasians in the U.S. carries the cystic fibrosis allele. More than a thousand human genetic disorders are known, so it is likely that most individuals carry the gene for some genetic disorder. It is unlikely that two individuals carrying recessive genes for the same disorder will meet and have children, so the disorders are much less common than the genes.

3. The genes for sex-linked disorders are on the *X* chromosome. If both parents are normal but have a child with a sex-linked disorder, the mother must carry the allele for the disorder on one of her *X* chromosomes, and she will pass it on to half of her children. Half of her daughters will carry the allele for coloboma iridis, but they will be of normal phenotype because they will inherit the normal allele from their father. Half of her sons will also inherit the coloboma iridis allele. Because they inherit *Y* chromosomes from their father (and no normal allele that could mask the coloboma iridis allele), half the sons will exhibit coloboma iridis. Thus, one-fourth of their children—all boys—will have coloboma iridis.

4. Both homozygous dominant and heterozygous individuals have the gene that codes for the enzyme and are therefore able to make the normal enzyme (although heterozygous individuals might make some defective enzyme as well). Because enzymes are normally needed in relatively small amounts, both homozygous dominant and heterozygous individuals will be of normal—dominant—phenotype. Only those individuals with two defective recessive genes will be unable to make any normal enzyme, and only they will be of recessive phenotype.

5. If Diane's brother has hemophilia (and pre-

sumably her father does not), her mother must be a carrier of the hemophilia allele, and she would pass this allele on to half her daughters. The probability that Diane is a carrier is 1/2. If Diane is a carrier, there is a probability of 1/4 that she and Craig will have a child—a son—with the disease. The probability both that Diane is a carrier and that Diane and Craig will have a child with hemophilia is $1/2 \times 1/4 = 1/8$.

6. The remains are probably those of Manuel, because the DNA fingerprint of the remains contains bands matching some of those in the father's DNA fingerprint and some of those in the mother's DNA fingerprint, but no bands that are not found in either of the parents' DNA fingerprints.

7. The lack of blood enzyme is a better candidate for gene therapy because bone marrow cells can be removed from the patient; their genes can be altered; and the cells can be reintroduced to the patient. The reintroduced cells can continue to proliferate through life, curing the disease. At this time it is not possible to genetically engineer specialized cells such as nerve cells that do not continue to proliferate, because it is not possible to get the corrected gene into the cells and have them function properly.

Chapter 14

Organizing Your Knowledge

Exercise 1: 1. D 2. E 3. G 4. F 5. A 6. B 7. C

Exercise 2: 1. comparative embryology 2. molecular biology 3. fossil record 4. comparative anatomy 5. molecular biology 6. biogeography 7. fossil record

Exercise 3: 1. Charles Darwin 2. evolution 3. vary 4. heritable 5. overproduction 6. resources 7. fit 8. survive 9. reproduce 10. fit 11. natural selection 12. snails 13. differential 14. population 15. population 16. isolated 17. gene pool 18. fit 19. genes 20. microevolution 21. population

Exercise 4: B. *RR, Rr, rr* C. 36, 48, 16 D. 36%, 48%, 16% E. 72, 48, 0 F. 0, 48, 32 G. $p = 0.6$, $q = 0.4$ H. 0.6, 0.4 J. *RR, Rr, rr* K. 36%, 48%, 16% L. 72, 48, 0 M. 0, 48, 32 N. $p = 0.6$, $q = 0.4$ O. The genotype and allele frequencies stayed the same in the second generation and will continue to do so in succeeding genera-

tions. This illustrates that the frequency of each allele in the gene pool will remain constant unless acted on by other agents. Sexual reproduction alone does not lead to microevolution—the so-called Hardy-Weinberg principle.

Exercise 5: 1. The population is no longer isolated; there is gene flow—migration of individuals with yellow alleles into the population. 2. A mutation alters the frequency of red and yellow alleles in the gene pool. 3. Because the population is small, a chance event such as a landslide can significantly alter the gene pool. This is called genetic drift. 4. Individuals differ in reproductive success, changing the relative frequencies of red and yellow alleles in the gene pool. This is natural selection, and it results in adaptation to the environment. 5. Mating is not random because a yellow plant does not stand an equal chance of mating with every other plant in the population. It is more likely to mate with a yellow plant.

Exercise 6:

```
 F
RECOMBINATION
   T  CONTINUOUSLY
   I  H    G          U
   L  R    E         U
   I  O    G        U
   Z  M  BOTTLENECK
   A  O    G         I
   T  S    R  P      F     A
   I  O    A  O      O     D
POLYMORPHIC    R     V     M
 N        A    H  N  MUTATION
          L    I  T  I     N     R
        C  C      R  T     T  Y  T    A  P
        L  A      E        Y  A    H
       DIPLOID             G
        N         U  GENETIC
        E         C
     BACTERIA
```

Exercise 7: 1. struggle 2. fittest 3. fitness 4. phenotype 5. phenotype 6. genotype 7. gene pool 8. population 9. genome 10. environment 11. bell 12. stabilizing 13. directional 14. change 15. diversifying 16. intermediate

Testing Your Knowledge

Multiple Choice: 1. e 2. b 3. d 4. d 5. d 6. a 7. c 8. e 9. b 10. b 11. e 12. d

Essay: 1. Because organisms produce more offspring than the resources in their environment can support, only a fraction of these offspring survive and reproduce themselves. Individuals in a population vary in their characteristics, and many of these varying traits are inherited. Those individuals with traits that fit them best to their environment are most likely to survive and reproduce, and thus these individuals leave more offspring than less fit individuals. Over time, the fitter types make up a larger fraction of the population, and through this process of natural selection the population adapts to its environment.

2. Biogeography, the geographical distribution of species, suggests that species change when they move into new habitats. Animals on islands resemble those on the nearest land mass more than those in more distant localities. The fossil record shows sequential changes in living things over time. Comparative anatomy and comparative embryology show the relationships among organisms descended from a common ancestor. Molecular biology shows that organisms thought to be more closely related on the basis of other criteria also have a greater proportion of their DNA and proteins in common.

3. A population is a group of individuals of a particular species living in a particular place at a particular time. A species is all the populations whose individuals can potentially interbreed. The white-tailed deer on an island constitute a population; all white-tailed deer are members of a particular species.

4. Darwin would have said that both horses and zebras are descended from an ancestral species that lived in the past. As the descendants of this species spread into different environments, they adapted in different ways to different conditions. Horses and zebras eventually became distinct species, but their similarities show that they are related.

5. In a small population, a chance event might alter the frequency of alleles in the gene pool so as to increase the frequency of an allele that would otherwise be disadvantageous. In a diploid population, a recessive allele affects the phenotype only when two copies of it are present in a homozygous individual. Many harmful recessive alleles in a population might be hidden in heterozygous individuals, not exposed to the effects of natural selection. Sometimes natural selection actually favors heterozygotes—a phenomenon known as heterozygote advantage. In this case, heterozygotes might have greater reproductive success than either homozygote, further increasing the frequency of an allele that is harmful in the homozygous recessive state.

6. All these phenomena alter allele and genotype frequencies in the gene pool but in different ways. In a small population, a chance event could kill some individuals and alter the gene pool. If individuals prefer some types of potential mates to others—that is, if mating is nonrandom—certain alleles are more likely to be passed on than others, and the gene pool can change. Migration of individuals in or out of the population can add new alleles or delete alleles from the gene pool. Mutation creates a new allele, altering allele frequencies. If a certain allele enables individuals with that allele to better survive and reproduce, that allele is more likely to be passed on to the next generation than others, and its frequency in the gene pool will increase. This is natural selection.

Applying Your Knowledge

Multiple Choice: 1. e 2. a 3. a 4. c 5. b
6. d 7. c 8. b 9. e 10. d

Essay: 1. Lamarck would have said that the ancestral butterfly had a shorter proboscis. It stretched the proboscis to reach the nectar deep in the flower, and this change was passed on to its offspring. In this way, the proboscis got longer and longer. Darwin would have said that the ancestral butterflies varied—some had longer probosci than others—and that proboscis length is inherited. Those butterflies with the longest probosci were able to get the most food from the flowers and were therefore better able to survive and reproduce. A greater proportion of butterflies in the next generation had longer probosci, and over time, this survival of the fittest resulted in an increase in the length of the proboscis.

2. Darwin saw that by selecting individuals with desired characteristics as breeding stock, humans play the role of the environment and bring about differential reproduction. If this artificial selection can bring about great change in pigeons in a few hundred years, natural selection must be able to cause even greater changes over millions of years.

3. An individual cannot adapt in an evolutionary sense. Each plant has particular traits; it can either stand heat or it cannot. Trina recognized that individual plants probably varied in their ability to withstand heat. Some of the plants are more fit than others. Less fit individuals die, and more fit individuals are better able to survive and reproduce, passing their genes for the traits that enable them to withstand the heat on

to the next generation. In the next generation of wildflowers, a greater proportion of individuals will be of the more fit type. Thus individual plants do not adapt, but the population does.

4. Use the formulas $p + q = 1$ and $p^2 + 2pq + q^2 = 1$ to answer these questions. 9% of the individuals in the population are homozygous recessive (white). Thus, $q^2 = .09$, and $q = 0.3$. Since $p = 1 - q$, $p = 0.7$. Frequencies of homozygous dominant, heterozygous, and homozygous recessive individuals are 49%, 42%, and 9%, respectively. Frequency of the green allele is 0.7. Frequency of the white allele is 0.3. The genotype and gene frequencies will stay the same in the next generation if the population is at Hardy-Weinberg equilibrium.

5. The population could be made to deviate from Hardy-Weinberg equilibrium in several ways. One could randomly scoop out some fish. In such a small population, this could probably alter allele and genotype frequencies by chance—genetic drift. One could cause gene flow by adding white fish to the aquarium. Subjecting the fish to X rays might cause mutations that would alter the makeup of the gene pool. One could segregate the fish so that mating is non-random—only allowing fish of the same color to mate with each other, for example. Putting in a predator that can see and catch the white fish more easily than the green ones would result in natural selection in favor of the green fish and adaptation of the fish population to their environment.

6. This is similar to the snail example in Module 14.5. The data indicate that the owls are catching a larger proportion of the dark mice than of the light mice. Perhaps the dark mice are easier to see against the white sand. If coloration is inherited, the light mice are more likely to pass their genes on to future generations, so a larger proportion of the mice will be light colored. Directional selection is occurring, and the mouse population is adapting to its environment.

Chapter 15

Organizing Your Knowledge

Exercise 1: 1. Because both are extinct, we must rely on appearance rather than the biological species concept. 2. Tigers and lions would probably not interbreed under natural conditions. 3. Appearance can be deceptive; they can interbreed. But would a chihuahua really be able to mate with a St. Bernard? 4. The criterion of in-terbreeding is useless for asexual organisms; we have to classify them by appearance and biochemical features. 5. A and C are becoming separate species, but there is still some gene flow between them via B. 6. If it is difficult to distinguish species on the basis of appearance, it might be difficult to know whether isolated populations are of the same or different species. 7. Are the birds the same species if they look alike? If the populations are separated, it might not be possible to determine whether they could interbreed in nature.

Exercise 2: 1. prezygotic, temporal isolation 2. postzygotic, hybrid breakdown 3. postzygotic, hybrid sterility 4. prezygotic, mechanical isolation 5. prezygotic, behavioral isolation 6. prezygotic, gametic isolation 7. prezygotic, habitat isolation 8. postzygotic, hybrid inviability

Exercise 3: There are many possible answers for each of these questions, for example: 1. ocean, mountain range 2. mountain range, river 3. separate river basins, dry land between lakes 4. mountain range, ocean 5. ocean, mountain range 6. land bridge, deep water between coral reefs

Exercise 4: There are many possible examples, such as the following: 1. As a large lake dries up, fish or frogs could be isolated in smaller ponds or springs. 2. Wildflowers or beetles could evolve on isolated mountains surrounded by lowlands. 3. Insects could evolve in an isolated cave.

Exercise 5: 1. yes, yes, 36 2. no, no, 33 3. no, no, 35 4. yes, yes, 36 5. yes, yes, 28 6. no, no, 21

Exercise 6: 1. smooth, jumpy 2. same, different 3. gradual transitions, sudden appearance and disappearance 4. slow, fast 5. continuous, quick change and stability

Exercise 7: 1. population 2. species 3. geographical barrier 4. genetic drift 5. natural selection 6. interbreed 7. pre 8. behavioral isolation 9. post 10. hybrid inviability 11. species 12. isolating 13. allopatric 14. punctuated equilibrium 15. polyploid 16. tetraploid 17. nondisjunction 18. isolated 19. species 20. sympatric

Testing Your Knowledge

Multiple Choice: 1. b 2. b 3. d 4. d 5. c 6. a 7. d 8. a 9. e 10. b

Essay: 1. It is not possible to see whether fossil species could have interbred. Isolated populations may not naturally have the opportunity to interbreed. The interbreeding criterion does not apply to asexual organisms.

2. A splinter population is cut off from the parent population by a geographical barrier such as a mountain range, a canyon, or a stretch of ocean. Changes in allele frequencies occur in the splinter population, caused by mutation, selection, and genetic drift. These changes are undiluted by gene flow from the parent population. Finally, the splinter population has diverged so much from the parent population that individuals from the two populations can no longer interbreed. The presence of prezygotic or postzygotic barriers tells us that they have become separate species.

3. Imagine two species of flowers: species A, with a diploid number of 10 ($2n = 10$), and species B, with a diploid number of 12 ($2n = 12$). Two things have to happen to produce a fertile hybrid—hybridization and then nondisjunction in meiosis. Pollen from species A ($n = 5$) fertilizes an egg from species B ($n = 6$). This produces a zygote with 13 chromosomes. This zygote can develop into a hybrid plant, but the hybrid would be sterile because its chromosomes cannot pair during meiosis. However, a failure during meiosis—nondisjunction—could produce gametes with 13 chromosomes. Self-fertilization among the gametes would produce a zygote with 26 chromosomes ($2n = 26$). This hybrid would be fertile, because all the chromosomes would be able to pair in meiosis.

4. A small, isolated population is more likely to undergo speciation, because it is more likely to be affected by natural selection and genetic drift in a local area and less affected by gene flow that would make it more like other populations.

5. According to the gradualist model, isolated populations gradually evolve differences as they adapt to local environments. There is a smooth, gradual process of change from one species to the next, through an incremental accumulation of microevolutionary differences. The punctuated equilibrium model suggests that evolution occurs in spurts. New species rapidly diverge from the parent stock and then remain unchanged for long periods. The fossil record more often shows the sudden appearance, persistence, and sudden disappearance of species predicted by the punctuated equilibrium model, rather than the gradual transitions predicted by the gradualist model.

Applying Your Knowledge

Multiple Choice: 1. a 2. e 3. e 4. d 5. a
6. e 7. b 8. c 9. b 10. c

Essay: 1. The western United States has much more rugged terrain than the east. There are many mountain ranges and canyons that would be formidable geographical barriers, isolating populations and promoting speciation.

2. A cell from a mule would contain 63 chromosomes, 32 from the horse gamete and 31 from the donkey gamete. A mule is sterile because, in meiosis, its chromosomes are unable to pair up, and then separate to form haploid cells.

3. This is a gray area, because dogs and coyotes probably would not interbreed under natural conditions in the wild. If dogs and coyotes are different species, this shows us that they are closely related, having recently evolved from a common ancestor species.

4. Dry stretches of desert between streams constitute geographical barriers that isolate fish populations in desert streams, much as stretches of ocean separate land animals on islands. Isolated fish populations might undergo different mutations and genetic drift and be subjected to different kinds of natural selection. They might evolve into different species, like animals isolated on oceanic islands.

5. Flies in newer populations might be less picky because they have not been separated from other populations as long, so they still share some characteristics (courtship signals, and so on) and can be "fooled" by flies of other species. Perhaps females are pickier than males because females only mate once and males can mate repeatedly. Females who make mistakes are not as successful at reproducing, so there is strong selection in favor of females who can tell the difference between their own species and other species.

6. According to the punctuated equilibrium model, species evolve rapidly and then remain unchanged for long periods. If the banana-eating moths are native to Hawaii, and bananas have only been there for 1000 years, this suggests that a new species of moth evolved there in 1000 years or less. This is a short time by evolutionary standards and is a bit of evidence in favor of the punctuated equilibrium model.

7. This criticism is not valid. Formation of a new species by allopatric speciation might take thousands of years. We have not been able to study species for long enough to see the whole process happen, but there is a lot of circumstantial evidence—Darwin's finches, desert pupfish, and so on. More important, we have seen many species of plants appear virtually overnight, via sympatric speciation—hybridization and polyploidy. Hugo de Vries saw this happen in primroses, for example.

Chapter 16

Organizing Your Knowledge

Exercise 1:

Exercise 2: A. 7 Mesozoic B. 10 Cenozoic
C. 2 Precambrian D. 3 Paleozoic E. 9 Cenozoic
F. 8 Mesozoic G. 4 Paleozoic H. 5 Paleozoic
I. 1 Precambrian J. 6 Paleozoic

Exercise 3: 1. The mammoth is 11,200 years old.
(Half the carbon would disappear in the first
5600 years, leaving 8 grams. Half of that would
disappear in the next 5600 years, leaving 4
grams.) The minimum and maximum ages of
the fossil are 10,080 to 12,320 years.

2. Half the potassium-40 would disappear in the
first 1.3 billion years, leaving 0.5 grams. Half of
this would disappear by 2.6 billion years, leaving 0.25 grams. At 3.9 billion years, half of that
would remain—about 0.12 grams.

Exercise 4: 1. L 2. J 3. G 4. M 5. N 6. O
7. A 8. P 9. C 10. D 11. E 12. F 13. H
14. K 15. I 16. B

Exercise 5: 1. Cretaceous 2. 65 3. 150 4. six
5. Pangaea 6. 90 7. cooling 8. iridium 9. meteorites 10. meteorite 11. asteroid 12. volcanoes
13. hair 14. milk 15. preadaptations 16. titanoth-
eres 17. speciation 18. species 19. regulatory
20. coding 21. neoteny 22. childhood 23. learn

Exercise 6: 1. D 2. F 3. B 4. C 5. G 6. E
7. A

Exercise 7: 1. species 2. genus 3. family 4. order
5. class 6. phylum 7. kingdom

Exercise 8: 1. B or F 2. F or B 3. D 4. A or C
5. C or A 6. E

Exercise 9: 1. both 2. cladistic 3. cladistic
4. classical 5. cladistic 6. classical 7. cladistic
cladistic 9. both 10. cladistic 11. classical
12. classical 13. cladistic 14. classical

Exercise 10: 1. Protista 2. Animalia 3. Fungi
4. prokaryotic 5. eukaryotic 6. eukaryotic 7. eukaryotic 8. single-celled 9. both 10. multicellular
11. multicellular 12. multicellular 13. autotrophic
14. heterotrophic; ingestion and digestion

Testing Your Knowledge

Multiple Choice: 1. d 2. b 3. a 4. b 5. d
6. a 7. a 8. b 9. b 10. b

Essay: 1. When the land masses fused, the total
amount of shoreline and shallow coastal areas
was reduced, and many marine species that inhabit shallow waters may have died as their
habitats shrank. Ocean current would have
shifted, altering climate. Less land would have
been near the moderate coastal climate, so the
climate in the interior may have become more
harsh, leading to extinction of land species.

2. These events occurred in the following order:
first prokaryotes, origin of eukaryotes, origin of
animals, appearance of first vertebrates, movement of plants and animals onto land, dominance of dinosaurs and cone-bearing plants,
origin of flowering plants, diversification of
mammals, appearance of humans.

3. Boundaries between geological eras and periods are marked by major transitions in the
forms of life present in the fossil record. For example, many forms of marine life died out at
the end of the Paleozoic era, and dinosaurs first
appeared in the Mesozoic. These major changes
in the history of life have apparently occurred
at irregular intervals, so the geological timescale
is uneven.

4. When plates collide, their edges may crumple
and form mountain ranges. The Himalayas
formed this way when India collided with
Eurasia. Volcanoes, such as Mount St. Helens,
often arise where plates grind past each other.

5. Animalia—Kingdom, Chordata—Phylum,
Vertebrata—Subphylum, Mammalia—Class,
Primates—Order, Hominidae—Family, *Homo*—
Genus, *sapiens*—species.

Applying Your Knowledge

Multiple Choice: 1. d 2. c 3. a 4. b 5. c
6. e 7. c 8. e 9. d 10. c

Essay: 1. There would be 200 milligrams after
5600 years, 100 milligrams after 11,200 years,
and 50 milligrams after 16,800 years. The tree
was buried 16,800 years ago.

2. Figure 16.5B in the text shows that Australia
(with Antarctica) broke away from Gondwana
while the other continents were still joined. This

might have happened after the marsupials appeared but before placental mammals evolved, so the placentals were never able to reach Australia and displace the marsupials.

3. Late in the Cretaceous, the climate cooled and shallow coastal seas receded. Many food plants used by the dinosaurs died out before the dinosaurs did. Continental drift may have caused climatic changes, but were these changes alone enough to cause the extinction of the dinosaurs? Rocks in India indicate that at the same time, massive volcanic activity released particles into the atmosphere, perhaps contributing to the cooling. Sediments from the end of the Cretaceous contain a layer of iridium, an element rare on Earth but common in meteorites. The iridium layer may be fallout from a huge cloud of dust produced when a large meteorite or asteroid hit the Earth. This could have blocked light and disrupted the climate for an extended period, killing off many plants and the dinosaurs that depended on them. Finally, it is possible that a combination of factors is responsible for the demise of the dinosaurs. Many forms of marine life also perished; any of these hypotheses could account for the extinction of both dinosaurs and marine species.

4. One way to determine how closely related the birds are would be to compare the amino acid sequences of proteins from the birds. Samples of hemoglobin, for example, could be sequenced; the more the sequences are alike, the closer the relationship between the birds. Similarly, samples of DNA could be sequenced. The greater the similarity in nucleotide sequences, the closer the relationship. A simpler way to compare DNA is DNA-DNA hybridization. DNA from two species is cut up and heated to separate the complementary strands. The single-stranded DNA from two species is mixed and cooled to re-form double-stranded DNA. The more similar the DNA of two species, the greater the number of complementary nucleotides and hydrogen bonds between the strands. Heat is used to test for complementarity; the more heat needed to part the strands, the more similar they are, and the more closely related the birds are.

5. Apparently, the similarities between the two groups of vultures are analogous, not homologous. The birds probably have similar lifestyles and live in similar environments. They have apparently been similarly shaped by natural selection, even though they are not closely related. This is called convergent evolution.

Chapter 17

Organizing Your Knowledge

Exercise 1: A. 6 B. 4 C. 2 D. 3 E. 8 F. 1 G. 9 H. 10 I. 5 J. 7 K. 11

Exercise 2: 1. simple molecules 2. energy from lightning, UV radiation 3. small organic molecules 4. polymerization 5. proteinoids 6. simple RNA replication 7. proteinoid microspheres 8. RNA-protein "co-ops" 9. enclosure of co-ops by microspheres 10. natural selection 11. the first cells

Exercise 3: 1. smaller—1 to 10 μm 2. no membrane-bound nucleus or other organelles 3. one large DNA molecule in a ring 4. no true sexual reproduction 5. binary fission 6. nearly all have cell walls of peptidoglycan 7. appeared perhaps 4 billion years ago

Exercise 4: 1. heterotroph 2. autotroph 3. chemoheterotroph 4. photosynthetic autotroph

Exercise 5:

```
                S                           E
C Y A N O B A C T E R I A           T       N       S
            R               P               D       T
  M E T H A N O G E N S       P     E   A   O       R
        E                   P       U   T   O       O
    A N T I B I O T I C S           U   T   P       M
        E                   O       B   O   O       A
        R           A R C H A E B A C T E R I A     T
B L O O M                   O       C   R   E   S   O
A       T       C O C C I           T   O   P       L
C       R                   C       E   P   H       I
I       O       S P R I N G S       R   H   S       T
L       P               E           I           F   E
L       H       E                   A           L   S
I     S U N L I G H T                           A
          E             R           P           G
          R         G L Y C O L Y S I S         E
          Y             L                       L
                        S P I R A L             A
```

Exercise 6: 1. prokaryotes 2. eukaryotes 3. oxygen 4. pathogens 5. 50 6. tissues 7. poisons (toxins) 8. exotoxins 9. endotoxins 10. *Salmonella* 11. water-treatment 12. sewage 13. antibiotics 14. postulates 15. pure culture 16. inoculate 17. experimental hosts 18. nutrients 19. legumes 20. nitrogen 21. cyanobacteria 22. organic 23. wastes 24. dead organisms 25. treatment 26. anaerobic 27. sludge 28. aerobic 29. lead (mercury, and so on) 30. tick

Exercise 7: 1. mitochondrion and chloroplast 2. Mitochondria and chloroplasts are like bacteria in several ways: They have their own DNA,

RNA, and ribosomes, which are similar to those of bacteria. They transcribe their own RNA and make their own proteins. They replicate their own DNA and reproduce by a process like binary fission. They have two membranes, and the inner one resembles a bacterial membrane. 3. It is possible that organelles first arose as infoldings of a prokaryote's plasma membrane. 4. nucleus and endomembrane system 5. Some organelles may be derived from symbiotic prokaryotes and others from the host cell membrane.

Exercise 8: 1. Protista 2. unicellular 3. protozoans 4. amoebas 5. pseudopodia 6. parasitic 7. malaria 8. cilia 9. *Paramecium* 10. flagellates 11. sleeping sickness 12. cellular 13. colonies 14. slime molds 15. nuclei 16. plant 17. photosynthetic 18. green 19. cellulose 20. colonial 21. multicellular 22. stems 23. roots 24. red 25. warm water 26. coralline algae 27. green 28. brown 29. largest 30. kelp 31. diatoms 32. silica 33. oil 34. cellulose 35. spinning

Exercise 9: 1. H 2. F 3. E 4. A 5. C 6. D 7. G 8. B

Testing Your Knowledge

Multiple Choice: 1. d 2. a 3. b 4. c 5. a 6. c 7. e 8. e 9. c 10. b 11. a 12. b 13. d 14. d 15. c

Essay: 1. Present-day conditions are different from conditions on Earth when life first appeared. Earth's present atmosphere is rich in oxygen, which tends to disrupt chemical bonds. Early Earth probably had a reducing atmosphere, instead of an oxidizing one. It would have added electrons to molecules, causing chemical bonds to form and causing simple molecules to combine to form more complex ones.

2. Miller's experiment showed how simple organic compounds could have formed from inorganic chemicals on the primitive Earth. A flask containing warm water simulated the sea, while CH_4, H_2, and NH_3 simulated the early atmosphere. Electrical sparks introduced into the flask simulated lightning. A condenser cooled water vapor and caused "rain" that washed dissolved compounds into the "sea." As material circulated through the apparatus, the sparks triggered chemical reactions that formed a variety of organic compounds, including a mixture of amino acids.

3. Small nucleic acid molecules—probably RNA at first—may have developed the ability to replicate. Proteinoids—simple proteins—might have acted as enzymes to catalyze this process, while the RNA "genes" may have come to code for the simple proteins. Proteinoids tend to self-assemble into fluid-filled microspheres, which could have acted as a protective, semipermeable membrane, enclosing these gene-enzyme "co-ops."

4. Bacteria cause tuberculosis, cholera, bubonic plague, and typhoid fever. Some bacteria cause disease by actually destroying tissues. Most cause illness via poisons—exotoxins secreted from cells or endotoxins that are components of cell walls.

5. Bacteria are important in the cycling of chemical elements between living things and the nonliving environment. Cyanobacteria restore oxygen to the atmosphere. Some bacteria take nitrogen from the air and convert it to forms in the soil that can be used by plants to make proteins and nucleic acids. Bacteria also act as decomposers, breaking down organic matter in dead organisms and body wastes and converting them to inorganic forms that are available to organisms for reuse. Some bacteria might be useful in similarly breaking down spilled petroleum and toxic wastes.

6. Primitive cyanobacteria, such as those that formed fossil stromatolites, dominated the Earth from about 3 billion to about 1 billion years ago. Their photosynthesis released enough oxygen (O_2) gas to make the atmosphere aerobic, setting the stage for the evolution of aerobic organisms such as animals.

7. The endosymbiotic hypothesis suggests that eukaryotic cells may have started out as cooperative communities of prokaryotic cells. An ancestral prokaryote may have ingested small heterotrophic bacteria that were able to use O_2 to release energy from organic molecules via cellular respiration. These cells remained alive inside the host cell and continued to perform respiration. They eventually became mitochondria. Similarly, small photosynthetic prokaryotes may have taken up residence inside host cells and become chloroplasts.

8. Multicellularity probably arose several times in Kingdom Protista, with today's seaweeds, plants, animals, and fungi evolving from different kinds of protists. Unicellular protists may have formed colonies of interconnected cells. Such a colony might form when a cell divided and its offspring remained connected. The cells gradually became more specialized and interdependent, with specialized cells coming to perform specific tasks, such as locomotion, feeding, gas exchange, and reproduction.

Applying Your Knowledge

Multiple Choice: 1. c 2. c 3. b 4. c 5. a 6. e 7. a 8. c 9. b 10. c

Essay: 1. This may be a problem for theories about the origin of life, because many different organic compounds make up a living thing. If a given set of starting conditions produce only certain amino acids, but not ATP or sugars, it is hard to see how this could lead to the origin of life. On the other hand, it is possible that different chemicals could have been formed under different conditions at different times or in different locations. Given enough time, this mixture of chemicals might have collected in one spot, giving rise to life.

2. There is no reason to believe that the chemical reactions that gave rise to organic molecules on the primitive Earth are restricted to our own neighborhood. It is possible that similar reactions could happen elsewhere, even in clouds of gas and dust in space. Organic matter in meteorites leads scientists to believe that the reactions that form organic molecules might be quite likely to happen in space or on other planets. Organic matter from space may have had a role in the origin of life on Earth, and it may indicate that life has arisen on other planets. (Of course, it is possible that organic compounds originating on the Earth could have contaminated the meteorites after they got here.)

3. The fact that UV light breaks chemical bonds poses some problems for the hypothesized role of UV in the origin of organic compounds. If UV light provided energy for the reactions that caused inorganic compounds to combine and form simple organic compounds, why did it not then cause these same organic compounds to break apart?

4. If the diphtheria bacterium is like most bacteria, it causes illness by producing a toxin (in this case probably an exotoxin secreted by the bacteria) that can cause damage or interfere with body functions at places removed from the site of infection.

5. The number of bacteria would double every 20 minutes, so they would double 18 times in 6 hours. After 20 minutes, there would be 200, after 40 minutes 400, after an hour 800, and so on. There would be more than 26 million bacteria at the end of 6 hours!

Chapter 18

Organizing Your Knowledge

Exercise 1: Differences: 1. Algae are supported by water; leaves and stems of plants contain rigid supporting elements. 2. Whole body of alga exposed to water, light, minerals; only aboveground parts of plant exposed to light and roots in contact with soil nutrients and water. 3. Alga does not have true organs with specialized functions; plant has discrete organs—roots, stems, and leaves. 4. Algae lack mycorrhizae; plants have mycorrhizae. 5. Algae have no cuticle; plants have a cuticle that prevents drying. 6. Algae lack stomata; gas exchange occurs via stomata in plants. 7. Algae lack vascular tissues; plants have vascular tissues that transport nutrients and water. 8. Algae disperse gametes in water; plant gametes are protected in gametangia. The egg is fertilized, and the zygote develops into an embryo there. Similarities: 1. Cells contain chloroplasts. 2. Have similar photosynthetic pigments. 3. Have cell walls made of cellulose. 4. Store carbohydrates in the form of starch. 5. Golgi vesicles form a cell plate that divides cytoplasm in cell division.

Exercise 2: A. 3, gymnosoperms, angiosperms B. 1, bryophytes (mosses) C. 4, angiosperms D. 2, bryophytes, ferns

Exercise 3: 1. sporophytes 2. haploid 3. diploid 4. gametophyte 5. gametes 6. gametangia 7. sperm 8. flagella 9. Fertilization 10. zygote 11. sporangium 12. spores 13. meiosis

Exercise 4: 1. meiosis 2. spores 3. mitosis and development 4. gametophyte 5. sperm 6. egg 7. fertilization 8. zygote 9. mitosis and development 10. sporophyte 11. sporangia 12. haploid phase 13. diploid phase

Exercise 5: 1. pine tree 2. male cones, female cones 3. male cones, pollen grains 4. female cones, ovules 5. ovule 6. ovule 7. eggs, sperm 8. pollen grain, sperm, egg 9. zygote, embryo, seed 10. seed, embryo 11. pine tree

Exercise 6: 1. D 2. E 3. B 4. A 5. F 6. C 7. G

Exercise 7: 1. meiosis 2. pollen grain (male gametophyte) 3. female gametophyte 4. egg 5. pollination 6. stigma 7. pollen tube 8. sperm 9. fertilization 10. zygote 11. seed 12. embryo 13. seed coat 14. food supply 15. fruit 16. sporophyte 17. ovary 18. ovule 19. anther 20. diploid phase 21. haploid phase

Exercise 8: 1. cones 2. flowers 3. wind 4. insects, and so on 5. a year or more 6. 12 hours

or so 7. more than a year 8. days or weeks
9. seeds unprotected 10. seeds protected and
dispersed by fruit 11. usually needle-shaped
12. usually broad and flat 13. thinner cell walls
14. thicker and stronger cell walls

Exercise 9: 1. A 2. G, A 3. F 4. F 5. G 6. M
7. A 8. M 9. M, F 10. G 11. A 12. F 13. A
14. G, A 15. F 16. G 17. M 18. G 19. A
20. M, F 21. A 22. A

Exercise 10:

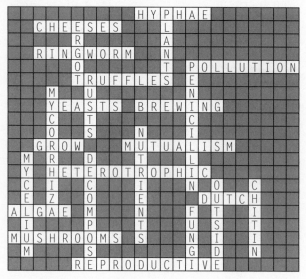

Exercise 11: 1. three 2. spores 3. hyphae
4. mycelium 5. mating types 6. nuclei 7. dikaryotic 8. fruiting body 9. zygotes 10. meiosis

Testing Your Knowledge

Multiple Choice: 1. a 2. d 3. b 4. d 5. a
6. b 7. c 8. b 9. b 10. e 11. d 12. e 13. d
14. c 15. c

Essay: 1. Algae are supported by water, and
they lack supporting tissues. The leaves and
stems of plants contain rigid supporting elements. The whole body of an alga is exposed to
water, light, and minerals, while only aboveground parts of the plant are exposed to light
and only the roots are in contact with soil nutrients and water. Plants have discrete organs—
roots, stems, and leaves—while algae do not
have true organs with specialized functions.
Plants have mycorrhizae, but algae do not.
Plants are covered by a cuticle that prevents
drying, pierced by stomata for gas exchange,
but algae have no need for such a covering or
stomata. Plants have vascular tissues that transport nutrients and water, but algae do not.
Algae disperse their gametes in water; plant ga-

metes are protected in gametangia. The egg is
fertilized, and the zygote develops into an embryo there.

2. Many people think they are plants because
they are rigid, "rooted" in one spot, and cannot
move. Seaweeds lack the specialized interdependent tissues and organs seen in plants. Fungi
are heterotrophic, while plants and seaweeds are
autotrophic—capable of making their own food
via photosynthesis.

3. Gymnosperms and angiosperms—the cone-
bearing and flowering plants—both produce
seeds. Seed plants do not require pools of water
for fertilization. Instead of sperm that swim to
the eggs, they produce pollen grains that protect
sperm and carry them through the air to the female parts of the plant. The seed is a survival
packet for the embryo—a protective coat and a
food supply that enable the embryo to get a
good start.

4. Gymnosperms—conifers such as pines and
firs—supply lumber for construction, wood pulp
for making paper, and most other forest products. Forests are also important for wildlife habitat, watersheds, and recreation. We can reduce
the cutting of forests by recycling lumber and
paper and by cutting trees no faster than they
can be replanted.

5. See Figure 18.12 in the text.

6. Animals assist angiosperms by carrying
pollen from anthers to carpels and by dispersing
seeds. Flowers produce nectar and pollen, which
are attractive to animals, and advertise their
presence with showy shapes, colors, and scents.
Fruits sometimes cling to the fur of animals or
attract animals to eat them and thereby disperse
seeds in their droppings.

7. The threadlike hyphae of fungi grow through
their food and secrete enzymes that digest the
food outside their bodies. The fungal cells then
absorb the digested nutrient molecules through
their cell membranes.

8. A lichen consists of a fungus growing in a
mutualistic relationship with green algae or
cyanobacteria. Their exact roles are not completely understood, but it is known that the
fungus receives food from the algae or cyanobacteria, and the photosynthetic cells receive
protection, water, and certain minerals from the
fungus.

Applying Your Knowledge

Multiple Choice: 1. e 2. d 3. c 4. a 5. b
6. e 7. b 8. b 9. c 10. d

Essay: 1. Like seed plants, ferns have true roots, stems, and leaves; vascular tissues that transport water and nutrients; and a life cycle in which the sporophyte stage predominates. Like mosses, ferns have flagellated sperm that swim to the egg.

2. Mosses and ferns are dispersed in the form of spores. Gymnosperms and angiosperms are dispersed in the form of seeds. Spores are very small and can be carried by the wind for long distances, but they are relatively vulnerable and short-lived. Seeds are bigger and cannot travel as far as fast (at least without help from animals), but they are protected by a seed coat and carry their own food supply that the embryo can draw on for a good start in life.

3. What probably happened is that the farmer killed beneficial mycorrhizal fungi along with the harmful parasites. Without the help of the mycorrhizae, the trees are not able to absorb water and soil minerals effectively.

4. Fungi can actually grow into other cells, where they are protected. They grow very rapidly, which enables them to come back quickly if not completely killed off. (Also, since they are eukaryotes, their metabolism is more like our own cells than is the metabolism of pathogenic bacteria. This makes them harder to kill than bacteria, because chemicals that injure fungi might injure human cells as well.)

5. Naming a lichen is tricky because the lichen consists of two organisms—a fungus and an alga or cyanobacterium. Does the lichen get one name or two?

Chapter 19

Organizing Your Knowledge

Exercise 1: The black line goes between bacteria and protists; bacteria are prokaryotic, and protists are eukaryotic. The blue line goes above protists; they are primarily unicellular, and those above them multicellular. The green line goes between plants and fungi; plants are autotrophs, and fungi and animals are heterotrophs. The red line goes between fungi and animals; fungi and plants have cell walls, and animals do not.

Exercise 2: See Module 19.2 in the text. A colony of cells may have become a hollow sphere. Cells of the colony then specialized, performing various functions.

Exercise 3: 1. Porifera 2. Cnidaria 3. sea 4. radial 5. pores 6. two 7. flagella 8. enters 9. exits 10. food particles 11. oxygen 12. nerves 13. digestive tract 14. budding 15. regeneration 16. jellyfishes 17. digestive 18. nervous 19. tissues 20. organ systems 21. polyp 22. medusa 23. jellyfish 24. *Hydra* 25. tentacles 26. cnidocytes 27. underneath 28. gastrovascular 29. wastes 30. incomplete

Exercise 4: 1. flatworms, mollusks, insects, chordates, and so on 2. parts arranged in a circle around a central axis 3. no dorsal and ventral surfaces 4. anterior and posterior ends 5. distinct head 6. immobile or slow-moving

Exercise 5: See Figure 19.7 in the text.

Exercise 6: 1. Cnidaria; Platyhelminthes; Nematoda 2. jellyfish, *Hydra*; planarian, tapeworm; *Caenorhabditis*, *Trichinella* 3. radial; bilateral; bilateral 4. medusa or polyp; ribbon or leaf shape; cylinder 5. none; none; pseudocoelom 6. incomplete; incomplete or none; complete 7. fresh water and salt water; water and parasites in animals; water, soil, and parasites in animals and plants 8. build coral reefs; human and animal parasites; human, animal, and plant parasites

Exercise 7: 1. E 2. E 3. C 4. G 5. D 6. C 7. B 8. F 9. G 10. E 11. D 12. A 13. C 14. D 15. D 16. E 17. E 18. G

Exercise 8: 1. Annelida 2. segments 3. sea 4. fresh water 5. walls 6. nervous 7. excretory 8. digestive 9. movement 10. flexibility 11. mobility 12. feces 13. polychaetes 14. ocean 15. tubes 16. appendages 17. feeding 18. leeches 19. blood 20. carnivores 21. fresh water 22. teeth 23. heart attacks 24. insects 25. vertebrae 26. muscles

Exercise 9:

```
C R U S T A C E A N S
    R                     T
    C H I T I N       E   I   A
    H                 X   C   B
    T H R E E         O   K   D
A           H O R S E S H O E
L           P         K       M
L           O   A N T E N N A E
E           D         L       N
R     B A R N A C L E S
G           N         T
W I N G S   N     J O I N T E D
C         E   W E B N   H   M
          G   L   N   S C O R P I O N
M O L T I N G         R       L
          E   D       A       L
P I N C E R S         X       I
          T                   P
I N S E C T S                 E
                  C E N T I P E D E S
                              E
                              S
```

Exercise 10: Circle the following: 1, 4, 6, 8, 9, 10, 12, 13, 14, 15, 16.

Exercise 11: See Figure 19.15A in the text.

Exercise 12: 1. D 2. F, G 3. E 4. D 5. D 6. G 7. A 8. C 9. A, B, C 10. G 11. C 12. D 13. G 14. F, G 15. B 16. A 17. D 18. C 19. G 20. F 21. B, C 22. A 23. G 24. C 25. G

Exercise 13: 1. Porifera 2. radial 3. bilateral 4. Cnidaria 5. Platyhelminthes 6. pseudocoelom 7. unsegmented 8. segmented 9. Echinodermata 10. exoskeleton 11. no exoskeleton 12. Arthropoda

Testing Your Knowledge

Multiple Choice: 1. c 2. b 3. b 4. c 5. a 6. a 7. b 8. b 9. a 10. c 11. e 12. c 13. e 14. d 15. c

Essay: 1. Animals are eukaryotic, multicellular heterotrophs that lack cell walls. They ingest their food and usually digest it in a digestive tract. Animals are diploid, except for their haploid eggs and sperm. The animal zygote usually goes through a blastula stage and a gastrula stage, and there may be a larval stage that metamorphoses into the adult.

2. Some characteristics important in determining an animal's phylum are number of cell layers, body symmetry, presence of tissues and organs, presence and type of body cavity, embryonic development of the body cavity, and presence or absence of segmentation.

3. All animal phyla discussed in this chapter have some kind of body cavity, except for sponges, cnidarians, and flatworms. The body cavity aids in movement, cushions internal organs, and may stiffen the body and help in circulation.

4. The mantle secretes the shell, but it can have other functions. In some bivalves, such as scallops, sensory structures such as eyes are arranged along the edge of the mantle. It functions in waste disposal in some mollusks. The mantle cavity houses the gills, and it is modified as a kind of lung in land snails. In squids, the mantle cavity can expel water for jet propulsion. Snails have a one-piece coiled shell. The two-part hinged shell of a clam can be closed for protection or opened for feeding. Squids have a small internal shell that stiffens the body.

5. Insects are arthropods, so they are covered by a tough exoskeleton that protects them from injury and drying out. They have jointed appendages that are modified for various functions—locomotion, sensing the environment, and feeding. Many insects have highly specialized mouthparts. Insects are small and can reproduce rapidly. They are the only invertebrates with wings, and they use their flying ability to find food, escape predators, and disperse to new habitats.

6. Reptile skin is covered by scales made of keratin, a waterproof protein that keeps the body from drying out. Reptiles are also able to lay their amniotic eggs on land. The reptile embryo develops within a hard shell and a protective, fluid-filled sac called the amnion. The young reptile does not go through an aquatic larval stage but rather emerges from the egg as a miniature adult, ready to cope with life on dry land. Amphibians have thin, moist skin that dries out easily. They must lay their jelly-covered eggs in water and their tadpole larvae must develop in water.

7. Birds have feathers, which form a lightweight covering and shape the wings. They have many modifications to reduce their weight: no teeth, a reduced tail, hollow bones and feathers, and air sacs in the body. They have large flight muscles attached to a keeled breastbone. They are endothermic, maintaining a high constant temperature and metabolic rate. The bird's circulatory system and lungs are highly efficient in delivering food and oxygen for the rapid metabolism needed for flight.

Applying Your Knowledge

Multiple Choice: 1. d 2. c 3. a 4. e 5. c 6. a 7. c 8. b 9. a 10. a

Essay: 1. Like all other animals, sponges are eukaryotic, multicellular heterotrophs that lack cells walls, ingest their food, and are mostly diploid.

2. The temporary digestive cavity of *Trichoplax* suggests a hypothesized stage in animal evolution when a relatively unspecialized colony of cells may have folded inward to form a digestive cavity, much as a blastula becomes a gastrula during animal development.

3. Pigs and people are infected when they eat undercooked meat. If meat is thoroughly cooked, the worms will be killed. Pigs can also become infected by ingesting the feces of infected humans or animals. A way to prevent the spread of the worms might be to use good sanitation and keep the pigs away from sewage.

4. Your criteria might include number of species or individuals or how widespread the animals are. By these criteria, arthropods and chordates are very successful. You might discuss particular structural, functional, or behavioral features or adaptations that contribute to their success.

5. The larvae suggest that mollusks and annelids are related. *Peripatus* is a possible link between annelids and arthropods. These three phyla share a major branch of the animal phylogenetic tree. They are all bilaterally symmetrical animals with a coelom that forms within masses of cells in the middle tissue layer.

6. A dog is a chordate because it has a dorsal hollow nerve cord (the brain and spinal cord), a flexible supporting rod called a notochord, gill structures behind the mouth, and a postanal tail (a tail posterior to the anus). It is a vertebrate because it has an internal skeleton made of bone, consisting of a skull, a backbone of vertebrae, and bones supporting its appendages—the legs. In the adult dog, the notochord is replaced by the vertebral column and the gill structures disappear, but these structures are present in a dog embryo.

7. Birds and mammals are endothermic. Unlike reptiles and amphibians, they maintain a constant high rate of metabolism and body temperature, enabling them to function in a cold environment.

Chapter 20

Organizing Your Knowledge

Exercise 1: 1.The tail is flattened into broad flukes, which propel the whale through the water. 2. The hummingbird's long, thin beak enables it to sip nectar from tubular flowers. 3. The sensitive nerve endings and finely controlled muscles of the hand enable you to grasp and manipulate objects. 4. The frog's long legs enable it to leap to safety, and its webbed feet make it a good swimmer. 5. The mosquito's mouthparts allow it to pierce the skin and suck blood. 6. The stomach chambers carry out various steps in breaking down hard-to-digest grass.

Exercise 2: 1. cells 2. tissue 3. epithelial 4. connective 5. nervous 6. organ 7. system 8. organism 9. organism 10. systems 11. organs 12. tissues 13. cells

Exercise 3: 1. stratified squamous epithelium 2. multiple layers of flattened cells 3. lining of esophagus, epidermis of skin 4. covers and pro-

tects surfaces subject to abrasion 5. simple cuboidal epithelium 6. single layer if cube-shaped cells 7. tubular passageways where urine forms in kidneys 8. simple squamous epithelium 9. linings of lungs and blood vessels 10. Exchange of materials by diffusion 11. simple columnar epithelium 12. single layer of elongated, cylindrical cells 13. secretes and absorbs in respiratory tubes and walls of digestive tract

Exercise 4:

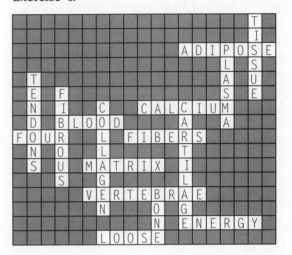

Exercise 5: 1. B 2. D 3. C 4. B 5. D 6. C 7. D 8. C 9. A 10. A 11. A 12. D 13. C 14. B 15. A

Exercise 6: 1. nervous 2. muscular 3. skeletal 4. digestive 5. respiratory 6. cardiovascular 7. lymphatic 8. integumentary 9. immune 10. excretory 11. endocrine 12. reproductive

Exercise 7: 1. X rays 2. Examination of hard tissues—bones, cartilage, dense tumors 3. Does not show soft tissues clearly, produces flat image, X rays can cause cancer 4. X-ray cross sections combined by computer 5. cross sections and 3-dimensional views of normal and abnormal tissues in brain, elsewhere 6. X rays cause cancer (but these are lower doses than standard X rays), expense 7. magnets, radio pulses, and computer 8. visualizes soft tissue such as brain, but does not visualize bone 9. expense 10. radiation from radioactive isotope picked up by detector 11. measuring metabolic activity, especially of brain 12. expense, radiation

Exercise 8: See Figure 21.11B in the text.

Exercise 9: 1. increase in temperature 2. eating a meal 3. hypothalamus (brain) 4. pancreas 5. increase in blood temperature 6. increase in blood sugar level 7. nerve impulses 8. hormone—insulin 9. sweat glands and blood vessels in skin 10. cells 11. sweat glands secrete sweat, and blood vessels dilate and fill with warm blood, cooling the body 12. cells take up sugar and lower blood sugar level 13. around 37°C 14. 70 to 110 milligrams of sugar per 100 milliliters of blood

Testing Your Knowledge

Multiple Choice: 1. b 2. d 3. c 4. d 5. b 6. e 7. b 8. a 9. d 10. e 11. b

Essay: 1. A wing is light—mostly feathers—but strong. Powerful breast muscles move the wing. Its broad surface catches the air and generates lift that keeps the bird aloft.

2. Epithelial tissue consists of sheets of closely packed cells that cover and line body surfaces. It protects, secretes, and absorbs. Connective tissue consists of sparse cells scattered in a matrix, which usually is a web of fibers embedded in a liquid, jelly, or solid. Most connective tissues form a framework that supports and protects the organs of the body. Muscle tissue consists of bundles of elongated cells called fibers. Muscle cells can contract and are responsible for body movement. Nervous tissue consists of cells called neurons and other supporting cells that form a communication and coordination system within the body. Neurons have elongated extensions that transmit signals to other cells.

3. The digestive system ingests food, breaks it down, and absorbs it into the bloodstream. The respiratory system supplies the blood with oxygen and disposes of carbon dioxide. The cardiovascular system transports nutrients and oxygen to cells and carries away wastes. The lymphatic system returns fluid to the blood and plays a role in body defense. The immune system protects the body from disease and foreign substances. The excretory system disposes of metabolic wastes. The endocrine system and nervous system control and coordinate body activities via hormones and nerve signals. The male and female reproductive systems produce gametes and support the growth of the developing embryo. The muscular system moves the body and its parts, the skeletal system supports and protects the body, and the integumentary system covers the body.

4. Computerized tomography uses less powerful (and therefore less potentially damaging) X rays than conventional X rays. A CT scan reveals more of the fine detail of soft tissues. It yields cross-sectional views, which can reveal features that might block one another on conventional X rays, and these cross sections can be combined by computer to give three-dimensional views.

5. In a house, the stimulus is a change in room temperature, and in the body it is a change in the temperature of the blood. The control center is the thermostat in a house, and in the body it is a "thermostat" in the brain. The home thermostat sends a signal to the furnace—the effector—which responds to an increase in room temperature by turning off and allowing the house to cool a bit. When blood temperature goes up, the brain sends a signal to body effectors—sweat glands and blood vessels in the skin. The sweat glands respond by secreting sweat, which increases evaporative cooling. The blood vessels dilate to bring warm blood to the surface, where it loses its heat and cools the body.

6. Small animals have a larger surface-to-volume ratio than larger animals, and most of their cells may be in direct contact with their surroundings, so their body surfaces may be sufficient to exchange materials with the environment. Larger animals have smaller surfaces in relation to overall volume, so they require specialized surfaces for exchange of materials with the environment.

Applying Your Knowledge

Multiple Choice: 1. b 2. c 3. c 4. e 5. c 6. c 7. e 8. c

Essay: 1. A woodpecker displays many features that illustrate the correlation between structure and function. It has sharp ears that enable it to detect insects inside a tree. It uses its sharp, chisel-like bill to cut a hole in the tree and its long tongue to probe for insects. Two toes on each foot face forward and two backward, enabling the woodpecker to cling to a vertical tree trunk, and its stout tail is used as a brace as it drills for insects.

2. An anatomist studies the structure of the body and so might be interested in the shape of the fish's fins and body, the structure and arrangement of the penguin's feathers, and how the insect's exoskeleton protects it from its enemies. A physiologist studies body function and so might be interested in how the fish's muscles propel it through the water, how the penguin generates body heat, and how the insect's senses warn it about potential predators.

3. Bone cells are embedded in a matrix of rope-like collagen fibers and hard calcium salts. This combination makes bone hard without being brittle. Simple squamous epithelium is a single layer of thin cells, forming the lining of the

lungs and blood vessels. Gases and liquids can easily diffuse through this thin layer, facilitating exchange. Blood cells are small and round and suspended in a liquid matrix, enabling them to be pumped around the body to transport oxygen and defend against infection.

4. The digestive system takes in, breaks down, and absorbs food into the blood. The respiratory system draws in air, and oxygen enters the blood in the lungs. The food and oxygen are transported to the brain by the cardiovascular system—the heart, blood, and blood vessels. The nervous and endocrine systems control these activities.

5. When blood calcium gets too high, the parathyroid glands slow absorption of calcium by the intestine and speed up excretion of calcium by the kidneys. When blood calcium gets too low, the parathyroids increase absorption by the intestine and slow excretion by the kidneys. This illustrates negative feedback because a change in the concentration of calcium triggers the control mechanisms to counteract further change in the same direction.

Chapter 21

Organizing Your Knowledge

Exercise 1: 1. herbivore, large piece feeder 2. omnivore, substrate feeder 3. herbivore, fluid feeder 4. carnivore, filter feeder 5. omnivore, large piece feeder 6. carnivore, absorptive feeder 7. carnivore, large piece feeder 8. carnivore, fluid feeder 9. omnivore, suspension feeder 10. herbivore, large piece feeder

Exercise 2: Label as in Figure 21.3B in the text. Color as follows: *Earthworm:* mouth, pharynx, esophagus, and crop yellow; gizzard red; intestine red and green; anus blue. *Grasshopper:* mouth, esophagus, and crop yellow; gizzard red; stomach red and green; gastric pouches green; intestine green and blue; anus blue. *Bird:* mouth, esophagus, and crop yellow; stomach and gizzard red; intestine red and green; anus blue.

Exercise 3: See Figure 21.4 in the text.

Exercise 4: 1. incisor 2. liver 3. circular 4. anus 5. gallbladder 6. pyloric sphincter 7. pharynx 8. liver 9. molars 10. longitudinal 11. duodenum (small intestine) 12. pancreas 13. tongue 14. large intestine 15. canines 16. oral cavity 17. stomach 18. bolus 19. salivary glands 20. small intestine 21. large intestine 22. epiglot-

tis 23. stomach 24. esophagus 25. rectum 26. oral cavity

Exercise 5: 1. esophagus 2. peristalsis 3. acid chyme 4. gastric glands 5. gastric juice 6. gastrin 7. mitosis 8. hydrochloric acid 9. pepsin 10. proteins 11. small intestine 12. protects 13. mucous 14. mucus 15. heartburn 16. gastric ulcer 17. mucus 18. caffeine (and so on) 19. pyloric sphincter 20. small intestine 21. duodenal ulcer 22. bile 23. pancreas 24. neutralizes 25. liver 26. fats 27. emulsification 28. lipase 29. proteins 30. amino acids 31. starch 32. monosaccharides 33. absorption 34. villi 35. microvilli 36. lymph 37. capillaries 38. diffuse 39. glycogen 40. large intestine 41. water 42. feces

Exercise 6: 1. B 2. E 3. A 4. C or F 5. D 6. C or F

Exercise 7: Here's a possible answer, in exactly 25 words: Every animal must obtain from its diet fuel for metabolism, raw materials for making the animal's own molecules, and substances the animal cannot make itself.

Exercise 8: Answers for this exercise will depend on your body weight.

Exercise 9:

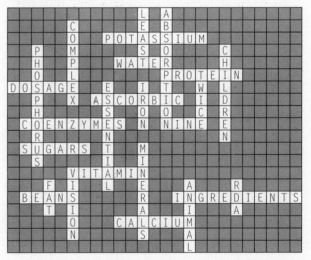

Exercise 10: 1. ↓ 2. ↑ 3. ↓ 4. ↓ 5. ↑ 6. ↑ 7. ↓ 8. ↓ 9. ↑ 10. ↑ 11. ↑ 12. ↓ 13. ↑ 14. ↓ 15. ↓

Testing Your Knowledge

Multiple Choice: 1. d 2. a 3. b 4. c 5. c 6. e 7. d 8. c 9. a 10. e 11. b 12. b 13. c 14. c 15. c

Essay: 1. The large polymer molecules in food—proteins, carbohydrates, and fats—are too large to pass through cell membranes. They cannot be taken up by cells unless they are digested first.

Also, an animal needs to break down these large molecules to obtain monomers to make its own polymers.

2. Both a grasshopper and a human have alimentary canals with a mouth and an anus. In both, food is cut and chewed by the mouth before it passes through the esophagus. In a grasshopper, chewed food is stored in the crop, ground by the gizzard, digested largely by the stomach, and absorbed in the stomach and gastric pouches. The intestine mainly compacts wastes. Humans have no crop or gizzard. The human stomach begins digestion, but most digestion and absorption occur in the small intestine. The human large intestine has much the same functions as the grasshopper intestine.

3. Saliva contains a slippery glycoprotein that lubricates food. Buffers neutralize food acids and help prevent tooth decay. Antibacterial agents kill food bacteria, and an enzyme in saliva begins the digestion of starch.

4. Herbivores generally have longer, more complex alimentary canals than carnivores, because plants are harder to digest. Horses and elephants house cellulose-digesting microbes in the colon and a large cecum. Some of the nutrients are absorbed by the cecum and colon, but some are lost. Rabbits compensate for this loss by reingesting their fecal pellets, extracting additional nutrients as the already-digested material passes through the alimentary canal a second time. Ruminant mammals such as cattle have four-chambered stomachs. Food is digested by bacteria in two of the compartments, then the softened material (the "cud") is regurgitated and rechewed, then swallowed and passed to another compartment where digestion is completed. The nutrients absorbed by the cow's intestine come from both plant material and the rapidly reproducing bacteria in the stomach.

5. Fad diets—low-carbohydrate diets, low-fat diets, and formula diets—may be inadequate in essential nutrients, may result in loss of body protein, may cause health problems such as salt imbalance and thinning hair, and may ultimately prove unsatisfying and result in bingeing and regaining lost weight. A balanced diet of 1200 calories or more, combined with moderate exercise, is best for healthy, gradual weight loss.

6. Food labels provide two kinds of information—a list of ingredients and nutrients present in the food. Such information is useful in planning a healthy, balanced diet—for example, obtaining the proper balance of vitamins, minerals, and other nutrients, or avoiding foods high in calories, sugar, fat, and sodium.

Applying Your Knowledge
Multiple Choice: 1. c 2. b 3. b 4. a 5. e 6. a 7. b 8. b 9. a 10. e

Essay: 1. Muscle is meat, so meat protein will supply James with the right mix of amino acids needed by the body for building muscle. Meat is also high in saturated fats, however, which can contribute to weight gain and heart disease. Plant foods are low in saturated fats, but because no single plant contains all the amino acids needed to build muscle, Michael must be careful to consume a variety of plants to get all the amino acids his body needs.

2. It would be most difficult to live without the small intestine, because it is the portion of the digestive system where most food molecules are digested and where nearly all absorption of nutrients takes place.

3. If trypsin and chymotrypsin were manufactured in their active form, they would digest proteins in the cells where they are made and stored. Like pepsin in the stomach, they are produced in an inactive form that does not digest protein, and they are converted to active, protein-digesting enzymes once they enter the intestine.

4. 4.83 kcal of food energy are liberated for every liter of oxygen consumed. This means that an astronaut would consume 2500 kcal/4.83 kcal/L = 518 liters of oxygen per day.

5. He could try eliminating all scandium from the diet and looking for ill effects. This might be a problem because this might harm human experimental subjects. He could experiment with animals, but they might not require scandium as he thinks humans do. If he eliminates scandium from the diet, he might accidentally eliminate other elements as well, and then it would be difficult to determine what effects are due to scandium alone. Finally, if scandium is required, it must be in minute quantities, or this requirement probably would have been discovered already. If this mineral is required in tiny concentrations, how do we make sure none is present?

6. Calculate by using the conversion factors in Module 21.14 in the text.

Chapter 22

Organizing Your Knowledge

Exercise 1: 1. transport 2. breathing 3. breathing 4. transport 5. exchange 6. breathing 7. exchange 8. exchange

Exercise 2: 1. C, P 2. A, R 3. D, Q 4. B, S 5. A, R 6. A, R 7. B, S

Exercise 3: 1. skin 2. oxygen 3. carbon dioxide 4. respiratory 5. water 6. moist 7. large 8. thin 9. gills 10. moist 11. less 12. lamellae 13. blood vessels 14. ventilate 15. opposite 16. blood 17. countercurrent exchange 18. oxygen 19. more 20. more 21. diffusion 22. air 23. higher 24. easier 25. dry 26. tracheal 27. tracheal 28. circulatory 29. membranes 30. oxygen 31. carbon dioxide 32. lungs 33. diaphragm 34. nose 35. filtered 36. nasal 37. pharynx 38. larynx 39. vocal cords 40. tense 41. less tense 42. trachea 43. bronchi 44. bronchioles 45. mucus 46. cilia 47. alveoli 48. millions 49. blood capillaries 50. carbon dioxide

Exercise 4: See Figure 22.6A in the text.

Exercise 5:

```
                    B R O N C H I
            M       A             E
            O       D     B L A C K
            N       O           R
            O Z O N E           T
            X
          F I V E
      E     D
C   E M P H Y S E M A
A   I
N   S T
C O U G H I N G
E   E   E
R   F   L   M U C U S
    U   I           M
    R   A S B E S T O S
        L           K
          C I L I A N
            L U N G
```

Exercise 6: 1. I 2. I 3. E 4. I 5. I 6. I 7. E 8. E 9. I 10. E 11. E 12. E

Exercise 7: 1. ↑ 2. ↓ 3. ↑ 4. ↓ 5. ↑ 6. ↑ 7. ↓ 8. ↑

Exercise 8: 1. circulatory 2. O_2 3. CO_2 4. low 5. high 6. alveoli 7. brain 8. drop 9. CO_2 10. diaphragm 11. ribs 12. CO_2 13. O_2 14. O_2 15. CO_2 16. CO_2 17. more 18. less 19. O_2 20. soluble 21. hemoglobin 22. red 23. iron 24. O_2 25. muscles 26. higher 27. more 28. out 29. into 30. red 31. hemoglobin 32. water 33. bicarbonate 34. plasma 35. buffering 36. rises 37. CO_2

Exercise 9: A. 5 B. 6 C. 1 D. 11 E. 2 F. 4 G. 9 H. 7 I. 3 J. 8 K. 10

Testing Your Knowledge

Multiple Choice: 1. b 2. d 3. a 4. e 5. a 6. c 7. b 8. d 9. a 10. e 11. b

Essay: 1. Air contains a higher concentration of oxygen than water, and it is easier to move across the respiratory surface, but it tends to dry the respiratory surface. There is no danger of the respiratory surface drying out in water, but water contains less oxygen than air and is harder to move.

2. Gills and lungs are both divided into small units that greatly enlarge their surface area for gas exchange. They both have a thin, moist surface through which oxygen passes into numerous blood vessels. They usually have a mechanism for ventilating the respiratory surface. The primary difference between gills and lungs is that gills project outward from the body and lungs are folded inward.

3. There are millions of alveoli in each lung, forming a large respiratory surface. Oxygen dissolves easily in the moisture inside an alveolus and passes easily through the thin wall of the alveolus and into the network of capillaries that cover its surface.

4. The two vocal cords are located in the larynx, or voice box, which is between the pharynx and the trachea. When you exhale, air rushes past the vocal cords. If they are tensed, they vibrate and produce sounds. The more the muscles in the larynx tense the vocal cords, the faster they vibrate, and higher-pitched sounds are produced. When the cords are less tense, they produce lower-pitched sounds.

5. When you inhale, the diaphragm contracts and moves downward and muscles between the ribs contract and expand the rib cage. This enlarges the chest cavity, and the lungs expand with it. This reduces the pressure in the alveoli to less than that of the atmosphere. Air rushes in from the area of higher pressure outside to the area of lower pressure in the alveoli. Thus, the lungs expand first, causing air to rush in.

6. Water passes through the gill in a direction opposite to the flow of blood within each of the many thin lamellae that make up the gill's respiratory surface. As blood picks up more and

more O_2, it comes into contact with water that has more and more O_2 available. Thus, a concentration gradient is maintained that favors diffusion of O_2 along the entire length of the lamella. This allows the gill to extract a large percentage of O_2 from the water flowing through it.

Applying Your Knowledge

Multiple Choice: 1. b 2. c 3. b 4. a 5. b 6. d 7. c 8. d 9. b 10. d

Essay: 1. When you inhale, air passes through your nose; into the nasal cavity; through the pharynx, larynx, and trachea; and into the bronchi that lead into each lung. The air continues through smaller and smaller bronchioles, which end in tiny alveoli. An O_2 molecule dissolves in the moisture on the surface of an alveolus and diffuses through the thin epithelium of the alveolus and into a capillary. The O_2 molecule dissolves in blood plasma, diffuses through the membrane of a red blood cell, and attaches to one of the iron atoms in a hemoglobin molecule. The blood returns to the heart, which pumps it through a blood vessel to a capillary in the brain. There the O_2 molecule detaches from the hemoglobin molecule and diffuses out of the red blood cell, into the plasma, through the wall of the capillary, and into a brain cell.

2. The respiratory surface should have a large area through which gas exchange can occur. This could be accomplished in a relatively small space by folding or dividing the surface into many smaller tubes, plates, or sacs—like the alveoli of the lungs or lamellae of fish gills. The surface must be moist so that gas molecules will dissolve readily, and it must be thin so that gases can diffuse through it easily and rapidly. The respiratory surface should be protected in some way—perhaps folded inside the robot's body—to keep it from drying out or being damaged. Finally, there should be some mechanism for ventilating the respiratory surface—similar to breathing or movement of a fish's gill covers—to move fresh air over the surface to maximize gas absorption.

3. The carbon monoxide molecules would attach to hemoglobin and reduce its ability to carry oxygen. Cells cannot use carbon monoxide as they use oxygen, so they might be starved for oxygen.

4. 20 cigarettes/day × 5 minutes/cigarette × 365 days/year × 40 years = 1,460,000 minutes. There are 365 × 24 × 60 = 525,600 minutes in one year, so this will shorten the smoker's life by 1,460,000/525,600 = 2.8 years.

5. The respiratory centers in the brain normally regulate breathing rate by monitoring blood pH. CO_2 combines with water in the blood to form carbonic acid, so if the concentration of CO_2 in the blood increases, the blood becomes more acidic, and pH drops. The brain detects this change and speeds up breathing, which removes carbon dioxide and returns blood pH to the acceptable range. In the submarine, CO_2 removal was functioning normally, so there was no buildup of CO_2 (and no drop in pH) in the blood of the sailors—and no response on the part of their respiratory centers—even though oxygen was getting dangerously low.

Chapter 23

Organizing Your Knowledge

Exercise 1: 1. G 2. F 3. H 4. J 5. I 6. A 7. K 8. B 9. D 10. C 11. L 12. E

Exercise 2: A: a. 1 b. 3 c. 4 d. 2 B: a. 1 b. 3 c. 2 d. 6 e. 5 f. 4

Exercise 3: *Labels:* a. superior vena cava b. capillaries of upper body c. pulmonary artery d. pulmonary artery e. capillaries of right lung f. aorta g. capillaries of left lung h. pulmonary vein i. pulmonary vein j. right atrium k. left atrium l. inferior vena cava m. aorta n. right ventricle o. left ventricle p. capillaries of lower body. *Colors and blood flow:* See Figure 23.4B in the text.

Exercise 4: 1. superior vena cava 2. right atrium 3. right ventricle 4. pulmonary arteries 5. pulmonary veins 6. left atrium 7. left ventricle 8. aorta 9. capillaries 10. inferior vena cava 11. right 12. pulmonary 13. left 14. systemic 15. artery

Exercise 5: 1. arteries 2. heart 3. capillaries 4. veins 5. capillaries 6. heart 7. thin 8. thick 9. in between 10. epithelium 11. epithelium, smooth muscle, connective tissue 12. epithelium, smooth muscle, connective tissue 13. no 14. no 15. yes

Exercise 6: 1. AV 2. SL 3. SL 4. AV 5. SL 6. AV

Exercise 7: Answers will depend on your heart rate.

Exercise 8: 1. chest, shoulder, arm, neck, jaw 2. pacemaker (SA node) 3. heart attack 4. AV node 5. coronary artery 6. atherosclerosis 7. bypass 8. brain

Exercise 9: 1. G 2. C 3. B 4. A 5. A or B
6. D 7. D 8. A or B 9. D or E 10. A 11. A
or B 12. F 13. B 14. F or G 15. F and G

Exercise 10: Some possible changes that might
cause increased blood flow through muscle capillaries are increased temperature, increased CO_2
concentration, increased concentration of other
waste products, decreased O_2 concentration, and
nerve impulses from the brain.

Exercise 11: The red arrow is oxygen-rich blood
flowing into the capillary; the blue arrow is
oxygen-poor blood leaving. At the arterial end
of the capillary, the blood pressure arrow is bigger than the osmotic pressure arrow, so the net
flow is out. At the venous end, the osmotic
pressure arrow is bigger than the blood pressure
arrow, so the net flow is in.

Exercise 12: 1. in between 2. blood clotting
3. erythrocytes 4. medium-sized 5. most numerous 6. largest 7. defense and immunity

Exercise 13: 1. high 2. pulse 3. arterioles
4. erythrocytes 5. oxygen 6. leukocytes 7. phagocytes 8. plasma proteins 9. platelets 10. capillary
11. epithelium 12. interstitial 13. connective tissue 14. platelets 15. prothrombin 16. fibrin
17. coronary arteries 18. heart attack 19. venule
20. vein 21. valve

Exercise 14: 1. atrium (The others are all blood
vessels.) 2. fibrin (The others are blood cells.)
3. fibrinogen (The others relate to blood pressure.) 4. aorta (The others are parts of the pulmonary circuit.) 5. leukocyte (The others are
involved in blood clotting.) 6. semilunar valve
(The others are parts of the heart's electrical system.) 7. leukocyte (The others relate to heart
contraction and relaxation.) 8. epithelium (The
others are concerned with delivery of oxygen to
cells.) 9. pulmonary artery (The others relate to
the systemic circuit.)

Testing Your Knowledge

Multiple Choice: 1. d 2. e 3. e 4. a 5. d
6. b 7. b 8. e 9. a 10. a 11. a 12. b 13. b

Essay: 1. A fish has a single circuit of blood
flow. The fish's two-chambered heart receives
and pumps only oxygen-poor blood from body
tissues. After leaving the heart, the blood passes
through capillaries in the gills, where it picks
up oxygen. In the gills, the blood slows down
considerably, then it continues on to capillaries
in various body tissues. A mammal, a terrestrial
vertebrate, has two blood circuits instead of one.
The right side of the heart receives oxygen-poor
blood returning from the tissues and pumps it
through capillaries in the lungs. The left side of
the heart receives this oxygen-rich blood and
pumps it to capillaries in body tissues. Blood
passing through the lung capillaries loses speed
and pressure, but the second trip through the
heart gives it a boost, so the oxygen-rich blood
is rapidly and efficiently delivered to the tissues.
This supports the higher metabolic rate characteristic of many land animals.

2. When you are exercising, cardiac output can
be increased by increasing the heart rate and
the amount of blood the heart pumps with each
beat.

3. Elevated blood pressure caused by reduced
flexibility of the artery walls can strain the
heart, because the heart has to work harder to
pump blood against the greater resistance in the
arteries. High blood pressure can also cause
small tears in the lining of blood vessels and increase the tendency for vessel-blocking plaques
to form there.

4. Blood pressure is highest in the aorta and
only slightly lower in the arteries, because the
heart is pushing on the blood with full force in
these vessels. As blood passes through the narrower arterioles, pressure drops sharply, mainly
due to increased resistance to blood flow caused
by friction between the blood and the extensive
inner surface of the many arterioles. Pressure
continues to drop as blood flows through the
capillaries, due to the resistance and huge surface area of the capillary walls. By the time
blood gets to the veins, its pressure is near zero.
Reduced pressure in the chest due to breathing,
squeezing action of skeletal muscles, and venous
valves help keep blood moving back to the
heart.

5. The large number of capillaries and their
minute size give them a tremendous combined
surface area through which materials can be exchanged between blood and interstitial fluid.
Capillary walls consist of a single thin layer of
epithelial cells, and there are gaps between these
cells, also facilitating exchange.

Applying Your Knowledge

Multiple Choice: 1. c 2. a 3. c 4. b 5. b
6. a 7. d 8. e 9. d 10. d

Essay: 1. Cardiac output is equal to the amount
of blood pumped by each heartbeat times the
number of beats per minute. The runner's cardiac output is 90 mL per beat × 160 beats per
minute = 14,400 mL, or 14.4 L, per minute.

2. In an artery, blood pressure increases during systole, when the heart is contracting and pushing blood into the artery. Pressure drops during diastole, when the heart relaxes. An arterial pressure of 130/80 tells you that the systolic pressure is 130 mm of mercury, and the diastolic pressure is 80 mm of mercury. The numerous arterioles and capillaries offer much resistance to blood flow. By the time blood gets to a vein, the force of the beating heart no longer propels it, so there are no pulses of pressure, just a slow, steady flow at a low, constant pressure.

3. At the capillaries, blood pressure tends to force fluid out of the tissues and into the tissues. The osmotic pressure of blood causes fluid to leave the tissues and enter the blood. Blood pressure tends to win out at the arterial end of a capillary, and osmotic pressure is greater at the venous end, but overall these forces are just about in balance. In a person with high blood pressure, blood pressure would overcome osmotic pressure, and there would be a net movement of fluid into the tissues, causing swelling.

4. If the blood vessels were reversed as described, the right ventricle would pump blood to the tissues, and this blood would soon return to the right atrium and then to the right ventricle, where this oxygen-poor blood would again be pumped to the tissues, without being sent to the lungs to pick up oxygen. Similarly, the left atrium and left ventricle would keep sending the same blood to the lungs, without ever pumping this oxygen-rich blood to the tissues that need it. Once the baby is born and the lungs are functioning, there would be no way for oxygen-rich blood to get to the tissues, which would die from lack of oxygen.

Chapter 24

Organizing Your Knowledge

Exercise 1: 1. H 2. D 3. E 4. I 5. F 6. A 7. C 8. B 9. G

Exercise 2: A. bacteria B. histamine C. swelling D. fluid E. leakiness F. pathogens G. cell debris

Exercise 3: 1. An antigen is a foreign molecule that elicits an immune response. An antibody is a protein that attaches to and counters a particular antigen. 2. T cells mature in the thymus and carry out cell-mediated immunity. B cells mature in bone marrow and carry out humoral immunity. 3. Antibodies secreted by B cells carry out humoral immunity, mainly against bacteria and viruses in body fluids. In cell-mediated immunity, T cells attack bacteria and viruses that have already entered body cells. 4. The lymphatic system consists of lymph vessels, lymph nodes, and other organs such as the thymus. It returns a fluid called lymph to the blood vessels and helps defend the body. The circulatory system consists of the heart, blood vessels, and blood. 5. Active immunity occurs when the body is stimulated by antigens and secretes antibodies in its own defense. In passive immunity, the body does not secrete its own antibodies but rather borrows antibodies temporarily. 6. Both blood capillaries and lymphatic capillaries are narrow tubes with thin walls, but blood capillaries are open at both ends, whereas lymphatic capillaries end in dead-ends. 7. Nonspecific defenses—the skin, digestive juices, antimicrobial proteins, and so on—act against any threat. Specific defenses—the immune system—target specific invaders.

Exercise 4: 1. nonspecific 2. immune 3. antigens 4. antibodies 5. antigen-binding sites 6. lymph 7. selection 8. clone 9. effector 10. antibodies 11. primary 12. days 13. immune 14. secondary 15. memory 16. decades

Exercise 5: 1. antigens 2. B cells 3. primary immune response 4. plasma cells 5. antibodies 6. antigens 7. memory B cells 8. secondary immune response 9. plasma cells 10. antibodies

Exercise 6: 1. activation of complement 2. precipitation of antigens 3. neutralization 4. agglutination of cells 5. cell lysis 6. phagocytosis

Exercise 7: For example, monoclonal antibodies might be used to identify human blood proteins in criminal investigations. Wildlife managers might use them to differentiate between wild and hatchery fish. Monoclonal antibodies might also be used to identify particular materials in archaeological digs—skins, animal remains, or food, for example.

Exercise 8: 1. H 2. D 3. F 4. A 5. E 6. B 7. C 8. G

Exercise 9: 1. In immunodeficiency diseases, components of the immune system are defective and the body is susceptible to infection. 2. An allergy is an overreaction by the immune system to normally harmless antigens, such as those in cat hair. This triggers the inflammatory response, which produces allergy symptoms. 3. Cancer may alter cancerous cells in such a way that the immune system usually recognizes the cells as foreign and destroys them. Cancer cells may sometimes fool or suppress the immune system, enabling a tumor to grow to fatal proportions. 4. The immune system of a heart transplant recipient recognizes the antigens of the transplanted heart as foreign and attacks the organ. 5. In rheumatoid arthritis, some of the patient's own antibodies form complexes with each other and complement. These complexes accumulate in joints, where they cause inflammation and joint damage. 6. HIV, the AIDS virus, attacks helper T cells and other immune cells, disabling the immune system and leaving the victim defenseless against infections and certain cancers. 7. Nerve fibers penetrate tissues that produce lymphocytes, and lymphocytes have receptors for chemicals produced by nerve cells. Thus psychological stress could affect immune function. It has been shown that students' natural killer cells are less effective and produce less interferon during exams.

Exercise 10: 1. HIV 2. condoms 3. blood 4. 200,000 5. 10 million 6. helper T cells 7. RNA 8. DNA 9. cancer 10. 3 years 11. AZT

Testing Your Knowledge

Multiple Choice: 1. d 2. e 3. d 4. e 5. a 6. d 7. e 8. c 9. c 10. b 11. a 12. c 13. e 14. d 15. a 16. e

Essay: 1. A vaccine contains a harmless form of a disease-causing microbe. It cannot make you sick, but it stimulates lymphocytes to become effector cells and mount a primary immune response. At the same time, it stimulates some lymphocytes to become memory cells, which are responsible for active immunity and are able to mount a rapid and massive secondary immune response if you are ever exposed to the "real" harmful microbe, even years later.

2. B cells carry out humoral immunity against microbes in body fluids. T cells carry out cell-mediated immunity against body cells infected by microbes or altered by cancer. When B cells are activated, they produce antibodies that circulate in body fluids and attach to microbe antigens. This clumps and inactivates the microbes and tags them for destruction by complement and phagocytes. Receptors on activated cytotoxic T cells enable them to bind to antigens on infected body cells. The T cells discharge a substance called perforin that destroys infected cells.

3. Immune system cells do not analyze antigens and produce antibodies to fit. Instead, the body contains small numbers of many types of immune cells, each type capable of making antibodies that fit a certain antigen. When an antigen appears, only the clone of cells capable of responding to that antigen are selected—a mechanism known as clonal selection. The cells that respond have the ability to make an antibody that fits the antigen before the antigen appears. Antibodies are not made to order.

4. They are lymph nodes. Lymph carries the microbes responsible for your sore throat to the lymph nodes, where T and B lymphocytes are activated. The activated B cells proliferate and produce antibodies that are carried by body fluids to sites of infection. T cells also travel to sites of infection. The rapid production of lymphocytes makes your lymph nodes swollen and tender.

5. When tissues are damaged, cells release chemical alarm signals such as histamine, which trigger various defensive mechanisms. Histamine causes blood vessels to dilate and become leakier. Blood flow increases and plasma passes out of the blood vessels and into the damaged tissues. Other chemicals attract white blood cells, which engulf bacteria and debris. Blood clotting proteins and platelets wall off the damaged area and allow repair to begin. Local increases in blood flow, fluid, and cells produce the redness, heat, and swelling characteristic of inflammation. Inflammation disinfects and cleans injured tissues and allows healing to begin.

Applying Your Knowledge

Multiple Choice: 1. a 2. b 3. d 4. c 5. e 6. b 7. b 8. a 9. a 10. d

Essay: 1. The primary immune response is slow and gradual. In the case of a bee sting, by the time the body is able to mount a strong response, the bee venom is gone. But the primary immune response triggers the development of memory cells, which are able to mount a quicker, more massive secondary immune response the second time the body is exposed to bee venom. This is what leads to anaphylactic shock.

2. It is possible that each time you catch a cold it is a different virus. You get sick for a while, but eventually the primary immune response fights off the cold virus. Memory cells make you immune to that virus but not to other cold viruses, which probably have different antigens. A two-year-old gets lots of colds, because a two-year-old is not yet immune to many of the cold viruses. By the age of 80, an individual has been exposed to and developed immunity to most cold viruses, so colds are less frequent.

3. The injection contained antibodies against hepatitis, so it conferred passive immunity to the disease. The antibodies soon disappeared from John's blood. Because he was not exposed to hepatitis antigens, his body never developed memory cells and active immunity against the disease.

4. A virus may have several different antigens on its surface. In addition, antigens are large molecules, and antibodies usually recognize localized regions, called antigenic determinants, on the surface of antigen molecules. It is possible that the same virus could activate lymphocyte clones that respond to different antigens or antigenic determinants in different people.

5. An antibody can attach to two different antigen molecules or microbes at once, and more than one antigen could attach to a given antigen or microbe. In this way, antibodies stick antigens and microbes together—processes called precipitation (antigens) and agglutination (microbes). The clumps of antigens or microbes formed in this way are easily engulfed by phagocytes.

6. The body directs the cell-mediated immune response, carried out by T cells, at transplanted organs, whose antigens the body senses as foreign. Transplant recipients are given drugs that suppress immunity, particularly cell-mediated immunity. This makes recipients more prone to infections, especially viral infections, because cell-mediated immunity is primarily directed against virus-infected cells.

Chapter 25

Organizing Your Knowledge

Exercise 1: 1. insect, great white shark, frog, snake, most fish 2. bird, polar bear 3. insect, frog, snake, most fish 4. all 5. great white shark, bird, polar bear 6. great white shark, bird, polar bear 7. insect, frog, snake, most fish

Exercise 2: 1. D 2. B 3. B 4. C 5. D 6. A 7. D 8. A 9. D 10. B 11. A 12. C 13. D 14. C

Exercise 3: 1. osmoconformer 2. osmoregulator 3. osmoregulator 4. osmoregulator 5. neither 6. gain 7. lose 8. lose 9. lose 10. lose 11. gain 12. lose 13. actively transports ions into cells 14. kidneys excrete dilute urine; digestive system and gills take up ions 15. drinks seawater; expels salts through gills and concentrated urine

16. drinks water and consumes salts; behavior and waterproof skin and eggs save water

Exercise 4: 1. most aquatic animals, including many fishes 2. mammals, amphibians, some fishes 3. birds, insects, many reptiles, snails 4. readily diffuses into water, no energy expended to make it 5. highly soluble in water, less toxic than ammonia 6. not toxic, requires little water for disposal 7. highly toxic, cannot be stored in body, requires much water for disposal 8. energy expended to make it, requires some water for disposal 9. energy required to make it

Exercise 5: See Figure 25.9 in the text.

Exercise 6: See Figure 25.9 in the text.

Exercise 7: 1. A 2. F 3. B 4. E 5. D 6. C

Exercise 8: 1. B 2. B 3. C 4. D 5. A 6. B 7. A 8. D 9. B

Exercise 9: 1. Dialysis is better if kidney damage is expected to be temporary. There are no problems with rejection or infection. Over the long term, dialysis is expensive and not always available, and it is no substitute for a real kidney because wastes build up between treatments. 2. A transplanted kidney works continuously, with no buildup of wastes as between dialysis treatments. The donor can live with one kidney, and over the long run a transplant is less expensive than dialysis. There is a chance of transplant rejection and infection due to drugs that suppress rejection.

Exercise 10: 1. hepatic portal vessel 2. glycogen 3. glucose 4. inactivated 5. ammonia 6. urea 7. inferior vena cava 8. renal 9. nephrons 10. glomerulus 11. Bowman's capsule 12. filtrate 13. reabsorption 14. secretion 15. excreted 16. active transport 17. water 18. loop 19. renal vein 20. secretion 21. urine 22. ureters 23. urinary bladder

Testing Your Knowledge

Multiple Choice: 1. a 2. d 3. b 4. c 5. b 6. a 7. c 8. e 9. c 10. e 11. b 12. e 13. c 14. c 15. a

Essay: 1. When an animal is too warm, nerves signal blood vessels in the skin surface to dilate, so more blood flows from the warm body core to the surface, and its heat can escape to the surroundings. Heat loss can be further increased by evaporative cooling, such as sweating or panting. When the body is too cold, sweating and panting are reduced, and nerves signal blood vessels in the skin to constrict, decreasing movement of heat from the warm body core to the surface and reducing heat loss.

2. A goose's legs must contain countercurrent heat exchangers. In such a heat exchanger, warm blood from the core of the body and cold blood from the foot flow in opposite directions in adjacent blood vessels. Heat from the body warms the blood from the foot. Cold blood is thus warmed as it moves up the leg, so it does not chill the body, and warm blood is cooled as it flows to the foot, so there is little heat left to lose when it gets there.

3. The internal fluids of a freshwater fish have a higher solute concentration than the surrounding water. This causes water to enter the fish by osmosis, mainly through the gills. The fish does not drink water (except with its food). To get rid of the excess water, the fish's kidneys produce large amounts of dilute urine. Some solutes are lost with the urine. To replace them, the digestive system absorbs ions from food, and the gills take up salt from the surrounding water. A saltwater fish has the opposite problem. The solute concentration of its body fluids is less than that of the surrounding water, so it tends to lose water by osmosis. To compensate, it drinks salt water and disposes of the excess salts via its gills and its kidneys, which excrete small amounts of concentrated urine.

4. An animal can bask in the sun or huddle together with other animals to warm up. It can turn in a direction that exposes less surface to the sun, move into the shade, hide in a damp burrow, or bathe to cool off.

5. In filtration, water and other small molecules are forced by blood pressure from the porous glomerulus into Bowman's capsule, the first part of the nephron tubule. The process of reabsorption is reclaiming valuable solutes and water from the filtrate. Solutes are reabsorbed by active transport mainly in the proximal and distal tubules, and water is reabsorbed by osmosis in the loop of Henle and the collecting duct. Some substances, such as H^+ ions, are secreted by active transport from the blood into the filtrate in the proximal and distal tubules. After the filtrate is refined by the processes of absorption and secretion, it is called urine. It leaves the kidney and is expelled from the body—a process called excretion.

Applying Your Knowledge

Multiple Choice: 1. a 2. c 3. c 4. e 5. d 6. c 7. c 8. b 9. e 10. d

Essay: 1. The windchill factor takes into account not only the temperature but also the movement of air, which accelerates heat loss. The body loses heat by conduction, convection, radiation, and evaporation. In comparison with still air, moving air increases heat loss by convection and evaporation. Thus, on a windy day, the body loses heat faster. The windchill factor reflects the fact that a windy day seems colder than a day without wind.

2. A horned lizard is an ectotherm. It warms itself by absorbing heat from its surroundings, so it expends a lot of effort moving in and out of the sun to maintain the proper body temperature. It does not generate heat from metabolism, so it does not eat much. A mouse is an endotherm. It gets most of its heat from its own metabolism, so it does not need to expend much effort changing its position or location relative to the sun or outside energy sources. But because it maintains a constant high temperature, it needs to expend more effort than the horned lizard hunting for food.

3. The liver occupies a critical location in the body—between the intestine and the heart. Substances that are ingested (or inhaled—the blood from the lungs eventually circulates through the liver) pass first through the liver before being pumped by the heart throughout the body. The liver has the metabolic machinery for altering and inactivating a large number of poisonous substances. Unfortunately, the liver also tends to take in substances that are poisonous to the liver itself. Thus, the liver is the first organ to be damaged by many harmful substances, such as mushroom poison or solvents in glue.

4. Mammals and birds are endotherms, which generate body heat via their own metabolism. They are able to stay warm enough to function even when there is little heat available from the outside environment. Reptiles such as crocodiles are ectotherms, which obtain heat from their surroundings. There is little heat available in the polar regions; an antarctic crocodile would not be able to obtain enough heat to function in this environment.

5. After the subject eats a candy bar, blood entering the liver would be high in glucose. The liver would store up some of this glucose as glycogen, so blood leaving the liver would have a lower glucose content. The liver removes and alters toxic substances such as alcohol. After a drink, blood leaving the liver would be lower in alcohol than blood entering the liver. Between

meals, the liver breaks down its stored glycogen and adds glucose to the blood, so the blood leaving the liver would contain more glucose than the blood entering the organ.

Chapter 26

Organizing Your Knowledge

Exercise 1: 1. neurosecretory cell 2. blood vessel 3. target cell 4. endocrine cell 5. secretory vesicles 6. blood vessel 7. target cell 8. nerve cell 9. nerve signals 10. nerve cell. For hormones and neurotransmitter molecules, see Figure 26.1 in the text.

Exercise 2: A: 1. steroid hormone 2. plasma membrane 3. receptor 4. DNA 5. new protein B: 1. non-steroid hormone 2. receptor 3. plasma membrane 4. enzyme 5. second messenger 6. effect on cellular function

Exercise 3: See Figure 26.3A and Table 26.3B in the text.

Exercise 4: 1. Thyroid gland 2. T_3 3. T_4 4. Lowers blood calcium 5. Estrogens 6. Stimulate uterus; female sex characteristics 7. Promotes uterine lining growth 8. Adrenal medulla 9. epinephrine 10. norepinephrine 11. Adrenal cortex 12. Increases blood sugar 13. Mineralocorticoids 14. Raises blood sugar 15. Insulin 16. Lowers blood sugar 17. Parathyroid glands 18. Parathyroid hormone 19. Pineal gland 20. Involved in body rhythms 21. Anterior lobe of pituitary 22. Growth hormone 23. Stimulates thyroid gland 24. Stimulates milk production 25. ACTH 26. Stimulates ovaries and testes 27. Follicle-stimulating hormone 28. Stimulates contraction of uterus and mammary gland 29. Antidiuretic hormone 30. Thymosin 31. Stimulates T lymphocytes 32. Testes 33. Support sperm formation; development of male characteristics

Exercise 5: 1. P 2. H 3. A 4. H 5. P 6. A 7. A 8. P 9. A (or H) 10. H 11. P 12. A 13. H 14. A

Exercise 6: 1. E 2. C 3. B 4. D 5. A 6. C 7. A 8. D

Exercise 7: 1. blood clotting 2. parathyroid glands 3. parathyroid hormone 4. increased 5. increased 6. release of calcium from 7. calcitonin 8. thyroid gland 9. decreased 10. decreased 11. deposition of calcium in

Exercise 8: 1. increase 2. increases 3. increase 4. increase 5. decrease 6. decrease 7. decreases 8. increase 9. increases 10. decrease 11. decreases 12. increases 13. decrease 14. decrease

Exercise 9: 1. E 2. E 3. M or G 4. M or G 5. M 6. E 7. E 8. G 9. M or G 10. M or G 11. G 12. M 13. E 14. E 15. G

Exercise 10: There is no single correct answer for a concept map; you learn by constructing it; many arrangements are possible.

Exercise 11: 1. All are sex hormones. 2. Both are produced by the posterior lobe of the pituitary. 3. All act to raise blood glucose. 4. ACTH stimulates the adrenal cortex to secrete glucocorticoids. 5. All target the kidney. 6. All are produced by the thyroid gland. 7. Both are secreted by the pancreas and have opposing effects on blood glucose. 8. All are secreted in response to stress. 9. All are secreted by the anterior lobe of the pituitary. 10. They have opposing effects on blood calcium. 11. Both affect milk secretion.

Testing Your Knowledge

Multiple Choice: 1. d 2. e 3. e 4. d 5. b 6. c 7. a 8. b 9. c 10. b 11. e 12. c 13. b 14. c 15. d

Essay: 1. The pituitary gland controls many other glands. But it does not control all other glands, and the pituitary itself is controlled by the hypothalamus, which is the real "master gland."

2. The adrenal medulla secretes epinephrine and norepinephrine, which trigger the fight or flight response to short-term stress. The adrenal cortex secretes mineralocorticoids and glucocorticoids in response to prolonged stress.

3. Hormones and neurotransmitters are both chemical signals between cells. Hormones are carried by the blood to distant cells, while neurotransmitters affect nearby cells directly.

4. For example: Gigantism—abnormally large size—is caused by oversecretion of growth hormone during development. Type I diabetes results from undersecretion of insulin. In type II diabetes, insulin levels are normal, but cells are unable to respond to it. In either type of diabetes, cells are unable to take up and use glucose and so burn fats and proteins instead. Glucose builds up in the blood, and numerous complications result.

5. Glucocorticoids are used in treating injuries because they reduce swelling and pain. But because they decrease pain, their use can result in further injury. They also suppress the inflammatory response and immunity to infection, so they may slow healing.

6. Insulin lowers blood sugar, and glucagon raises it. Parathyroid hormone raises blood calcium, and calcitonin lowers it.

Applying Your Knowledge

Multiple Choice: 1. c 2. c 3. d 4. b 5. c
6. d 7. b 8. e 9. a 10. e

Essay: 1. When a person with hypoglycemia eats sugar, hyperactive B cells in the pancreas secrete too much insulin, which causes blood sugar to drop too low. Eating sugar would just make this problem worse.

2. In diabetes, cells break down fat for energy because they have not received the insulin signal to take in and use glucose.

3. The enlarged thyroid gland might secrete more T_3 and T_4 than normal, boosting metabolic rate and body temperature. A pituitary tumor might cause the pituitary to secrete too much TSH, causing the thyroid gland to secrete too much T_3 and T_4.

4. An organ is affected by a hormone only if it has the proper receptor. Many cells have insulin receptors, but only kidney cells have receptors for ADH.

5. Receptors could vary in various tissues in different animals. An animal that lacks receptors that respond to dioxin might not be affected at all. In another animal with a large number of receptors in liver, cells might sustain serious liver damage, while a different animal whose immune cells have some receptors might suffer slight immune system impairment.

Chapter 27

Organizing Your Knowledge

Exercise 1: 1. hermaphroditism 2. sperm 3. asexual reproduction 4. zygote 5. gamete 6. regeneration 7. external fertilization 8. ovum 9. copulation 10. budding 11. internal fertilization 12. fragmentation

Exercise 2: 1. oviduct 2. ovary 3. uterus 4. vagina 5. follicles 6. corpus luteum 7. wall of uterus 8. endometrium 9. cervix 10. seminal vesicle 11. prostate gland 12. bulbourethral gland 13. epididymis 14. testis 15. scrotum 16. vas deferens 17. erectile tissue 18. urethra 19. glans 20. prepuce

Exercise 3: 1. seminiferous tubules of testes 2. follicles in ovary 3. daily and continuously

4. before birth 5. two secondary spermatocytes 6. one secondary oocyte and a first polar body 7. four immature sperm cells 8. one ovum and a second polar body 9. four 10. one 11. millions per day 12. one per month

Exercise 4: 1. hypothalamus 2. follicle-stimulating hormone (FSH) 3. luteinizing hormone (LH) 4. follicle 5. ovary 6. estrogen 7. endometrium 8. uterus 9. ovum 10. follicle 11. negative 12. FSH 13. LH 14. preovulatory 15. pituitary 16. FSH 17. LH 18. ovulation 19. fourteenth 20. oviduct 21. postovulatory 22. corpus luteum 23. progesterone 24. follicles 25. progesterone 26. fertilized 27. endometrium 28. menstruation

Exercise 5: A. 3, R B. 4, P C. 1, S D. 2, Q

Exercise 6:

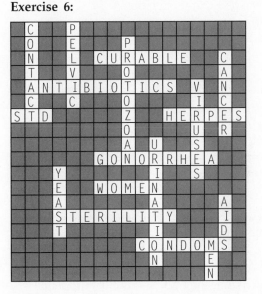

Exercise 7: A. prevents release, 1 or 2 B. prevents meeting, 8 C. prevents meeting, 7 D. prevents release, 1 or 2 E. prevents implantation, 4 F. prevents meeting, 6 G. prevents meeting, 5 H. prevents release, 3 I. prevents meeting, 9

Exercise 8: 1. A sperm swims to the egg, propelled by a flagellum powered by ATP. 2. The acrosome in the sperm head bursts on contact, releasing enzymes that digest an opening in the egg's coat. 3. Species-specific protein molecules on the surface of the sperm head bind with proteins on the egg. 4. If an egg were fertilized by more than one sperm, the resulting zygote nucleus would contain too many chromosomes and could not develop normally. 5. Fusion of sperm and egg membranes causes the egg membrane to quickly become impenetrable to other sperm, and then another barrier, the fertilization membrane, forms, which also keeps other sperm out. 6. The egg's metabolism speeds up and the egg and sperm nuclei fuse, forming the diploid zygote nucleus.

Exercise 9: A. 8 B. 5 C. 6 D. 4 E. 11 F. 1 G. 2 H. 9 I. 7 J. 10 K. 3 L. blastocoel M. digestive cavity (archenteron) N. blastocoel O. blastula P. neural tube Q. coelom R. somite S. zygote T. digestive cavity (archenteron) U. gastrula V. blastopore W. notochord X. ectoderm Y. mesoderm Z. endoderm

Exercise 10: 1. endoderm 2. mesoderm 3. endoderm 4. ectoderm 5. mesoderm 6. mesoderm 7. endoderm 8. ectoderm

Exercise 11: 1. induction 2. pattern formation 3. cell migration 4. cell aggregation 5. changes in cell shape 6. differentiation

Exercise 12: 1. I 2. M 3. J 4. G 5. A 6. K 7. P 8. T 9. R 10. S 11. B 12. H 13. N 14. Q 15. D 16. O 17. E 18. F 19. C 20. L

Exercise 13: There are many possible combinations; for example, a man's sperm could fertilize a donated egg in vitro. This egg could then be planted in his partner's uterus. A woman's eggs could be fertilized with her partner's sperm and the embryo implanted in the uterus of a surrogate mother.

Testing Your Knowledge
Multiple Choice: 1. e 2. c 3. d 4. c 5. b 6. c 7. b 8. a 9. b 10. d

Essay: 1. When food is abundant and the environment is unchanging, asexual reproduction would enable a single aphid to produce many offspring quickly. The populations could expand rapidly and exploit available resources. Sexual reproduction increases genetic variability among offspring. This might make the species more adaptable when the environment is changing and food supplies are less predictable.

2. The outermost extraembryonic membrane, the chorion, forms the embryo's part of the placenta. Extensions of the chorion, called chorionic villi, absorb oxygen and nutrients from the mother's blood. The amnion forms a fluid-filled sac around the embryo. The fluid absorbs shock and keeps the embryo from drying out. The yolk sac produces the embryo's first blood cells and germ cells, which later give rise to gamete-producing cells in the gonads. The allantois forms part of the umbilical cord and part of the embryo's urinary bladder.

3. Primary oocytes form in a female embryo and begin dividing to form eggs before birth. At maturity, FSH and LH cause a single follicle containing an egg to mature and rupture each month, and fertilization triggers the final meiotic cell division that completes the egg's development. Sperm production does not begin until a male reaches maturity, but after that point primary spermatocytes divide and mature at a rate of millions per day, stimulated by FSH. A primary oocyte divides to form a single ovum and some polar bodies, which are discarded. A primary spermatocyte divides to form four sperm.

4. See Figures 27.10 and 27.11 in the text.

5. The pill is associated with an increased risk of cardiovascular problems, especially among women who smoke or have a history of such problems. Long-term use may increase the risk of liver tumors. The pill reduces the risk of ovarian and endometrial cancers and the risks associated with pregnancy. Overall, death rates associated with pregnancy are twice those from use of the pill.

Applying Your Knowledge
Multiple Choice: 1. c 2. d 3. b 4. b 5. e 6. a 7. a 8. d 9. e 10. c

Essay: 1. The fish that reproduce sexually would probably adapt most successfully to a changing environment, because sexual reproduction produces more variation among offspring than does asexual reproduction. Among a variety of offspring there is more likelihood of some individuals being suited to changed environmental conditions.

2. Sperm cannot develop at body temperature. The testes are located in the scrotum, a sac outside the abdominal cavity, where the sperm-forming cells can be kept cool enough to function normally. Tight jeans hold the testes closer to the body, where the temperature is warmer and sperm cannot develop normally.

3. The eggs of aquatic animals are usually fertilized externally, in the water. The male might court the female for her to lay her eggs, but it is not necessary for a male and a female to be close to one another for fertilization to occur. Land animals usually use internal fertilization. The male and the female must copulate so that sperm are deposited inside the body of the female. This requires closer cooperation between the male and the female, which is accomplished by more elaborate courtship behaviors.

Chapter 28

Organizing Your Knowledge

Exercise 1: See Figure 28.2 in the text.

Exercise 2:

```
          I
  N E U R O N   C   M
          T   E   O U T P U T
          E   N   T
  I N T E G R A T I O N   G
    E       N   R       L
  P E R I P H E R A L     A
    V   I   U   L   S E N S O R Y
    E   N   R       C   D
    O   O       C L   S   G   K
    D E N D R I T E S   B R A I N
      E   S       R         N   O
      S       S U P P O R T I N G
              S         N   L
          M Y E L I N   P   I   B
              S       U   A X O N
                      T       D
                              Y
```

Exercise 3: A. 3, T   B. 1, P   C. 2, Q   D. 4, R   E. 5, S

Exercise 4: 1. neurons 2. electrical 3. neurotransmitters 4. cleft 5. axon 6. plasma membrane 7. receptor 8. receiving cell 9. channels 10. action potentials 11. enzyme

Exercise 5: 1. Y 2. N 3. Y 4. Y 5. N 6. N 7. N

Exercise 6: 1. Reduced level of serotonin would result in lower level of stimulation and depressed mood. 2. Increased inhibition would result in lower level of stimulation and depressed mood. 3. This would simulate stimulation by serotonin, elevating mood. 4. Increased inhibition would result in lower level of stimulation and depressed mood. 5. Decreased inhibition by GABA would elevate mood. 6. Increased stimulation by serotonin would elevate mood.

Exercise 7: 1. brain 2. sensory division 3. autonomic nervous system 4. ventricles 5. motor division 6. somatic nervous system 7. white matter 8. cerebrospinal fluid 9. peripheral nervous system 10. gray matter 11. central nervous system 12. spinal cord 13. spinal nerves

Exercise 8: 1. stimulates activity 2. inhibits activity 3. constricts 4. relaxes 5. stimulates erection 6. promotes ejaculation 7. inhibits 8. accelerates 9. stimulates saliva production 10. inhibits saliva production 11. constricts 12. dilates

Exercise 9: See Figure 28.14 in the text. Note that there is no interneuron in the knee-jerk reflex.

Exercise 10: 1. cerebral cortex, E 2. thalamus, F 3. hypothalamus, D 4. pons and medulla, B 5. midbrain, C 6. cerebellum, A

Exercise 11: 1. rapid eye movement (REM) 2. midbrain 3. motor 4. frontal 5. motor 6. cerebellum 7. limbic 8. right 9. split 10. sensory 11. spinal cord 12. thalamus 13. somatosensory 14. parietal 15. left 16. corpus callosum 17. short 18. long 19. synapses 20. amygdala 21. hippocampus 22. left 23. association 24. left 25. medulla oblongata 26. pons

Testing Your Knowledge

Multiple Choice: 1. b 2. d 3. d 4. d 5. a 6. c 7. c 8. d 9. a 10. c

Essay: 1. See Figure 28.7 in the text. The transmitting and receiving neurons are separated by a narrow synaptic cleft. A chemical neurotransmitter is contained in small vesicles in the synaptic knobs at the end of the axon of the transmitting cell. When an action potential arrives at the end of the transmitting cell's axon, the vesicles fuse with the plasma membrane. Neurotransmitter molecules are released into the synaptic cleft, rapidly diffuse across, and bind to receptor molecules in the membrane of the receiving cell. This opens ion channels, allowing ions to diffuse through the membrane and initiating new action potentials in the receiving cell. Then the neurotransmitter is broken down by an enzyme, and the ion channels close.

2. The parasympathetic and sympathetic divisions are both parts of the autonomic nervous system, which exerts motor control over many involuntary (mostly internal) body functions. In general, the two systems oppose each other. The sympathetic system prepares the body for intense activities by speeding up the heart, dilating breathing passages, stimulating glucose release by the liver, but slowing down activities of the digestive tract. The parasympathetic system primes the body for rest and digestion by slowing the heart, constricting the breathing passages, and stimulating the digestive system.

3. The relative size of the vertebrate brain increased in birds and mammals. The three original regions of the brain (forebrain, midbrain, and hindbrain) became subdivided into subregions with specific functions, such as the medulla, cerebellum, and cerebrum. The third trend was an increase in the power of the forebrain, specifically the cerebrum, which is much larger in birds and mammals than in other vertebrates.

4. The three functions of the nervous system are sensory input, integration, and motor output. As you read the question, sensory receptors in your eyes send signals to processing centers in the brain—sensory input. The brain interprets the signals, comparing them with memory and determining their symbolic and conceptual meaning—integration. Finally, the brain sends signals to the muscle cells in your arm and hand, directing their movements as you write an answer—motor output.

5. Some transmitting neurons deliver excitatory neurotransmitters, which open Na^+ channels and thus trigger action potentials in the receiving cell. Other transmitting cells deliver inhibitory neurotransmitters, which open channels for Cl^- and decrease the receiving cell's tendency to transmit action potentials. Only if excitatory signals outweigh inhibitory signals will the receiving cell transmit action potentials.

Applying Your Knowledge

Multiple Choice: 1. e 2. a 3. e 4. c 5. a 6. d 7. b 8. a 9. c 10. d

Essay: 1. a. Acetylcholine stimulates receiving neurons to transmit action potentials. If the receptors for acetylcholine are blocked, acetylcholine will not be able to stimulate receiving neurons and the nerve signals will be stopped at the synapse. b. If acetylcholine release is inhibited, transmitting cells will not be able to signal receiving cells, and nerve signals will be stopped at the synapse. c. If DF prevents acetylcholine from being broken down, acetylcholine will remain in receptors in the receiving cell, and the receiving cell will continue to transmit signals, without being stimulated by the sending cell.

2. When dominoes topple, each domino affects the next one in the row. The dominoes do not travel along the row, but the topple is relayed along the row, one domino to the next. Similarly, when sodium gates open in a neuron, movement of sodium ions into the neuron triggers gates in the next portion of the membrane to open, letting sodium in at that point, which triggers gates to open farther down, and so on. Sodium does not travel along the neuron, just the change in the neuron membrane. Thus, the action potential travels along the neuron, much the way the topple travels along the row of dominoes. The biggest difference between the dominoes and the neuron is that the dominoes can only topple once. The neuron membrane

can repolarize after each action potential and thus quickly transmit many action potentials in succession.

3. Nerve signals from the brain pass through the spinal cord on their way to the nerves leading to the arms and legs. Apparently, Andrew's injury damaged the spinal cord between the brain and the place where nerves to the legs originate, so that signals to the legs were interrupted. His injury must have been below the connection between the spinal cord and nerves to the arms, so nerve signals to and from the arms were not affected. The brain is not involved in the knee-jerk reflex. Signals travel to the spinal cord via a sensory neuron, across a synapse, and back to leg muscles via a motor neuron. His spinal cord is injured above these neurons, so impulses from sensory neurons do not reach the brain, and Andrew cannot feel his legs. But he doesn't have to feel his legs for the reflex to function.

4. The cerebrum is responsible for higher mental functions such as interpreting sensory information, language, and control of voluntary movements. Centers in the brainstem and midbrain, such as the medulla and hypothalamus, control involuntary vital functions such as heart rate, breathing, and body temperature. These centers may continue to do their jobs even if the cerebrum is damaged.

Chapter 29

Organizing Your Knowledge

Exercise 1: A. 2, M B. 4, P C. 3, N D. 5, L E. 1, Q F. 6, O

Exercise 2: 1. Electromagnetic receptor 2. Near eyes of pit viper (snake) 3. Mechanoreceptor 4. Skin and hairs 5. Electroreceptor 6. Electromagnetic receptor 7. Skin of electric fish 8. Photoreceptor 9. Electromagnetic receptor 10. Light 11. Pain receptor 12. Hair cell 13. Sound 14. Thermoreceptor 15. Heat 16. Skin 17. Stretch receptor 18. Mechanoreceptor 19. Skeletal muscle 20. Chemoreceptor 21. Chemicals in external environment

Exercise 3: 1. sclera 2. muscle, G 3. cornea, B 4. iris, E 5. pupil 6. lens, F 7. aqueous humor, C 8. choroid 9. retina, D 10. optic nerve, A 11. vitreous humor

Exercise 4: See Figure 29.6 in the text.

Exercise 5: 1. Farsightedness 2. Nearsightedness 3. Astigmatism 4. Distant 5. Nearby 6. Varies

7. Nearby 8. Distant 9. Varies 10. Eyeball too short or lens less elastic 11. Eyeball too long 12. Misshapen lens or cornea 13. Corrective lens that curves more sharply on inside than outside 14. Corrective lens that curves more sharply on outside than inside 15. Lens that compensates for asymmetry in eye

Exercise 6: *Rods:* 3, 5, 6, 8, 9, 10, 12, 13, 15, 16, 17 *Cones:* 1, 2, 4, 5, 7, 8, 11, 13, 14, 15

Exercise 7: 1. stirrup, G 2. anvil, G 3. hammer, G 4. auditory canal, F 5. semicircular canals, H 6. auditory nerve, D 7. eardrum, E 8. pinna, C 9. oval window 10. cochlea, A 11. eustachian tube, B 12. outer ear 13. middle ear 14. inner ear

Exercise 8: 1. B 2. F 3. E 4. G 5. H 6. C 7. D 8. A 9. D 10. E

Testing Your Knowledge

Multiple Choice: 1. c 2. b 3. a 4. e 5. c 6. d 7. b 8. b 9. a 10. a

Essay: 1. Snakes called pit vipers have infrared receptors near their eyes that enable them to detect the heat given off by warm-blooded prey. Some birds may be able to detect the magnetic field of the Earth and use this information to find their way when migrating. Certain fishes find their prey by sensing the electrical currents produced by the prey's muscles.

2. Sensory receptors send action potentials to the brain, and the brain is made aware of sensory stimuli. This is called sensation. Perception is meaningful interpretation of these sensory data.

3. Farsightedness is a vision defect that occurs when the eyeball is too short, and the lens cannot bend light rays sharply enough to bring an image into focus on the retina. It can also occur when the lens loses its elasticity and cannot curve enough to bring an image into focus. It can be corrected with a lens that curves more sharply on the inside than the outside, making light rays converge slightly before entering the eye.

4. A stimulus changes the permeability of the receptor cell membrane by causing ion gates to open. Positively charged ions flow into the cell, altering the membrane potential, increasing it to a level called the receptor potential. If the receptor potential is big enough, it triggers action potentials that are sent to the central nervous system.

5. The senses of smell and taste both rely on chemoreceptors. There are different kinds of smell and taste receptors, each stimulated by molecules of particular types. The particular combination of receptors stimulated gives rise to our perception of different tastes and smells. Smells are carried by the air, and their molecules dissolve in the mucus lining the nasal cavity before stimulating smell receptors. Taste receptors are usually stimulated by direct contact between a substance and receptors in taste buds.

Applying Your Knowledge

Multiple Choice: 1. d 2. c 3. e 4. c 5. a 6. c 7. a 8. c 9. c 10. d

Essay: 1. Receptors for both the sense of hearing and balance are located in the inner ear. Sound receptors are in the cochlea, and balance receptors are in the semicircular canals, saccule, and utricle. Meniere's disease must impair the inner ear in some way.

2. Blue cones are stimulated when we see blue; yellow cones are stimulated when we see yellow. White stimulates all three types of cones, and black stimulates none of them.

3. One method that has been tried is to blindfold people, drive them around an unfamiliar area, spin them around to make sure they are completely mixed up, then ask them to point toward their starting point. Two groups of people were tested: One group wore helmets containing powerful magnets (which would presumably interfere with a magnetic sense); the other group wore nonmagnetic helmets. (The results were tantalizing but inconclusive.)

4. Changes in the environment tend to be more important than ongoing "background" stimuli. Natural selection has shaped the brain to sense a sudden change more readily than a continuing stimulus. Pain is an important message about danger, injury, or disease. Natural selection would not favor an animal that gets used to pain. It is important for the animal to be reminded of pain and deal with its cause.

5. Hair cells are the receptors for hearing; if they were damaged, they would not be able to respond to vibrations in the cochlea. A brain injury that affects the auditory portion of the cerebral cortex would impair the brain's ability to sense or perceive sounds. Arthritis of the middle ear bones might interfere with their movement and prevent them from transmitting vibrations to the cochlea. A torn eardrum might not be

able to pick up vibrations from the air and transmit them to the ear bones. Earwax in the auditory canal would block sound waves from setting the eardrum in motion.

Chapter 30

Organizing Your Knowledge

Exercise 1: 1. friction 2. gravity 3. tripod 4. momentum 5. peristalsis 6. friction 7. gravity 8. streamlined 9. up and down 10. airfoils 11. greater 12. lift 13. gravity

Exercise 2: 1. earthworm, *Hydra* 2. insect, crab, clam 3. sponge, sea star, human 4. fluid under pressure in a body compartment 5. armorlike covering or shell over outside 6. hard supporting elements embedded inside soft tissues 7. fluid (water) 8. insect: chitin; clam: calcium carbonate 9. sponge: calcium carbonate or silica spicules and protein fibers; human: bone and cartilage 10. cushions, gives body shape, peristaltic movement 11. protection, support, movement 12. support, movement, protection 13. cannot support locomotion in which body is held off ground; offers little protection 14. insect exoskeleton does not grow with animal; must be molted, leaving animal vulnerable 15. does not offer strong protection of exoskeleton

Exercise 3: See Figure 30.3A in the text.

Exercise 4: 1. B, E 2. C, F 3. A, D

Exercise 5: 1. G 2. I 3. H 4. B 5. C 6. F 7. B 8. E 9. D 10. A 11. B 12. H 13. J

Exercise 6: See Figure 30.8 in the text.

Exercise 7: 1. capillaries 2. tendon 3. nerves 4. motor unit 5. neuromuscular junction 6. fiber 7. acetylcholine 8. action potential 9. tubules 10. myofibrils 11. Z 12. sarcomeres 13. thin 14. actin 15. thick 16. myosin 17. slide 18. calcium 19. heads 20. ATP 21. shortening 22. calcium 23. endoplasmic reticulum

Exercise 8: For example: "Tubular robot cleans out drains" or "Robot with digger arms tunnels for trapped miners" or "Vacuum robot cleans up oil spills."

Testing Your Knowledge

Multiple Choice: 1. c 2. b 3. a 4. d 5. c 6. d 7. b 8. d 9. e 10. b

Essay: 1. A wing is an airfoil, a shape that is thicker at the leading edge and more curved on top than underneath. When a bird flaps its wings, the air passing over the wings has to travel a longer distance than the air moving under the wings. Air molecules are spaced farther apart above a wing than below, resulting in greater air pressure below than above, which lifts the wing.

2. A skeleton supports the body against gravity; this function is illustrated by the vertebral column. The skeleton protects organs, as the ribs protect the lungs and heart. The skeleton works with muscles in movement; this is illustrated by the leg bones and their opposing sets of muscles.

3. Arthropods—animals such as insects, crabs, and spiders—have jointed exoskeletons. An exoskeleton is strong and effective at protecting the organs inside. A drawback of an exoskeleton is that it does not grow with the animal and must be molted periodically. This is wasteful and leaves the animals temporarily vulnerable.

4. The two main forces that must be overcome by a moving animal are friction and gravity. Aquatic animals are buoyed up by the water and are not affected much by gravity, but the frictional resistance of water is a problem for them. Land animals must expend energy supporting themselves against the pull of gravity, but air offers much less resistance than water, so friction is not a serious problem for them.

5. Check your drawings against Figure 30.9A in the text.

Applying Your Knowledge

Multiple Choice: 1. c 2. e 3. d 4. a 5. a 6. c 7. b 8. d 9. b 10. b

Essay: 1. The frog's long fingers and toes support webs that enable the frog to swim. Its hind legs, much longer than its front legs, are used for hopping. The frog's short, rigid backbone stiffens its body for hopping. Flexible cartilage in its joints and sternum absorb shock on landing.

2. An earthworm is segmented, and each of its segments is filled with fluid. Muscles in different segments can perform different movements simultaneously. So, for example, circular muscles can contract in anterior segments, lengthening them and pushing the front of the worm forward. At the same time, longitudinal muscles can shorten posterior segments, drawing the rear of the worm up behind. This enables the worm to make varied crawling and turning movements. The whole body of the roundworm is one big fluid-filled compartment, so the

whole worm has to move the same way at once.

3. The short wings of a songbird are easy to flap and enable the bird to fly fast and make tight turns. An eagle uses its broad wings to lift its prey from the ground. The elongated wings of a gull are more suited to gliding long distances with minimal effort.

4. A motor unit is a neuron and the muscle fibers it controls. In the hand, the motor units are small; one neuron might control only one muscle fiber. In the leg, the motor units are bigger; one neuron might control hundreds of muscle fibers. Thus, even though the hand is smaller than the leg, it takes more neurons to control the hand, because of its one-to-one arrangement of neurons and muscle fibers. Therefore, more area in the brain is devoted to controlling the hands than the legs.

5. Walking with the legs splayed to the sides is hard work. A lot of energy is devoted simply to fighting gravity and keeping the body off the ground. Animals that walk this way are slower than those whose legs are directly under the torso. If dinosaurs' legs were under the body, like those of horses and elephants, they probably were a lot faster than alligators or lizards.

Chapter 31

Organizing Your Knowledge

Exercise 1: There are many possible answers to this question. As I write this, I can see landscape plants such as Douglas firs, rhododendrons, and roses that shade the house, produce oxygen, hold soil in place, and beautify my surroundings. There are lots of papers, made from trees, on my oak desk. There is a pack of chewing gum on the desk; chicle for making gum comes from a tropical tree. I am wearing a cotton shirt and jeans and drinking a cup of tea.

Exercise 2: *Monocots:* D—complex array of vascular bundles, E—floral parts usually in multiples of three, F—veins usually parallel, H—one cotyledon, I—fibrous root system. *Dicots:* C—vascular bundles in ring, G—flower parts usually in multiples of four or five, B—veins usually netlike, J—two cotyledons, A—taproot usually present

Exercise 3:

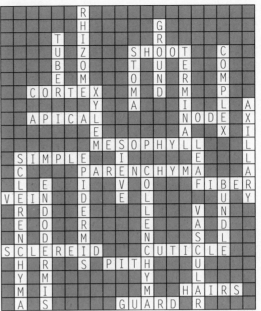

Exercise 4: See Figures 31.6B, C, and D in the text. a. monocot stem. b. root. c. leaf. d. dicot stem

Exercise 5: 1. Q 2. I 3. J 4. P 5. M 6. L 7. N 8. A 9. D 10. F 11. O 12. E 13. G 14. K 15. B 16. H 17. C

Exercise 6: See Figure 31.7B in the text.

Exercise 7: 1. apical meristem 2. outer cylinder 3. inner cylinder 4. cortex 5. primary phloem

Exercise 8: 1. bark 2. secondary phloem 3. cork (with cambium) 4. vascular cambium 5. sapwood 6. heartwood 7. secondary xylem (colors from inside out: red, pink, green, light blue, dark blue)

Exercise 9: See Figure 31.9 in the text. The carpel (stigma, style, and ovary) should be pink and the stamens (anther and filament) blue.

Exercise 10: 1. water 2. germination 3. seed coat 4. embryo 5. cotyledons 6. endosperm 7. sporophyte 8. meiosis 9. spores 10. gametophytes 11. sepals 12. petals 13. anthers 14. stamens 15. stigmas 16. carpels 17. male 18. sperm 19. Pollination 20. ovary 21. sperm 22. an ovule 23. embryo sac 24. gametophyte 25. egg 26. zygote 27. endosperm 28. fertilization 29. seed 30. embryo 31. cotyledons 32. ovary 33. an aggregate fruit 34. sugar 35. seed coats

Exercise 11: 1. E 2. A 3. D 4. B 5. C

Testing Your Knowledge

Multiple Choice: 1. e 2. c 3. a 4. d 5. e
6. b 7. b 8. a 9. b 10. c 11. c

Essay: 1. The root anchors the plant, absorbs water and minerals, transports water and minerals to other parts of the plant, and stores food. The epidermis, especially root hairs, functions mainly in absorption; the vascular cylinder carries out transport; and the cortex stores food.

2. See Figure 31.6C in the text. Epidermis is one system, xylem and phloem the vascular tissue system, and pith the ground tissue system. The vascular cylinder is in the center of the root, surrounded by a thick cortex and the epidermis. The center of a stem is filled by pith, with the vascular tissues forming a ring, and a thin cortex on the outside, covered by epidermis.

3. The diploid sporophyte stage is the full-grown flowering plant, with roots, stems, and leaves. Its flowers contain anthers and ovules, in which cells undergo meiosis to produce haploid spores. These spores divide by mitosis to develop into pollen grains (the male gametophytes) and embryo sacs (female gametophytes) —both haploid. The gametophytes produce haploid gametes—sperm and eggs—by mitosis.

4. A meristem consists of unspecialized parenchyma cells that divide and generate new cells and tissues. Meristems occur in the tips of roots and the apical and axillary buds of shoots. These are responsible for primary—lengthwise—growth. An apple tree also has cylinders of vascular cambium and cork cambium just inside its trunk, which enable the tree to grow in diameter—secondary growth.

5. Sclerenchyma cells have thick, rigid secondary walls hardened with lignin, which enables them to form a sturdy framework for the stem. The water-conducting xylem cells in a root die when mature and form long, hollow pipes that carry water. Epidermal cells are thin and transparent, letting light into the leaf. They secrete a waxy cuticle, which keeps the leaf from drying out.

6. The vascular cambium forms a cylinder just underneath the bark of a tree, starting out between the primary xylem and the primary phloem. The vascular cambium is a meristem; its cells divide and give rise to xylem cells and phloem cells. As cambium cells divide and differentiate, new xylem cells are added to the inside of the cambium and new phloem cells are added to the outside of the cambium. This causes the tree to grow in diameter.

7. Pollination is transfer of a pollen grain from a stamen to the stigma of a carpel. The pollen grain germinates, and its tube cell gives rise to a pollen tube that grows down into the ovary to an ovule. Its generative cell divides to form two sperm cells, which enter the embryo sac. One of the sperm cells fertilizes the egg, forming the plant zygote. The other sperm fertilizes a large diploid cell, which divides to form the triploid endosperm, food stored in the seed for the embryo. Fertilization of the haploid egg by one sperm and a diploid cell by another sperm is called double fertilization, and it occurs only in plants. Fertilization in animals involves only a single haploid egg and a single haploid sperm.

8. Some plants, such as turnips, have large roots that store food. Stems are modified in many ways. Horizontal runners enable grasses and strawberries to reproduce and spread vegetatively. Enlarged horizontal stems called tubers, such as potatoes, store food. A bulb is also a stem that functions in storage and reproduction by fragmentation. The leaves of some plants, such as celery, have enlarged petioles that store food and water. Leaves may be modified to form tendrils, which help plants climb. Cactus spines are modified leaves that protect the plant from grazing animals.

Applying Your Knowledge

Multiple Choice: 1. a 2. c 3. c 4. d 5. c
6. d 7. a 8. d 9. a 10. d

Essay: 1. Most of a tree trunk is xylem, and xylem cells are dead at functional maturity, so the cells that make up most of a tree are not alive.

2. The spike would pass through layers of cork, cork cambium, and secondary phloem, which make up the tree's bark. Then it would penetrate the vascular cambium, the functional secondary xylem of the sapwood, and the nonfunctional secondary xylem of the heartwood.

3. Garlic—bulb (modified stem); walnut—seed; cabbage—leaves; carrot—root; cauliflower—flower buds; artichoke—leaves and flower bud; asparagus—stem; cucumber—fruit and seeds; broccoli—flower buds; rice—fruit (mostly seed); potato—tuber (underground stem).

4. Apical meristems at the tips of roots and shoots contain dividing, elongating, differentiating cells that are responsible for plant growth. After Grandpa nailed the horseshoe to the tree, the tips of its branches continued to grow, and

other meristems caused the tree to expand in girth, but the part of the trunk where he nailed the horseshoe, 5 feet from the ground, no longer contained embryonic cells capable of elongation.

5. It would be difficult to kill a palm by girdling it. An oak tree is a dicot. Its vascular tissues are arranged in a ring, with the phloem just under the cork layer of the bark, vulnerable to damage. The palm is a monocot, with its vascular bundles scattered throughout the diameter of the trunk, and therefore less vulnerable to damage.

6. The vascular cylinder is in the middle of a root. Lateral roots must start growing from the inside of the main root so that the vascular cylinders of the main root and lateral roots will connect.

7. Seeds contain all the nutrients needed to sustain the plant embryo until its leaves and roots are functional. Plant cells and human cells need much the same nutrients, so seeds are a good source of nutrition.

Chapter 32

Organizing Your Knowledge

Exercise 1: 1. carbon dioxide (CO_2) 2. water (H_2O) 3. oxygen (O_2) 4. inorganic ions (minerals) 5. CO_2 6. air 7. H_2O 8. soil 9. nitrogen 10. nitrogen 11. magnesium 12. nitrogen 13. phosphorus 14. minerals 15. O_2

Exercise 2: See Figure 32.2B in the text.

Exercise 3: A. 4, diffuse (evaporate) B. 1, xylem, osmosis C. 5, cohesion, adhesion D. 2, root pressure E. 6, hydrogen F. 3, guard, osmosis, stomata, transpiration G. 7, transpiration-adhesion-cohesion

Exercise 4: 1. source 2. sink 3. source 4. source 5. source 6. source 7. source 8. source 9. sink 10. source 11. sink 12. sink 13. sink 14. sink

Exercise 5: 1. C, H, O, N, S, P, Ca, P, Mg 2. C, H, O, N, S, P 3. Fe, Cl, Cu, Mn, Zn, Mo, B 4. N 5. Ca 6. K 7. Fe 8. P 9. Mg

Exercise 6: 1. B 2. H 3. G 4. J 5. D 6. I 7. C 8. A 9. E 10. F

Exercise 7: 1. Humus content of the soil can become depleted, reducing its fertility, and fertilizer runoff can pollute water. 2. This can reduce soil erosion, but herbicides can pollute the soil. 3. This could channel runoff, reduce penetration of water into the soil, and increase soil erosion.

4. Salts that are left behind when irrigation water evaporates can make soil too salty for crops. 5. This adds humus, restoring the natural texture and fertility of soil and reducing dependence on inorganic fertilizers and pesticides. 6. Drip irrigation uses less water, allows plants to absorb most of the water, reduces salt buildup, and reduces water loss from evaporation and drainage.

Exercise 8: 1. Nitrogen 2. proteins 3. nucleic acids 4. air 5. ammonium 6. nitrate 7. nitrogen-fixing 8. ammonifying 9. clay 10. nitrifying 11. nitrate 12. mycorrhizae 13. mutualistic 14. acid 15. photosynthetic 16. dodder 17. mistletoe 18. Carnivorous 19. Venus flytrap 20. nitrogen 21. insects 22. legume 23. nodules 24. nitrogen-fixing 25. organic compounds 26. N_2 27. rotate 28. corn 29. protein 30. vegetarian 31. fertilizer 32. bacteria 33. genetic engineering 34. genes

Testing Your Knowledge

Multiple Choice: 1. b 2. a 3. e 4. d 5. d 6. a 7. c 8. e 9. c 10. b 11. c 12. c

Essay: 1. Root cells actively pump mineral ions into the xylem. Water follows by osmosis and may exert a slight push, called root pressure, that may move water a few meters up a tree. The primary force is a pull exerted by the transpiration of water molecules from the tree's leaves. The water molecules cling to each other and cell walls by hydrogen bonds—phenomena known as cohesion and adhesion. As each water molecule evaporates from a leaf, it pulls on the next, which pulls on the next. This pull, exerted through a column of water molecules in xylem extending all the way to the roots, is what moves water from roots to leaves.

2. If you grew plants in hydroponic culture, bathing them in solutions of known composition, you could compare plants that received a complete nutrient solution with plants receiving the same solution minus molybdenum.

3. Phloem transports sugars, via a pressure-flow mechanism, from sugar sources to sugar sinks. In the summer, a leaf is a sugar source and a potato is a sugar sink. At the leaf, where sugar is being made, it is pumped into the phloem by active transport. Water follows by osmosis, raising the water pressure at this end of the phloem "pipeline" and pushing phloem sap toward the underground potato tubers. In a potato, sugar is transported out of the phloem and stored, and water follows. This lowers the

water pressure at this end of the phloem tube, maintaining the pressure gradient that keeps phloem sap moving from leaves to potatoes. Early in the spring, when the potato is full of carbohydrates and the leaves are developing but not yet performing photosynthesis, their roles are reversed. The potato is a sugar source and the leaf a sugar sink, and phloem sap flows upward from potato to leaf.

4. Clay particles are negatively charged, and because opposite charges attract, positively charged nutrient ions tend to cling to them. Root hairs obtain these ions by positive-ion exchange. The root hairs release hydrogen (H^+) ions into the soil solution. The H^+ ions take the place of the nutrient ions on the clay particles, and the root hair can then absorb the free ions.

5. Humus holds water but keeps the topsoil porous enough to allow aeration of roots. The nutrients in humus are relatively insoluble and released gradually, so they are not washed from the soil and do not pollute water.

6. Agricultural scientists have bred new strains of crop plants that produce more and better-quality protein. They may be able to develop crop plants and nitrogen-fixing bacteria that are able to obtain more nitrogen for making protein from the air, and therefore require less fertilizer. They may be able to insert genes for nitrogen fixation into the cells of crop plants, so the plants will be able to fix their own nitrogen, without help from mutualistic bacteria.

Applying Your Knowledge

Multiple Choice: 1. c 2. e 3. d 4. e 5. d
6. e 7. b 8. d 9. e 10. d

Essay: 1. Spraying with antibiotics probably killed beneficial nitrogen-fixing bacteria growing in nodules on the roots. Without their help, the bean plants became nitrogen deficient. The plants and root fungi probably formed a mutually beneficial mycorrhizal association. The fungi helped the plants absorb water and nutrients and may have protected them from disease. Without the fungi, the plants suffered.

2. Bubbles in the xylem would interfere with transport of water to the leaves. Transpiration of water from the leaves pulls water through the xylem from the roots. Bubbles in the xylem tubes would disrupt the cohesion of water molecules and reduce this transpiration-adhesion-cohesion pull.

3. Most plants open their stomata during the day. This allows them to take in the CO_2 used in photosynthesis and transpire water, which

pulls water and minerals up from the roots. In its hot, dry environment, the jade plant might lose too much water if its stomata remained open during the day. By opening its stomata at night, it reduces water loss.

4. Potassium is needed as an activator of several enzymes, and it is also the main solute that regulates water balance in a plant—the opening and closing of guard cells, for example. The plant does not build molecules containing potassium, but it needs potassium as a regulator of its activities.

5. Acid and heavy rain would be most likely to deplete the nitrogen in the soil. Organic matter decays slowly in acid soil, making little nitrogen available. Unlike phosphate or magnesium ions, nitrate is negatively charged. The rain would be most likely to wash these negative ions from the soil, because they would not cling to negatively charged clay particles. Denise and Roger could enrich the soil by planting and plowing under legumes, such as clover.

Chapter 33

Organizing Your Knowledge

Exercise 1: 1. left 2. left 3. straight 4. right (then left to light) 5. left 6. straight 7. left 8. straight 9. no growth 10. no growth 11. straight 12. left 13. straight 14. right 15. no growth

Exercise 2: 1. auxins 2. Cytokinins 3. IAA 4. apical meristem 5. roots 6. embryos 7. fruits 8. xylem sap 9. IAA 10. cell walls 11. inhibits 12. stimulates 13. inhibits 14. axillary 15. bushy 16. Cytokinins 17. auxins 18. cytokinins 19. inhibit 20. stimulate 21. lower 22. cytokinins 23. root 24. shoot

Exercise 3: 1. ABA 2. G 3. ABA 4. ABA 5. G 6. G 7. ABA 8. ABA 9. G 10. ABA 11. G 12. G 13. ABA

Exercise 4: 1. ethylene 2. aging 3. drying 4. ripening 5. tomatoes 6. CO_2 7. ethylene 8. deciduous 9. water 10. pigments 11. chlorophyll 12. shorter 13. cooler 14. auxin 15. abscission

Exercise 5: 1. gibberellins 2. ethylene 3. auxins 4. cytokinins

Exercise 6: 1. C 2. A 3. D 4. E 5. B

Exercise 7: 1. Growth of a shoot toward light. 2. Auxin migrates from light to dark side of shoot tip, causing cells on dark side to elongate faster. 3. Gravitropism 4. Gravity may pull organelles downward, affecting auxin distribution and rate of growth on opposite sides of root or shoot. 5. Thigmotropism 6. Growth movement in response to touch

Exercise 8: 1. yes 2. no 3. no 4. no 5. yes 6. no 7. yes 8. yes 9. yes 10. no

Testing Your Knowledge

Multiple Choice: 1. e 2. a 3. e 4. b 5. c 6. b 7. c 8. b 9. b 10. e

Essay: 1. Responses called tropisms enable a plant to alter its pattern of growth to adjust to environmental changes. Phototropism is bending toward light. Hormones called auxins migrate from the light side to the dark side of a shoot tip, causing cells on the dark side to elongate faster. Gravitropism, response to gravity, also appears to be controlled by auxins. Gravity may pull organelles downward, redistributing auxins in the process. This causes the lower side of a shoot and the upward side of a root to grow faster, even in the dark.

2. Circadian rhythms continue even when a plant is sheltered from environmental changes—kept in constant darkness, for example. A biological clock does not keep perfect time, so circadian rhythms tend to drift somewhat in relation to the external environment. Environmental cues such as the daily cycle of light and dark enable plants to reset the clock and keep in sync with the environment.

3. As fruit ripens and ages, it produces ethylene gas, which can diffuse through the air from fruit to fruit. Ethylene accelerates ripening, so the ethylene from damaged or overripe fruit can speed up the softening of crisp apples.

4. The signals for leaf drop are the shorter, cooler days of autumn. Auxin prevents leaf drop, but as a leaf ages, it produces less auxin. Cells in the abscission layer, at the base of the leaf stalk, begin to produce ethylene. The ethylene causes enzymes to digest cell walls in the abscission layer. Eventually, the stalk breaks away, and the leaf falls.

5. Long-day plants flower when the night is shorter than a critical length. Short-day plants flower when the night is longer than a critical length. Phytochrome, a colored protein, absorbs light and helps the plant set its biological clock so that it can measure the length of the night. Light changes a form of phytochrome called P_r to P_{fr}. P_{fr} is converted back into P_r during the night. The biological clock may measure the time between the beginning of the P_{fr} to P_r change (sunset) and the end of the P_{fr} to P_r change (sunrise).

Applying Your Knowledge

Multiple Choice: 1. b 2. d 3. b 4. a 5. e 6. d 7. c 8. b 9. e 10. d

Essay: 1. The gas probably contained ethylene, which causes changes in the abscission layer at the base of a leaf that cause the leaf to drop from the stem.

2. Different concentrations of plant hormones can have quite different effects on the same plant. Relatively low auxin concentrations stimulate elongation of stems, but higher concentrations inhibit stem growth.

3. Normally, production of auxins by stems would be balanced by production of cytokinins by roots. Auxins tend to inhibit axillary buds, while cytokinins stimulate them. Grafting a shoot to a large root system would alter this balance. The excess of cytokinins from the roots would stimulate development of axillary buds, which would grow into lateral stems. This would make the plant more bushy, producing a shoot system proportional to the root system.

4. Auxin, produced by the leaf, normally offsets the effects of ethylene, produced by the stalk. Ethylene causes the stalk to drop off. Smearing auxin on the stalk should offset the effects of ethylene and cause the stalk to remain attached to the stem.

5. Apparently, the age or size of soybean plants has little influence on whether or not they flower. Soybeans must be short-day (long-night) plants, which flower in the fall, when the night exceeds a critical length. Once the plants are mature, they wait until the nights are a certain length to flower.

6. Abscisic acid (ABA) acts as a growth inhibitor. It causes seed dormancy by inhibiting germination. ABA might be a good place to start.

7. You could grow a shoot straight up, then turn the plant so that the shoot is horizontal. Place a light under the horizontal shoot. Will it grow toward the light or away from gravity?

Chapter 34

Organizing Your Knowledge

Exercise 1: 1. biosphere 2. ecosystems 3. abiotic factors 4. biotic factors 5. nutrients 6. populations 7. habitats 8. ecology 9. interactions 10. groups belonging to same species 11. organisms

Exercise 2: 1. scarcity of water 2. penguin 3. high body temperature 4. pronghorn antelope 5. cold 6. fire 7. heat and dryness 8. cold and wind 9. dissolved oxygen

Exercise 3: 1. directly 2. North Pole 3. tilt 4. summer 5. winter 6. warms 7. moisture 8. rises 9. rain 10. descends 11. 30 12. desert 13. moisture 14. rotation 15. prevailing 16. trade winds 17. northeast 18. currents 19. winds 20. rotation 21. continents 22. Gulf Stream 23. rainfall 24. rain shadows

Exercise 4:

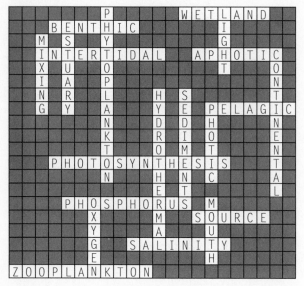

Exercise 5: See Modules 34.9–34.18 in the text.

Exercise 6: 1. desert 2. temperate grassland 3. temperate forest 4. tropical forest 5. taiga 6. tundra

Exercise 7: 1. E 2. C 3. G 4. F 5. B 6. A 7. B 8. D 9. G 10. H 11. A 12. B 13. E 14. G 15. C 16. H 17. D 18. E 19. F 20. C

Exercise 8: See Figure 34.9 in the text.

Testing Your Knowledge

Multiple Choice: 1. d 2. b 3. a 4. a 5. c 6. c 7. d 8. b 9. d 10. c 11. b 12. a

Essay: 1. There are many possible examples. Beavers dam streams and create ponds. Coral animals build coral reefs, which create habitats for many other species. Overgrazing might create a desert. Trees create shade that might alter the temperature and soil moisture.

2. During the postwar era, new technologies such as chemical fertilizers and pesticides came into widespread use. These increased food production and promised to wipe out insect-borne diseases such as malaria. In the 1960s it became clear that technology had a down side. Pesticide residues threatened wildlife, and pests became resistant to chemicals. Now many people are concerned about such negative effects of human activities on the biosphere.

3. The sun's rays strike the equator directly, warming the air and evaporating moisture. Warm air rises, and as it rises, it cools. Its moisture condenses and falls as rain, watering tropical forests. Cool air is heavier than warm air, so it descends, 30° north and south of the equator, creating deserts.

4. In North America, the temperate forest and temperate grassland have been most extensively altered, mostly by agriculture and urban development.

5. The intense cold, relatively light precipitation, and the short growing season of the far north and high mountains prevent the growth of trees. The ground is permanently frozen—permafrost. This kind of environment is characterized by the tundra biome—shrubs, grasses, mosses, and lichens—species that can live under these extreme conditions. Tropical rain forests are characteristic of warm, humid equatorial areas where there is abundant rainfall and little seasonal variation in climate.

Applying Your Knowledge

Multiple Choice: 1. e 2. d 3. a 4. d 5. d 6. b 7. d 8. b 9. b 10. c

Essay: 1. Temperature, rainfall, soil nutrients, wind, and fire are all abiotic factors that would affect a tree. Predation by herbivorous insects and competition with other trees are biotic factors that might affect the tree. Day to day, the tree might adapt to changes in these factors by adjusting its rate and pattern of growth. For example, it might grow more slowly in a colder location, or it might drop its leaves in the winter. Over many generations, the tree has become adapted to a particular range of environmental factors via natural selection.

2. Water absorbs and stores heat, so the range of temperatures in the ocean is much narrower than on land. Since most of the ocean is within the same narrow temperature range, temperature is not an important factor shaping the distribution of life in the sea. Temperature is more variable on land, so it has a bigger impact on the distribution of land life. Most of the ocean is without light, but most places on land are light for about half of each day. Thus, light is

not a factor in where living things can exist on land, but it is an important factor in the sea.

3. The routes of sailing ships depend on the prevailing winds. On their way to North America, ships from Europe first sailed south to catch the northeast trade winds. They first sailed north to catch the westerlies on the return to Europe.

4. The biomes characteristic of warm environments, such as most deserts and grasslands, will probably expand toward the poles. Those characteristic of colder climates, such as taiga and tundra, will probably shrink to the north.

5. The photic zone, where light is available for photosynthesis, is only a thin layer at the surface of the sea. Here there is plenty of light but few nutrients needed for the growth of phytoplankton. Sediments at the bottom of the sea, the benthic zone, are rich in nutrients, but here there is no light for photosynthesis. Life is abundant in the shallow waters of the continental shelves, where sunlight reaches the bottom, but the continental shelves make up only a small portion of the oceans.

Chapter 35

Organizing Your Knowledge

Exercise 1: 1. 2 hectares 2. This figure will vary but will probably be about 30 shrubs. 3. This figure will vary but will probably be about 15 shrubs per hectare. 4. This figure will vary but will probably be about 250 shrubs. 5. One could take more samples, or count every shrub, but this would take more time and cost more money. 6. The shrubs are clumped. This could be due to variations in moisture or soil nutrients, or perhaps because offspring sprout close to parent plants. 7. Mice are mobile; they do not hold still to be sampled. 8. Using the equation in Module 35.2, $N = (40 \times 40)/10 = 160$ mice. 9. $160/16.8 =$ about 10 mice per hectare. 10. You have to assume that a marked mouse has the same chance of being caught as an unmarked individual. You could be wrong. If the mice are territorial and you put the traps in the same locations both times, you are more likely to recapture the same mice. If the captured mice learn to avoid the traps, you are then less likely to recapture them.

Exercise 2: 1. 6 aphids 2. 20 aphids 3. 8, 28 4. 11, 39 5. 55, 77, 108, 151, 211, 295 6. The graph should look a bit like Figure 35.3A in the text. 7. It is J-shaped. 8. This kind of growth could not continue. Eventually, some environmental factor, such as limited food, would slow the growth of the aphid population. 9. The graph is S-shaped. 10. It looks as though the population briefly exceeded carrying capacity, then dropped a bit. 11. The population leveled off at carrying capacity—the number of individuals the environment could support over the long term as food or space became limited. 12. around 700 deer 13. logistic growth

Exercise 3: 1. DI 2. DD 3. DI 4. DD 5. DI 6. DD 7. DI 8. DD 9. DD 10. DI 11. DI 12. DD 13. DD 14. DI 15. DD 16. DI

Exercise 4: 1. rapid population growth when conditions are favorable 2. small 3. large 4. many 5. few 6. older 7. density-independent 8. density-dependent 9. quantity 10. dandelions, many insects 11. coconuts, polar bears

Exercise 5: 1. B 2. A 3. E 4. A 5. E 6. D 7. D 8. C 9. C

Exercise 6: 1. five 2. birth 3. death 4. agriculture 5. birth 6. Black Death 7. 500 8. Industrial 9. death 10. 1.7 11. slow 12. reproductive 13. postreproductive 14. baby boom 15. fewer 16. developing 17. age structure 18. prereproductive 19. carrying capacity 20. birth 21. death 22. food 23. jobs 24. social security 25. health care 26. birth 27. death

Testing Your Knowledge

Multiple Choice: 1. e 2. c 3. d 4. d 5. d 6. a 7. a 8. c 9. c 10. d 11. e 12. b

Essay: 1. The exponential growth curve is J-shaped and describes a population growing at an ever-accelerating rate under conditions that do not restrain population growth. The rate of growth depends only on how big the population is, and as the population increases, it grows faster. The logistic growth curve is S-shaped and describes the growth of a population that is limited by its environment. When the population is small, it grows slowly. Growth accelerates as the population increases, but growth slows down and eventually ceases as the population approaches and reaches carrying capacity, the number of organisms that the environment can support.

2. Carrying capacity depends on species and habitat. The carrying capacity of a desert for foxes might be quite different from the carrying capacity of a forest for mice. The animals are different sizes, they require different kinds and

amounts of food, and desert and forest habitats produce different amounts of food.

3. Density-independent factors include temperature, rainfall, fire, and floods. Density-dependent factors include food supply and predation. Density-independent factors are not affected by the density of the population. Density-dependent factors affect a greater fraction of the population as the population increases.

4. Human culture, not biology, is responsible for our rapid and sustained population growth. Nutrition, health care, and sanitation have enabled us to maintain a birth rate that exceeds our death rate. Agricultural and industrial technology have enabled us to more effectively exploit the Earth's resources and inhibit the mechanisms that would normally inhibit population growth. In effect, our population has continued to grow because we have increased the Earth's carrying capacity.

5. Some populations, such as the lynx and the snowshoe hare, undergo regular ups and downs in population density, called population cycles. The lynx population cycle probably depends on the ups and downs of their prey, the hare. Hare population cycles are probably caused by changes in the availability and quality of their plant food. The plants produce unpalatable chemicals when the hares graze on them. It is also possible that stress from overcrowding may reduce fertility, leading to population cycles.

6. Organisms with opportunistic life histories are usually small in size, mature quickly, produce many offspring, and do not invest much parental care in their offspring. Examples are dandelions and aphids. They usually live in unpredictable environments and multiply exponentially when conditions are favorable. They emphasize producing large numbers of offspring quickly. Equilibrial species are large in size, mature later, and produce fewer offspring, but give their offspring more care. Their populations may be quite stable, held near carrying capacity by density-dependent factors. They emphasize production of better-endowed offspring that can compete in a well-established population. Polar bears and chimpanzees are equilibrial species.

Applying Your Knowledge

Multiple Choice: 1. e 2. b 3. e 4. a 5. d 6. a 7. c 8. b 9. b 10. c

Essay: 1. Using the equation in Module 35.2, there are 2000 smallmouth bass in the lake.

2. The total size of the population is different from population density. Density is the number of individuals of a species per unit of area. The population of the United States is bigger than the population of Japan, but the people of Japan are packed into a much smaller area, so Japan has a higher population density.

3. A population that grows logistically grows slowly when the population is either large or small, and grows fastest when the population is at an intermediate level. (The S-shaped growth curve is steepest at around half of carrying capacity.) The salmon population will grow fastest when it is kept at 50,000.

4. Temperature is a density-independent limiting factor. Its effect alone on the robin population is not influenced by the size of the population. One would expect the same percentage of robins to be killed whether the population is large or small. Competition for shelter is a density-dependent limiting factor. It affects a bigger percentage of the robins when the population is larger.

5. These populations were introduced to environments where they are not subjected to their normal limiting factors, such as predators and parasites. Without these limits, the populations grew larger than would otherwise be the case.

6. Populations on islands or in lakes are easier to define than in other areas. The organisms are limited by natural geographical boundaries, and there is no mixing, interbreeding, or confusion with organisms from elsewhere. They might also be easier to find and sample than populations with less-defined boundaries.

Chapter 36

Organizing Your Knowledge

Exercise 1: 1. D 2. V 3. V 4. V 5. D 6. T 7. T 8. T 9. D 10. T 11. S 12. S 13. D 14. S

Exercise 2: 1. commensalism 2. parasitism 3. predator/prey 4. predator/prey 5. parasitism 6. commensalism 7. mutualism 8. competition 9. mutualism 10. parasitism 11. competition 12. predator/prey

Exercise 3: 1. black widow spider 2. barnacle 3. polar bear 4. sea star 5. redwood 6. poison-arrow frog 7. nitrogen-fixing bacterium

Exercise 4: 1. early 2. late 3. early 4. late 5. early 6. early 7. early

Exercise 5: 1. PP 2. 1C 3. 1C 4. 2C 5. 2C

6. PP 7. 1C 8. 1C 9. 2C 10. 2C 11. 2C
12. 3C 13. 3C 14. 3C 15. 3C 16. 4C 17. 4C
18. D. Your food web should look something like Figure 36.9 in the text, with primary producers on the bottom, quaternary consumers on the top, and arrows pointing up to show direction of energy flow.

Exercise 6: 1. production energy 2. tertiary consumers 3. energy used in cellular respiration 4. secondary consumers 5. energy in wastes 6. primary consumers 7. primary producers 8. sunlight

Exercise 7: 1. N_2 2. nitrogen-fixing bacteria 3. protein 4. bacteria 5. ammonia 6. nitrifying 7. nitrate 8. ammonia 9. nitrate 10. assimilation 11. amino acid 12. denitrifying

Exercise 8: 1. Burning of wood and fossil fuels adds CO_2 to the air faster than it is used, causing a buildup of CO_2 in the atmosphere and possible global warming. 2. When a forest is clear-cut, vegetation holds less moisture, and the increased runoff of water washes mineral nutrients from the soil. 3. Runoff from deforested land, fertilizer from farmland, animal wastes from pastures and stockyards, or sewage from cities can add nutrients to a lake, causing rapid eutrophication. The lake can become choked with vegetation and algae, and their decomposition can deplete oxygen, kill fish, and reduce species diversity. 4. Uncontrolled exploitation of natural ecosystems, such as the taiga biome of North America and Siberia and the tropical rain forests, threatens the survival of thousands of species and may alter climate. 5. Destruction of tropical forests reduces transpiration and may alter the amount of water vapor in the air, changing weather patterns.

Testing Your Knowledge

Multiple Choice: 1. c 2. a 3. c 4. d 5. a
6. b 7. d 8. c 9. b 10. d 11. e 12. c

Essay: 1. At each trophic level in an ecosystem, organisms fail to ingest some food, and some of the food they do eat passes through their systems undigested. Much of the food they do assimilate is burned in cellular respiration, releasing energy that is used to carry out body activities and is eventually given off as heat. This leaves little energy—an average of 10%— available to the next trophic level. Ninety percent of the energy is lost to the next level. If humans eat grain, they get 10 to 100 times the energy that would be available if they fed the grain to cattle and then ate beef, because most of the energy in the grain is lost in passing through the additional trophic level.

2. Decomposition is the breakdown of organic materials—plant litter, animal wastes, dead organisms, and so on—into inorganic compounds. Bacteria and fungi are an ecosystem's most important decomposers. They recycle organic material and release inorganic nutrients such as ammonia, nitrate, and phosphate, which are then available for plants to use for growth.

3. The speed of antelope is an adaptation to pursuit by fast predators. The cooperative hunting of a lion pride is an adaptation to their speedy prey. Predator and prey have shaped each other's evolution. Symbiotic relationships are also shaped by coevolution. For example, plants and their pollinators often coevolve. The shape and color of a flower might adapt to the behavior of a particular insect, and the life cycle and mouthparts of the insect might adapt to the flower.

4. Squirrel—primary consumer; oak tree—primary producer; mosquito—secondary or higher consumer; great white shark—secondary or higher consumer; moose—primary consumer; cheetah—secondary or higher consumer; mushroom—detritivore; spider—secondary or higher consumer; phytoplankton—primary producer; grass—primary producer; vulture—detritivore

5. Water evaporates from oceans and other bodies of water and becomes water vapor. Transpiration by plants also adds water vapor to the air. Winds blow water vapor over the land, and the water condenses and forms rain. Precipitation runs off the land, forming streams and rivers, which return water to the sea, and the cycle is repeated.

6. A fern captures a CO_2 molecule in photosynthesis and uses the carbon atom it contains to build a glucose molecule, which ends up in a leaf. An insect eats the leaf and burns the glucose molecule in cellular respiration, releasing CO_2 to the air. A tree captures the CO_2 and uses it to build a protein molecule in its bark. The carbon atom is released in CO_2 when the tree burns in a forest fire. A tree a hundred miles away again captures the CO_2 in photosynthesis and uses it to build a cellulose molecule. When the tree topples and dies, decomposers such as bacteria could break down the cellulose in cellular respiration and release the carbon in CO_2, but this tree does not decompose. It is buried and forms coal. Millions of years later, the coal is burned in a power plant, and the carbon atom again ends up in CO_2.

Applying Your Knowledge

Multiple Choice: 1. e 2. d 3. e 4. c 5. b
6. c 7. e 8. a 9. b 10. d 11. d 12. b

Essay: 1. You could remove A—the predator species—from some of the ponds and see if this had any effect on the abundance of other species. If A is a keystone predator, you would expect its removal to alter the relative sizes of its prey populations. For example, species B might crowd out the other species.

2. Various species of trees are the prevalent form of vegetation in the forest, but grass is the prevalent form of vegetation in the park. The number of species would probably be reduced in the park, so its species diversity would be lower. The trophic structure of the park would be altered, with fewer large primary producers (trees) and more small primary producers (grass); fewer primary consumers that feed on trees and more primary consumers that feed on grass; and a different mix of higher consumers that feed on them. Even the detritivores would probably be different. The forest community is probably more stable; left alone it would probably persist unchanged for many years. The park is less stable; left alone, ecological succession would turn it into a forest. Also, the park is less stable in terms of disruption by disease or catastrophe. A disease that affected one of the few species in the park would have more impact than a disease that affected one of the many species in the forest.

3. *P. aurelia* and *P. caudatum* have similar niches; their needs and abilities are similar, so they compete with each other. *P. bursaria* apparently has a different niche from *P. aurelia*; they do not compete as much. If *P. aurelia* and *P. caudatum* are similar, and *P. aurelia* can coexist with *P. bursaria*, then *P. caudatum* could probably coexist with *P. bursaria* if they were cultured together.

4. Much of the energy in the food consumed by the inhabitants of the shelter is used in cellular respiration to power body activities; this energy is lost as heat. Thus, the body wastes of the inhabitants contain only a fraction of the energy that was originally in their food. The mushrooms, like the shelter inhabitants, use much of the energy in these wastes in their own cellular respiration, storing little of it in their tissues. As energy is lost from the system, the mushroom crops would rapidly dwindle, and the inhabitants would quickly starve. Energy flows through an ecosystem; it is degraded to heat and dissipated, not recycled. Without energy from outside—light (from the sun or electric lights) to grow crops, canned food, and so on—the system quickly runs out.

5. If 10% of the energy on one trophic level is transferred to the next level, 20 kilograms of algae will grow 2 kilograms of crustaceans, and this will grow 0.2 kilograms of fish. If 1 square meter of pond grows 0.2 kilograms of fish and you want 1000 kilograms of fish, the pond will have to have an area of 1000/0.2 = 5000 square meters.

6. The least chipmunk can live in a broader range of habitats—both the sagebrush and the sagebrush-pinon pine zones. The yellow pine chipmunk is narrower in its requirements and is confined to the sagebrush-pinon pine zone. The two species have somewhat similar niches because they compete. If both species are present, the yellow pine chipmunk is apparently a better competitor, and it excludes the least chipmunk from the sagebrush-pinon pine zone—an example of competitive exclusion.

7. Coevolution shapes host-parasite relationships. The deadliest strains of viruses do not spread as readily because they die with their hosts. Less deadly strains are more successful and displace the more deadly strains. The parasites thus become less deadly over time. Among the hosts, those individuals with the greatest natural resistance to the viruses survive the epidemic and pass their resistance on to their children, increasing the resistance of the host population over time. This will probably happen among the Indians, but at great human cost.

Chapter 37

Organizing Your Knowledge

Exercise 1: 1. *Proximate:* Movement of a small object. *Ultimate:* Fitness of frogs has been enhanced by their ability to detect and capture insects. The fact that it is "automatic" suggests it is genetic. 2. *Proximate:* Shape and pattern of egg. *Ultimate:* Reproductive success of gulls has been enhanced by an ability to recognize and retrieve their own eggs. The pattern of spots must vary, so the pattern on a particular egg must be learned by experience, but looking for it could be genetic. 3. *Proximate:* The appearance and location of the hive. *Ultimate:* Bees that could find their way back to the hive were more successful than those who couldn't. The location of the hive can vary, so finding it each day results from experience. 4. *Proximate:* Red wing patches. *Ultimate:* Those males that identify themselves and defend their territories have greater reproductive success than those who do not. Based on the information given, this behavior could be genetic or learned. 5. *Proximate:* The locked cage. *Ultimate:* Those monkeys able to use their intelligence to solve these kinds of problems have the most offspring. The monkey has to figure out the lock, so it has not seen it before. The ability to open the lock must be a result of experience. 6. *Proximate:* The appearance or sound of a bee. *Ultimate:* Toads that could not learn how to avoid harmful animals soon died out. The behavior of the toad is modified by experience. 7. *Proximate:* Colder weather or shorter days of autumn. *Ultimate:* Monarchs that flew south for the winter had greater reproductive success. Because the butterflies have not made the trip before, the behavior must be genetically programmed. 8. *Proximate:* Moving "hawk" silhouette. *Ultimate:* Chicks that run for cover live to reproduce. Because newly hatched chicks do it, the behavior must be genetic.

Exercise 2: 1. host eggs 2. ejects eggs from nest 3. chicken 4. runs for cover 5. bird 6. stuffs food into mouth 7. red-winged blackbird 8. threatens and attacks intruder 9. frog 10. moving insect 11. human infant 12. adult face 13. gull or graylag goose 14. rolls egg into nest

Exercise 3: 1. association 2. imprinting 3. imprinting 4. habituation 5. innovation 6. imprinting 7. innovation 8. association 9. imitation 10. habituation (probably association too) 11. imprinting 12. imitation

Exercise 4: See Chapter 3, Exercise 10, for a discussion of constructing concept maps.

Exercise 5: 1. Catch some turtles off Brazil and tag them in some way. Later catch some turtles laying eggs on Ascension Island and look for the tags. 2. See which way worms move when placed on the boundary between dark and light areas. 3. First see if the crabs retain their rhythmic behavior if moved into a laboratory away from the ocean. If so, move some crabs to a lab in a different time zone, where tides would be different. See if they stay on the schedule of their original habitat or the new location. 4. Blindfold a homing pigeon and see if it can still get home. 5. See whether birds in the lab, at a constant temperature, still become restless as fall approaches. 6. See whether the rat retains a circadian rhythm of activity under constant light.

Exercise 6: 1. specialists 2. dominant 3. search image 4. efficient 5. optimal foraging 6. territory

7. scent 8. food 9. offspring 10. dominance 11. control 12. fighting 13. communication 14. scent 15. visual 16. sounds 17. agonistic 18. ritual 19. courtship 20. threatening 21. altruism 22. kin 23. genes 24. reciprocal

Exercise 7: Honeybee—near the center, but closer to movements. Ring-tailed lemur—in the center. Wolf—in the center. Cricket—near sounds. Human—between sounds and movements, but closer to sounds. Domestic dog—in the center, but closer to chemicals. Common loon—between movements and sounds, but closer to movements. White-crowned sparrow—between movements and sounds, but closer to sounds.

Exercise 8: 1. C 2. M 3. I 4. G 5. D 6. L 7. B 8. A 9. E 10. F 11. H 12. K 13. J

Testing Your Knowledge

Multiple Choice: 1. a 2. c 3. c 4. a 5. d 6. d 7. b 8. b 9. b 10. e 11. b 12. d

Essay: 1. Newly hatched geese are genetically programmed to learn, during a short critical period after hatching, to follow a moving object (usually their mother). Similarly, a songbird is programmed to learn its species' song during a critical period, then sing that song later. Other birds imprint on their own eggs or offspring. Imprinting provides potential mates, parents, and offspring with a quick way to identify each other, saving time and energy. The energy saved can be used in finding food and rearing young, increasing the chances of reproductive success.

2. A dominance hierarchy assures access to food, water, mates, and nesting or roosting sites for higher-ranking individuals. It establishes order within the group, so individuals can devote energy to finding food, defending against predators, locating mates, or caring for young, instead of fighting each other. This contributes to fitness.

3. Courtship rituals signal that potential mates are not threats to each other. They also confirm that individuals are of the correct species and sex and are in condition to breed.

4. Social behavior provides organization within animal populations, but for animals to coordinate their activities, they must be able to signal each other about their intentions and actions. Bees communicate the location of nectar by "dancing." Wolves and cheetahs mark their territories with scents. Loon courtship involves a complex series of movements. Lemurs wave

their scented tails to communicate aggression. A frog attracts a mate by croaking.

5. Some birds use the position of the sun as a reference point, adjusting for its motion across the sky. Other birds that migrate at night use the fixed north star as a directional reference. Some birds may be able to sense the Earth's magnetic field.

6. Fixed-action patterns enable animals to perform behaviors correctly the first time, with no time required for learning. FAPs seem to be most important in animals with short life spans, such as insects. In birds and mammals, FAPs are important in mating and parental care, where there is little chance for practice and little margin for behavioral error.

7. Wolves, chimpanzees, and bees cooperate in social groups. Social cooperation aids in hunting and foraging for food, mating, care of the young, and defense.

Applying Your Knowledge

Multiple Choice: 1. d 2. b 3. c 4. c 5. b 6. d 7. a 8. e 9. e 10. b

Essay: 1. Group hunting allows animals to capture prey that are too fast, too wary, or too big for one predator to bring down alone. The biggest disadvantage of group hunting is that each predator must share food with others in the group.

2. Mating behaviors in cuckoos must be largely innate—genetically programmed. One way to test this would be to raise cuckoos from the time of hatching in isolation from other birds, then, when they are mature, to observe whether they are able to court and mate.

3. The proximate cause of the diving behavior is an environmental cue—the sound of a bat's sonar. The ultimate cause of the diving behavior has to do with natural selection. The fitness (reproductive success) of moths is enhanced by diving to the ground when they hear bat sonar.

4. This is a question of optimal foraging. Alone, a wagtail does not catch as many insects, so it does not obtain as much energy, but it does not go as far, so it does not expend as much energy either. In a flock, a wagtail gets more insects and energy, but it must expend more energy covering more territory to get them. Foraging alone or in a flock depends on which provides the most energy gain for energy expended.

5. This is an example of association. The patients learn to associate particular foods eaten just before treatment with the "punishment" of chemotherapy, and subsequently they want to avoid those foods. This kind of learning would help an animal quickly learn to avoid harmful foods.

6. Biological clocks in most animals tend to run a little slow. In humans, the clock naturally runs on about a 25-hour cycle, so it is naturally easier for a person to stay up later than to get up earlier.

7. Altruism might be fatal for an individual bird, but if it saves the lives of its relatives, its genes will live on through them. An individual might enhance its fitness by helping its relatives. This idea is called kin selection. Alternatively, reciprocal altruism might be involved here. A favor extended today might be repaid by the recipient tomorrow. In this way, unrelated birds might enhance each other's fitness.

8. A sociobiologist might say that this behavior is at least partially a consequence of our evolution. Adults are genetically programmed to be attracted to and take good care of babies. An individual is likely to be around his or her own babies the most. Individuals who like their babies enhance their reproductive success, and liking babies is passed on via their offspring.

Chapter 38

Organizing Your Knowledge

Exercise 1: For example: Human intellect and culture have given us an unequaled ability to manipulate nature, which makes us the most successful and destructive species on the Earth.

Exercise 2: 1. sensitive hands with opposable digits 2. eyes positioned for depth perception, limber shoulder joints, and opposable digits 3. limber shoulder joints and opposable digits 4. eyes positioned for depth perception and sensitive hands with opposable digits 5. limber shoulder and hip joints and opposable digits 6. eyes positioned for depth perception and sensitive hands with opposable digits 7. eyes positioned for depth perception and limber shoulder joints

Exercise 3: 1. primates 2. prosimians 3. anthropoids 4–5. lemurs, loris 6. monkeys 7. hominids 8. apes 9. New World monkeys 10. Old World monkeys 11. tamarins 12–13. baboon, langurs 14. australopithecines 15. modern human 16–20. orangutan, chimpanzee, bonobo, gorilla, gibbons

Exercise 4: 1. C 2. F 3. B 4. F 5. E 6. E
7. F 8. B 9. C 10. A 11. D or G 12. D or G
13. E

Exercise 5: 1–2. *A. robustus* and *A. africanus* 3.
H. habilis 4. *H. erectus* 5. Neanderthal 6. *H. sapiens* 7–8. *A. robustus* and *A. africanus* 9. *H. habilis*
10. *H. erectus* 11. Neanderthal 12. *H. sapiens*

Exercise 6: 1. agriculture 2. machine age
3. scavenging-gathering-hunting 4. scavenging-gathering-hunting 5. agriculture 6. machine age
7. agriculture 8. machine age 9. scavenging-gathering-hunting 10. agriculture

Exercise 7: 1. Acid precipitation 2. Increase efficiency in use of fossil fuels; burning cleaner
fuels 3. CO_2 buildup from burning of fossil
fuels to power industry, vehicles, home heating;
combined with deforestation 4. Reducing fossil-fuel use; renewable energy sources; replanting
forests 5. Destruction of marine life 6. Reducing
oil consumption; safer tankers 7. Pollution and
biological magnification of chemical pesticides
such as DDT 8. Reducing use of chemical pesticides 9. Reduction of biodiversity; changes in atmosphere and climate; loss of potential useful
products 10. Conservation of forests; extractive
reserves to use forests without destroying them
11. CFC pollutants released from air conditioners, refrigerators, and so on 12. Damage to the
atmosphere's protective ozone layer 13. Over-hunting 14. Whaling ban

Exercise 8: 1. Each person in an industrialized
country uses a much greater share of the resources and produces more pollution than an
individual in a developing country. 2. Tropical
forests may contain 80% of the world's species
of plants and animals. Besides, most of the forest land in the United States has already been
cut. 3. The rain forest produces oxygen, uses
CO_2, and affects the water cycle. Destroying the
forest could alter global climate. 4. The greenhouse effect warms the Earth enough to make
life possible. But the CO_2 added by human activities might increase the greenhouse effect and
make it too warm. 5. It is pollution of the atmosphere by CFCs from air conditioners and refrigerators that destroys the ozone layer. Combustion of fossil fuels intensifies the greenhouse
effect but has little effect on ozone.

Testing Your Knowledge

Multiple Choice: 1. c 2. a 3. c 4. d 5. c 6. e
7. c 8. d 9. b 10. a

Essay: 1. Primates have limber hip and shoulder
joints and opposable digits on their hands and

feet, useful for climbing, grasping, and manipulating objects. They have an acute sense of
touch in their hands and feet. Their snouts are
short, and their eyes are set close together on
the front of their face, which enhances depth
perception.

2. Chimpanzees (and closely related bonobos)
are the living primates most closely related to
humans. They are apes and share numerous
anatomical similarities with humans. Chimpanzees and humans share behavioral similarities as
well. Chimps are very intelligent. They can
make and use simple tools; they raid other social groups of their own species; they can learn
sign language; and their behavior in front of
mirrors indicates that they are, to some degree,
self-aware. Biochemical evidence also shows the
close relationship between chimps and humans;
human DNA differs from chimpanzee DNA by
less than 3%.

3. a. Anthropoids have relatively larger brains
and depend more on sight and less on sense of
smell than prosimians. b. Old World monkeys
have nostrils that are narrow and close together,
and many have a tough seat pad. New World
monkeys lack the seat pad, and their nostrils
are wide open and far apart. They also have
prehensile (grasping) tails. c. Monkeys have
tails; apes do not. Monkeys have front limbs
that are equal in length to their hind limbs; the
front limbs of apes are longer than their hind
limbs. d. Apes walk with their knuckles on the
ground; humans are bipedal, walking upright.
(There are other less distinctive differences, such
as language and use of tools, but chimps share
these traits to some degree.)

4. The two most important physical changes in
the evolution of humans were upright walking
and development of a larger brain.

5. Early African hominids may have driven several species of saber-toothed cats to extinction.
Homo sapiens may have been responsible for
wiping out many large animals in Europe,
Australia, and the New World. Intensive farming and overgrazing in the Middle East depleted
the soil and exposed it to erosion, leaving much
of the area a desert.

6. Humans have not changed much physically
in the last 100,000 years. But our culture—accumulated knowledge, customs, beliefs, arts, and
especially technology—has grown and changed
enormously. We alter our environment to suit
our needs and desires, increasing the carrying
capacity of the Earth for humans but changing

the Earth much faster than we or other living things can adapt to biologically. The human population has grown exponentially, and all life on Earth is threatened by the changes we have wrought.

Applying Your Knowledge

Multiple Choice: 1. b 2. a 3. c 4. b 5. c
6. b 7. d 8. b 9. a 10. e

Essay: 1. Top predators depend on all the species below them in the food web. If a disturbance alters the vegetation, or affects a primary or secondary consumer, the disturbance also affects the top predators who depend on the lower trophic levels for food. A good example is biological magnification of pesticides.

2. a. Conversion to solar and hydroelectric power would decrease global warming, because this would reduce CO_2 emissions. b. Growing trees use CO_2, so this would reduce global warming. c. Burning and decomposition of logs add CO_2 to the air, and trees remove it, so deforestation would increase global warming.
d. More efficient cars would burn less gasoline and produce less CO_2 per mile, so this would reduce global warming. e. Fewer people would use less fossil fuels and cut forests at a slower rate, slowing CO_2 buildup and global warming.

f. Decomposition produces CO_2, so this would increase global warming.

3. Hominids are thought to have diverged from other apes 4 to 5 million years ago and started walking upright. A large brain evolved later. A small-brained bipedal hominid 4.0 million years old fits current theory. A large-brained hominid that retained its knuckle-walking ways 2.5 million years ago would be harder to explain.

4. Given enough iron, the algae might grow faster, speed up photosynthesis, and take CO_2 from the atmosphere at a faster rate. Since CO_2 traps heat in the atmosphere, the drop in CO_2 would reduce the rate of global warming. But trying this might have unintended effects, such as disrupting ocean food webs.

5. When did humans diverge from the apes? What was the common ancestor of apes and humans like? Are the australopithecines our ancestors, or simply another branch of the hominid family tree? Did different races of humans evolve from different populations of *Homo erectus*, or can all modern humans be traced back to a single origin? What is the role of the Neanderthals in human evolution, and what happened to them? What triggered the rapid cultural evolution of the last 100,000 years? What next?